D1560031

Organic Syntheses with Noble Metal Catalysts

This is Volume 28 of
ORGANIC CHEMISTRY
A series of monographs
Editors: ALFRED T. BLOMQUIST and HARRY WASSERMAN

A complete list of the books in this series appears at the end of the volume.

Organic Syntheses with Noble Metal Catalysts

PAUL N. RYLANDER

Engelhard Industries Division
Engelhard Minerals and Chemicals Corporation
Menlo Park, Edison, New Jersey

ACADEMIC PRESS New York and London 1973

A Subsidiary of Harcourt Brace Jovanovich, Publishers

ACADEMIC PRESS, INC.
111 Fifth Avenue, New York, New York 10003

United Kingdom Edition published by
ACADEMIC PRESS, INC. (LONDON) LTD.
24/28 Oval Road, London NW1

Library of Congress Cataloging in Publication Data

Rylander, Paul Nels, DATE
Organic syntheses with noble metal catalysts.

Includes bibliographies.
1. Chemistry, Organic–Synthesis. 2. Platinum group catalysts. I. Title
QD262.R9 547′.2 72-13608
ISBN 0–12–605360–X

PRINTED IN THE UNITED STATES OF AMERICA

Contents

Preface ix

Chapter 1. Dehydrogenation

Catalysts 1
Experimental Procedures 3
Aromatization 4
Formation of New Bonds 17
Dehydrogenation of Ketones to Phenols 26
Dealkylation and Isomerization 29
Hydrogen Exchange Processes 35
Loss of Functional Groups 43
References 53

Chapter 2. Homogeneous Hydrogenation

Catalysts 60
Olefins 61
Sulfur Compounds 68
Labeling 69
Diminished Disproportionation 70
Asymmetric Hydrogenations 71
Miscellaneous 73
References 74

Chapter 3. Oxidation

Oxidation of Olefins in Aqueous Systems 77
Oxidation of Olefins in Nonaqueous Systems 80

Oxidation of Complex Olefins 85
Oxidative Coupling 88
Acetoxylation 95
Oxidation of Alkylaromatics 97
Oxidation of Alcohols 99
Oxidative Dehydrogenation 112
N-Dealkylation 113
References 115

Chapter 4. Osmium and Ruthenium Tetroxides as Oxidation Catalysts

OSMIUM TETROXIDE

Metal Chlorates as Oxidants 122
Peroxides as Oxidants 125
Air as Oxidant 128
Periodate as Oxidant 129
Hypochlorite as Oxidant 132
Hexacyanoferrate as Oxidant 132
Disproportionation 133

RUTHENIUM TETROXIDE

Oxidation of Aromatics 134
Oxidation of Alcohols 136
Oxidation of Olefins 138
Oxidation of Acetylenes 140
Ruthenium Tetroxide–Sodium Hypochlorite 140
References 141

Chapter 5. Isomerization

Double-Bond Migration 145
Configurational Changes 160
Skeletal Isomerization 163
Aromatizations 166
Valence Isomerization 168
References 171

Chapter 6. Oligomerizations, Telomerizations, and Condensations

Oligomerization of Olefins 175
Oligomerization of Substituted Olefins 178
Telomerization of Olefins 181
Oligomerization of Dienes 185
Addition of Dienes to Olefins 189

Oligomerization of Acetylenes 191
Telomerization of Dienes 194
Condensations 208
References 211

Chapter 7. Carbonylation and Hydroformylation

Catalysts 215
Acetylenes 217
Amines 222
Azides 224
Alcohols 224
Halogen Compounds 225
Olefins 226
Carbonylation of Dienes 241
Carbonylation of Olefinic Compounds 247
Nitro Compounds 252
References 254

Chapter 8. Decarbonylation and Desulfonylation

DECARBONYLATION
Aliphatic Aldehydes 261
Unsaturated Aldehydes 262
Aromatic Aldehydes 264
Acid Halides 265
Aroyl Cyanides 267
Ketones 268
Anhydrides 268
Alcohols 269
Ethers 270
Esters and Acids 270
β-Ketoamides 270
Formate Esters 271
DESULFONYLATION
References 272

Chapter 9. Silicon Chemistry

Catalysts for Hydrosilylation of Olefins 274
Addition of Silanes to Olefins 277
Addition of Silanes to Diolefins 278
Addition of Silanes to Substituted Olefins 279
Hydrosilylation of Acetylenes 282
Addition to the Nitroso Function 283
Addition of Aminosilicon Hydrides to Olefins 284

Preface

The field of catalysis of organic reactions by noble metals is in its infancy. Despite a long history, most noble metal-catalyzed reactions have been developed only within the last decade, a consequence of a burgeoning research effort. Impetus for intensive research stems from the realization that platinum group metal catalysts can be adapted readily to the synthesis of low cost, large volume compounds and to an increasing awareness of the versatility and uniqueness of these catalysts in organic synthesis.

This book is an attempt to cull much of the literature on catalysis by noble metals and to present it in such a form as to be of use to those interested in organic synthesis. This aim determined the style and contents of the book, and much that might be of interest to others has been omitted. My approach in this work is descriptive rather than mechanistic. Experimental conditions are given frequently and the effect of process variables is stressed. Mechanistic aspects of reactions are confined to brief statements of the authors' views and citation of the original literature. Hopefully this orientation and scope will give the reader convenient access to the enormous possibilities inherent in organic synthesis with noble metal catalysts.

I am indebted and grateful to Mrs. Irene Tafaro for her skillful typing of a not easily read manuscript and for her care in checking references.

PAUL N. RYLANDER

CHAPTER 1

Dehydrogenation

Catalytic dehydrogenation over platinum group metals provides a convenient, frequently used method for synthesis of organic compounds. As a rule, catalytic dehydrogenation must be carried out at elevated temperatures to displace the equilibrium in favor of the dehydrogenated product. If a hydrogen acceptor, such as oxygen or nitrobenzene, is present, the reaction may be carried out under milder conditions. The usually vigorous conditions of dehydrogenation promote various side reactions in proportions that depend on the substrate structure as a whole, the reaction conditions, and the catalyst. Most successful dehydrogenations involve establishment or enlargement of an aromatic system, or oxidation of a particular function that offers a preferential point of attack. Much industrial interest evolves around dehydrogenation of paraffins to olefins and diolefins.

CATALYSTS

All platinum metals have been used as dehydrogenation catalysts, but most reactions have been carried out over palladium and, to a lesser extent, platinum. Very few comparisons of more than two platinum metal catalysts have been made and it is difficult to arrive at generalities concerning the most efficient metal. Limited work suggests that rhodium and iridium may prove especially useful when hydrogenolysis is desired as a concomitant side reaction (Rylander and Steele, 1968). Dehydrogenation activity depends on the metal and support (Lago *et al.*, 1956) and also on the method of catalyst preparation (Linstead *et al.*, 1937).

An inverse relationship between hydrogenation activity and dehydrogenation activity has been noted, and it was suggested that a typical hydrogenation catalyst can be converted to a dehydrogenation catalyst by poisoning (Packendorff and Leder-Packendorff, 1934).* One excellent hydrogenation catalyst was converted to a dehydrogenation catalyst by treating it at 300°C with a benzene solution of phosphorus trichloride. In another example, sulfur increased the dehydrogenating efficiency of rhodium in contrast to its poisonous effect in hydrogenation (Hernandez and Nord, 1948). The observation sometimes made that in dehydrogenation the yield of product increases with re-use of the catalyst may be related to this effect as well as to a diminution in side reactions.

Palladium-on-carbon catalysts prepared by alkaline formaldehyde reduction or by hydrogen reduction of palladium chloride (Horning, 1955) have proved useful in various batch type dehydrogenations (Rabjohn, 1963; Horning *et al.*, 1948a). A preparation of a palladium-on-granular carbon for use in continuous processing featured exclusion of oxygen from all stages of the preparation and from finished catalysts. This catalyst was immediately satisfactory for dehydrogenation of decalin or tetralin, but dehydrogenation of certain heterocyclic compounds was satisfactory only after several uses of the catalysts (Anderson *et al.*, 1963). Palladium black has been used with success in the preparation of phenanthrol, whereas under the same conditions, palladium oxide and palladium-on-carbon were unsatisfactory. Platinum oxide also was useful in these reactions (Mosettig and Duvall, 1937). Rhodium has not been much used in dehydrogenations, but it has given some very satisfactory results, especially in exchange reactions (Anderson and Anderson, 1957; Newman and Lednicer, 1956).

As with hydrogenation catalysts, a considerable mystique surrounds the preparation of dehydrogenation catalysts. For instance, platinum-on-alumina catalysts used for dehydrogenation of hydronaphthalene derivatives are said to be improved both in efficiency and stability by incorporation of small amounts of neodymium (Hiser, 1968). Other examples of unexpected synergism are cited later. But despite the complexity of the problem, it is usually not too difficult to find a dehydrogenation catalyst that will give quite satisfactory results; this is most easily done by using a catalyst that had proven satisfactory in similar situations in the past.

* Hydrogenation and dehydrogenation reactions are carried out generally under much different temperatures: the statement does not, therefore, imply that the catalyst shifts the equilibrium.

Homogeneous dehydrogenation catalysts have been studied very little, but the work does suggest that they may prove a supplement to heterogeneous procedures. Blum and Biger (1970) compared $RhCl(Ph_3P)_3$, $RhCl_3(Ph_3As)_3$, $IrCl(CO)(Ph_3P)_2$, $RuCl_2(Ph_3P)_3$, and 10% palladium-on-carbon for the dehydrogenation of a variety of hydrocarbons. The rates generally were comparable for all catalysts, but the homogeneous reactions were more selective. For example, dehydrogenation of 9,10-dihydroanthracene over 10% palladium-on-carbon afforded up to 50% 1,2,3,4-tetrahydroanthracene, whereas over $RhCl_3(Ph_3As)_3$ and $RuCl_2(Ph_3P)_3$ quantitative yields of anthracene were obtained. Alcohols can be dehydrogenated to ketones with a rhodium trichloride–stannous chloride catalyst (Charman, 1970). Tin is necessary to prevent precipitation of rhodium metal (Charman, 1967). No synthetic application of this system seems to have been reported. Palladium complexes used stoichiometrically are effective in converting enone systems to dienones (Howsam and McQuillin, 1968).

Experimental Procedures

Catalytic dehydrogenations in batch operation are usually carried out by heating a mixture of substrate and catalyst, with or without solvent, to reflux. Dehydrogenation–hydrogenation are reversible reactions and dehydrogenation is facilitated greatly if liberated hydrogen is expelled from the system by boiling or by sweep of an inert gas. Boiling, which is essential for the highest reaction rates, may be achieved without excessive temperatures by diminution of pressure or use of appropriate solvent (Linstead and Michaelis, 1940). For instance, the difficult aromatization of the B ring of 3-methoxyestra-1,3,5(10)-trien-17-one was achieved by refluxing the compound in anisole for 23 days with 5% palladium-on-carbon catalyst. The solvent was chosen with a boiling point such that dehydrogenation would occur but not thermal degradation (Scaros and Bible, 1968). Alcohol proved an effective solvent for dehydrogenation of complex diacids and anhydrides without epimerization of the acid function (Bachmann and Controulis, 1951).

If carbon dioxide is used as a sweep gas, the volume of hydrogen produced can be measured after passing the off-gas through an absorbent such as a sodium hydroxide solution (Eastman and Detert, 1951). Reactions producing water may be carried out conveniently in an ordinary distilling flask that allows removal of water as formed and thus prevents violent sputtering (Newman and Zahm, 1943).

Dehydrogenations are also suitably carried out in vapor phase over supported platinum metal catalysts in a tubular reactor (Anderson *et al.*, 1963; Doering *et al.*, 1953). For example, cyclohexanol and cyclohexanone are each converted to phenol by continuous dehydrogenation over platinum-on-granular carbon at 300°C. Over ruthenium-on-carbon, on the other hand, cyclohexanol is converted to cyclohexanone at 320°C at a weight hourly space velocity of 0.70 (Rylander and Kilroy, 1958; British Patent 849,135). In dehydrogenation in tubular reactors attention must be paid to the equilibrium constant for the reaction; operation at too low a temperature will result in incomplete conversions. In the above example, substantially complete conversion to phenol cannot be obtained at temperatures much less than 300°C; in contrast, any temperature that gives a convenient rate can be used in batch dehydrogenations where the liberated hydrogen is removed as formed. Improved results may sometimes be obtained in dehydrogenations by employing hydrogen as a sweep gas despite its adverse effect on the equilibrium. For instance, phenol is formed from dehydrogenation of a cyclohexanol–cyclohexanone mixture by passing the mixture over 2% platinum-on-carbon granules at 340°–385° with 4 to 8 moles of hydrogen at a liquid hourly space velocity of 0.7 to 0.8. Under these conditions, little or no benzene or cyclohexene is formed and catalyst life is prolonged (Feder and Silber, 1968). The use of acidic supports in this type of reaction promotes dehydration and they should be avoided if the oxygen function is to be preserved.

AROMATIZATION

Dehydrogenation of hydroaromatic systems is facilitated by the resonance energy derived in forming an aromatic system. For example, cycloheptane and hexamethylenimine are relatively stable under conditions by which cyclohexane and piperidine are converted readily to benzene and pyridine (Ehrenstein and Marggraff, 1934). A second, and perhaps more important, feature of aromatization is the delimiting of dehydrogenation to defined, predictable portions of the molecule. This together with the thermal stability of the resulting aromatic system allows fair to excellent yields in many reactions. Selected examples of the formation of a variety of carbocyclic and heterocyclic aromatic compounds derived by dehydrogenation are given in the following sections with the aim of illustrating the diversity of types of compounds, of catalysts, and of reaction conditions. Partially hydrogenated latent aromatic systems are much more readily aromatized than the fully hydrogenated compound. It is possible using palladium-on-carbon as a catalyst to completely dehydrogenate the former in a mixture of the two without any dehydrogenation of the latter (Harris *et al.*, 1971). Aromatic dehydrogenation is endothermic and the reaction has been studied for use as a heat sink in connection with travel in space vehicles (Ritchie and Nixon, 1967).

Carbocyclic Aromatic Systems

Dehydrogenation has been used with considerable success as a key step in the synthesis of many complex aromatic hydrocarbons. Pyracene (**II**) was obtained in 67% yield by heating 4.55 gm of **I** with 200 mg of 10% palladium-on-carbon at 300°–320°C under a nitrogen atmosphere for 90 minutes (Anderson and Wade, 1952). Later it was found that dehydrogenation proceeded more cleanly by the use of a 5% rhodium-on-alumina catalyst. A mixture of 0.24 gm of **I**, 15 ml of anhydrous benzene, and 0.1 gm of 5% rhodium-on-alumina heated in a sealed tube at 290°C for 18 hours afforded pyracene in 81% yield (Anderson and Anderson, 1957).

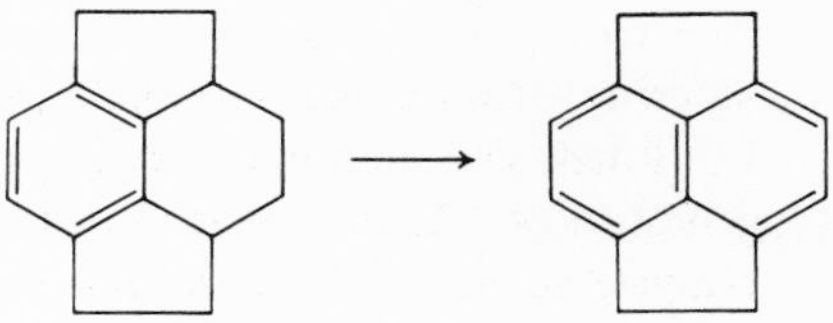

(**I**) (**II**)

Dehydrogenation of 0.3 gm of **III** in 20 ml of refluxing cymene over 0.2 gm 10% palladium-on-carbon gave quantitative yields of **IV** (Bergmann and Szmuszkovicz, 1951).

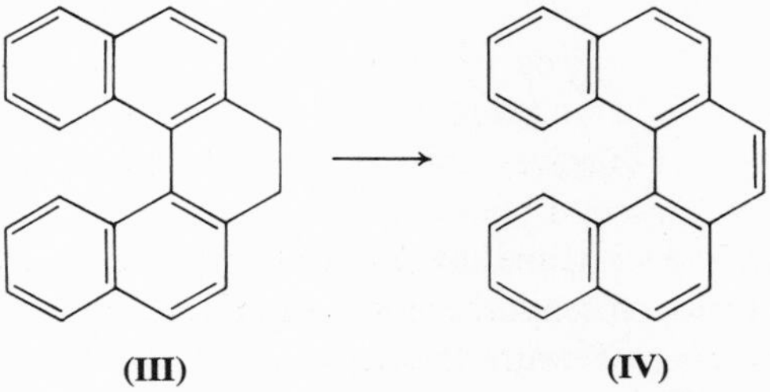

(**III**) (**IV**)

A general synthesis of linear condensed polynuclear aromatic hydrocarbons involves a dehydrogenation over palladium-on-carbon at 340°–375°C (Bailey and Liao, 1955).

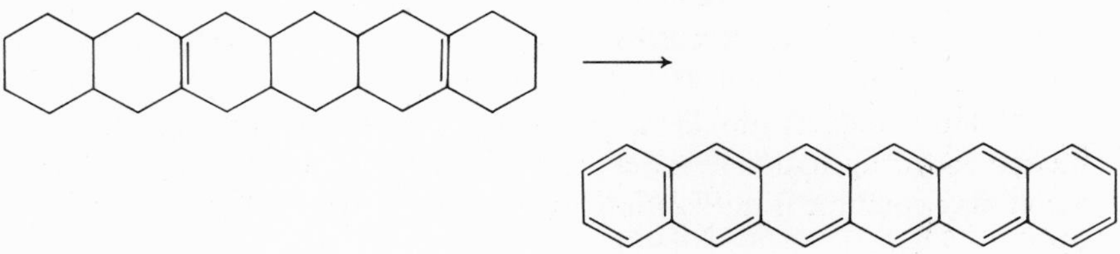

Dehydrogenation of **V** in boiling cymene over a 10% palladium-on-carbon catalyst with a nitrogen sweep affords 1,1′-biisoquinoline (**VI**) in good yield (Nielsen, 1970).

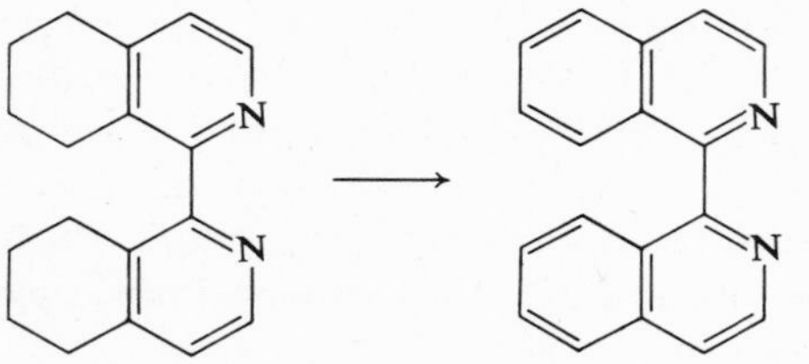

(V) (VI)

A convenient synthesis of 1,5-dimethylnaphthalene involves dehydrogenation of 3,4-dihydro-1,5-dimethylnaphthalene over palladium-on-carbon (Butz, 1940). Dehydrogenation of **VII** over platinum black at 300°C affords **VIII**, the reaction proceeding so as to form the more stable phenanthrene system (Rahman *et al.*, 1970).

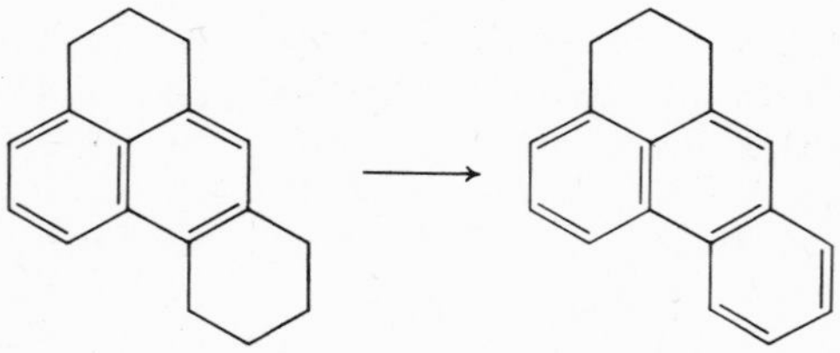

(VII) (VIII)

Dehydrogenation over palladium proved to be a useful technique in synthesis of benzocyclobutenes. An intimate mixture of **IX** and 5% palladium-on-carbon heated together for 6 to 7 hours affords **X** in 52–72% yield (Garrett, 1969).

(IX) (X)

R = H, CH_3

Diels-Alder adducts provide a convenient source of compounds that may undergo dehydrogenation to the corresponding aromatic derivatives. Aromatization may combine isomerization and dehydrogenation (Woods and Viola, 1956).

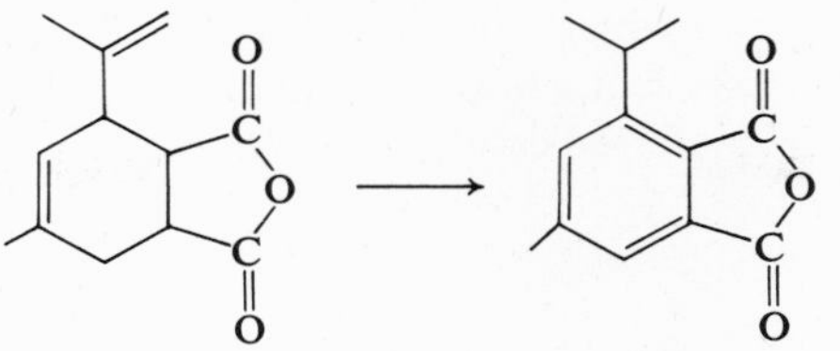

Under suitable conditions, dehydro dimers may be formed in a single reaction. For example, butadiene is dimerized to vinylcyclohexene and this compound isomerized and dehydrogenated, all reactions occurring with a single reactor. Passage of 100 ml/minute of a mixture consisting of 15% butadiene in inert gas over 5 gm of 2% ruthenium-on-carbon at 250°C affords 75% ethylbenzene and 25% vinylcyclohexene and isomers (Williamson, 1970).

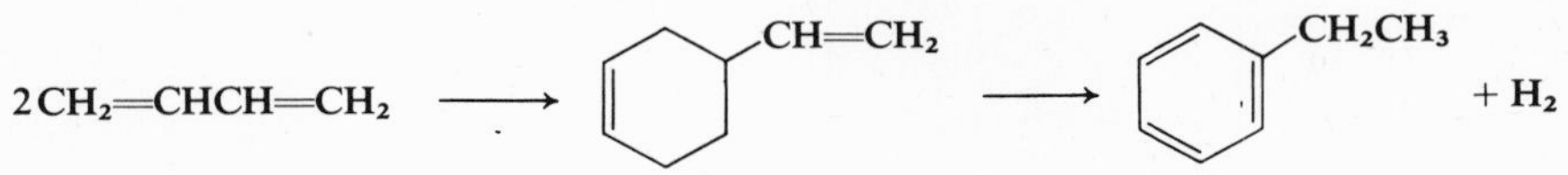

$2\,CH_2{=}CHCH{=}CH_2 \longrightarrow$ (vinylcyclohexene, $CH{=}CH_2$) $\longrightarrow$ (ethylbenzene, CH_2CH_3) $+ H_2$

Nitrogen-Containing Compounds

Hydrogenation of compounds containing or forming a basic nitrogen atom sometimes proceeds with difficulty because of poisoning of the catalyst by the strongly adsorbed nitrogen atom. Dehydrogenation, on the other hand, seems to be little affected by the presence of basic nitrogen and many such compounds have been successfully dehydrogenated. Palladium-on-carbon has been the most used catalyst (Galat, 1951; Ritter and Murphy, 1952; Burnett and Ainsworth, 1958; Yakhontov *et al.*, 1969; Ninomiya *et al.*, 1969; Terashima *et al.*, 1969; Wani *et al.*, 1970).

Indazole is obtained in 62% yield by refluxing 5 gm of 4,5,6,7-tetrahydroindazole with 3.5 gm of 5% palladium-on-carbon in 100 ml of dry decalin for 24 hours. Similarly, 4,5,6,7-tetrahydro-3(1*H*)-indazolone (**XI**) affords 3(1*H*)-indazolone (**XII**). Dehydrogenation of 4,5-trimethylenepyrazole under the above conditions fails; starting material is recovered unchanged (Ainsworth, 1957).

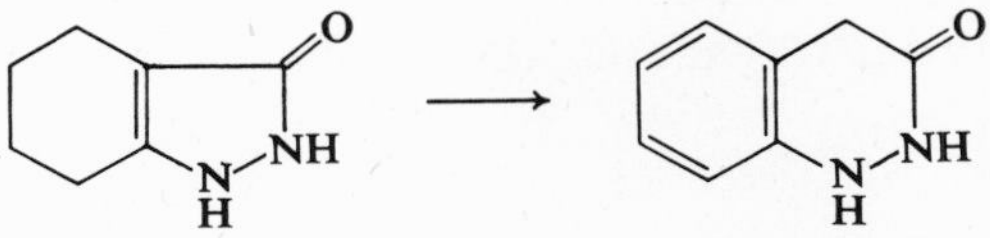

(XI) → (XII)

Dehydrogenation of **XIII** over palladium-on-carbon gives the corresponding indole, **XIV**, without difficulty, but the reaction fails when applied to the chloro compound, **XV**, due to hydrogenolysis of the halogen (Hester *et al.*, 1970). Palladium makes an excellent catalyst for dehydrohalogenation of aromatic halogen and is not the preferred catalyst when halogen is to be preserved.

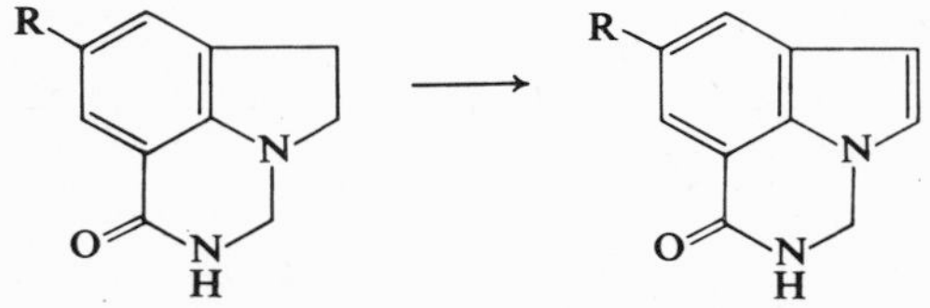

(XIII) R = H
(XV) R = Cl

(XIV)

Dehydrogenation of 1,11 α-iminoestrone-3-methyl ether (**XVI**) over 10% palladium-on-carbon affords the indole derivative of estrone, **XVII**, in 90% yield (Cantrall *et al.*, 1967).

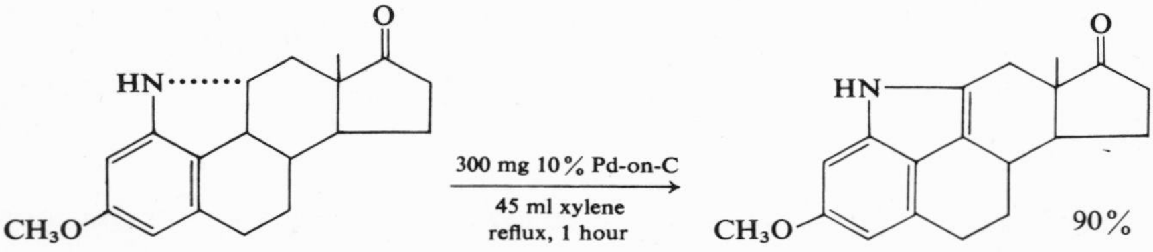

300 mg 10% Pd-on-C
45 ml xylene
reflux, 1 hour

(XVI)
892 mg

(XVII) 90%

Various tetrahydrocarbazoles are dehydrogenated easily to the corresponding carbazoles by refluxing in xylene over a 30% palladium-on-carbon catalyst. In most cases, the carbazoles are isolated in excellent yield and in a high state of purity merely by evaporation of the solvent after removal of the catalyst (Campaigne and Lake, 1959). Good results are obtained also in this reaction over palladium-on-carbon in refluxing triethylbenzene (Horning *et al.*, 1948a). Cinnamic acid may be used both as solvent and hydrogen acceptor in this dehydrogenation (Hoshino and Takiwa, 1936).

Heterocyclics containing a piperidine or pyrrolidine ring are converted easily to the corresponding pyridines and pyrroles by dehydrogenation. A mixture of 0.2 gm of **XVIII** and 0.45 gm of 5% palladium-on-carbon in 4 gm of diphenyl at 250°–260°C affords **XIX** (Sargent and Agar, 1958).

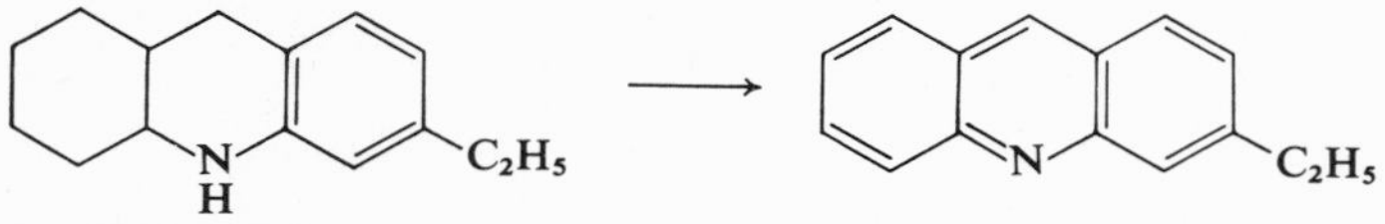

(XVIII)

(XIX)

Dihydroisoquinolines may be converted to isoquinolines by refluxing in *p*-xylene over palladium black (White and Dunathan, 1956). Dehydrogenation of complex tetrahydroisoquinolines may be carried out also over 10% palladium-on-carbon at 225°–235°C (Locke and Pelletier, 1959). A commercially attractive synthesis of papeverine involves dehydrogenation of 3,4-dihydropapeverine (Pal, 1958). The dehydrogenation can be carried out in boiling tetralin over 10% palladium-on-carbon (Mozingo, 1955) affording papeverine in 95% yield.

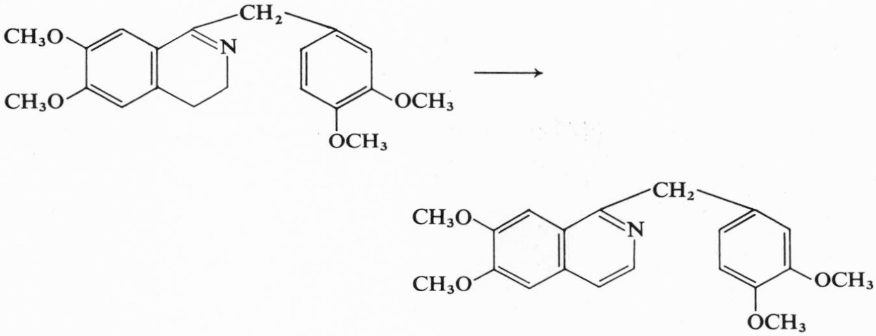

Dehydrogenation of octahydro-1,5-pyrindine (**XX**) over 30% palladium-on-carbon in the vapor phase with hydrogen as a carrier gas affords 6,7-dihydro-1,5-pyrindine (**XXI**) in 77% yield (Lochte and Pittman, 1960). The dehydrogenation was expected to yield **XXI** and not 1,5-pyrindine (**XXII**) since earlier attempts to dehydrogenate **XII** to 1,5-pyrindine had all met with failure (Prelog and Szpilfogel, 1945).

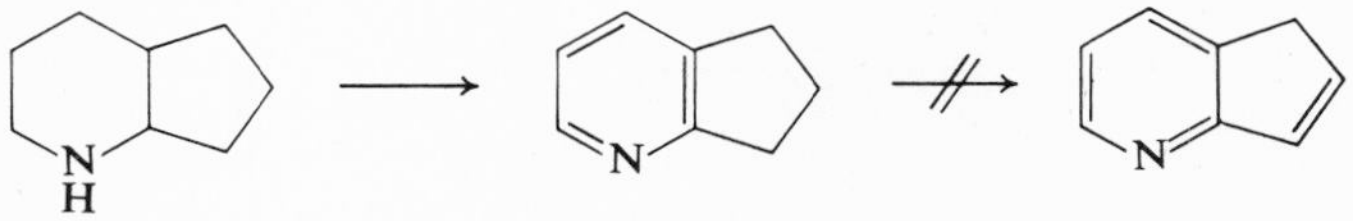

(**XX**) (**XXI**) (**XXII**)

Saturated and unsaturated valerolactams can be dehydrogenated over palladium to provide a variety of substituted pyridones that would be difficult to prepare by other routes (Shamma and Rosenstock, 1961).

An attractive, but little used, entry to indoles involves dehydrogenation of 4,5,6,7-tetrahydroindoles. For example, dehydrogenation of **XXIII** over 10%

palladium-on-carbon in refluxing mesitylene affords 1-(*p*-hydroxyphenyl)-5-methoxy-2-phenylindole (**XXIV**) in 78% yield, the methoxy group being largely retained (Bell *et al.*, 1970).

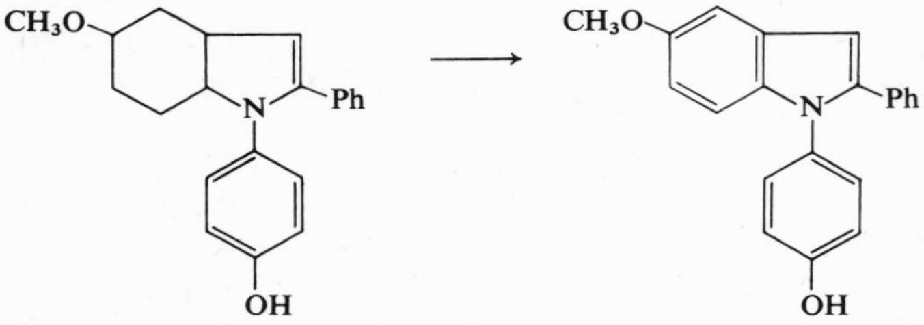

(XXIII) → (XXIV)

Formation of 5-trimethylsilylindole by dehydrogenation of the corresponding indoline proceeds smoothly in boiling xylene over palladium-on-carbon, but the reaction is unsatisfactory in boiling toluene (Belskii *et al.*, 1968).

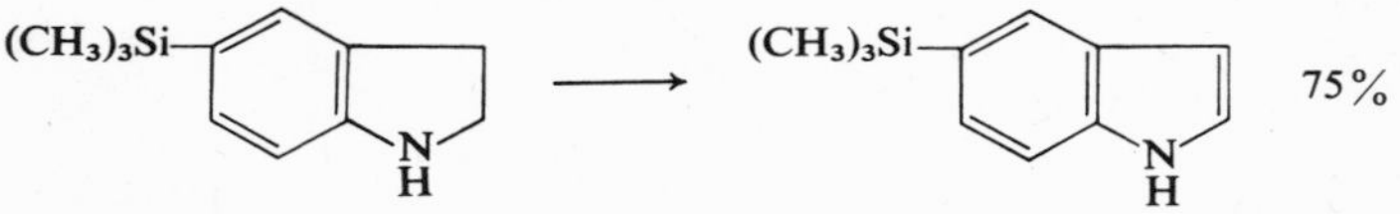

75%

A new route to quinolines and isoquinolines involves dehydrogenation of the products obtained through condensation of piperidone enamines and methyl *β*-vinylacrylate (Danishefsky and Cavanaugh, 1968).

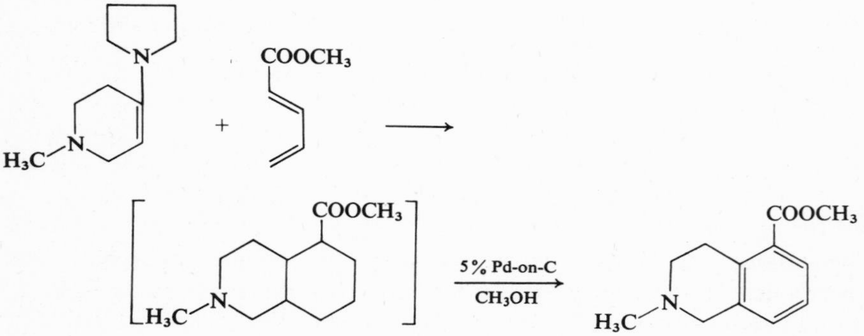

5% Pd-on-C, CH_3OH

The intermediate bicyclic condensation product obtained in 94% yield is a mixture of three compounds at various levels of unsaturation. Dehydrogenation of 1.95 gm of this mixture over 0.3 gm of 5% palladium-on-carbon in 10 ml of refluxing methanol for 6 hours affords 2-methyl-5-carbomethoxy-1,2,3,4-tetrahydroisoquinoline in 93% yield. A similar sequence begins with *N*-benzyl-3-piperidone, but ends with debenzylation of the isoquinoline and dehydrogenation of both rings, affording 5-carbomethoxyisoquinoline. Dehydrogenation of 4.8 gm of the condensation mixture in 100 ml of refluxing *p*-cymene containing 15 gm of *trans*-stilbene over 2.0 gm of 5% palladium-on-carbon affords the product in 56% overall yield from the piperidone.

Quinolines are formed by beginning with 3-piperidones. The dehydrogenation steps have some remarkable aspects. Dehydrogenation of 2.5 gm of the

O NCH₂Ph N H NCH₂Ph N COOCH₃ COOCH₃ N CH₂Ph (XXV) Pd/C COOCH₃ N (XXVII) Pd/C COOCH₃ N CH₂Ph (XXVI)

mixture of condensation compounds, **XXV**, in 10 ml of refluxing toluene over 1 gm of 5% palladium-on-carbon for 26 hours affords 1-benzyl-5-carbomethoxy-1,2,3,4-tetrahydroquinoline (**XXVI**) in 51% yield. This compound undergoes a surprisingly facile debenzylation and dehydrogenation under very mild conditions to afford 5-carbomethoxyquinoline (**XXVII**). A solution of 102 mg of **XXVI** in 5 ml of absolute alcohol with 500 mg of 5% palladium-on-carbon, held for 2 hours at atmospheric pressure and room temperature, evolves gas and affords 5-carbomethoxyquinoline in 81% yield (Danishefsky and Cavanaugh, 1968). This example is noteworthy because of the unusually mild conditions employed. Under more vigorous conditions, nitrogen substituents such as alkyl (Fujita *et al.*, 1968), acyl (Taub *et al.*, 1967; Wendler *et al.*, 1969a, b), benzoyl (Chemerda and Sletzinger, 1970), or carbethoxy (Rapoport and Willson, 1962) usually are lost readily to permit aromatization of the nitrogen ring.

Conversion of **XXVIII** to **XXIX** provides another example of facile aromatization which in this case may be achieved merely by refluxing in xylene for 14.5

hours with activated carbon without metal. This and other compounds of this series are readily dehydrogenated over palladium-on-carbon (Capps and Hamilton, 1953).

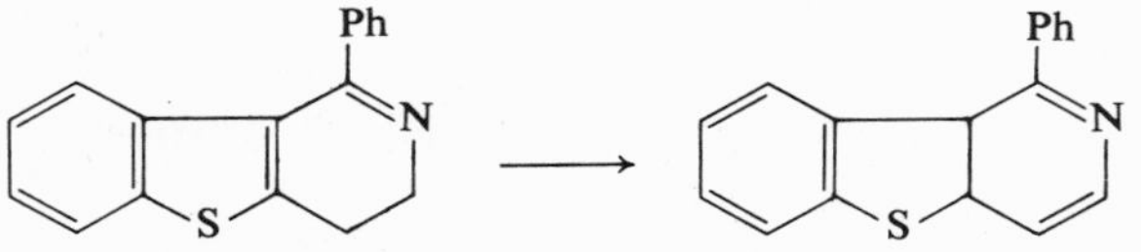

(XXVIII) → (XXIX)

Ease of dehydrogenation may be markedly dependent on stereochemistry. A striking example of differences in reactivity of stereoisomers is illustrated by behavior of *cis*- and *trans*-decahydroquinoxalines under dehydrogenation conditions. The *cis* isomer is dehydrogenated readily to tetrahydroquinoxaline in refluxing *p*-cymene under a slow stream of nitrogen in the presence of rhodium-on-alumina, whereas the *trans* isomer is unaffected. Moreover, the *cis* isomer is dehydrogenated merely by dissolving it in phenyl ether at 60°C; other solvents such as chloroform, ethyl ether, benzene, or isooctane exhibit no dehydrogenation properties (Broadbent *et al.*, 1960).

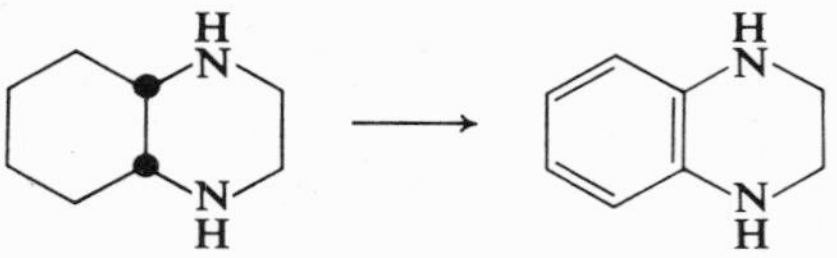

Certain dehydrogenations may occur even under hydrogenation conditions. On treatment of **XXX** with 3 atm of hydrogen in the presence of platinum oxide, the benzyl group suffers hydrogenolysis and additionally a hydrogen is lost, forming a 5,6-iminium bond which may be stabilized by conjugation with the nitrogen atom at the bridgehead affording **XXXI** (Leonard *et al.*, 1967).

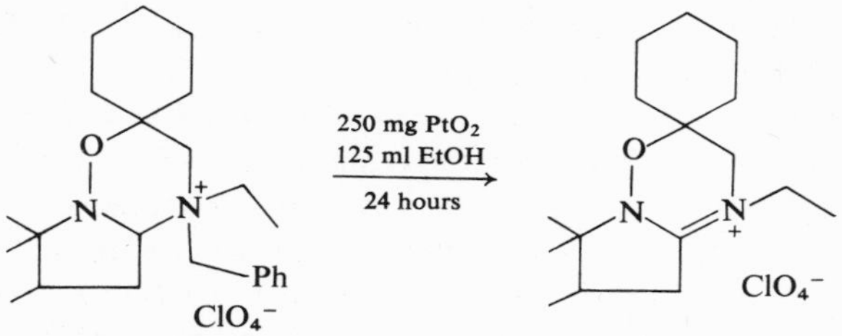

250 mg PtO_2
125 ml EtOH
24 hours

(XXX) 1 gm → (XXXI)

Palladium-on-carbon proved satisfactory for the dehydrogenation of **XXXII** to the new heteroaromatic compound, 14,16,18-tribora-13,15,17-

triazatriphenylene (**XXXIII**). The yields were low but all chemical attempts to achieve this reaction such as treatment with sulfur, selenium, or manganese dioxide, were uniformly unsatisfactory.

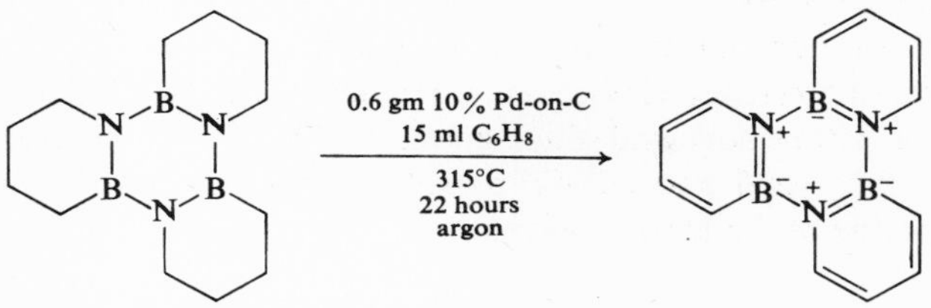

0.6 gm 10% Pd-on-C
15 ml C_6H_8
315°C
22 hours
argon

(**XXXII**) 3 gm (**XXXIII**)

In similar manner, dehydrogenation of 1-benzyl-2-phenyltetrahydro-borazarene (**XXXIV**) affords a mixture of 2-phenylborazarene (**XXXV**) and 1-benzyl-2-phenylborazarene (**XXXVI**) (Davies *et al.*, 1967).

(**XXXIV**) (**XXXV**) (**XXXVI**)

Pyrroles

A variety of pyrrolidine derivatives have been dehydrogenated both in liquid and vapor phase to afford pyrroles. Excellent yields of pyrrole are obtained by dehydrogenation of pyrrolidine over a palladium-on-silica gel catalyst at 420°C with a weight hourly space velocity of 0.5 and hydrogen as a carrier gas. The silica gel is characterized by having a silanol group density of 5–10 $SiOH/10^{-18}$ m^2. Silanol densities beyond these limits results in catalysts that deactivate readily (Guyer and Fritze, 1970). Pyrroles are also obtained by dehydrogenation of the corresponding pyrrolidine over a palladium chloride-on-alumina catalyst. For example, 4500 lbs of *N*-methylpyrrolidine passed over 200 lb of catalyst at 175°–200°C with a weight hourly space velocity of 3, affords 3300 lb of *N*-methylpyrrole together with some unchanged starting material. Spent catalysts are reactivated by oxidation with low concentrations of oxygen (Zellner, 1961). Pyrrolidine is converted to pyrrole by passage over 0.5% rhodium-on-alumina pellets at 650°C and a liquid hourly space velocity of 6 to 8. Shorter contact times give lower yields, longer contact times more decomposition (Patterson and Drenchko, 1959). 1-Pyrroline may be an intermediate in these dehydrogenations (Fuhlhage and Vanderwerf, 1958).

A general procedure for the synthesis of terpyrroles involves condensation of a 2,2′-bipyrrole and a 2-pyrrolidinone to give a pyrrolinylbipyrrole followed by dehydrogenation of the latter over palladium-on-carbon (Rapoport *et al.*, 1964). 2,2′-Bipyrrole and certain derivatives are obtained readily by dehydrogenation of an appropriate pyrrolidinylpyrrole over 5% palladium-on-carbon in refluxing xylene (Rapoport and Holden, 1962). Similar dehydrogenations reported (Rapoport and Castagnoli, 1962) afford a mixture of 2,2′-bipyrrole in 25% yield and 2,2′-(1′-pyrrolinyl)-pyrrole in 46% yield. Di-*n*-hexyl ether is one of the better solvents.

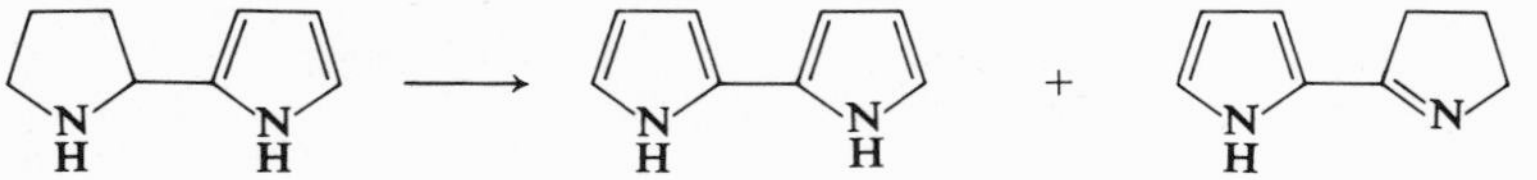

Bipyrroles can be obtained by dehydrogenation of 2,2′-(1′-pyrrolinyl)-pyrroles, but the yields are not good except when an ester group is present on the nucleus. The authors attributed the improved yields with ester-containing compounds to activation of the hydrogen atom α to the ester and stabilization of the resulting bipyrrole (Rapoport and Bordner, 1964).

Other Heterocyclics

A variety of heterocyclic compounds have been prepared by dehydrogenation over noble metals. Facile formation of dehydrogenated sulfur compounds is noteworthy inasmuch as the reverse reaction usually proceeds with difficulty and with severe catalyst inhibition by the sulfur present. The phenomenon is in accord with earlier comments that dehydrogenation catalysts may be improved at times by sulfiding. An example of facile dehydrogenation of sulfur containing compounds is the formation of thiazoles from the dihydro precursor. A mixture of 75 gm of 2-(2′-thiazolin-2′-yl)benzimidazole and 7.5 gm of 5% palladium-on-carbon refluxed under nitrogen in 500 ml of diphenyl ether for 32 hours affords 2-(2′-thiazolyl)benzimidazole in 77% yield (Ennis, 1969).

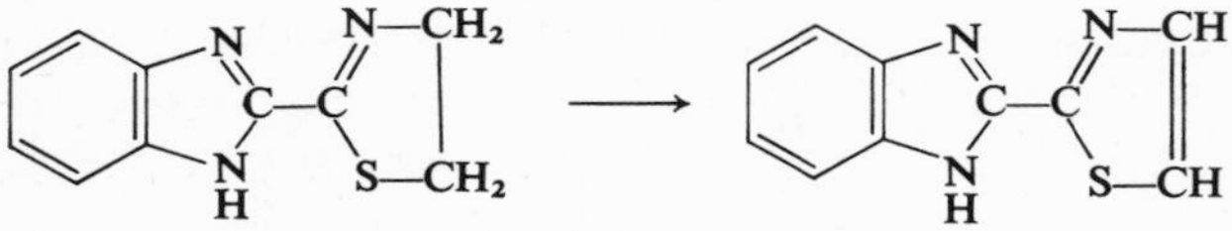

An interesting dehydrogenation also involving a rearrangement has been reported by Anderson *et al.* (1963). Continuous dehydrogenation of **XXXVII** over a palladium-on-carbon catalyst at 340°–360°C affords the octadehydro compound, **XXXVIII**, in 32% yield accompanied by benzthiophene (**XXXIX**) in 9% yield. Benzthiophene is apparently a rearrangement product of **XXXVIII**

and can be obtained from it by heating at 450°C. The catalyst used in these dehydrogenations improves with use.

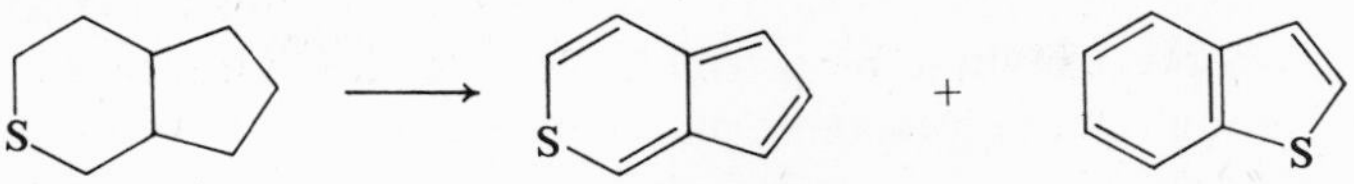

(XXXVII) → (XXXVIII) + (XXXIX)

Furan derivatives may be prepared by dehydrogenation of dihydrofurans (Chatterjee *et al.*, 1964) and coumarins by dehydrogenation of dihydrocoumarins (Esse and Christensen, 1960; Das Gupta and Chatterjee, 1968; Das Gupta *et al.*, 1969b; Yates and Field, 1970). Disproportionation may accompany dehydrogenation. 2-Methyl-4,5-dihydrofuran is converted over ruthenium, rhodium, osmium, or iridium-on-carbon to a mixture of α-methylfuran and tetrahydromethylfuran. Some methyl propyl ketone is also formed through hydrogenolysis or through isomerization of tetrahydromethylfuran (Shuikin *et al.*, 1962b).

The following example illustrates a convenient synthesis of certain coumarins which involves condensation of methyl methacrylate with the appropriate phenols in the presence of aluminum chloride and dehydrogenation of the product over palladium-on-carbon in refluxing diphenyl ether (Das Gupta *et al.*, 1969a).

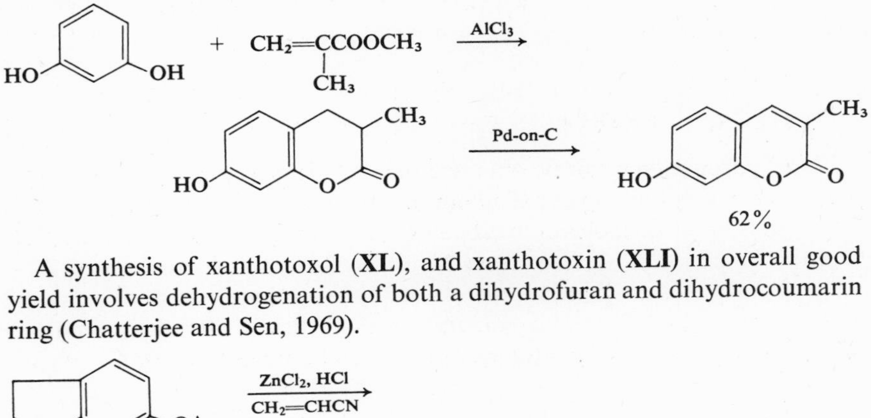

HO, OH + $CH_2{=}C(CH_3)COOCH_3$ —$AlCl_3$→ HO, CH_3, O, O —Pd-on-C→ HO, CH_3, O, O

62%

A synthesis of xanthotoxol (**XL**), and xanthotoxin (**XLI**) in overall good yield involves dehydrogenation of both a dihydrofuran and dihydrocoumarin ring (Chatterjee and Sen, 1969).

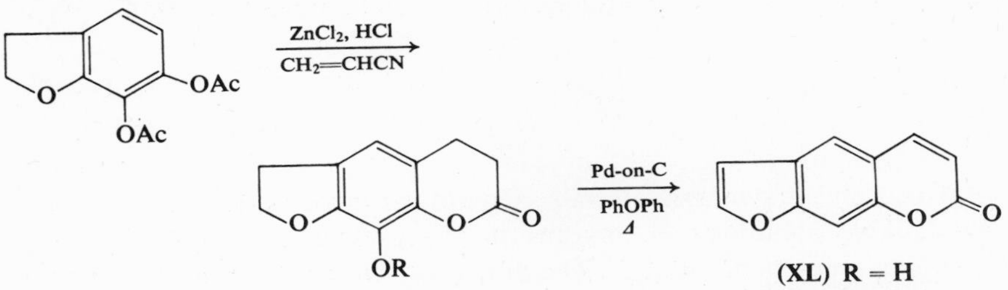

OAc, OAc —$ZnCl_2$, HCl, $CH_2{=}CHCN$→ OR —Pd-on-C, PhOPh, Δ→

(**XL**) R = H
(**XLI**) R = CH_3

Dehydrogenation of tetrahydroxanthyletin (**XLIII**) over palladium-on-carbon in boiling diphenyl ether affords only dihydroxanthyletin (**XLIV**) and not the expected xanthyletin (**XLII**). Failure of the chroman ring to dehydrogenate probably can best be attributed to steric hindrance by the *gem*-dimethyl group (Das Gupta and Das, 1969).

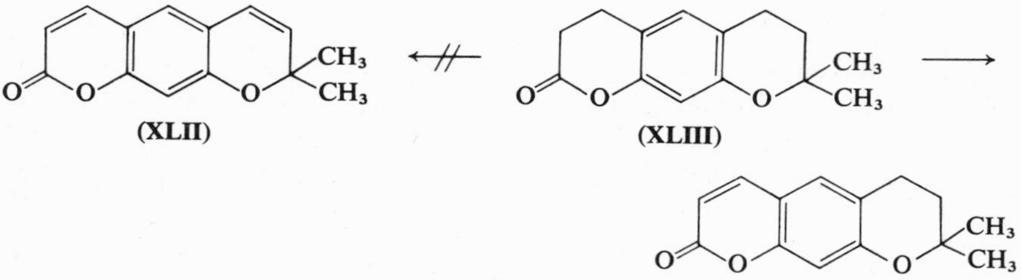

(XLII) (XLIII) (XLIV)

Various nitrogen heterocyclics can be obtained by a combination of amination and dehydrogenation of oxygen heterocyclics. Pyridine is obtained in yields up to 60% by interaction of ammonia and 2-hydroxymethyltetrahydrofuran over a palladium-on-alumina catalyst at 235° to 450°C. The reaction proceeds through a mixture of tetrahydropyridines and is accompanied by heavier products. The latter may be eliminated completely by conducting the reaction with 1 mole of oxygen present per mole of the furan (Butler and Laundon, 1970a).

$$+ \ NH_3 \longrightarrow \ + \ 2H_2O + 2H_2$$

Tetrahydropyridine can be obtained in 27% yield by passage of 2-aminomethyltetrahydrofuran over palladium-on-alumina or silica-alumina at 300°C in the presence of 1.5 moles of hydrogen. Contact time is about 0.9 second. The same product may be obtained from 2-hydroxymethyltetrahydrofuran in 21% yield by passage of this material over the catalyst with 6 moles each of ammonia and hydrogen at 0.7 second contact time. An acidic alumina support is necessary for ring expansion to occur (Butler and Laundon, 1970b).

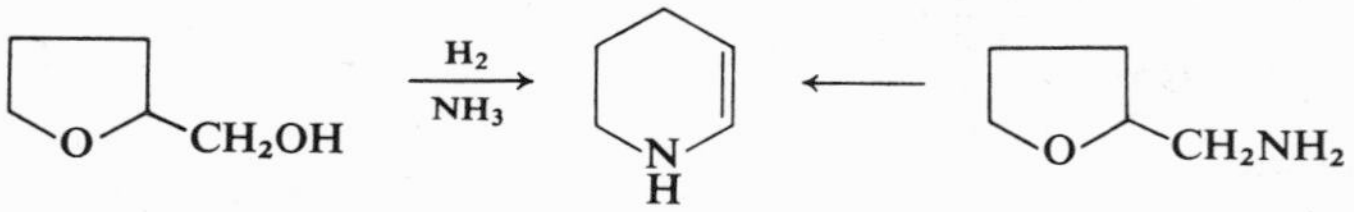

A convenient synthesis of 3-alkoxypyridines involves dehydrogenation of 2,3-dialkoxy-3,4-dihydro-1,2-pyrans in the presence of ammonia over a platinum-on-alumina catalyst. The dihydropyrans are formed by interaction of 1,2-dialkoxyethylenes and α,β-unsaturated carbonyl compounds.

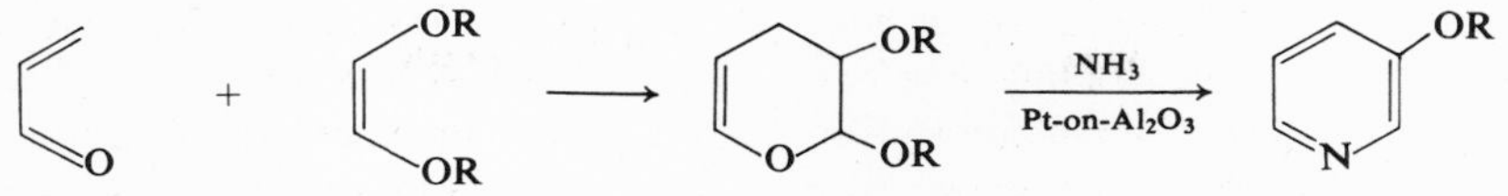

2,3-Diethoxy-3,4-dihydro-1,2-pyran, formed in 74% yield by interaction of acrolein and 1,2-diethoxyethylene, is passed over 0.5% platinum-on-alumina at 3 gm/hour per milliliter of catalyst with ammonia and water at 220°C to afford 3-ethoxypyridine in 67% yield. Similarly, 3-propoxypyridine is prepared from 2,3-dipropoxy-3,4-dihydro-1,2-pyran in 58% yield and 3-ethoxy-4-methylpyridine from 2,3-diethoxy-4-methyl-3,4-dihydro-1,2-pyran in 71% yield (Chumakov and Sherstyuk, 1967).

Formation of New Bonds

Under dehydrogenation conditions new carbon–carbon, carbon–nitrogen and carbon–oxygen bonds may form. These are considered separately in the sections that follow. New carbon–carbon bonds formed in isomerization of the carbon skeleton are discussed in the section on isomerization.

Carbon–Carbon Bond Formation: Intermolecular Processes

A useful synthesis of 2,2′-biquinolyls, compounds otherwise prepared with some difficulty, involves dehydrogenation of quinolines over palladium catalysts. The conversions are not large and most of the starting material is recovered unchanged. The general procedure for heterocyclic biaryls involves refluxing a stirred mixture of quinoline (or related compound) with 10% by weight of 5% palladium-on-carbon for 24 hours. There is little advantage to continuing heating beyond 24 hours. The presence of oxygen had no adverse effect on the yield (Rapoport *et al.*, 1960). Later workers found rhodium-on-carbon to be superior to palladium in this reaction (Jackson *et al.*, 1963); the overall yield of coupled products is 3 to 6 times greater over rhodium. However, over rhodium, unlike palladium, the coupled product contains about $\frac{2}{3}$ 2,2′-biquinolyl and $\frac{1}{3}$ 2,3′-biquinolyl.

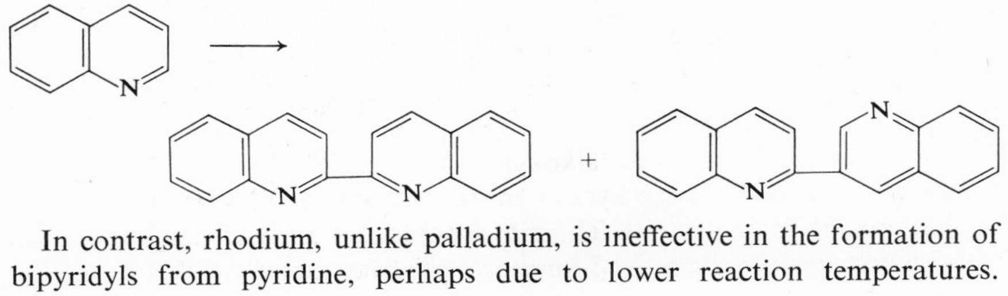

In contrast, rhodium, unlike palladium, is ineffective in the formation of bipyridyls from pyridine, perhaps due to lower reaction temperatures.

Palladium-on-carbon is about twice as effective as palladium-on-alumina (Jackson *et al.*, 1963). Dehydrogenation of pyridine over palladium results also in the formation of small amounts of pyrrole (Rylander and Karpenko, 1970). Ring contraction of pyridine was noted earlier when pyridine was treated with degassed Raney nickel (Sargeson and Sasse, 1958).

An unusual type of coupling occurs when 2-methylquinoline is employed in this reaction. No 4,4′-biquinolyl is formed but instead a small yield of 1,2-di(2-quinolyl)ethane is obtained. The reaction is similar to the coupling of toluene to afford dibenzyl.

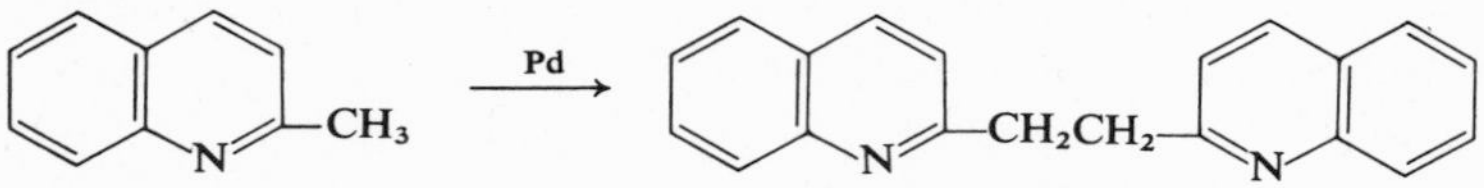

N CH_3 $\xrightarrow{Pd}$ N CH_2CH_2 N

Dehydrogenation of 2-methylquinoxaline (**XLV**) over palladium-on-carbon at 200°C takes a different course and affords **XLVI** in which new carbon–carbon and nitrogen–carbon bonds are established (Cheeseman and Tuck, 1968).

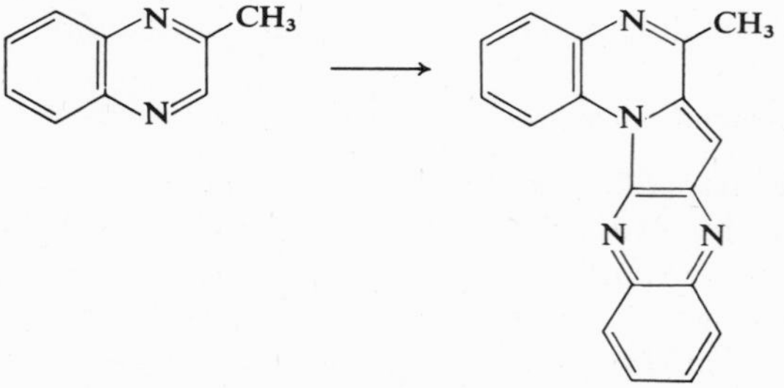

N CH_3 N ⟶ N CH_3 N N N

(XLV) **(XLVI)**

Acetylenes

An elegant method for synthesis of cycl(3,2,2)azine derivatives involves treatment of pyrrocoline with dimethyl acetylenedicarboxylate in boiling toluene containing a 5% palladium-on-carbon catalyst (Galbraith *et al.*, 1959).

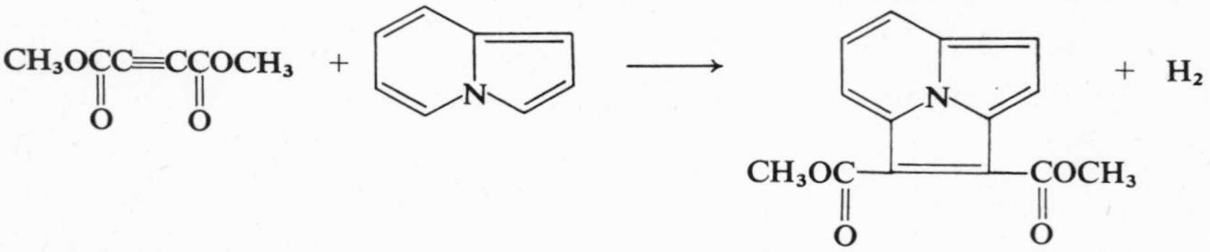

$CH_3OCC{\equiv}CCOCH_3$ (O O) + N ⟶ N CH_3OC—$COCH_3$ (O O) + H_2

The reaction has been applied also to the synthesis of **XLVIII** be refluxing **XLVII** under nitrogen in toluene containing diethyl acetylenedicarboxylate, 10% palladium-on-carbon, and a trace of hydroquinone. In this case, the addition is facilitated by recovery of the full resonance energy of two benzenoid rings (Godfrey, 1959).

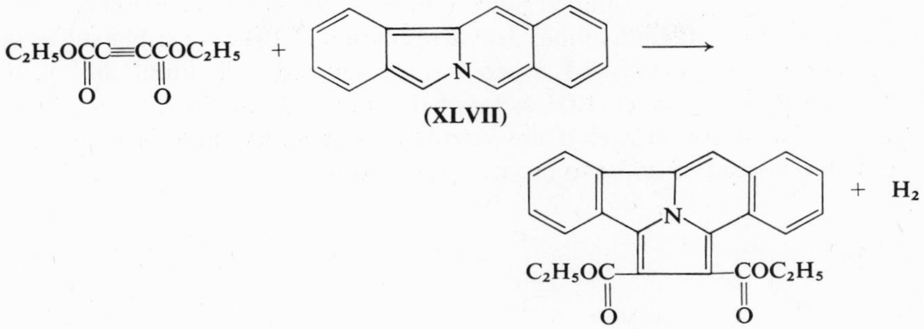

$C_2H_5OCC{\equiv}CCOC_2H_5$ (with C=O, C=O) + (XLVII) ⟶

C_2H_5OC(=O) ... COC_2H_5(=O) + H_2

(XLVIII)

Unsymmetrical acetylenes may undergo this dehydrogenation with a high degree of specificity. The cyclazine, **L**, was obtained in 50% yield by refluxing methyl propiolate and methyl 1-methoxycarbonylmethyl-6,8-dimethylindolizine-2-carboxylate (**XLIX**) in dry toluene over 5% palladium-on-carbon for 23 hours (Acheson and Robinson, 1968). The mode of addition had been established earlier for several cases (Boekelheide *et al.*, 1963).

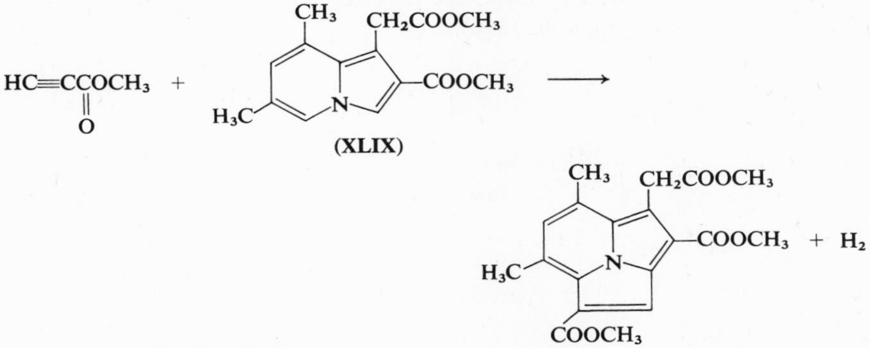

$HC{\equiv}CCOCH_3$ (C=O) + CH_3, CH_2COOCH_3, $COOCH_3$, H_3C (XLIX) ⟶

CH_3, CH_2COOCH_3, $COOCH_3$, H_3C, $COOCH_3$ + H_2

(L)

Carbon–Carbon Bond Formation: Intramolecular Processes

Formation of carbon–carbon bonds occurs rather frequently under dehydrogenation conditions, especially so as the temperature is raised. The most frequent examples are a result of an alkyl group migration. Others involve joining of aromatic nuclei or alkylation of an aromatic. The outcome may depend on both the catalyst and the temperature of reaction. Palladium or platinum is the usual catalyst. A novel variation is a supported sulfided palladium. The catalyst is superior to a mixture of palladium and sulfur in that the former avoids introduction of sulfur into the product (Crawford and Supanekar, 1970). The sulfided palladium may give different products than either sulfur or palladium alone; dehydrogenation of **LII** over either sulfur or palladium affords mostly **LI**, whereas over sulfided palladium, the cyclodehydrogenation product, **LIII**, is favored (Crawford and Supanekar, 1969). Arsenic and phosphorus, but not selenium or mercury, may be used with palladium instead of sulfur to promote cyclization.

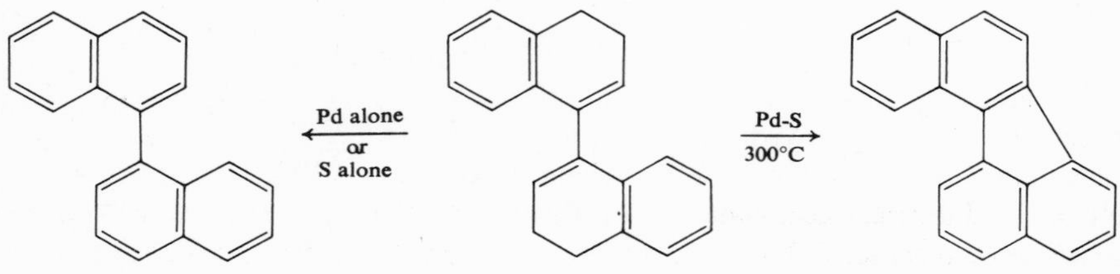

Pd alone or S alone ← ; Pd-S, 300°C →

(LI) **(LII)** **(LIII)**

An example of the effect of temperature on the product formed is the dehydrogenation of **LV** over palladium-on-carbon. At 250°C the aromatic, **LVI**, is formed, whereas at 320°C the product is the pentacyclic compound, **LIV** (Canonne and Regnault, 1969).

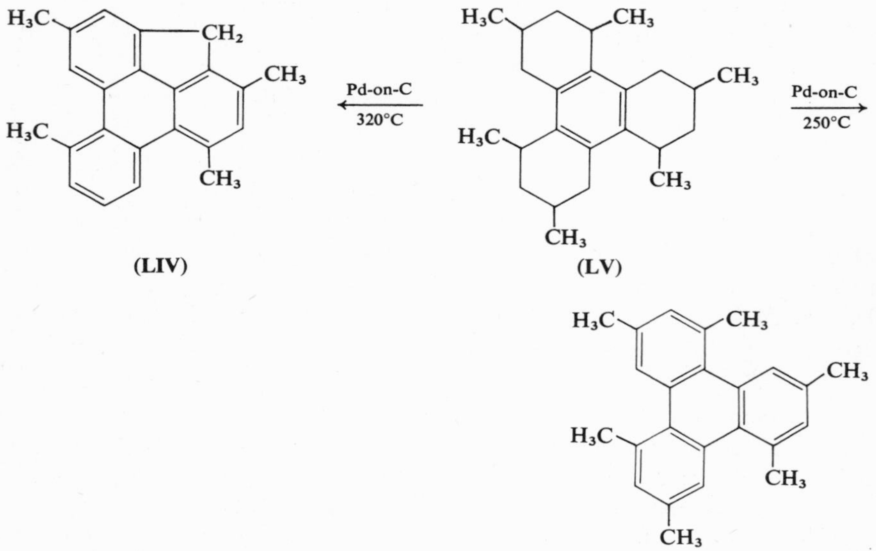

Pd-on-C, 320°C ← ; Pd-on-C, 250°C →

(LIV) **(LV)** **(LVI)**

Aromatics containing suitably orientated alkyl groups may undergo facile alkylation under dehydrogenation conditions. Acenaphthene and acenaphthylene are prepared in better than 90% combined yield by passage of α-ethylnaphthalene over 0.7% platinum-on-alumina containing lithium, sodium, or potassium carbonate as promoter. The yield of acenaphthylene increases with increasing temperature (Suld, 1967).

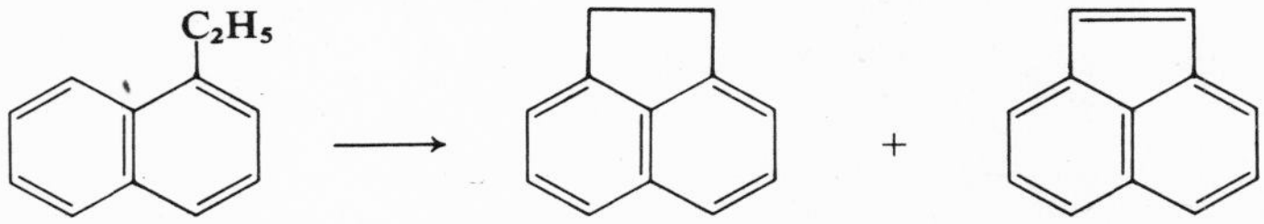

C_2H_5

+

The catalyst may have a profound effect on the products obtained in this type of reaction. Dehydrocyclization of *β-n*-butylnaphthalene over 10% platinum-on-carbon affords anthracene as the sole product, whereas over chromium-alumina catalysts phenanthrene is the major product (Shuikin *et al.*, 1960, 1962a).

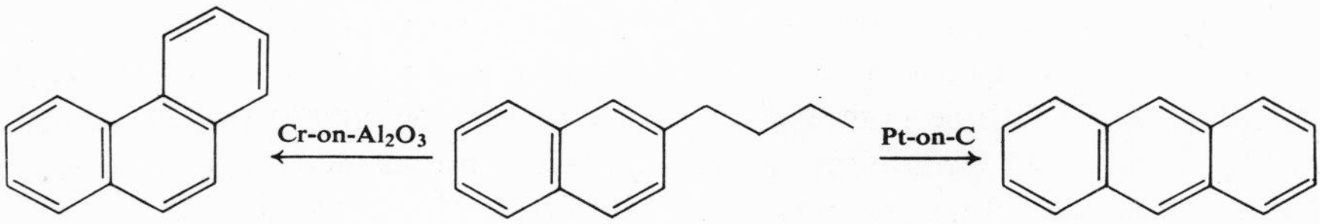

Cr-on-Al_2O_3 Pt-on-C

Azulene is formed in 20% yield together with naphthalene on heating cyclodecane at 340°C with palladium-on-carbon (Prelog and Schenker, 1953).

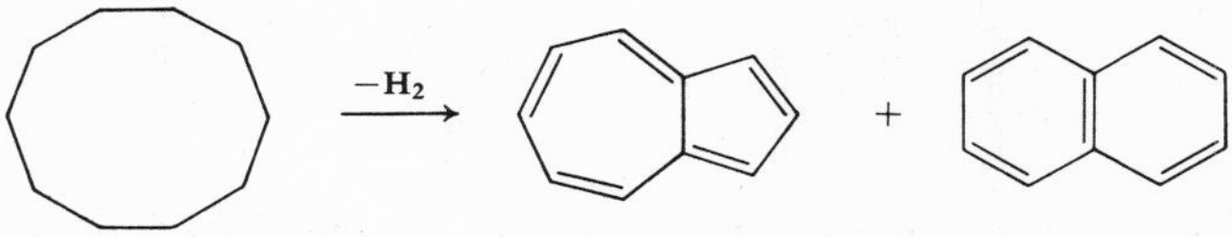

$-H_2$ +

A new synthesis of 3-ketobenzo(*d,e*) steroids involves cyclodehydrogenation of propargyl enol ethers of Δ^4-3 ketosteroids over palladium-on-carbon in refluxing solvent such as dimethylformamide, pyridine, or *o*-dichlorobenzene. Hydrogen acceptors such as ethyl cinnamate may be present. Yields are about 30% (Ercoli *et al.*, 1968).

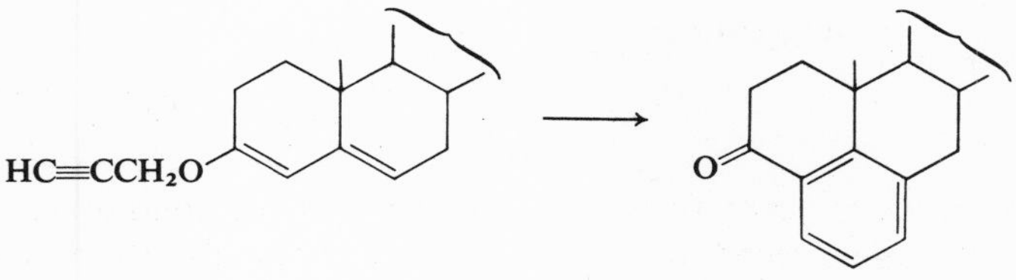

$HC{\equiv}CCH_2O$ O

Suitably disposed compounds may readily form a new carbon–carbon bond through transannular reactions. Dehydrogenation of linderane (**LVII**) over 10% palladium-on-carbon at 300°C for 30 seconds was accompanied by loss of the lactone and epoxide function and ring closure to form ujacazulene (**LVIII**) (Takeda *et al.*, 1964).

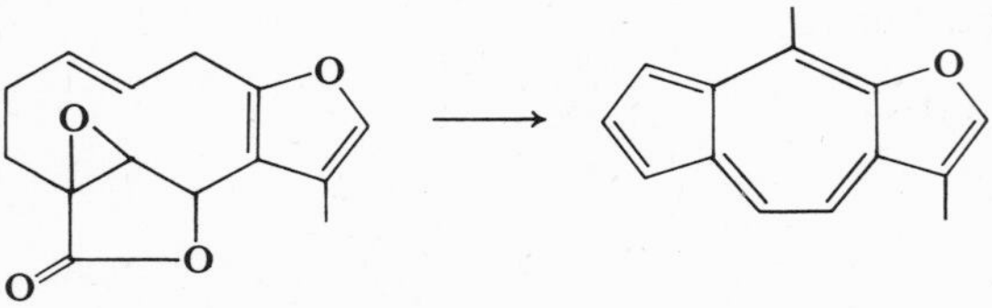

(LVII) **(LVIII)**

Formation of Carbon–Nitrogen Bonds

Ring closure with formation of a carbon–nitrogen bond is apt to occur in systems containing a carbonyl or incipient carbonyl function. Imidazole is formed by dehydrogenation of a gaseous mixture of formamide, ethylenediamine, and hydrogen over platinum-on-alumina at 340°–480°C (Green, 1966). Pyridine and quinoline compounds are formed by dehydrogenation of oxo nitriles over 0.5% palladium-on-alumina. Continuous dehydrogenation of 2-(2-cyanoethyl)cyclohexanone at 250°–300°C affords quinoline, and at lower temperatures, 200°–260°C, a mixture with 5,6,7,8-tetrahydroquinoline (Simpson *et al.*, 1959).

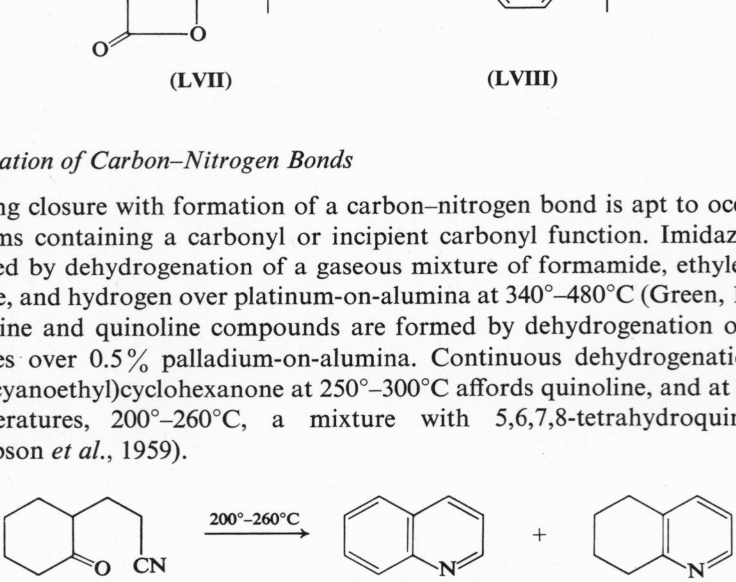

O CN —200°–260°C→ N + N

A useful synthesis of the pyrrocoline ring system involves dehydrogenation of 3-(2′-pyridyl)-1-propanol. The ring closure probably proceeds through an intermediate propionaldehyde. Pyrrocoline was obtained in 50% yield by refluxing, after removal of air, 12 gm of substrate over 0.5 gm of 10% palladium-on-carbon for 12 hours. 2-Phenyl-5-methylpyrrocoline was obtained similarly from 3-(6-methyl-2-pyridyl)-2-phenyl-1-propanol (Boekelheide and Windgassen, 1959).

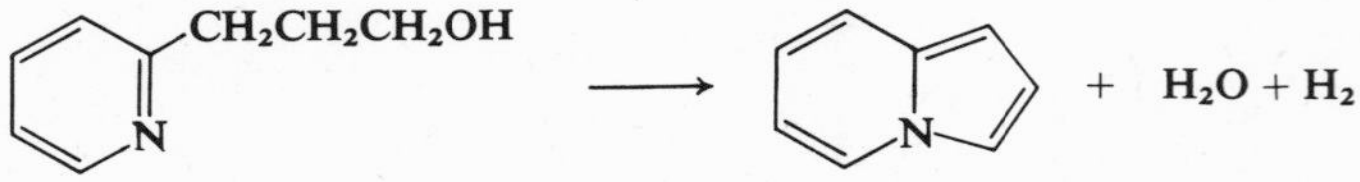

$CH_2CH_2CH_2OH$ (N) ⟶ (N) + $H_2O + H_2$

Similar reactions occur with hydrocarbons, but the conditions are more severe than those needed with oxygenated compounds. Pyrrocoline, 1,2-

benzopyrrocoline, and 2,3-benzopyrrocoline have been formed by continuous dehydrogenation over platinum-on-alumina at 490°C of 2-propylpyridine, 2-(*o*-tolyl)pyridine, and 2-benzylpyridine, respectively. Indoles are formed by dehydrogenation of *o*-ethylaniline over platinum-on-alumina at 480°C (Voltz *et al.*, 1959). At 550°C and a space velocity of 1, *o*-ethylaniline is dehydrogenated over platinum-on-alumina to afford high yields of indole together with lesser amounts of *o*-vinylaniline. Indole is obtained to the virtual exclusion of *o*-vinylaniline if the dehydrogenation is carried out in the presence of hydrogen, whereas in the presence of an inert gas such as nitrogen, the yield of *o*-vinylaniline is increased to 15% (Voltz and Weller, 1961).

Coupling of the aromatic rings in diphenylamines through dehydrogenation provides a convenient synthesis of carbazoles. The reaction is carried out continuously with both hydrogen and steam present to keep the catalyst clean (German Patent 1,203,785). Carbazoles may be formed also by dehydrogenation of *o*-aminobiphenyls.

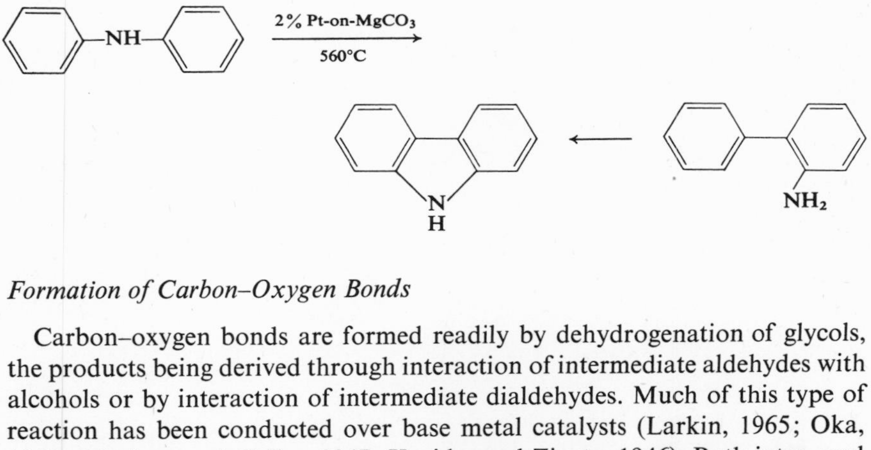

—NH— $\xrightarrow[560°C]{2\%\ Pt\text{-}on\text{-}MgCO_3}$

N
H ⟵ NH_2

Formation of Carbon–Oxygen Bonds

Carbon–oxygen bonds are formed readily by dehydrogenation of glycols, the products being derived through interaction of intermediate aldehydes with alcohols or by interaction of intermediate dialdehydes. Much of this type of reaction has been conducted over base metal catalysts (Larkin, 1965; Oka, 1962; Schniepp and Geller, 1947; Kyrides and Zienty, 1946). Both intra- and intermolecular reactions occur. Dehydrogenation of ethylene glycol over 0.2% palladium oxide-on-carbon granules at 310°–320°C or over 1% platinum-on-carbon granules at 240°–250°C affords 2,3-dihydro-*p*-dioxin in 76 and 70% yield, respectively (Guest and Kiff, 1964).

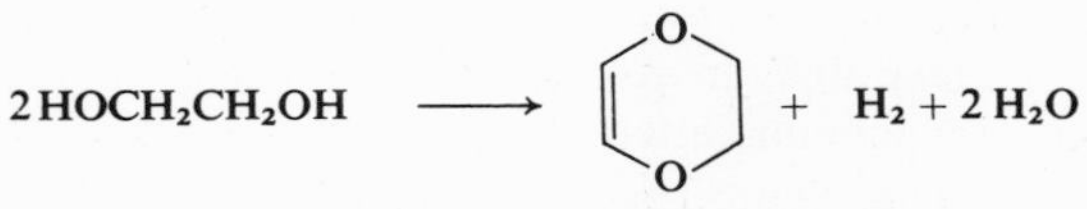

$$2\,HOCH_2CH_2OH \longrightarrow \text{(O, O ring)} + H_2 + 2\,H_2O$$

Ring closures occur readily with appropriately substituted phenols to afford benzofurans in a manner analogous to the formation of indoles from *ortho*-

substituted anilines. Benzofuran is formed by continuous dehydrogenation of *o*-ethylphenol over platinum-on-alumina at 480°C (Voltz *et al.*, 1959). Similarly, *o*-allylphenol is converted to 2-methylbenzofuran in 45% yield in passage over lithium hydroxide-treated 0.25% platinum-on-alumina spheres at a weight hourly space velocity of 0.2 at 550°C. *o*-Propylphenol may be used also, but the yield is only about 9% (Illingworth and Louvar, 1966). Alkalia treatment of alumina is a useful technique to minimize high-temperature cracking reactions.

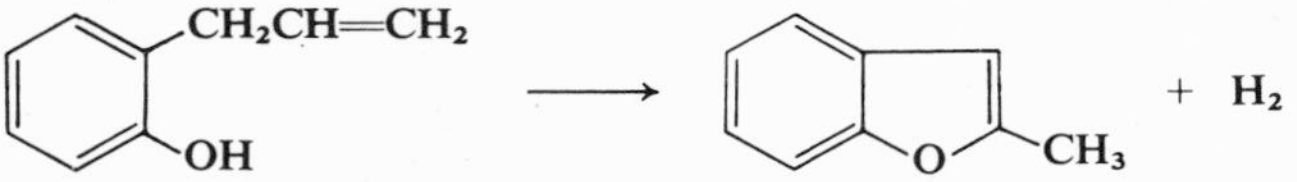

$CH_2CH{=}CH_2$, OH → O, CH_3 + H_2

An unusual rearrangement occurs during dehydrogenation of cyclohexylphenols to phenylphenols over palladium-on-carbon at 300°C. In addition to the expected phenylphenol, diphenyl ether is formed in appreciable quantities together with small amounts of dibenzofuran from the *ortho* isomer (Matsumura *et al.*, 1971).

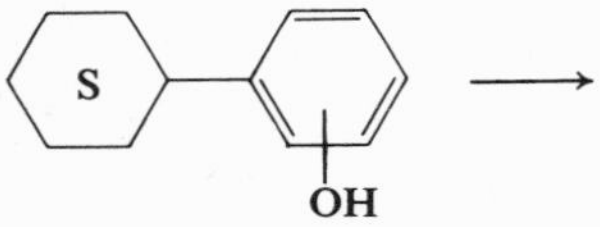

S, OH →

o-Cyclohexylphenol
p-Cyclohexylphenol

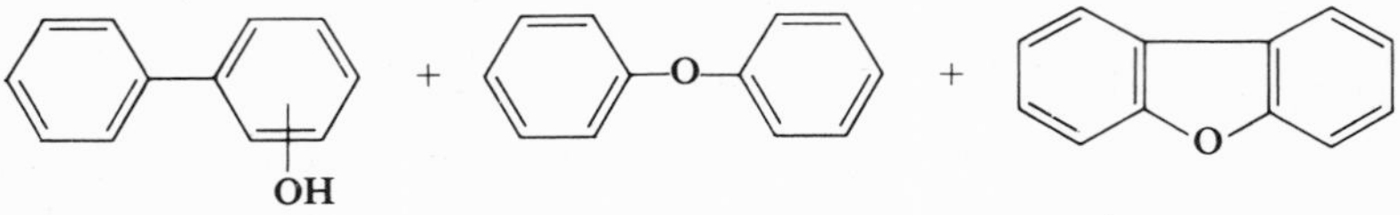

OH + O + O

72% *ortho*	23%	3%
79% *para*	17%	—

Most dehydrogenations occur only at elevated temperature unless a hydrogen acceptor is present. An unusual exception has been reported by Ling and Djerassi (1970). An extremely facile dehydrogenation of dichotine (**LIX**) occurs when this substrate is merely stirred with 10% palladium-on-carbon in absolute ethanol under nitrogen, resulting in ring closure and formation of a carbinolamine ether, **LX**. The reaction can be readily reversed.

Another unusually facile dehydrogenation occurs when the oxazolidines, **LXI**, are refluxed in an aqueous ethanol solution over 5% palladium-on-carbon or Raney nickel. The reaction proceeds cleanly to give either the amide, **LXII**, or hydroxyamide, **LXIII**, but not a mixture of both. The product depends strongly on the substituents at C-2.

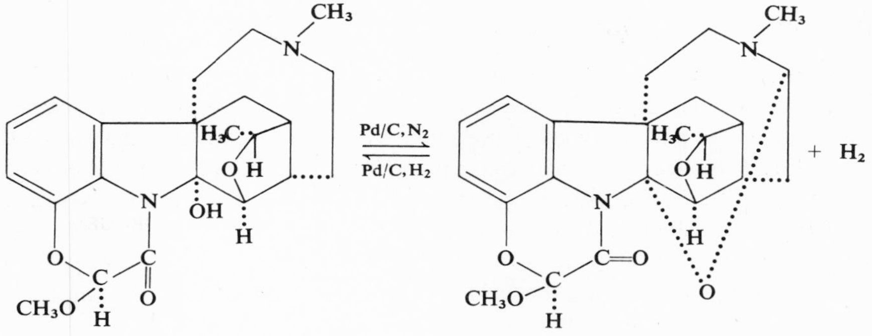

Pd/C, N_2 / Pd/C, H_2 + H_2

(LIX) (LX)

EtOH / H_2O

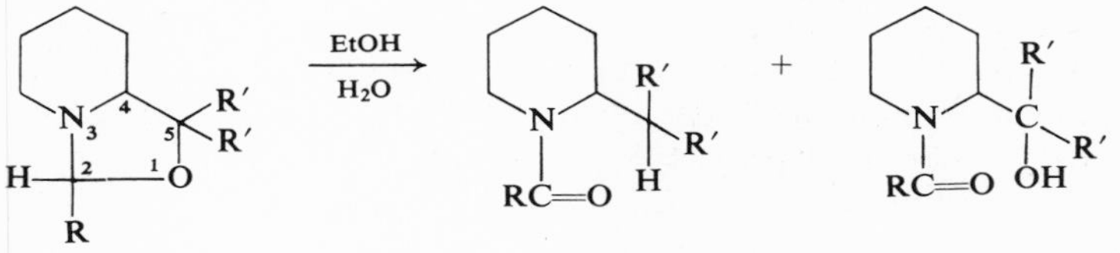

(LXI) (LXII) (LXIII)

When R′ = C_6H_5 quantitative yields of the amide, **LXII**, are obtained within 2 hours, whereas with R′ = CH_3 the reaction is much slower and after 24 hours, the hydroxyamide, is obtained in 40% conversion together with unchanged starting material. The authors tentatively suggested dehydrogenation proceeds by a hydride abstraction. When R′ = C_6H_5 the amide is formed in 20% yield even under hydrogenation conditions (Ghiringhelli and Bernardi, 1967).

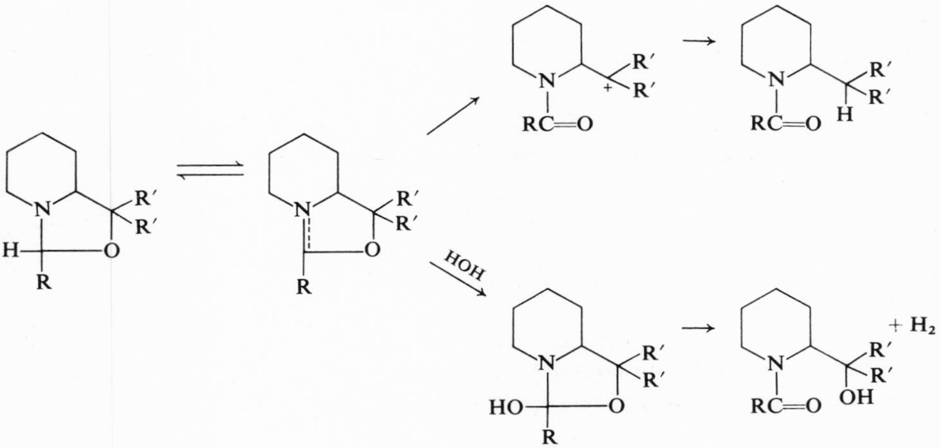

HOH + H_2

Dehydrogenation of Ketones to Phenols

Saturated or unsaturated ketones contained in an incipient aromatic system are dehydrogenated readily to the corresponding hydroxyaromatic compound. Some form of palladium or platinum is the catalyst usually used in these reactions. Hydrocarbons are used frequently as solvents, but materials such as phenyl ether, ethylene glycol, and ethylene glycol monoethyl ether may also prove quite satisfactory (Horning and Horning, 1947a). Improved results may be obtained if the feed is first alkaline-washed to remove traces of acid (Feder and Silber, 1968). At times, the yields are only fair because of a competing loss of the oxygen function, a reaction discussed further in the section on loss of functional groups. This side reaction is influenced by both the metal and support. Carbon may prove more satisfactory than alumina when dehydration is to be minimized. For instance, cyclohexanone is converted to phenol nearly quantitatively over 0.5% platinum-on-carbon granules at 300°C, whereas over 0.5% platinum-on-alumina, the phenol is contaminated with about 10% benzene (Rylander and Kilroy, 1958).

Mosettig and Duvall (1937) examined in some detail the effect of catalyst and solvent on the dehydrogenation of 1-keto- and 4-keto-1,2,3,4-tetrahydrophenanthrene to the corresponding phenanthrols.

O HO

0.33 gm Pd → naphthalene, 24 hours, reflux

86%

Naphthalene was a much more effective solvent for this reaction than either tetralin or xylene. Palladium black was the best catalyst, whereas palladium-on-carbon (in limited examination) was ineffective. Satisfactory yields were obtained also with platinum oxide. Palladium differed significantly from platinum in that over the latter catalyst yield loss was caused more by incomplete reaction than by side reaction, whereas over palladium, the reverse was true. Other workers found palladium-on-carbon quite satisfactory for the dehydrogenation of similar molecules (Newman and Blum, 1964; Fales *et al.*, 1955). Dehydrogenation in a stream of nitrogen of 5 gm of **LXIV** over 1.0 gm of 10% palladium-on-carbon in 40 ml of refluxing α-methylnaphthalene afforded the phenol **LXV**, in 74% yield after recrystallization (Turner *et al.*, 1956).

Dehydrogenation of ketones has also proved useful in the synthesis of hydroxyazulenes. Continuous dehydrogenation of **LXVI** over 30% palladium-on-carbon suspended on asbestos affords 4-hydroxyazulene, **LXVII** (Anderson and Nelson, 1951).

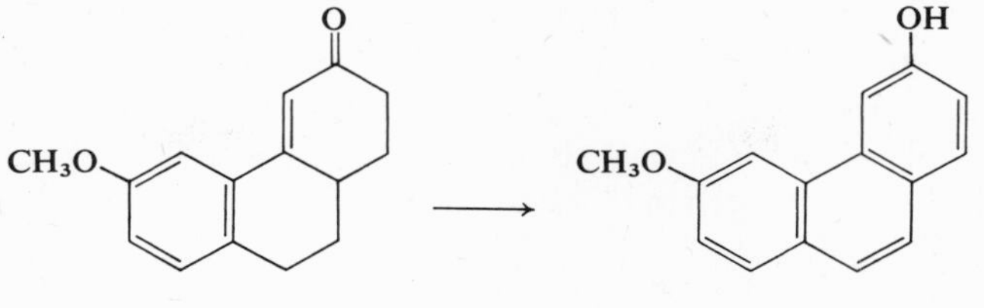

(LXIV) → (LXV)

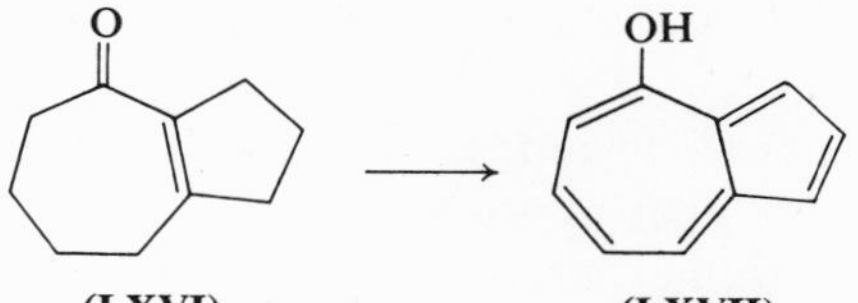

(LXVI) → (LXVII)

Secondary changes of the phenol may occur in certain structures (Walker, 1958). Dehydrogenation of 1.78 gm of **LXVIII** over 830 mg of 10% palladium-on-carbon in an open tube at 250°C for 1 hour affords 5,8-dimethylcoumarin (**LXIX**), a double bond being introduced in the new lactone ring. The reaction was also of diagnostic value, eliminating **LXX** as an alternative structure for the substrate, inasmuch as a vinylogous keto acid would be expected to undergo decarboxylation at these elevated temperatures (Wendler *et al.*, 1951). Lactonization does not occur when higher homologs are dehydrogenated (Bhandari and Bhide, 1970).

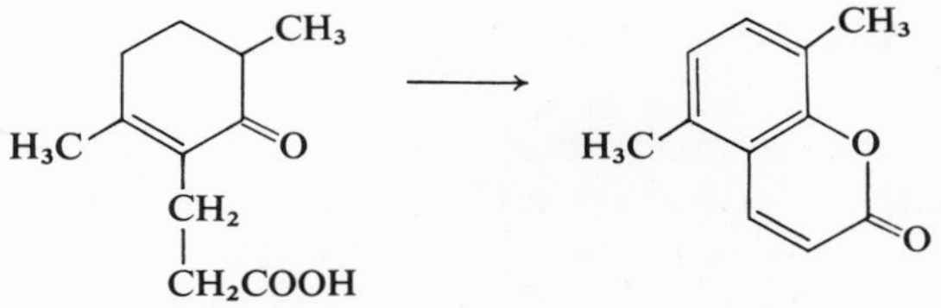

(LXVIII) → (LXIX)

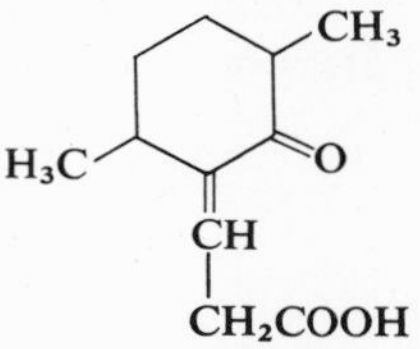

(LXX)

Aromatizations may be accompanied by hydrogenolysis reactions. A useful synthesis of 2-methylphenols and 2,6-dimethylphenol follows this course and involves hydrogenolysis of the Mannich bases of cyclohexanone. For example,

100 gm of 2,6-bis(dimethylaminomethyl)cyclohexanone is heated with 6 gm of 5% palladium-on-carbon for 2 hours at 210°C while a stream of nitrogen is passed through the reaction mixture. Steam distillation of the product affords 2,6-dimethylphenol in 62% yield (Bajer and Carr, 1968). Surprisingly efficient use is made of the hydrogen in this hydrogenolysis.

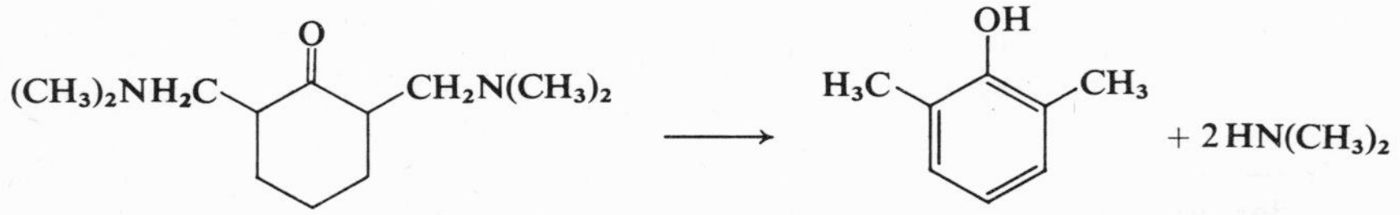

$(CH_3)_2NH_2C$ ... $CH_2N(CH_3)_2$ ⟶ H_3C ... CH_3 + $2\,HN(CH_3)_2$

A synthesis of 6-fluoroequilenin (**LXXII**) involves dehydrogenation of the enone, **LXXI**, over 10% palladium-on-carbon at 180°C. Aromatization is accompanied by dehydrofluorination; the initially formed benzylic fluoride was expected to be unstable under the conditions of the reaction. Lesser amounts of the completely dehydrofluorinated products, equilenin and 3-hydroxy-5,7,9(10)-estratrien-17-one are also formed (Boswell *et al.*, 1971).

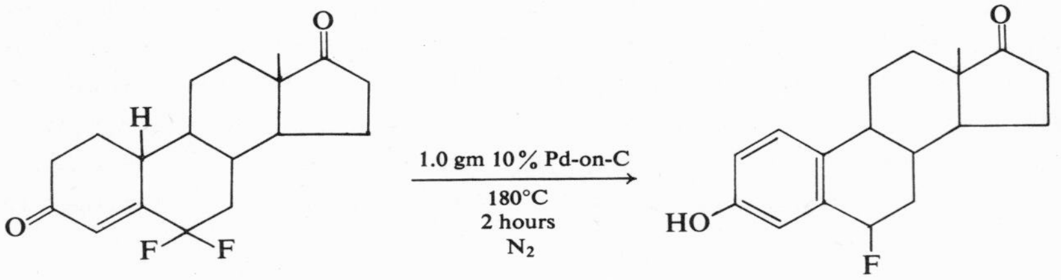

1.0 gm 10% Pd-on-C, 180°C, 2 hours, N_2

(LXXI) 0.5 gm **(LXXII)**

An interesting inhibition of phenol formation was reported in dehydrogenation of the indoline derivative, **LXXIII**, over palladium-on-carbon in *p*-cymene; the product was the benzindole, **LXXIV**. On the other hand, the *N*-benzoyl, **LXXV**, or the *N*-acetyl derivatives, **LXXVI**, are converted under similar conditions to the corresponding phenols, **LXXVII** and **LXXVIII**. The authors attributed the stabilization of the naphthalene system in the *N*-acylated compounds to suppression of the interaction between the nitrogen

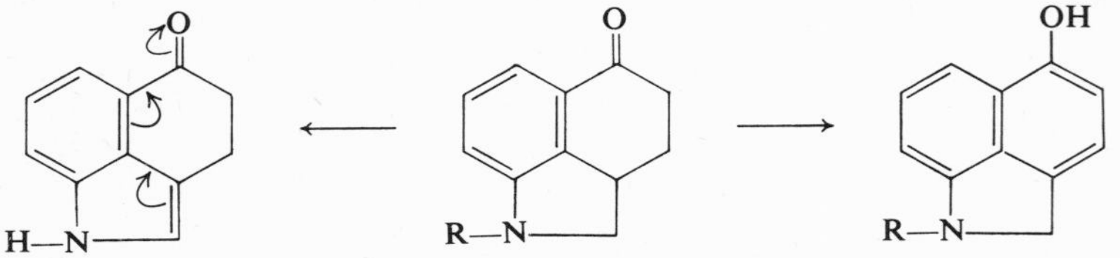

(LXXIV) R = H

(LXXIII) R = H
(LXXV) R = —COPh
(LXXVI) R = —$COCH_3$

(LXXVII) R = —COPh
(LXXVIII) R = —$COCH_3$

atom and the carbonyl group in the ketonic isomers, indicated by arrows (Kornfeld *et al.*, 1956).

DEALKYLATION AND ISOMERIZATION

Aromatization of alkyl-substituted hydroaromatics systems may be accompanied by breaking of the ring–alkyl bond with either elimination or migration of the split-off fragment. These latter reactions occur of necessity when the latent aromatic ring contains a quaternary carbon; they are apt to occur when the reactions relieve ring strain or crowding of the molecule. The final products may depend in large measure on the catalyst used.

Quaternary Carbons

Aromatization of rings containing a quaternary carbon atom occur under more vigorous conditions than simple dehydrogenations since a carbon–carbon bond must be broken. The ratio of migration to cleavage depends on the catalyst and the support (Linstead and Thomas, 1940), and to some extent the catalysts may determine which bond is broken. The relationship between the type of catalyst and the type of reaction is by no means always clear. Adkins and co-workers found in dehydrogenation through hydrogen transfer with benzene that nickel-on-kieselguhr favors migration, whereas platinum- or nickel-on-nickel chromite favors elimination (Adkins and Davis, 1949; Adkins and England, 1949). On the other hand, dehydrogenation of 1-methyl-1-phenylcyclohexane over chromia-alumina catalyst affords methane and biphenyl (Ipatieff *et al.*, 1950a, b) whereas over platinum-on-alumina at 340°C the product is apparently 2-methylbiphenyl derived by migration (Linsk, 1950). Similarly, dehydrogentaion of 1,1,3-trimethylcyclohexene over chromia-alumina afforded only *m*-xylene derived by elimination, whereas over platinum-on-alumina some trimethylbenzenes, derived by migration, were found as well (Pines *et al.*, 1953). Dehydrogenation of 1,1,3-triphenylindane over palladium-on-carbon affords 1,2,3-triphenylindene derived by migration (Hodgkins and Hughes, 1962).

Ph Ph Ph — 0.2 gm 5% Pd-on-C, 350°C, 20 minutes → Ph Ph Ph + H_2

1 gm

In dehydrogenations over noble metals, the results may depend on the method of preparation of the catalyst as well as the metal itself, as illustrated by dehydrogenation of 9-methyloctalin. With platinum catalysts prepared by reducing chloroplatinic acid with hydrogen at 135°C or palladium-on-carbon made by reduction with formaldehyde in alkaline solution, elimination of the methyl group predominates and the main product is naphthalene accompanied by some α-methylnaphthalene. But over platinum catalysts prepared by the method of Loew (1880), the migration of the methyl group becomes the main reaction, affording α-methylnaphthalene with little or no naphthalene (Linstead *et al.*, 1937).

In certain cases, the catalyst also determines to some extent which carbon–carbon bond is broken. Dehydrogenation of 1-methyl-1-ethyl-, or 1-methyl-1-*n*-propyl, or 1-methyl-1-isopropylcyclohexane over chromia-alumina affords toluene as the major product in each case, whereas over platinum-on-alumina, there is no selectivity in the removal of the alkyl groups (Pines and Marechal, 1955).

Adkins and England (1949) derived a number of useful generalities in dehydrogenation reactions of compounds containing quaternary carbons with benzene present as a hydrogen acceptor at 350°–375°C. Aromatization proceeds more cleanly over platinum than over nickel catalysts. Over platinum-on-carbon catalysts, aromatization is brought about chiefly through elimination of an alkyl group whereas the nickel(k) catalyst promotes aromatization through inducing alkyl group migration. Nickel-on-nickel chromite is intermediate between these two, inducing both elimination and migration of alkyl groups. Palladium-on-carbon is also an effective catalyst in promoting aromatization of compounds of this type (Adkins and Hager, 1949).

Dealkylation has been shown not to involve, in some cases at least, an olefin intermediate. Dehydrogenation of 1,1,3-trimethylcyclohexane over platinum-on-alumina affords *m*-xylene and methane, whereas over the same catalyst, dehydrogenation of the olefinic, 1,1,3-trimethyl-x-cyclohexene affords 62% *m*-xylene and 38% of 1,2,4-trimethylbenzene. Isomerization during dehydrogenation of the cyclohexane will occur over this catalyst, however, if the feed contains small amounts of *sec*-butyl chloride, used as a source of hydrogen chloride. Dehydrogenation of 1,1,3-trimethylcyclohexane containing 4 mole percent *sec*-butyl chloride afforded 32% 1,2,4- and 21% 1,2,3-trimethylbenzene (Pines *et al.*, 1953).

Strain Relief

Dealkylation of hydroaromatics is facilitated if ring strain is relieved by the reaction. Pinene, for instance, is converted readily over palladium or platinum catalysts to a mixture of *p*-cymene and pinane through cleavage of the cyclobutane ring and disproportionation (Linstead *et al.*, 1940).

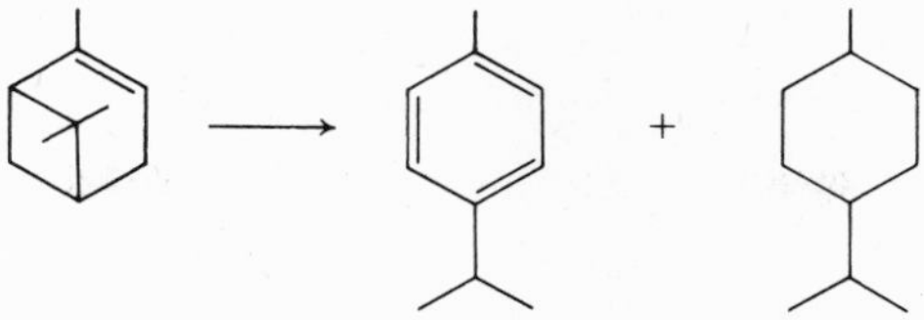

+

Some ring cleavage of this type may be thermally induced. Cleavage of the four-membered ring of pinane over platinum-on-alumina, -pumice, or -charcoal at 240°–300°C occurs apparently by two different reactions; a thermal cleavage favors formation of *p*-cymene and the hydrogen liberated in the aromatization step of this reaction favors hydrogenolysis of pinane and formation of 1,1,2,3- and 1,1,2,5-tetramethylcyclohexane and *o*-menthane. At higher temperatures, the tetramethylcyclohexanes undergo demethanation and dehydrogenation to the corresponding trimethylbenzenes (Pines *et al.*, 1948). 2,2,-Dimethylnorpinane on dehydrogenation over platinum-on-alumina at 322°C affords 94% isopropylbenzene and 5% *o*-xylene. The authors suggested *o*-xylene is formed through a *gem*-dimethylcyclohexene intermediate (Ipatieff *et al.*, 1951).

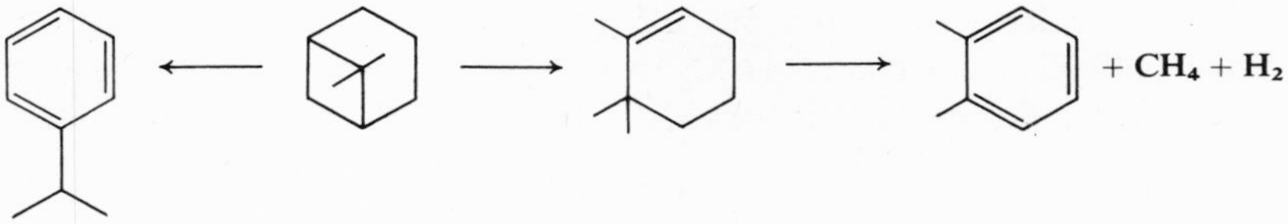

$+ CH_4 + H_2$

Release of ring strain by dealkylation facilitates aromatization, but the reaction still goes with more difficulty than those in which a carbon–carbon bond does not have to be broken. The cyclopropane, **LXXIX**, is stable to dehydrogenation under conditions in which **LXXX** was rapidly converted to the corresponding naphthalene (Newman *et al.*, 1958).

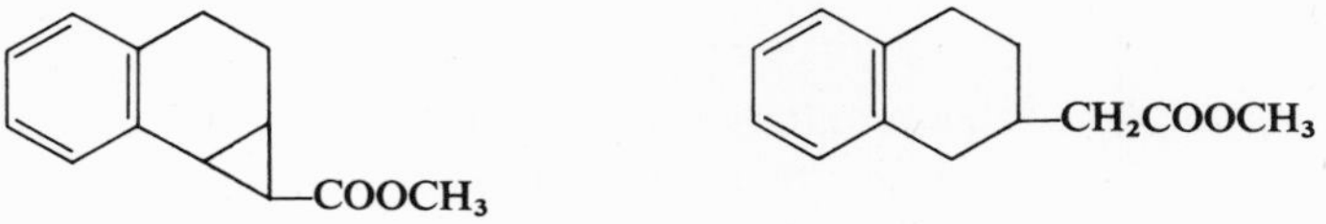

$COOCH_3$ CH_2COOCH_3

(LXXIX) **(LXXX)**

Aromatization of the cyclopropane, **LXXXI**, over 10% palladium-on-carbon at 325°C affords the aromatic, **LXXXII**, with the most highly substituted bond of the cyclopropane ring being broken in the process (Brown *et al.*, 1969). This course most relieves crowding and permits aromatization with the least movement. The side chain becomes saturated, presumably by migration of the double bond into the ring.

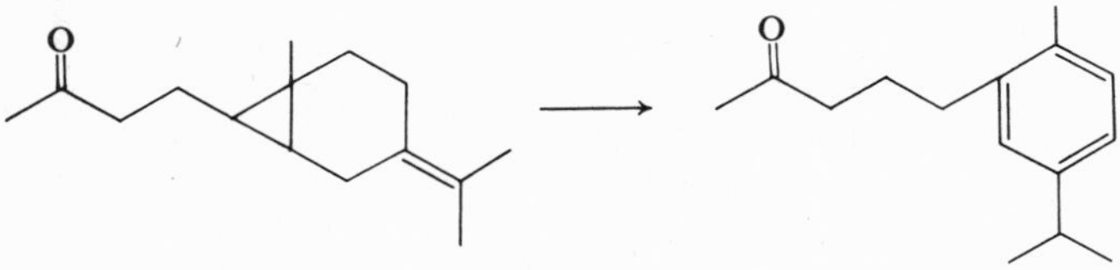

(LXXXI) (LXXXII)

Spiranes

Spiranes constitute a special class of quaternary compounds in which the fragment partially split off during dehydrogenation may undergo further reaction resulting in skeletal isomerization. Tetralins containing a methyl-substituted spirocyclopentane ring form methylphenanthrenes, the methyl group being preserved (Sengupta and Chatterjee, 1952, 1953, 1954a). On the other hand, an α-ethyl substituent is apt to be involved in the reaction and form a pyrene. The authors assumed a partially reduced 4-ethylphenanthrene intermediate which undergoes cyclodehydrogenation to pyrene (Sengupta and Chatterjee, 1954b). A propyl-substituted spiran, **LXXXIII**, affords 1-methylpyrene (**LXXXIV**), probably through cyclization of a partially reduced 4-*n*-propylphenanthrene (Chatterjee, 1955a). Similarly, a butyl substituent, **LXXXV**, affords 1-ethylpyrene (**LXXXVI**). Both 4-*n*-propyl- and 4-*n*-butyl-1,2-dihydrophenanthrene were in fact shown to be converted to 1-methyl- and 1-ethylpyrene, respectively.

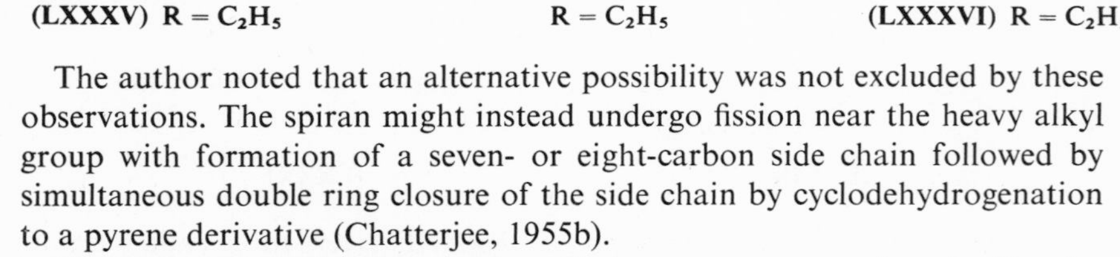

CH_2CH_2R

(LXXXIII) $R = CH_3$
(LXXXV) $R = C_2H_5$

$R = CH_3$
$R = C_2H_5$

(LXXXIV) $R = CH_3$
(LXXXVI) $R = C_2H_5$

The author noted that an alternative possibility was not excluded by these observations. The spiran might instead undergo fission near the heavy alkyl group with formation of a seven- or eight-carbon side chain followed by simultaneous double ring closure of the side chain by cyclodehydrogenation to a pyrene derivative (Chatterjee, 1955b).

The tendency of spiranes to undergo fission was put to good use by Freudewald and Konrad (1970) in the synthesis of *p*-phenylphenol (**LXXXIX**). Dehydrogenation of *p*-cyclohexenylphenol over palladium catalysts takes place smoothly in the presence of a hydrogen acceptor such as α-methylstyrene or nitrobenzene. For example, 174 gm of *p*-cyclohexenylphenol and 615 gm of nitrobenzene heated to 150°–180°C for 8 hours with 50 gm of 5% palladium-on-carbon affords 136 gm of *p*-phenylphenol (80% yield). Cyclohexenylphenol (**LXXXVIII**) is obtained by cleavage of 1,1-bis(*p*-hydroxyphenyl)cyclohexane (**LXXXVII**). The preferred mode of operating is to combine cleavage and dehydrogenation into a single step.

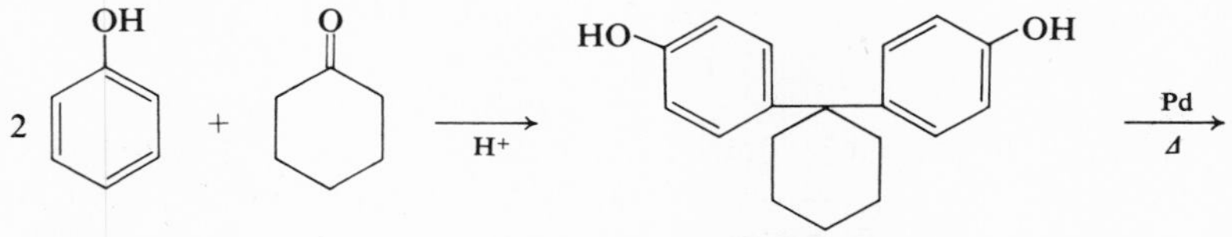

(LXXXVII)

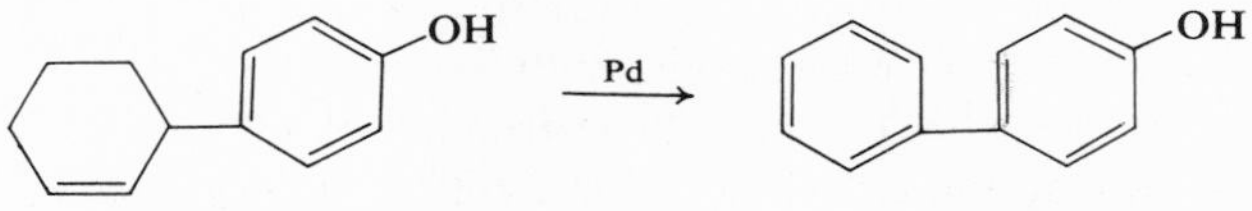

(LXXXVIII) (LXXXIX)

Crowded Compounds

Compounds with a number of substituents on the hydroaromatic ring may undergo aromatization only with difficulty. Relief of strain through dealkylation is facilitated by crowding of substituents, but at the same time, access of the ring to the catalyst surface is impeded. Such compounds usually undergo a slow aromatization with alkyl group migration or elimination. Dehydrogenation of **XC** over palladium-on-carbon proceeds with difficulty and affords 2,4-dimesitoyltoluene (**XCI**) in low yield. The difficulty in dehydrogenation of this compound was attributed to the steric effect of four vicinal substituents blocking approach to the catalyst (Fuson and Sauer, 1963). Demethylation proceeds so as to provide the greatest strain relief.

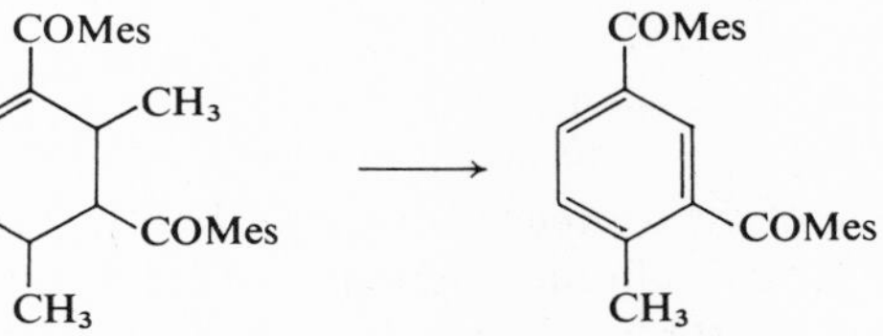

(XC) (XCI)

Similarly, crowding of the methyl groups in the diol, **XCII**, favors dealkylation, and dehydrogenation affords a mixture of mono- and dimethyl-1,2-benzanthracenes with the benzyl hydroxyl functions being readily eliminated in the process (Newman *et al.*, 1960).

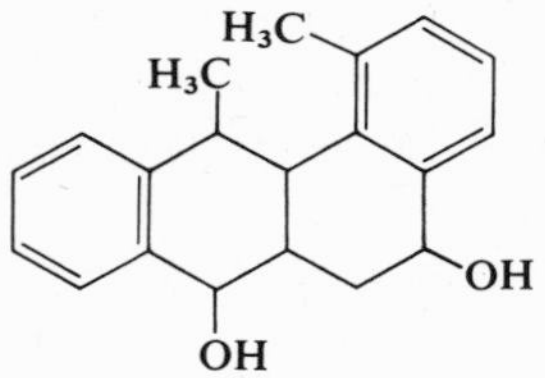

(XCII)

Polynuclear compounds may undergo aromatization only with difficulty and accompanied by considerable degradation. For example, the polynuclear compounds, **XCIII**, **XCIV**, and **XCV** were dehydrogenated only in very poor yield, the reaction being in each case accompanied by extensive fragmentation. Dehydrogenation of **XCIII** over 10% palladium-on-carbon at 300°C afforded only 10% of **XCVI**, the only hydrocarbon that could be isolated (Phillips and Chatterjee, 1958). Complete destruction of **XCV** occurred when dehydrogenation was attempted with selenium, sulfur, or platinum black (Fieser *et al.*, 1936).

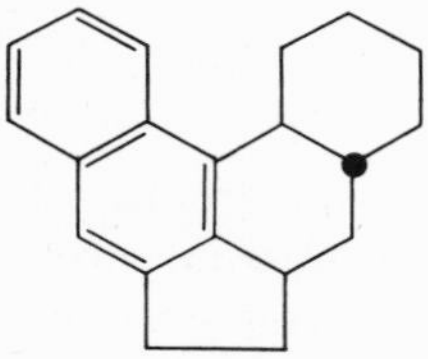

(XCIII) 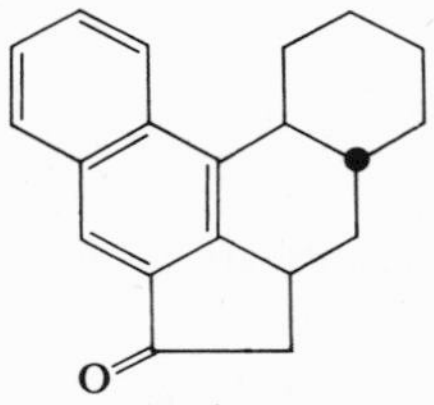 (XCIV)

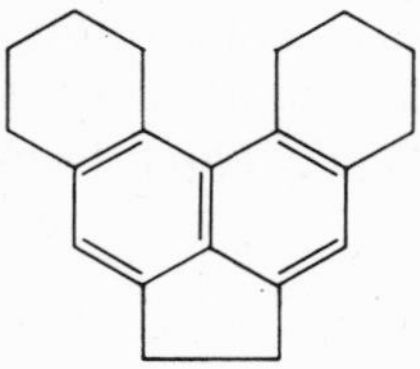

(XCV) 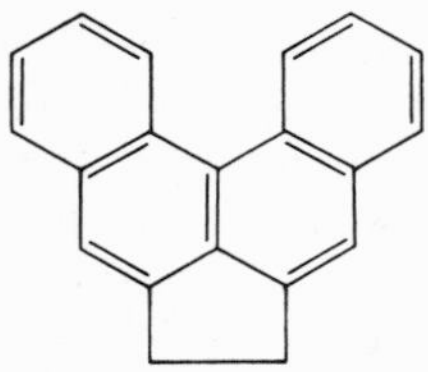 (XCVI)

Dehydrogenation of the tricyclic compound, **XCVII**, over 10% palladium-on-carbon at 150°–250°C afforded 4-amino-2-ethyl-pyrimidine (**XCVIII**) in 57% yield, presumably through dealkylation and deamination as indicated (Van Winkle *et al.*, 1966).

Rearrangement and degradation is favored by higher temperatures. Dehydrogenation of **XCIX** over palladium-on-carbon at temperatures in excess of 300°C affords only small yields of 1,8-diphenylnaphthalene (**C**) accompanied by substantial amounts of rearranged products, whereas in boiling cumene, **C**

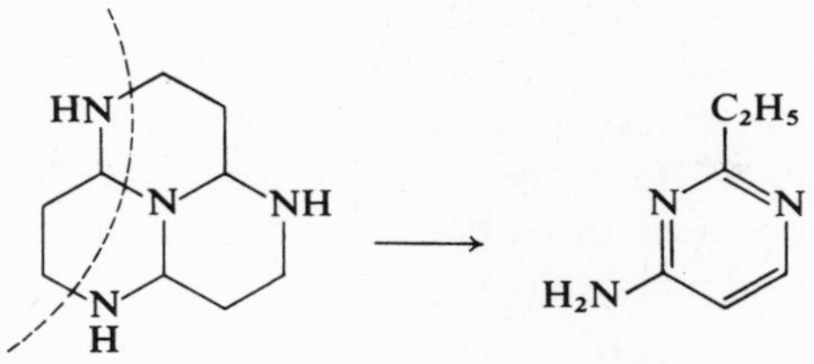

HN NH N N H
(XCVII)

C_2H_5 N N H_2N
(XCVIII)

is obtained in 54% yield with less contamination by rearranged products (House and Bashe, 1967).

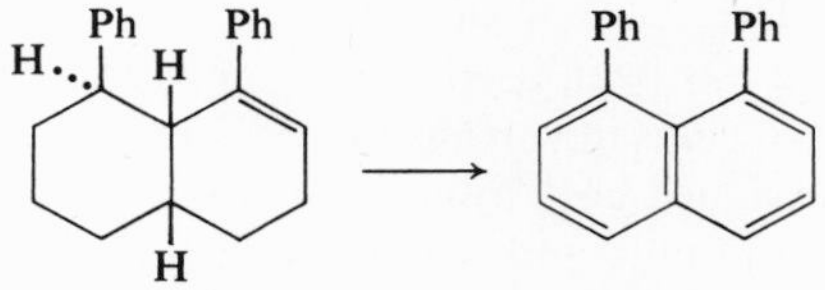

Ph Ph H H H
(XCIX)

Ph Ph
(C)

Hydrogen Exchange Processes

A useful technique for achieving dehydrogenation under relatively mild conditions involves carrying out the reaction in the presence of a hydrogen acceptor to remove hydrogen as it is formed. The method is particularly suited to heat-sensitive compounds. Since the acceptor is reduced, the exchange reaction is also a method of hydrogenation and at times is more selective and affords higher yields than direct catalytic hydrogenation (Braude *et al.*, 1954b). Some examples that might more properly be considered a hydrogenation are given in the sections on ketones and peroxides as acceptors.

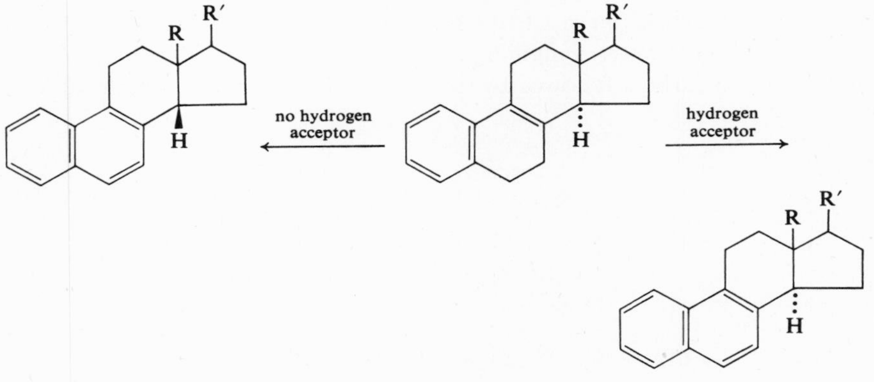

R′ R H ← no hydrogen acceptor — R′ R H — hydrogen acceptor → R′ R H

Dehydrogenations carried out in the presence of hydrogen acceptors usually give the same products as are obtained in its absence, but exceptions exist. For instance, dehydrogenation of 17-substituted, 13-alkylgona-1,3,5(10),8-tetraene over palladium affords the 14β-pentaene in the absence of a hydrogen acceptor, and 14α-pentaene in its presence. With 17-hydroxy materials, the 14β-pentaene is formed in either case (Buzby *et al.*, 1969).

Cinnamic and Maleic Acids as Acceptors

Cinnamic acid (Kubota, 1939; Suginome, 1959) and maleic acid have been used with success as hydrogen acceptors in exchange reactions mainly with palladium catalysts. The rate of dehydrogenation over palladium in the presence of maleic acid has been used as a diagnostic tool in determining the stereochemistry of fused rings in various indole alkaloids and their derivatives. Configurations were assigned both by comparison of the rates of dehydrogenations with known compounds and by considerations of the relative ease of approach of the oxidizable hydrogens to the catalyst surface (Wenkert and Roychaudhuri, 1958). Platinum catalysts have also been used successfully in this type of reaction. For example, yohimbic acid heated with maleic acid in water for 5 hours over platinum black affords tetradehydroyohimbic acid (Majima and Murahashi, 1935).

The terpenes *d*-α-phellandrene, *d*-limonene, and terpinolene are each converted efficiently to *p*-cymene by refluxing in ethanol or tetrahydrofuran over palladium-on-carbon in the presence of hydrogen acceptors such as maleic, fumaric, or cinnamic acids, or nitroaromatic compounds. Other terpenes are converted to *p*-cymene less efficiently, reflecting structural differences in the terpenes (Pallaud and Hoa, 1964). Certain sesquiterpenes are converted by dehydrogenation to azulenes, providing a diagnostic tool for structure elucidation (Pallaud and Hoa, 1965).

An example of selectivity in aromatization of a complex polynuclear ring system is the conversion of **CI** to **CII**. Dehydrogenation at reflux of 0.50 gm of **CI** in 150 ml of water containing 700 mg of maleic acid and 250 mg of 20% palladium-on-carbon, followed by treatment with hydrogen bromide, affords

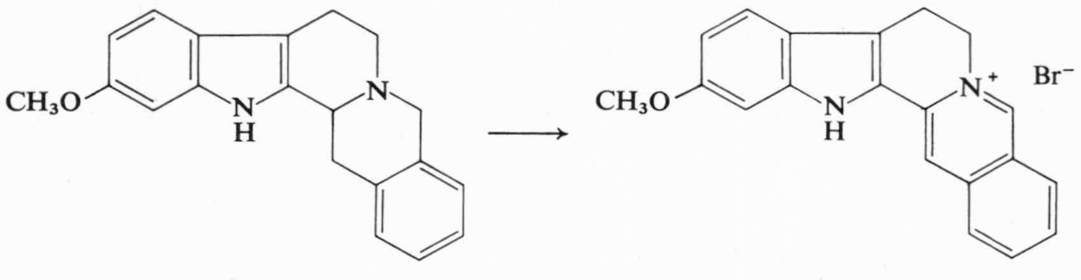

CH_3O N H N → CH_3O N H N^+ Br^-

(CI) **(CII)**

340 mg of the quaternary base, **CII**. The authors noted that apparently a tetrahydroisoquinoline ring system is dehydrogenated more readily to an isoquinoline than the tetrahydro-β-carboline is to a carboline (Elderfield *et al.*, 1958).

Hydroxyquinolines have been converted to the corresponding phenols in excellent yield by dehydrogenation in the presence of maleic acid as hydrogen acceptor. A mixture of 2 gm of **CIII**, 2 gm of palladium black, 0.2 gm of 30% palladium-on-carbon, and 30 ml of water refluxed for 24 hours affords **CIV** in 95% yield. A pH of 9–10 is necessary for this dehydrogenation, but compounds lacking the methyl substituent are dehydrogenated readily at pH 7 (Elderfield and Maggiolo, 1949). The reaction is noteworthy in that both dehalogenation and dehydration are unimportant.

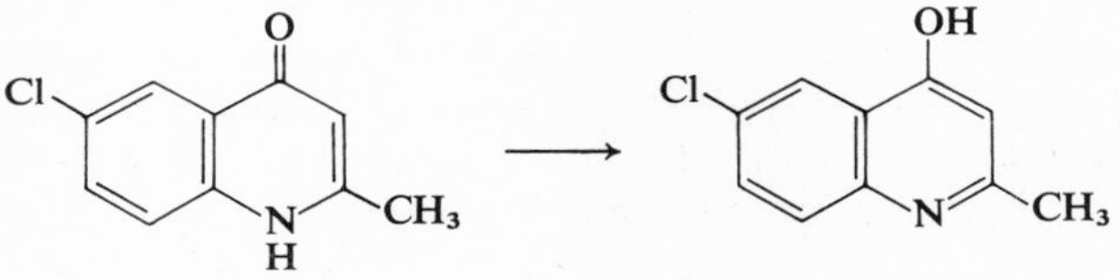

O Cl N H CH_3 → OH Cl N CH_3

(CIII) **(CIV)**

Benzene as Acceptor

Benzene has been used as a hydrogen acceptor in a variety of reactions and in the presence of diverse catalysts. The reactions are carried out under pressure at elevated temperatures. The preferred catalyst frequently depends on the nature of the functions present in the substrate. Supported nickel, for instance, proved better than platinum-on-carbon in converting hydroaromatics to phenols with benzene as acceptor at 300°–350°C. Platinum was frequently more active, but favored hydrogenolysis of the oxygen function (Adkins *et al.*, 1941). Rhodium-on-alumina is much more effective than palladium in promoting dehydrogenation of **CV** to **CVI** by transfer to benzene at 300°C. Dehydrogenation of 4.26 gm of **CV** was carried out by heating at 300°C with 45 ml of thiophene-free benzene and 2.13 gm of 5% rhodium-on-alumina for 10 hours. This substrate undergoes dehydrogenation with considerable difficulty and this

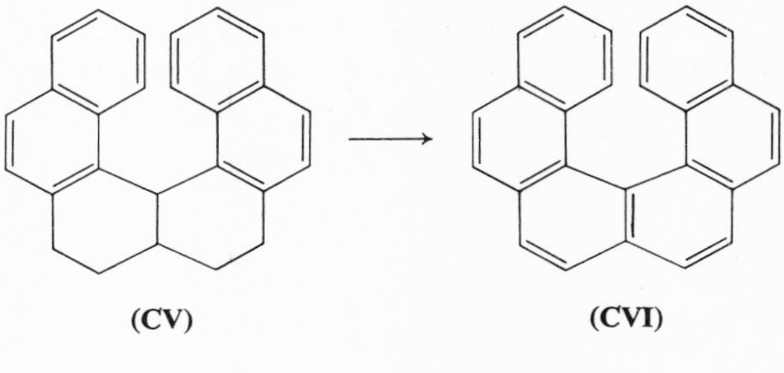

(CV) **(CVI)**

procedure was the best of a number tried. Other compounds in this series were dehydrogenated similarly (Newman and Lednicer, 1956).

Thiophene-free benzene was used with success in these exchange reactions, but other workers have noted that over nickel catalysts much improved results are obtained if suitable amounts of thiophene or diphenyl sulfide are added to the reaction (Adkins *et al.*, 1948). The same promotion by sulfur may also apply to noble metal catalysts.

Cyclohexenes as Acceptors

Cyclohexenes can function as both hydrogen acceptor and donor in hydrogen transfer reactions over noble metal catalysts. Molecules containing a cyclohexene or cyclohexadiene structure may therefore undergo disproportionation into a mixture of aromatic and saturated molecules. The tendency to disproportionation is sometimes so strong that disproportionation occurs readily even under hydrogenation conditions (Rylander, 1967). On the other hand, when cyclohexene is functioning as a hydrogen donor in the presence of another acceptor, disproportionation may be diminished or entirely eliminated. This phenomenon has been interpreted in terms of co-adsorption of the donor and acceptor on the catalyst surface (Braude *et al.*, 1954a). A variety of compounds have been effectively and selectively hydrogenated in the presence of cyclohexene donors (Braude *et al.*, 1954c). These include olefins, acetylenes, and azo and nitro compounds as well as certain activated types of carbonyl groups (Linstead *et al.*, 1952). The rate-determining step in disproportionation of cyclohexene over palladium is thought to involve hydrogen transfer between two associatively adsorbed cyclohexene molecules, producing a π-allylpalladium complex (Carrà *et al.*, 1964; Carrà and Ragaini, 1967). The ratio of hydrogenation to disproportionation is influenced by the solvent (Sedlak, 1966).

Disproportionation of cyclohexenes frequently affords mixtures of aromatic and saturated materials proportional to the extent of unsaturation in the substrate, that is, the reaction occurs with little or no loss of hydrogen. At higher temperatures, however, the entire substrate is converted to an aromatic compound with loss of appropriate hydrogen. Reaction temperature thus becomes an important determinant of product composition. For example, no reaction occurs when limonene is heated with platinum-on-carbon at 102°C, but at 140°C *p*-cymene and *p*-menthane are formed in approximately 2 to 1

3 ⟶ 2 +

ratio; at 305°C in the vapor phase *p*-cymene is formed in good yield (Linstead *et al.*, 1940).

Pinene, when heated with platinum-on-carbon at 156°C, is converted into an approximately equimolecular mixture of pinane and *p*-cymene, even though aromatization requires breaking a carbon–carbon bond. Only the 6:7 bond and none of the 4:7 bond is broken in the reaction, for no *o*-cymene is formed (Linstead *et al.*, 1940).

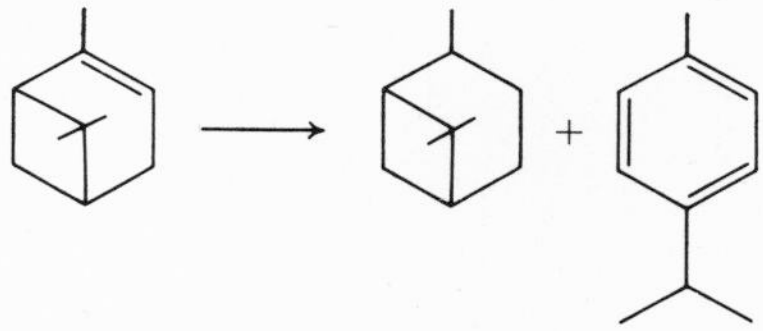

Even compounds containing a quaternary carbon in the incipient aromatic ring may undergo disproportionation if conditions are vigorous enough. Prolonged treatment of *β*-selinene (**CVII**) over palladium-on-carbon at 205°C afforded a mole each of eudalene (**CVIII**), tetrahydroselinene (**CIX**), and methane (Linstead *et al.*, 1940).

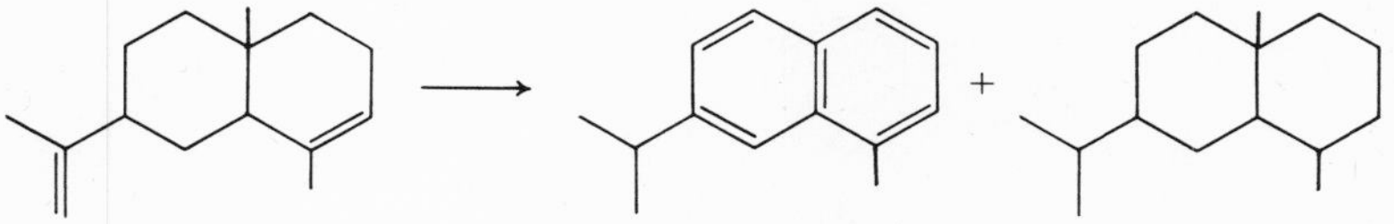

$+ CH_4$

(CVII) **(CVIII)** **(CIX)**

An order of palladium ≫ platinum > rhodium has been established for decreasing activity in the disproportionation of cyclohexene (Hussey *et al.*, 1968) and one would expect that palladium would generally make the most effective catalyst for disproportionation of similar compounds. For instance, the aromatic compound, **CXI**, and hydroaromatic compound, **CXII**, are prepared by disproportionation of **CX** by heating at 230°–240°C with palladium-on-carbon. Some dealkylated product, **CXIII**, is also formed (Mori and Matsui, 1968). Fleck and Palkin (1937) had earlier compared a number of catalysts in this type of reaction and found palladium-on-carbon by far the best. Palladium chloride has also been used in disproportionation reactions, but the true catalyst may be palladium metal (Karol and Carrick, 1966).

Disproportionation of Δ^4-cyclohexene-1,2-dicarboxylic acids follows an unusual course and affords a mixture of cyclohexane dicarboxylic acid, benzoic acid, and carbon dixoide. Methyl and phenyl substituents in the 3-position cause selective removal of the adjacent carboxyl group and afford *m*-methylbenzoic acid and *m*-phenylbenzoic acid, respectively. 3,6-Diphenyl-

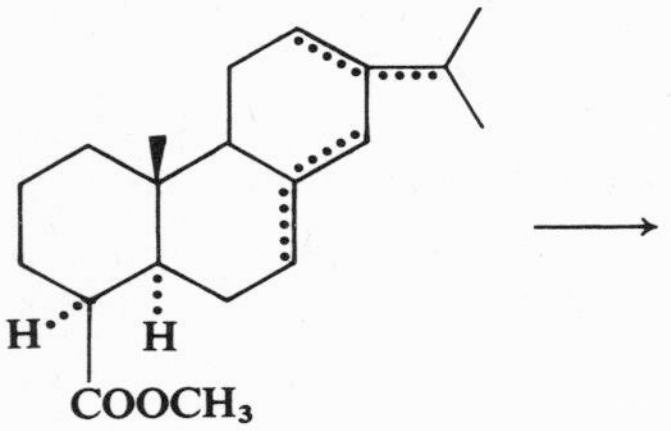

(CX)

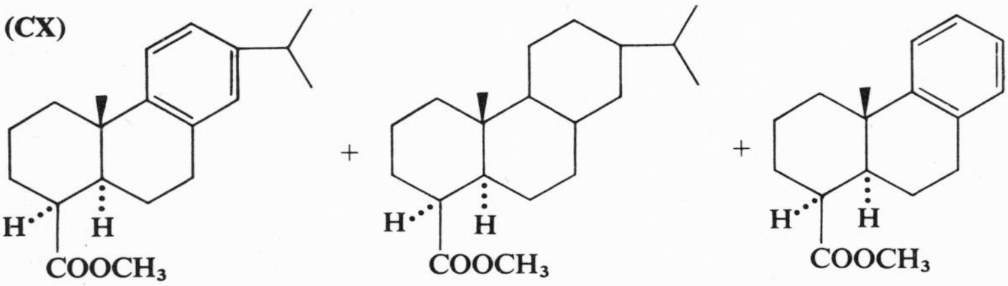

(CXI) + (CXII) + (CXIII)

tetrahydrophthalic acids lose both carboxyl groups and afford a mixture of terphenyl and 3,6-diphenylhexahydrophthalic acid (Jackman, 1960).

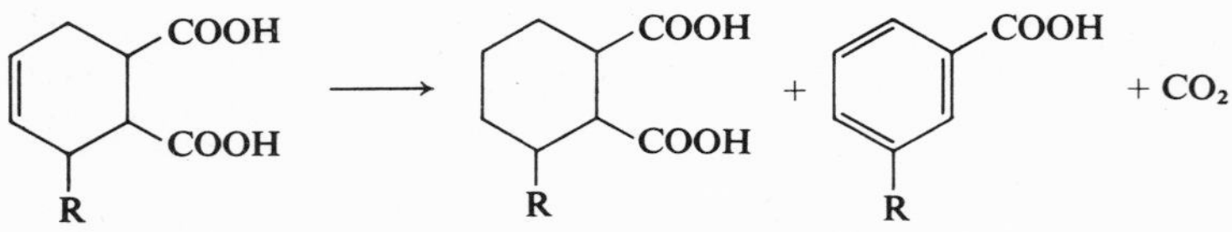

$R = CH_3, C_6H_5$

In some cases, cyclohexene apparently functions as both acceptor and donor during the course of the reaction. The sole product of a transfer reaction of **CXIV** and cyclohexene over palladium-on-carbon was the cyanoethylene compound, **CXVI**. Presumably, but not necessarily, **CXVI** arises from **CXIV** via the intermediate **CXV**, since **CXV** on treatment with palladium-on-carbon in ethanol yields **CXVI** (Irwin and Wibberley, 1968).

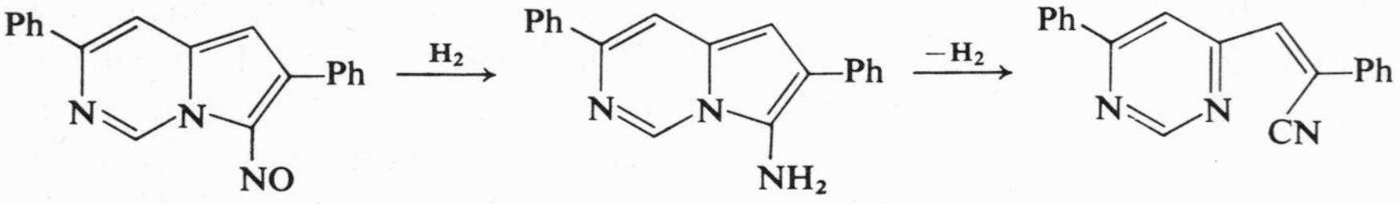

(CXIV) $\xrightarrow{H_2}$ (CXV) $\xrightarrow{-H_2}$ (CXVI)

Nitrobenzene as Acceptor

Nitrobenzene is an effective solvent for carrying out dehydrogenations; it provides both a convenient reaction temperature at its boiling point and a good hydrogen sink. For instance, a smooth dehydrogenation of 1-(*p*-trifluoromethylphenyl)-4-methoxycyclohexene (500 gm) is achieved in 22 hours over 166 gm of 10% palladium-on-carbon in refluxing nitrobenzene as a solvent and hydrogen acceptor. The authors did not report whether any attempt was made to dehydrogenate without prior dehydration (Bach *et al.*, 1968).

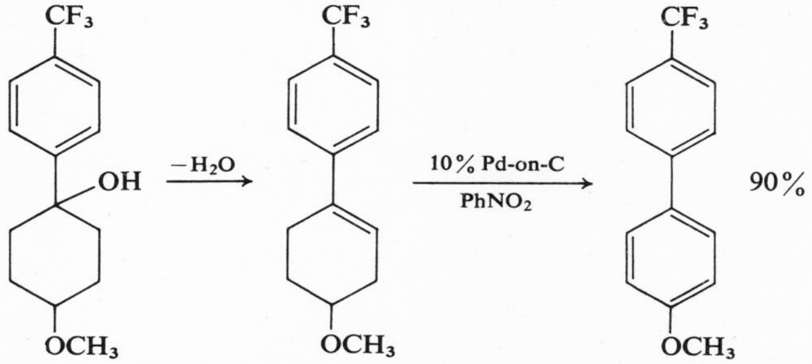

Nitrobenzene has been used with success as a solvent and hydrogen acceptor in the conversion of 4-imino-3-cyanopiperidine to 4-amino-3-cyanopyridine, a dehydrogenation carried out only with difficulty in inert solvents.

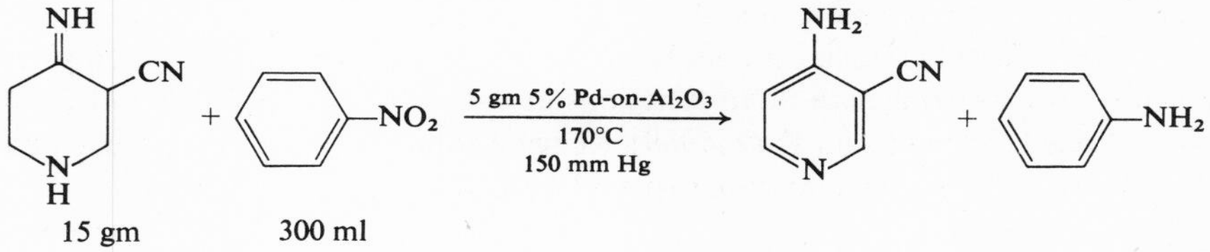

TABLE I

EFFECT OF ACETIC ACID ON YIELD OF 4-AMINO-3-CYANOPYRIDINE[a]

% Acetic acid based on weight of substrate	% Yield of isolated product
None	51
0.05	56.6
0.20	66.2
0.66	76.0
1.32	80.7
2.00	77.2
3.00	71.8
6.60	0

[a] Fifteen grams of 4-imino-3-cyanopiperidine, 300 ml of nitrobenzene, 5 gm of 5% Pd-on-Al_2O_3, 170°C, 160 mm Hg pressure.

Water formed in the reaction was removed continuously by a nitrogen sweep while the system was maintained under partial vacuum. The presence of small amounts of acetic acid improves the yield of product, whereas larger amounts completely suppress dehydrogenation (see Table I). This use of carboxylic acids in dehydrogenations is novel and seems worthy of further study (Marschik and Rylander, 1970). Palladium-on-alumina was clearly superior to palladium-on-carbon in these reactions; over the latter catalyst substantial quantities of dimeric product were formed (Marschik, 1964).

Ketones as Acceptors

Unique use has been made of ketones as acceptors in an exchange reaction involving aqueous isopropanol and trimethyl phosphite with soluble iridium compounds as catalysts. A feature of the reaction is that it provides exceptionally high proportions of axial alcohols. For instance, reduction of 3-*t*-butylcyclohexanone and 3,3,5-trimethylcyclohexanone afford the corresponding axial alcohol in yields of 98 and 99% (Haddad *et al.*, 1964). The presence of water avoids formation of ethers (Mleczak, 1965). The exchange has been applied with considerable success in the synthesis of steroids; reduction of 3-oxo steroids gives axial alcohols in the 5α- or 5β-series. The reaction is selective for 3-oxo groups; oxo functions at C-6, C-11, C-12, C-17, and C-20 do not react (Browne and Kirk, 1969). Reagent solutions which were preheated for about 14 hours gave more consistent behavior than fresh solutions.

Peroxide as Acceptor

Johns (1971) described an exchange reaction in which a steroid peroxide dehydrogenates ethanol, affording androsta-4,6,8(14)-triene-3,17-dione. Other

O O HO O O $\xrightarrow[\text{Pd, 25°C, 18 hours}]{C_2H_5OH}$ HO OH OH O $\xrightarrow[\text{Pd, 10 days}]{C_2H_5OH}$ O O routes to this conjugated system have been described but none match the efficacy of this approach. Prolonged treatment of either the epidioxide or primary product with palladium black in ethanol results in the formation of a trienone, obtained in 60% yield.

Loss of Functional Groups

Many dehydrogenations are carried out on molecules containing various functional groups, which may or may not survive the rigors of the reaction. The fate of the function depends on a number of factors that include the temperature, catalyst, solvent (if any), the type of function, and the substrate structure as a whole (Newman and Bye, 1952).

Hydroxyl Groups

Hydroxyl groups are lost easily during dehydrogenation reactions, especially if the hydroxyl is in a tertiary position (Reggel and Friedel, 1951; Barnes and Reinhold, 1952; Kakisawa *et al.*, 1968; Eisenbraun *et al.*, 1971). For example, an excellent procedure for preparing 1-*n*-propylphenanthrene (**CXVIII**) involves heating the allyl carbinol, **CXVII**, with palladium-on-carbon at 215°–220°C for ½ hour or at 320°C for only 10 minutes. The extracyclic double bond is saturated during the reaction, perhaps through double-bond migration (Bachmann and Wilds, 1938).

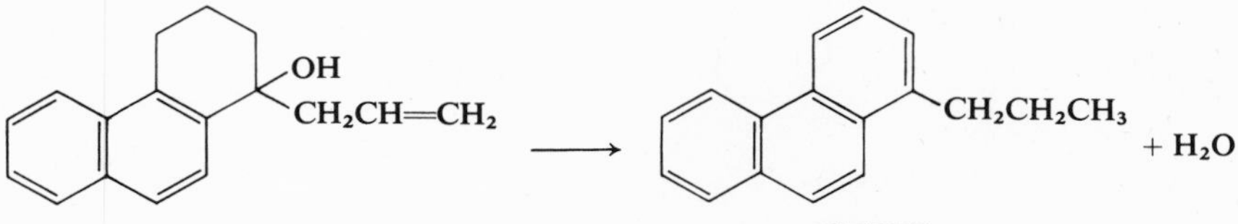

OH
$CH_2CH{=}CH_2$ → $CH_2CH_2CH_3$ + H_2O

(CXVII) **(CXVIII)**

Loss of tertiary hydroxyl may occur readily during dehydrogenation even in compounds that are relatively stable to dehydration. For example, **CXIX** is not dehydrated by steam distillation from 10% sulfuric acid or by distillation at atmospheric pressure, but it forms *o*-terphenyl (**CXX**) in good yield by heating with palladium-on-carbon at 300°C (Woods and Scotti, 1961).

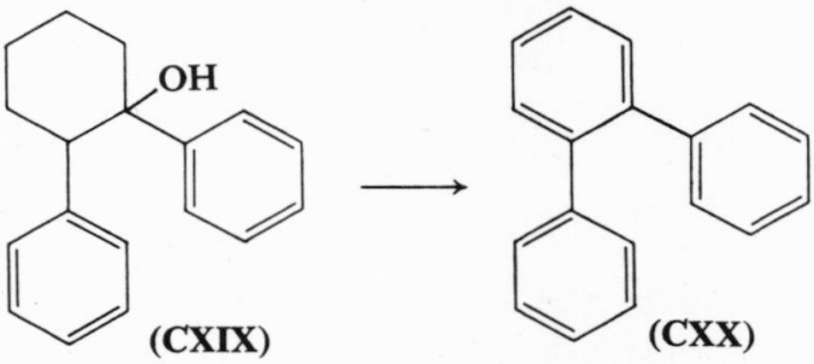

OH →

(CXIX) **(CXX)**

Some workers dehydrate carbinols before carrying out aromatization (Stiles and Sisti, 1961) and the prior dehydration at times results in considerably increased overall yields (Anderson and Nelson, 1951). On the other hand, although the alcohol, **CXXI**, could first be dehydrated to an olefin and the olefin dehydrogenated over palladium-on-carbon at 285°C to **CXXII** in 92% yield, it was found best for preparative purposes to omit isolation of the olefin, which was sensitive to air and to acids; the crude carbinol is most conveniently dehydrogenated directly (Peterson and Kloetzel, 1958).

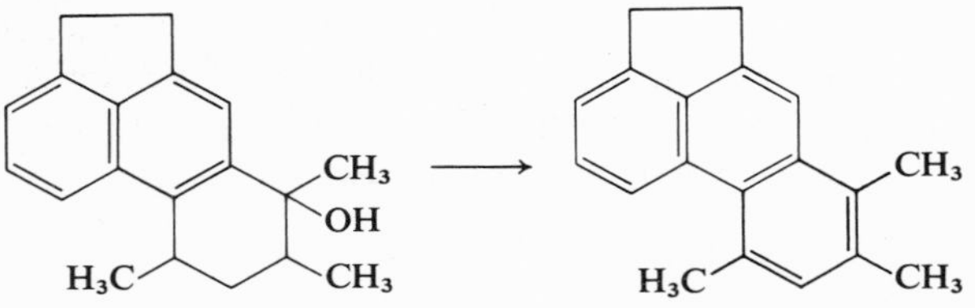

(CXXI) → (CXXII)

Benzyl alcohols are particularly apt to be lost during dehydrogenation reactions. Dehydrogenation of **CXXIII** over palladium-on-carbon at 270°–280°C proceeds with loss of water to afford **CXXIV** in high yield (Horton and Walker, 1952).

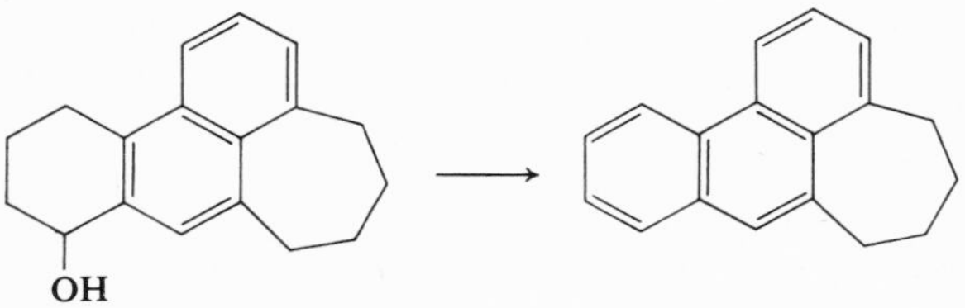

(CXXIII) → (CXXIV)

Similarly, complete loss of hydroxyl occurs on dehydrogenation of 1.3 gm of **CXXV** in 15 ml of boiling decalin over 0.3 gm of 30% palladium-on-carbon. Filtration and extraction produce 1.1 gm of 7,8-dimethoxyisoquinoline, **CXXVI** (Gensler *et al.*, 1968).

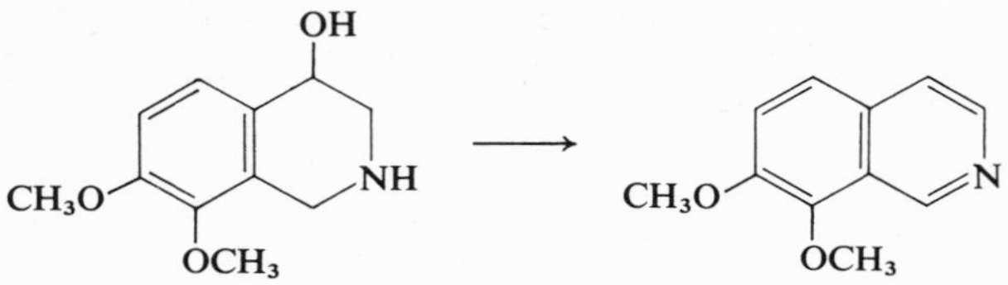

(CXXV) → (CXXVI)

A further illustration of the sensitivity to loss of functions in a benzyl position is provided by the work of Gardner and Horton (1952). Dehydrogenation of the ketone **CXXVII**, carbinol **CXXVIII**, and acetate **CXXIX**, over palladium-on-carbon results in each case in loss of the oxygen function and formation of **CXXX**. The acetate undergoes loss of acetic acid at temperatures as low as 160°C. The authors assumed the carbinol suffered dehydration rather than hydrogenolysis, but this interpretation was not supported experimentally.

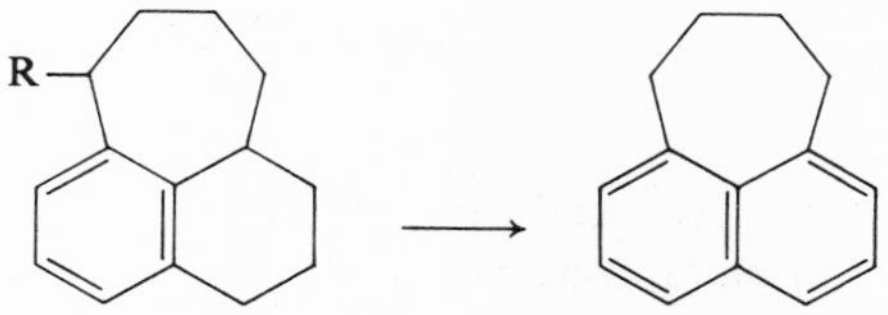

(CXXVII) R = =O
(CXXVIII) R = —OH
(CXXIX) R = —OAc

(CXXX)

Linstead and Michaelis (1940) examined the formation of naphthols from alcohols and ketones of the hydronaphthalene group and concluded the tendency for elimination of the oxygen atom is greatest for those substrates furthest removed from the aromatic type. In dehydrogenation over palladium-on-carbon, the following results were obtained:

Compound	% Yield of total aromatic material	% Yield of naphthol
ar-β-Tetralol	97	55
ac-β-Tetralol	96	60
trans-β-Decalol	51	17
cis-β-Decalol	56	12

Improved yields of naphthol were obtained by the use of a solvent. For example, equal amounts of naphthalene and β-naphthol were obtained from boiling (263°C) *ac*-β-tetralol, whereas dehydrogenation of the substrate in boiling (195°C) mesitylene solvent afforded twice as much naphthol as naphthalene.

The oxygen function may be lost under dehydrogenation conditions after being converted to the phenol. Hay (1968) heated a hydroxy-, methoxy-, or amino-substituted biphenyl with a cyclohexyl-substituted phenol and converted both compounds to biphenyls. The reaction provides an interesting way of achieving hydrogenolysis of substituents. Palladium is the preferred catalytic metal.

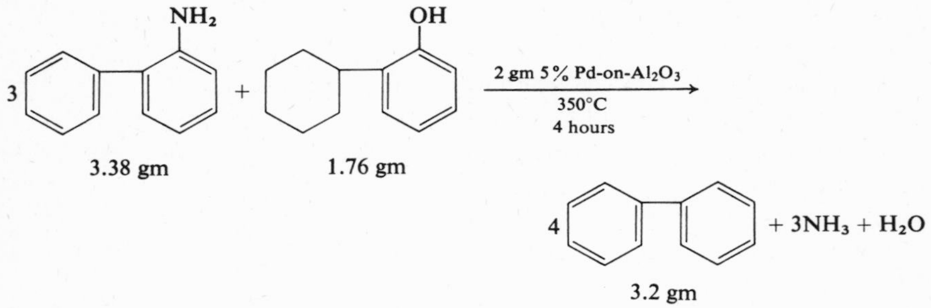

NH_2 OH

3 + $\xrightarrow[\text{350°C, 4 hours}]{\text{2 gm 5\% Pd-on-}Al_2O_3}$

3.38 gm 1.76 gm

4 $+ 3NH_3 + H_2O$

3.2 gm

Primary alcohols may be lost during dehydrogenation through decarbonylation probably by first being converted to an aldehyde (Newman and O'Leary, 1946). Protection of the hydroxyl by acetylation will prevent decarbonylation, but the oxygen function may be lost nonetheless. Carbinols, whether acetylated or not, will in general be lost under dehydrogenation conditions when attached to an aromatic or incipient aromatic ring (Newman and Mangham, 1949). The fate of the carbinol depends on its position.

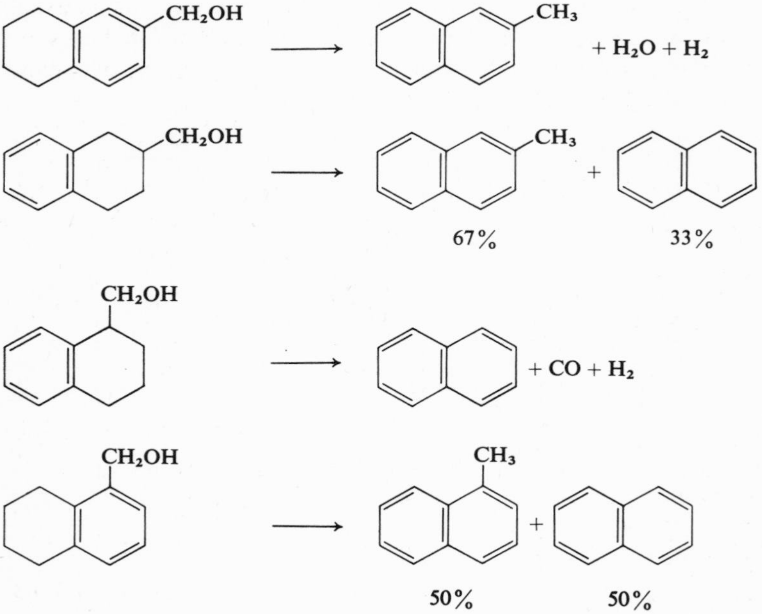

CH_2OH ⟶ CH_3 $+ H_2O + H_2$

CH_2OH ⟶ CH_3 +

67% 33%

CH_2OH ⟶ $+ CO + H_2$

CH_2OH ⟶ CH_3 +

50% 50%

These results are in accord with expectations from hydrogenation experiments. Palladium is an excellent catalyst for hydrogenolysis of aromatic carbinols, whereas it is poor for hydrogenolysis of aliphatic alcohols. The authors noted that very efficient use is made of the hydrogen in the hydrogenolysis reactions and they viewed the process as a sort of internal oxidation–reduction (Newman and Zahm, 1943).

Similarly, dehydrogenation of 3-hydroxymethyl-4,5,6,7-tetrahydroindazole is accompanied by decarbonylation with formation of indazole (Ainsworth, 1957).

CH_2OH — N — NH $\xrightarrow[\text{decalin reflux}]{\text{Pd-on-C}}$ N — NH $+ 3H_2 + CO$

Nitro Group

Nitro groups may or may not survive dehydrogenation reactions. Since these functions are very easily reduced, survival may be contingent on the quantity of hydrogen liberated during the reaction. Dehydrogenation of 1.78 gm of 5-nitro-7-azaindoline over 0.8 gm of 5% palladium-on-carbon in 50 gm of refluxing Dowtherm affords 5-nitro-7-azaindole in 50% yield. The quantity of hydrogen liberated in this case was insufficient for complete reduction (Robinson *et al.*, 1959). One might assume that a useful technique for preservation of a nitro function during dehydrogenation would be to carry out the reaction in nitrobenzene solvent as hydrogen acceptor.

O_2N — N — NH $\longrightarrow$ O_2N — N — NH

Anilines and alkylanilines have been prepared from nitrocyclohexanes by passage over a palladium-on-alumina catalyst at 300°–400°C and contact times of 3 to 4 seconds (LeMaistre and Sherman, 1969).

Carboxylic Acids and Esters

Carboxylic acids attached to hydroaromatic systems readily undergo decarboxylation under dehydrogenation conditions (Horning *et al.*, 1948b; Deno, 1950; Herz and Rogers, 1953; Lednicer and Hauser, 1958; House *et al.*, 1968a, b). Decarboxylation apparently accompanies aromatization and is favored by elevated temperatures; aromatization without decarboxylation is favored by lower temperatures (Walker, 1953; Horning and Walker, 1952). Under vigorous conditions, side-chain carboxylic acid functions may also be lost (Geissman and Turley, 1964). Aromatization of 7-carboxy-1,8(9)-*p*-menthadiene over 5% palladium-on-carbon at 215°C affords 7-carboxy-*p*-cymene, whereas at 310°–320°C *p*-cymene results (Oldroyd *et al.*, 1950).

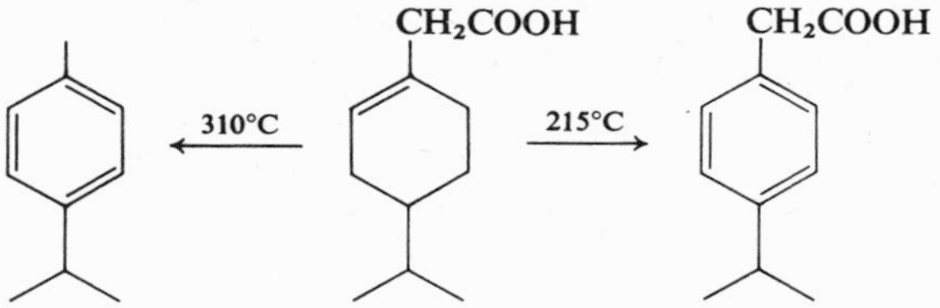

CH_2COOH CH_2COOH

310°C 215°C

Attempts to aromatize and decarboxylate **CXXXI** with formation of 1,2-diphenylbenzocyclobutene failed; instead a mixture was obtained in 76% yield consisting of 1 part 9-phenylanthracene (**CXXXII**) and 5 parts *o*-dibenzylbenzene (**CXXXIII**) (Blomquist and Meinwald, 1960).

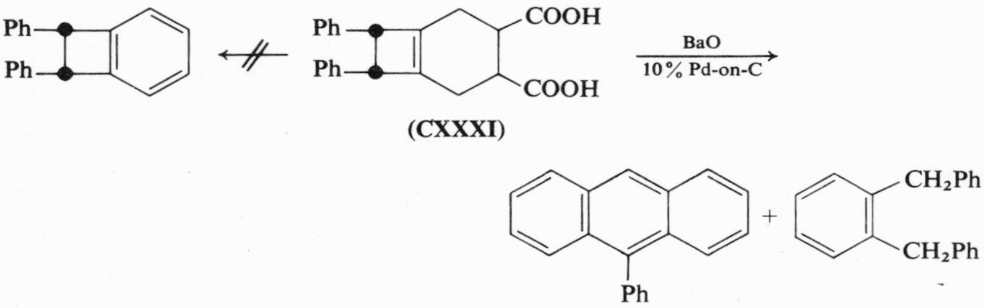

Ph Ph Ph Ph COOH COOH

BaO

10% Pd-on-C

(**CXXXI**)

CH_2Ph CH_2Ph

Ph

(**CXXXII**) (**CXXXIII**)

Esters may be more resistant to elimination than the corresponding free acid. Dehydrogenation of the acid, **CXXXIV**, over 10% palladium-on-carbon gave inconsistent results; an initial attempt at dehydrogenation afforded the desired α-methyl-7-ethyl-2-naphthenacetic acid (**CXXXVI**) in good yield, but attempts to repeat the experiment gave a decarboxylated product. Consistent yields of the desired acid were achieved by carrying out the dehydrogenation with the methyl ester, **CXXXV**, followed by saponification of the product (Locke and Pelletier, 1959).

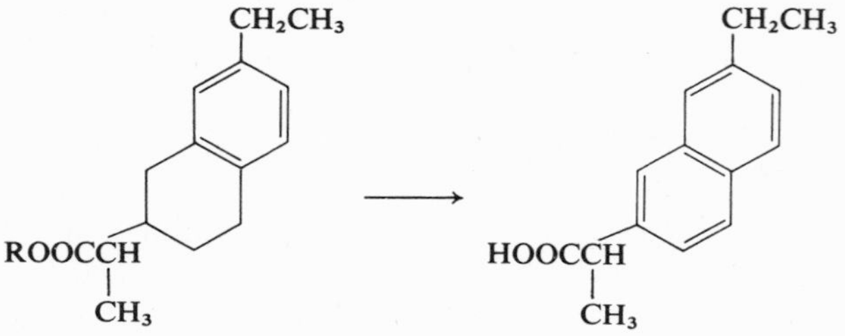

CH_2CH_3 CH_2CH_3

ROOCCH HOOCCH

CH_3 CH_3

(**CXXXIV**) R = H (**CXXXVI**)

(**CXXXV**) R = $—CH_3$

But esters too may be lost during dehydrogenations (Nielsen *et al.*, 1967). The ester **CXXXVII** was converted to **CXXXVIII** on heating with 30% palladium-on-carbon at 210°C. The ester function rather than the lactone underwent cleavage possibly through cracking of the ethyl radical (Taylor and Strojny, 1960). Elimination was probably facilitated by the quaternary carbon structure.

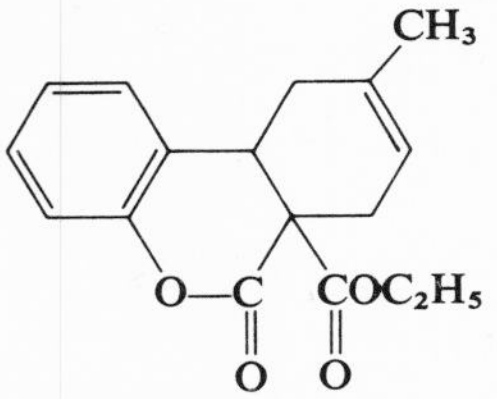

(CXXXVII) 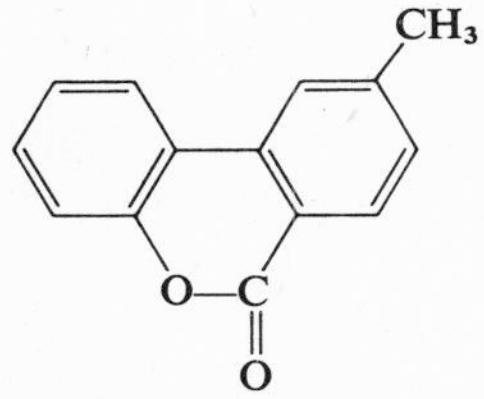 (CXXXVIII)

Hydroaromatic lactones may be easily decarboxylated (Minn *et al.*, 1956). Dehydrogenation of **CXXXIX** over palladium-on-carbon fails to give dehydropodophyllotoxin, but affords **CXL** instead (Gensler *et al.*, 1960).

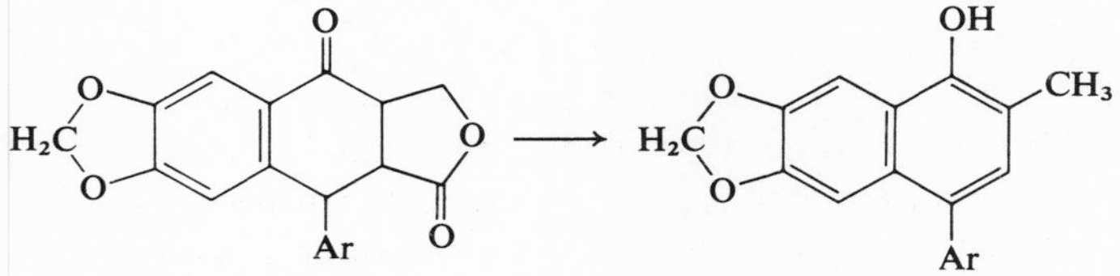

(CXXXIX) (CXL)

Anhydrides of hydroaromatic systems may also be lost during aromatization reactions. Anthracene (**CXLII**) is obtained by heating **CXLI** at 270°C for 3 hours with a mixture of 5% palladium-on-carbon and copper–chromite (Bailey *et al.*, 1962).

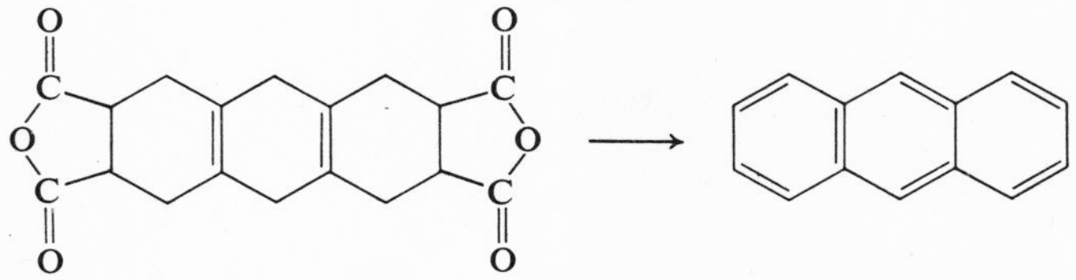

(CXLI) (CXLII)

Ketones

Ketonic functions may be lost readily during dehydrogenation. The reaction may be useful in establishing the carbon skeleton (Sarett *et al.*, 1952).

Facile loss seems limited to those ketones either part of or adjacent to an aromatic or hydroaromatic ring. The ketone function in **CXLIII** was converted to a methylene group, **CXLIV**, under dehydrogenation conditions,

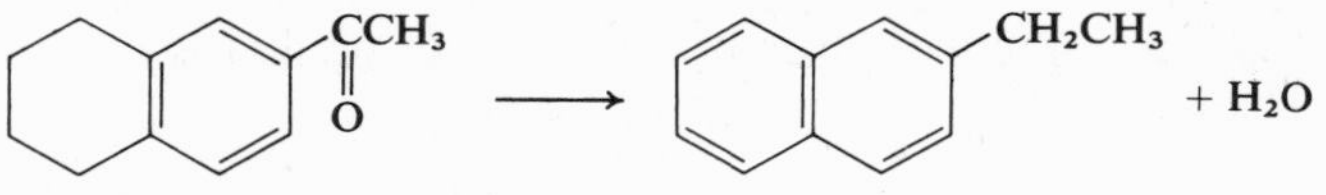

$+ H_2O$

(CXLIII) (CXLIV)

whereas **CXLV** was converted to **CXLVI** with the ketone function intact (Newman and O'Leary, 1946).

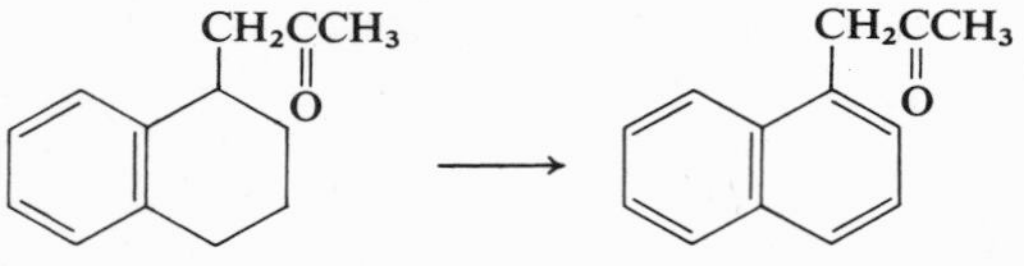

(CXLV) (CXLVI)

These results are in keeping with hydrogenation tendencies over palladium; palladium makes an excellent catalyst for reduction of aromatic ketones, whereas it is very poor for aliphatic ketones.

Ketones that are part of a hydroaromatic system give on dehydrogenation a mixture of phenols and deoxygenated compounds in ratios that depend on the catalyst, substrate, temperature, and solvent (Springer *et al.*, 1971). Loss of the oxygen atom, a reaction of synthetic usefulness, is favored by higher temperatures and lack of solvent. A convenient synthesis of *m*-terphenyl involves dehydrogenation of 2,6-di(1-cyclohexenyl)cyclohexanone (40 gm) over 5% palladium-on-alumina (40 gm) at 350°C. The mixture is heated in a metal bath with stirring until hydrogen evolution ceases (about 4 hours). At lower temperatures, large percentages of various phenols are formed, whereas at 350°C, they are negligible (Kahovec and Pospisil, 1969).

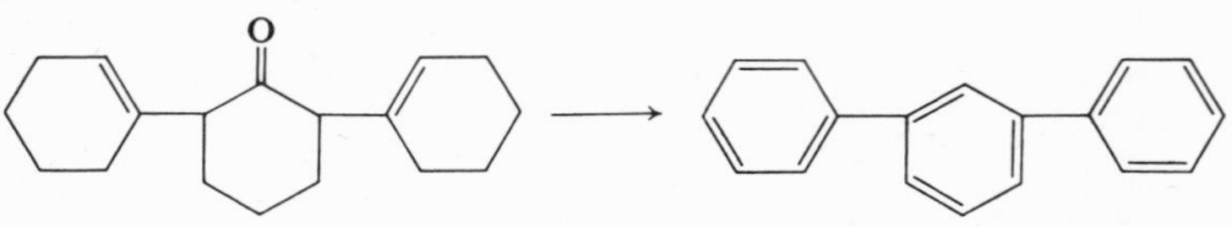

Deoxygenation has been applied with varying success to the synthesis of naphthalenes from tetralones and decalones (Linstead and Michaelis, 1940). The merit of the reaction lies in by-passing two of the usual steps, conversion to an alcohol and dehydration. The reaction products may include sizable amounts of dinaphthyl condensation products (Eisenbraun *et al.*, 1969). Inasmuch as loss of the oxygen atom consumes hydrogen, deoxygenation is favored in reactions carried out in a sealed tube. Dehydrogenation of the ketone, **CXLVII**, over palladium-on-carbon in a sealed tube proceeds with loss of the oxygen function to produce cadalene (**CXLVIII**) in good yield (Hayashi *et al.*, 1969).

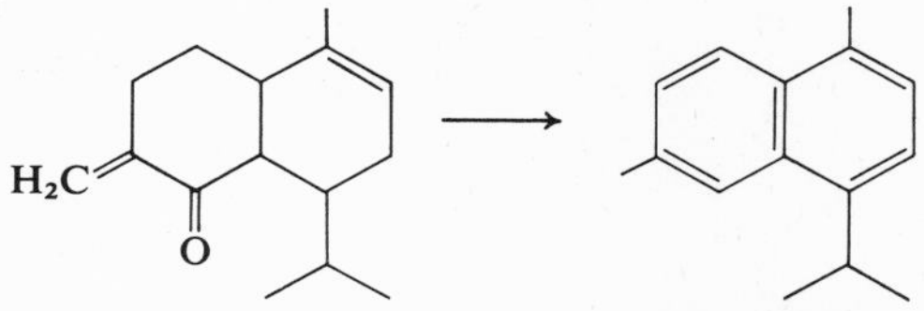

H_2C O

(CXLVII) **(CXLVIII)**

Ketones attached to a quaternary carbon are cleaved presumably with loss of carbon monoxide. For instance, dehydrogenation of **CLXIX** over palladium-on-carbon gives **CL** among other products (Dreiding and Voltman, 1954). Similar changes of the D ring have been noted by others (Dreiding and Pummer, 1953; Gentles *et al.*, 1958; Dreiding and Tomascewski, 1954, 1958). Methoxy groups attached to aromatic rings usually survive dehydrogenation (Bachmann and Dreiding, 1950; Walker, 1953; Barnes and Reinhold, 1952; Galat, 1951; Ritter and Murphy, 1952), but other examples of loss of methoxy have been recorded (Cocker *et al.*, 1950).

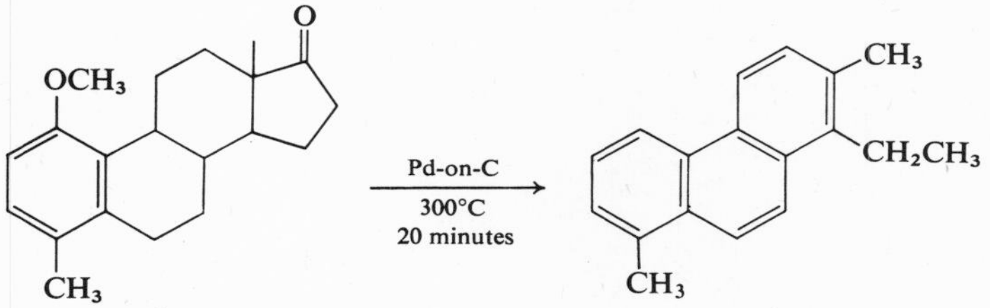

O OCH_3 CH_3 CH_3 CH_2CH_3 CH_3

Pd-on-C
300°C
20 minutes

(CXLIX) **(CL)**

Ring D cleavage in this type of molecule occurs at temperatures above 300°C. At temperatures around 250°C, aromatization without cleavage occurs (Bachmann and Dreiding, 1950).

Ethers

Ethers that undergo facile hydrogenolysis under hydrogenation conditions might be expected to be cleaved in dehydrogenation reactions. For instance, attempted dehydrogenation of **CLI** over 10% palladium-on-carbon in refluxing xylene or decalin affords 4-chlorophenol (**CLII**) and 1-methylisoquinoline (**CLIII**) (Tute *et al.*, 1970). This compound was successfully dehydrogenated by a novel reaction involving treatment with excess *m*-chloroperbenzoic acid.

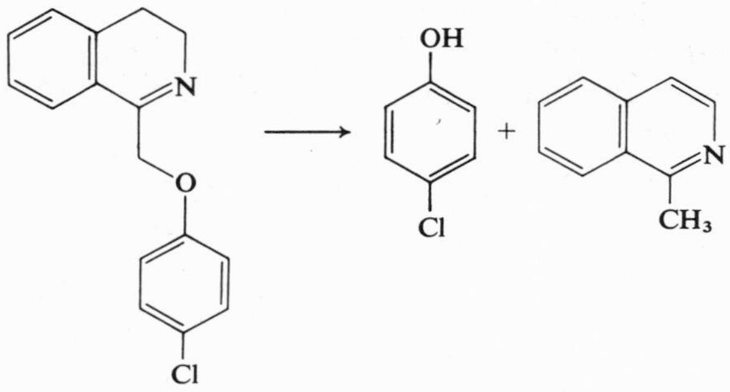

OH

N

O

Cl

Cl

CH_3

N

(CLI) (CLII) (CLIII)

Lower temperatures favor dehydrogenation without cleavage, higher temperatures favor cleavage. Dehydrogenation of homopterocarpin (**CLV**) over palladium-on-carbon in refluxing mesitylene affords 2,3-dehydrohomopterocarpin (**CLIV**), whereas at 300°C, the coumarin, **CLVI**, is obtained. The authors accounted for the formation of **CLVI** by a free-radical sequence (Bowyer *et al.*, 1964).

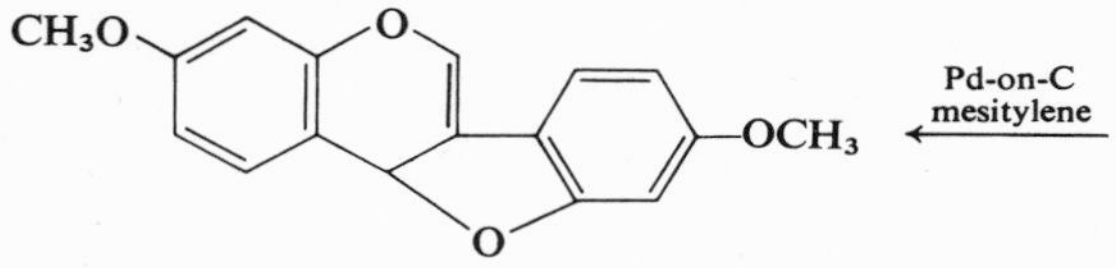

CH_3O O OCH_3 O

Pd-on-C
mesitylene

(CLIV)

CH_3O O OCH_3 O

Pd-on-C
300°C

(CLV)

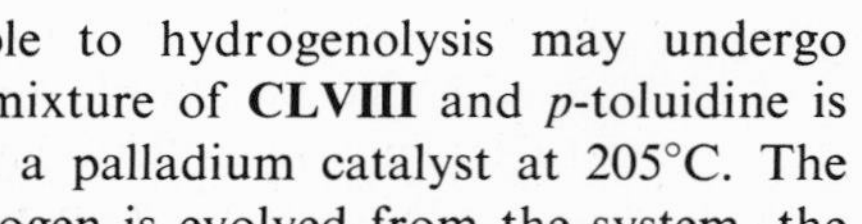

CH_3O O OCH_3

(CLVI)

Nitrogen–Nitrogen Bonds

Nitrogen–nitrogen bonds susceptible to hydrogenolysis may undergo cleavage during dehydrogenation. A mixture of **CLVIII** and *p*-toluidine is obtained when **CLVII** is heated with a palladium catalyst at 205°C. The reaction is interesting in that no hydrogen is evolved from the system, the whole process being an autohydrogenolysis (Eastman and Detert, 1951).

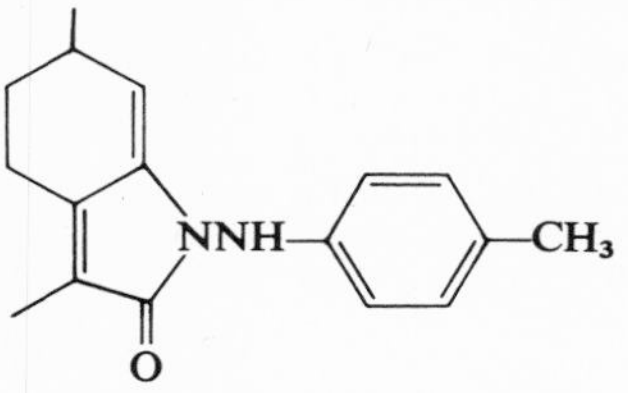

(CLVII) 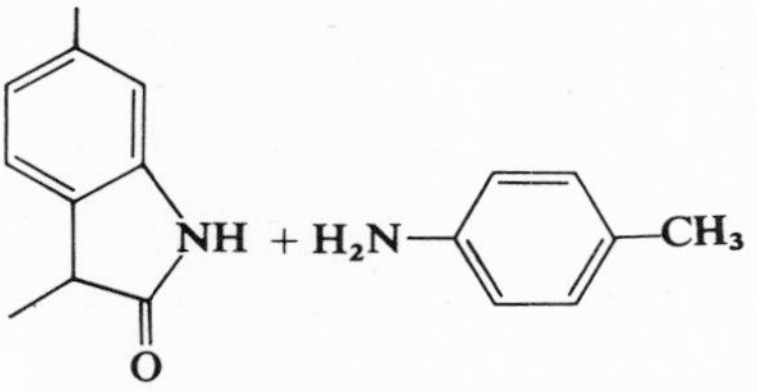 (CLVIII)

A similar type of cleavage is observed when the azine, **CLIX**, is refluxed in triethylbenzene over 5% palladium-on-carbon: 3,5-Dimethylaniline is obtained in about 50% yield. However, under similar conditions, no anilines are formed from **CLX** and **CLXI** (Horning and Horning, 1947b).

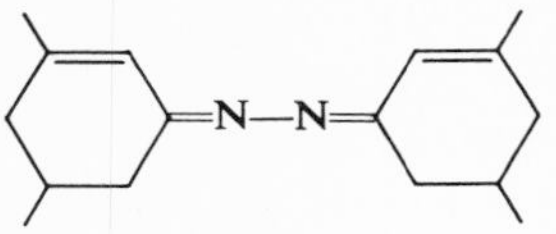

(CLIX) 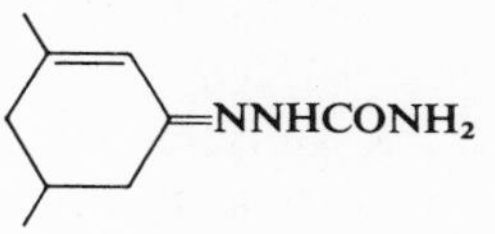 (CLX) 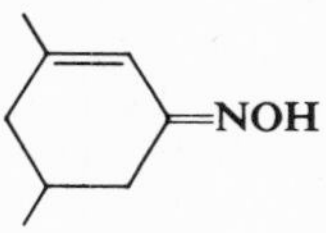 (CLXI)

Cyclohexanones may be converted to anilines instead of phenols by treatment with hydrazine followed by dehydrogenation of the resulting azine. Dehydrogenation in a nitrogen atmosphere of the azine of **CLXII** over 5% palladium-on-carbon in refluxing *p-t*-butyltoluene affords the aniline **CLXIII** (Robison *et al.*, 1966).

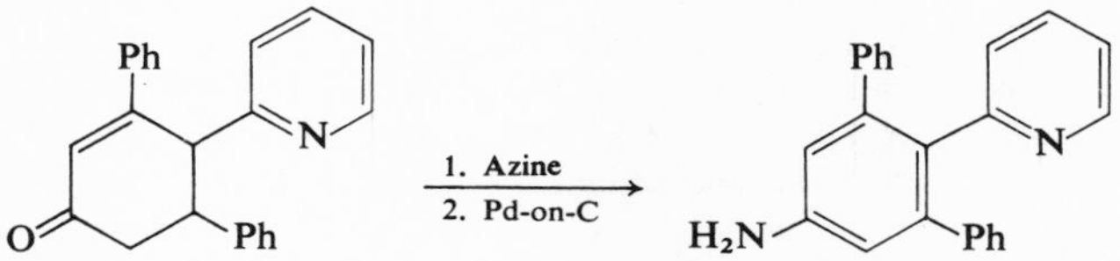

1. Azine
2. Pd-on-C

This reaction had been applied earlier to the synthesis of alkylanilines (Horning and Horning, 1947b) and aminobiphenyls and naphthylamines (Horning *et al.*, 1948a). The yield of amines obtained by refluxing the azines in triethylbenzene were generally the order of 50%. Although the reaction proceeded readily with the azines, similar attempts to convert oximes, oxime benzoates, and semicarbazones all meet with failure (Horning and Horning, 1947b).

REFERENCES

Acheson, R. M., and Robinson, D. A. (1968). *J. Chem. Soc., C* p. 1633.
Adkins, H., and Davis, J. W. (1949). *J. Amer. Chem. Soc.* **71**, 2955.
Adkins, H., and England, D. C. (1949). *J. Amer. Chem. Soc.* **71**, 2958.

Adkins, H., and Hager, G. F. (1949). *J. Amer. Chem. Soc.* **71**, 2962.
Adkins, H., Richards, L. M., and Davis, J. W. (1941). *J. Amer. Chem. Soc.* **63**, 1320.
Adkins, J., Rae, D. S., Davis, J. W., Hager, G. F., and Hoyle, K. (1948). *J. Amer. Chem. Soc.* **70**, 381.
Ainsworth, C. (1957). *J. Amer. Chem. Soc.* **79**, 5242.
Anderson, A. G., Jr., and Anderson, R. G. (1957). *J. Org. Chem.* **22**, 1197.
Anderson, A. G., Jr., and Nelson, J. A. (1951). *J. Amer. Chem. Soc.* **73**, 232.
Anderson, A. G., Jr., and Wade, R. H. (1952). *J. Amer. Chem. Soc.* **74**, 2274.
Anderson, A. G., Jr., Harrison, W. F., and Anderson, R. G. (1963). *J. Amer. Chem. Soc.* **85**, 3448.
Bach, F. L., Barclay, J. C., Kende, F., and Cohen, E. (1968). *J. Med. Chem.* **11**, 987.
Bachmann, W. E., and Controulis, J. (1951). *J. Amer. Chem. Soc.* **73**, 2736.
Bachmann, W. E., and Dreiding, A. S. (1950). *J. Amer. Chem. Soc.* **72**, 1323.
Bachmann, W. E., and Wilds, A. L. (1938). *J. Amer. Chem. Soc.* **60**, 624.
Bailey, W. J., and Liao, C-W. (1955). *J. Amer. Chem. Soc.* **77**, 992.
Bailey, W. J., Fetter, E. J., and Economy, J. (1962). *J. Org. Chem.* **27**, 3479.
Bajer, F. J., and Carr, R. L. K. (1968). U.S. Patent 3,394,399.
Barnes, R. A., and Reinhold, D. F. (1952). *J. Amer. Chem. Soc.* **74**, 1327.
Bell, M. R., Zalay, A. W., Oesterlin, R., Schane, P., and Potts, G. O. (1970). *J. Med. Chem.* **13**, 664.
Belskii, I. F., Gertner, D., and Zilkha, A. (1968). *J. Org. Chem.* **33**, 1348.
Bergmann, E. D., and Szmuszkovicz, J. (1951). *J. Amer. Chem. Soc.* **73**, 5153.
Bhandari, R. G., and Bhide, G. V. (1970). *Chem. Ind.* (*London*) p. 868.
Blomquist, A. T., and Meinwald, Y. C. (1960). *J. Amer. Chem. Soc.* **82**, 3619.
Blum, J., and Biger, S. (1970). *Tetrahedron Lett.* p. 1825.
Boekelheide, V., and Windgassen, R. J., Jr. (1959). *J. Amer. Chem. Soc.* **81**, 1456.
Boekelheide, V., Gerson, F., Heilbronner, E., and Meuche, D. (1963). *Helv. Chim. Acta* **46**, 1951.
Boswell, G. A., Jr., Johnson, A. L., and McDevitt, J. P. (1971). *J. Org. Chem.* **36**, 575.
Bowyer, W. J., Chatterjea, J. N., Dhoubhadel, S. P., Handford, B. O., and Whalley, W. B. (1964). *J. Chem. Soc., London* p. 4212.
Braude, E. A., Linstead, R. P., and Mitchell, P. W. D. (1954a). *J. Chem. Soc., London* p. 3578.
Braude, E. A., Linstead, R. P., and Wooldridge, K. R. H. (1954b). *J. Chem. Soc., London* p. 3586.
Braude, E. A., Linstead, R. P., Mitchell, P. W. D., and Wooldridge, K. R. H. (1954c). *J. Chem. Soc., London* p. 3595.
Broadbent, H. S., Allred, E. L., Pendleton, L., and Whittle, C. W. (1960). *J. Amer. Chem. Soc.* **82**, 189.
Brown, E. D., Sam, T. W., and Sutherland, J. K. (1969). *Tetrahedron Lett.* p. 5025.
Browne, P. A., and Kirk, D. N. (1969). *J. Chem. Soc., C* p. 1653.
Burnett, J. P., Jr., and Ainsworth, C. (1958). *J. Org. Chem.* **23**, 1382.
Butler, J. D., and Laundon, R. D. (1970a). *J. Chem. Soc., B* p. 716.
Butler, J. D., and Laundon, R. D. (1970b). *J. Chem. Soc., B* p. 1525.
Butz, E. W. J. (1940). *J. Amer. Chem. Soc.* **62**, 2557.
Buzby, G. C., Smith, R. C., and Smith, H. (1969). U.S. Patent 3,479,376.
Campaigne, E., and Lake, R. D. (1959). *J. Org. Chem.* **24**, 478.
Canonne, P., and Regnault, A. (1969). *Tetrahedron Lett.* p. 243.
Cantrall, E. W., Conrow, R. B., and Bernstein, S. (1967). *J. Org. Chem.* **32**, 3445.
Capps, D. B., and Hamilton, C. S. (1953). *J. Amer. Chem. Soc.* **75**, 697.

Carrà, S., and Ragaini, V. (1967). *Tetrahedron Lett.* p. 1079.
Carrà, S., Beltrame, P., and Ragaini, V. (1964). *J. Catal.* **3**, 353.
Charman, H. B. (1967). *J. Chem. Soc., B* p. 629.
Charman, H. B. (1970). *J. Chem. Soc., B* p. 584.
Chatterjee, D. K., and Sen, K. (1969). *Tetrahedron Lett.* p. 5223.
Chatterjee, D. K., Chatterje, R. M., and Sen, K. (1964). *J. Org. Chem.* **29**, 2467.
Chatterjee, D. N. (1955a). *J. Amer. Chem. Soc.* **77**, 414.
Chatterjee, D. N. (1955b). *J. Amer. Chem. Soc.* **77**, 513.
Cheeseman, G. W. H., and Tuck, B. (1968). *Tetrahedron Lett.* p. 4851.
Chemerda, J. M., and Sletzinger, M. (1970). U.S. Patent 3,509,172.
Chumakov, Yu. I., and Sherstyuk, V. P. (1967). *Tetrahedron Lett.* p. 771.
Cocker, W., Cross, B. E., Fateen, A. K., Lipman, C., Stuart, E. R., Thompson, W. H., and Whyte, D. R. A. (1950). *J. Chem. Soc., London* p. 1781.
Crawford, M., and Supanekar, V. P. (1969). *J. Chem. Soc., C* p. 832.
Crawford, M., and Supanekar, V. R. (1970). *J. Chem. Soc., C* p. 1832.
Danishefsky, S., and Cavanaugh, R. (1968). *J. Org. Chem.* **33**, 2959.
Das Gupta, A. K., and Chatterjee, R. M. (1968). *Chem. Commun.* p. 502.
Das Gupta, A. K., and Das, K. R. (1969). *J. Chem. Soc., C* p. 33.
Das Gupta, A. K., Chatterjee, R. M., and Das, K. R. (1969a). *J. Chem. Soc., C* p. 29.
Das Gupta, A. K., Chatterjee, R. M., and Das, K. R. (1969b). *J. Chem. Soc., C* p. 1749.
Davies, K. M., Dewar, M. J. S., and Rona, P. (1967). *J. Amer. Chem. Soc.* **89**, 6294.
Deno, N. C. (1950). *J. Amer. Chem. Soc.* **72**, 4057.
Doering, W. von E., Mayer, J. R., and DePuy, C. H. (1953). *J. Amer. Chem. Soc.* **75**, 2386.
Dreiding, A. S., and Pummer, W. J. (1953). *J. Amer. Chem. Soc.* **75**, 3162.
Dreiding, A. S., and Tomascewski, A. J. (1954). *J. Org. Chem.* **19**, 241.
Dreiding, A. S., and Tomascewski, A. J. (1958). *J. Amer. Chem. Soc.* **80**, 3702.
Dreiding, A. S., and Voltman, A. (1954). *J. Amer. Chem. Soc.* **76**, 537.
Eastman, R. H., and Detert, F. L. (1951). *J. Amer. Chem. Soc.* **73**, 4511.
Ehrenstein, M., and Marggraff, I. (1934). *Chem. Ber.* **67B**, 486.
Eisenbraun, E. J., Springer, J. M., Hinman, C. W., Flanagan, P. W., Hamming, M. C., and Linder, D. E. (1969). "The Reaction of Naphthalenones with Pd-on-C." Div. Petrol. Chem. Preprints, New York.
Eisenbraun, E. J., Hinman, C. W., Springer, J. M., Burnham, J. W., Chou, T. S., Flanagan, P. W., and Hamming, M. C. (1971). *J. Org. Chem.* **36**, 2480.
Elderfield, R. C., and Maggiolo, A. (1949). *J. Amer. Chem. Soc.* **71**, 1906.
Elderfield, R. C., Lagowski, J. M., McCurdy, O. L., and Wythe, S. L. (1958). *J. Org. Chem.* **23**, 435.
Ennis, B. C. (1969). U.S. Patent 3,481,947.
Ercoli, A., Gardi, R., and Brianza, C. (1968). U.S. Patent 3,419,582.
Esse, R. C., and Christensen, B. E. (1960). *J. Org. Chem.* **25**, 1565.
Fales, H. M., Warnhoff, E. W., and Wildman, W. C. (1955). *J. Amer. Chem. Soc.* **77**, 5885.
Feder, J. B., and Silber, A. D. (1968). U.S. Patent 3,391,199.
Fieser, L. F., Fieser, M., and Hershberg, E. B. (1936). *J. Amer. Chem. Soc.* **58**, 1463.
Fleck, E. E., and Palkin, S. (1937). *J. Amer. Chem. Soc.* **59**, 1593.
Freudewald, J. E., and Konrad, F. M. (1970). British Patent 1,205,944.
Fuhlhage, D. W., and Vanderwerf, C. A. (1958). *J. Amer. Chem. Soc.* **80**, 6249.
Fujita, E., Fuji, K., and Tanaka, K. (1968). *Tetrahedron Lett.* p. 5905.
Fuson, R. C., and Sauer, R. J. (1963). *J. Org. Chem.* **28**, 2323.
Galat, A. (1951). *J. Amer. Chem. Soc.* **73**, 3654.
Galbraith, A., Small, T., and Boekelheide, V. (1959). *J. Org. Chem.* **24**, 582.

Gardner, P. D., and Horton, W. J. (1952). *J. Amer. Chem. Soc.* **74**, 657.
Garrett, J. M. (1969). *Tetrahedron Lett.* p. 191.
Geissman, T. A., and Turley, R. J. (1964). *J. Org. Chem.* **29**, 2553.
Gensler, W. J., Johnson, F., and Sloan, A. D. B. (1960). *J. Amer. Chem. Soc.* **82**, 6074.
Gensler, W. J., Shamasundar, K. T., and Marburg, S. (1968). *J. Org. Chem.* **33**, 2861.
Gentles, M. J., Moss, J. B., Herzog, H. L., and Hershberg, E. B. (1958). *J. Amer. Chem. Soc.* **80**, 3702.
Ghiringhelli, D., and Bernardi, L. (1967). *Tetrahedron Lett.* p. 1039.
Godfrey, J. C. (1959). *J. Org. Chem.* **24**, 581.
Green, H. A. (1966). U.S. Patent 3,255,200.
Guest, H. R., and Kiff, B. W. (1964). U.S. Patent 3,149,130.
Guyer, P., and Fritze, D. (1970). U.S. Patent 3,522,269.
Haddad, Y. M. Y., Henbest, H. B., Husbands, J., and Mitchell, T. R. B. (1964). *Proc. Chem. Soc., London* p. 361.
Harris, L. E., Duncan, W. P., Hall, M. J., and Eisenbraun, E. J. (1971). *Chem. Ind.* (*London*) p. 403.
Hay, A. S. (1968). U.S. Patent 3,415,896.
Hayashi, S., Matsuo, A., and Matsuura, T. (1969). *Tetrahedron Lett.* p. 1599.
Hernandez, L., and Nord, F. F. (1948). *J. Colloid Sci.* **3**, 377.
Herz, W., and Rogers, J. L. (1953). *J. Amer. Chem. Soc.* **75**, 4498.
Hester, J. B., Jr., Rudzik, A. D., and Veldkamp, W. (1970). *J. Med. Chem.* **13**, 827.
Hiser, R. D. (1968). U.S. Patent 3,402,210.
Hodgkins, J. E., and Hughes, M. P. (1962). *J. Org. Chem.* **27**, 4187.
Horning, E. C., ed.-in-chief. (1955). "Organic Synthesis," Collective Vol. 3, p. 686. Wiley, New York.
Horning, E. C., and Horning, M. G. (1947a). *J. Amer. Chem. Soc.* **69**, 1359.
Horning, E. C., and Horning, M. G. (1947b). *J. Amer. Chem. Soc.* **69**, 1907.
Horning, E. C., and Walker, G. N. (1952). *J. Amer. Chem. Soc.* **74**, 5147.
Horning, E. C., Horning, M. G., and Platt, E. J. (1948a). *J. Amer. Chem. Soc.* **70**, 288.
Horning, E. C., Horning, M. G., and Walker, G. N. (1948b). *J. Amer. Chem. Soc.* **70**, 3935.
Horton, W. J., and Walker, F. E. (1952). *J. Amer. Chem. Soc.* **74**, 758.
Hoshino, T., and Takiwa, K. (1936). *Bull. Chem. Soc. Jap.* **11**, 218.
House, H. O., and Bashe, R. W. (1967). *J. Org. Chem.* **32**, 784.
House, H. O., Larson, J. K., and Muller, H. C. (1968a). *J. Org. Chem.* **33**, 957.
House, H. O., Sauter, F. J., Kenyon, W. G., and Riehl, J. J. (1968b). *J. Org. Chem.* **33**, 961.
Howsam, R. W., and McQuillin, F. J. (1968). *Tetrahedron Lett.* p. 3667.
Hussey, A. S., Schenach, T. A., and Baker, R. H. (1968). *J. Org. Chem.* **33**, 3258.
Illingworth, G. E., and Louvar, J. J. (1966). U.S. Patent 3,285,932.
Ipatieff, V. N., Meisinger, E. E., and Pines, H. (1950a). *J. Amer. Chem. Soc.* **72**, 2772.
Ipatieff, V. N., Appell, H. R., and Pines, H. (1950b). *J. Amer. Chem. Soc.* **72**, 4260.
Ipatieff, V. N., Czajkowski, G. J., and Pines, H. (1951). *J. Amer. Chem. Soc.* **73**, 4098.
Irwin, W. J., and Wibberley, D. G. (1968). *Chem. Commun.* p. 878.
Jackman, L. M. (1960). *Advan. Org. Chem.* **2**, 329.
Jackson, G. D. F., Sasse, W. H. F., and Whittle, C. P. (1963). *Aust. J. Chem.* **16**, 1126.
Johns, W. F. (1971). *J. Org. Chem.* **36**, 2391.
Kahovec, J., and Pospisil, J. (1969). *Chem. Ind.* (*London*) p. 919.
Kakisawa, H., Tateishi, M., and Kusumi, T. (1968). *Tetrahedron Lett.* p. 3783.
Karol, F. J., and Carrick, W. L. (1966). U.S. Patent 3,287,427.
Kornfeld, E. C., Fornefeld, E. J., Kline, G. B., Mann, M. J., Morrison, D. E., Jones, R. G., and Woodward, R. B. (1956). *J. Amer. Chem. Soc.* **78**, 3087.
Kubota, T. (1939). *J. Chem. Soc. Jap.* **60**, 604.

Kyrides, L. P., and Zienty, F. B. (1946). *J. Amer. Chem. Soc.* **68,** 1385.
Lago, R. M., Prater, C. D., and Weisz, P. B. (1956). *Amer. Chem. Soc., Div. Petrol. Chem., Prepr.* **1,** No. 1, 87.
Larkin, D. R. (1965). *J. Org. Chem.* **30,** 335.
Lednicer, D., and Hauser, C. R. (1958). *J. Amer. Chem. Soc.* **80,** 6364.
LeMaistre, J. W., and Sherman, A. H. (1969). U.S. Patent 3,427,355.
Leonard, N. J., Durand, D. A., and Uchimaru, F. (1967). *J. Org. Chem.* **32,** 3607.
Ling, N. C., and Djerassi, C. (1970). *J. Amer. Chem. Soc.* **92,** 6019.
Linsk, J. (1950). *J. Amer. Chem. Soc.* **72,** 4257.
Linstead, R. P., and Michaelis, K. O. A. (1940). *J. Chem. Soc., London* p. 1134.
Linstead, R. P., and Thomas, S. L. S. (1940). *J. Chem. Soc., London* p. 1127.
Linstead, R. P., Millidge, A. F., Thomas, S. L. S., and Walpole, A. L. (1937). *J. Chem. Soc., London* p. 1146.
Linstead, R. P., Michaelis, K. O. A., and Thomas, S. L. S. (1940). *J. Chem. Soc., London* p. 1139.
Linstead, R. P., Braude, E. A., Mitchell, P. W. D., Wooldridge, K. R. H., and Jackman, L. M. (1952). *Nature (London)* **169,** 100.
Lochte, H. L., and Pittman, A. G. (1960). *J. Amer. Chem. Soc.* **82,** 469.
Locke, D. M., and Pelletier, S. W. (1959). *J. Amer. Chem. Soc.* **81,** 2246.
Loew, O. (1880). *Chem. Ber.* **23,** 289.
Majima, R., and Murahashi, S. (1935). *Collect. Pap. Fac. Sci., Osaka Imp. Univ., Ser. C* **2,** 341; *Chem. Astr.* **30,** 3437 (1936).
Marschik, J. F. (1964). Unpublished observations from Engelhard Ind. Research Laboratories, Menlo Park, New Jersey.
Marschik, J. F., and Rylander, P. N. (1970). U.S. Patent 3,517,021.
Matsumura, H., Imafuku, K., Takano, I., and Matsuura, S. (1971). *Bull. Chem. Soc. Jap.* **44,** 567.
Minn, J., Sanderson, T. F., and Subluskey, L. A. (1956). *J. Amer. Chem. Soc.* **78,** 630.
Mleczak, W. (1965). *Wiad. Chem.* **19,** 633; *Chem. Abstr.* **64,** 551 (1966).
Mori, K., and Matsui, M. (1968). *Tetrahedron* **24,** 6573.
Mosettig, E., and Duvall, H. M. (1937). *J. Amer. Chem. Soc.* **59,** 367.
Mozingo, R. (1955). *Org. Syn., Collect. Vol.* **3,** 687.
Newman, M. S., and Blum, J. (1964). *J. Amer. Chem. Soc.* **86,** 503.
Newman, M. S., and Bye, T. S. (1952). *J. Amer. Chem. Soc.* **74,** 905.
Newman, M. S., and Lednicer, D. (1956). *J. Amer. Chem. Soc.* **78,** 4765.
Newman, M. S., and Mangham, J. R. (1949). *J. Amer. Chem. Soc.* **71,** 3342.
Newman, M. S., and O'Leary, T. J. (1946). *J. Amer. Chem. Soc.* **68,** 258.
Newman, M. S., and Zahm, H. V. (1943). *J. Amer. Chem. Soc.* **65,** 1097.
Newman, M. S., Sagar, W. C., and Cochrane, C. C. (1958). *J. Org. Chem.* **23,** 1832.
Newman, M. S., Sagar, W. C., and Georges, M. V. (1960). *J. Amer. Chem. Soc.* **82,** 2376.
Nielsen, A. T. (1970). *J. Org. Chem.* **35,** 2498.
Nielsen, A. T., Dubin, H., and Hise, K. (1967). *J. Org. Chem.* **32,** 3407.
Ninomiya, I., Naito, T., and Mori, T. (1969). *Tetrahedron Lett.* p. 3634.
Oka, S. (1962). *Bull. Chem. Soc. Jap.* **35,** 562.
Oldroyd, D. M., Fisher, G. S., and Goldblatt, L. A. (1950). *J. Amer. Chem. Soc.* **72,** 2407.
Packendorff, K., and Leder-Packendorff, L. (1934). *Chem. Ber.* **67B,** 1388.
Pal, B. C. (1958). *J. Sci. Ind. Res., Sect. A* **17,** 270.
Pallaud, R., and Hoa, H. A. (1964). *Chim. Anal. (Paris)* **46,** 501; *Chem. Abstr.* **62,** 6515 (1965).
Pallaud, R., and Hoa, H. A. (1965). *Chim. Anal. (Paris)* **47,** 22; *Chem. Abstr.* **62,** 16306 (1965).
Patterson, J. M., and Drenchko, P. (1959). *J. Org. Chem.* **24,** 878.

Peterson, R. C., and Kloetzel, M. C. (1958). *J. Amer. Chem. Soc.* **80**, 1416.
Phillips, D. D., and Chatterjee, D. N. (1958). *J. Amer. Chem. Soc.* **80**, 4364.
Pines, H., and Marechal, J. (1955). *J. Amer. Chem. Soc.* **77**, 2819.
Pines, H., Oblerg, R. C., and Ipatieff, V. N. (1948). *J. Amer. Chem. Soc.* **70**, 533.
Pines, H., Jenkins, E. F., and Ipatieff, V. N. (1953). *J. Amer. Chem. Soc.* **75**, 6226.
Prelog, V., and Schenker, K. (1953). *Helv. Chim. Acta* **36**, 1181.
Prelog, V., and Szpilfogel, S. (1945). *Helv. Chim. Acta* **28**, 1684.
Rabjohn, N., ed.-in-chief. (1963). "Organic Synthesis," Collective Vol. 4, p. 536. Wiley, New York.
Rahman, A., Vuano, B. M., and Rodriguez, N. M. (1970). *Chem. Ind.* (*London*) p. 1173.
Rapoport, H., and Bordner, J. (1964). *J. Org. Chem.* **29**, 2727.
Rapoport, H., and Castagnoli, N., Jr. (1962). *J. Amer. Chem. Soc.* **84**, 2178.
Rapoport, H., and Holden, K. G. (1962). *J. Amer. Chem. Soc.* **84**, 635.
Rapoport, H., and Willson, C. D. (1962). *J. Amer. Chem. Soc.* **84**, 630.
Rapoport, H., Iwamoto, R., and Tretter, J. R. (1960). *J. Org. Chem.* **25**, 372.
Rapoport, H., Castagnoli, N., Jr., and Holden, K. G. (1964). *J. Org. Chem.* **29**, 883.
Reggel, L., and Friedel, R. A. (1951). *J. Amer. Chem. Soc.* **73**, 1449.
Ritchie, A. W., and Nixon, A. C. (1967). *Amer. Chem. Soc., Div. Petrol. Chem., Prep.* **12**, 175.
Ritter, J. J., and Murphy, F. X. (1952). *J. Amer. Chem. Soc.* **74**, 763.
Robison, M. M., Robison, B. L., and Butler, F. P. (1959). *J. Amer. Chem. Soc.* **81**, 743.
Robison, M. M., Pierson, W. G., Dorfman, L., Lambert, B. F., and Lucas, R. A. (1966). *J. Org. Chem.* **31**, 3206.
Rylander, P. N. (1967). "Catalytic Hydrogenation Over Platinum Metals," p. 99. Academic Press, New York.
Rylander, P. N., and Karpenko, I. (1970). Unpublished observations of Engelhard Minerals and Chemicals Research Laboratory, Menlo Park, New Jersey.
Rylander, P. N., and Kilroy, M. (1958). *Engelhard Ind., News Bull.* published privately.
Rylander, P. N., and Steele, D. R. (1968). *Engelhard Ind. Tech. Bull.* **9**, 115.
Sarett, L. H., Lukes, R. M., Poos, G. I., Robinson, J. M., Beyler, R. E., Vandegrift, J. M., and Arth, G. E. (1952). *J. Amer. Chem. Soc.* **74**, 1393.
Sargent, L. H., and Agar, J. H. (1958). *J. Org. Chem.* **23**, 1938.
Sargeson, A. M., and Sasse, W. H. F. (1958). *Proc. Chem. Soc., London* p. 150.
Scaros, M. G., and Bible, R. H., Jr. (1968). U.S. Patent 3,379,744.
Schniepp, L. E., and Geller, H. H. (1947). *J. Amer. Chem. Soc.* **69**, 1545.
Sedlak, M. (1966). *Anal. Chem.* **38**, 1503.
Sengupta, S. C., and Chatterjee, D. N. (1952). *J. Indian Chem. Soc.* **29**, 438.
Sengupta, S. C., and Chatterjee, D. N. (1953). *J. Indian Chem. Soc.* **30**, 27.
Sengupta, S. C., and Chatterjee, D. N. (1954a). *J. Indian Chem. Soc.* **31**, 11.
Sengupta, S. C., and Chatterjee, D. N. (1954b). *J. Indian Chem. Soc.* **31**, 285.
Shamma, M., and Rosenstock, P. D. (1961). *J. Org. Chem.* **26**, 2586.
Shuikin, N. I., Erivanskaya, L. A., and Yang, A. H. (1960). *Dokl. Akad. Nauk SSSR* **133**, 1125.
Shuikin, N. I., Erivanskaya, L. A., and Yang, A. H. (1962a). *Zh. Obshch. Khim.* **32**, 823.
Shuikin, N. I., Belskii, I. F., and Karakhanov, R. A. (1962b). *Izv. Akad. Nauk SSSR, Otd. Khim. Nauk* p. 138.
Simpson, B. D., Schnitzer, A. N., and Cobb, R. L. (1959). U.S. Patent 3,007,931.
Springer, J. M., Hinman, C. W., Eisenbraun, E. J., Flanagan, P. W., Hamming, M. C., and Linder, D. E. (1971). *J. Org. Chem.* **36**, 686.
Stiles, M., and Sisti, A. J. (1961). *J. Org. Chem.* **26**, 3639.

Suginome, H. (1959). *J. Org. Chem.* **24**, 1655.
Suld, G. (1967). U.S. Patent 3,325,551.
Takeda, K., Minato, J., Hamamoto, K., Horibe, I., Nagasaki, T., and Ikuta, M. (1964). *J. Chem. Soc., London* p. 3577.
Taub, D., Kuo, C. H., and Wendler, N. L. (1967). *J. Chem. Soc., C* p. 1558.
Taylor, E. C., and Strojny, E. J. (1960). *J. Amer. Chem. Soc.* **82**, 5198.
Terashima, T., Kuroda, Y., and Kaneko, Y. (1969). *Tetrahedron Lett.* p. 2535.
Turner, R. B., Nettleton, D. E., Jr., and Ferebee, R. (1956). *J. Amer. Chem. Soc.* **78**, 5923.
Tute, M. S., Brammer, K. W., Kaye, B., and Broadbent, R. W. (1970). *J. Med. Chem.* **13**, 44.
Van Winkle, J. L., McClure, J. D., and Williams, P. H. (1966). *J. Org. Chem.* **31**, 3300.
Voltz, S. E., and Weller, S. W. (1961). U.S. Patent 2,967,202.
Voltz, S. E., Krause, J. H., Erner, W. E. (1959). U.S. Patent 2,891,965.
Walker, G. N. (1953). *J. Amer. Chem. Soc.* **75**, 3387.
Walker, G. N. (1958). *J. Org. Chem.* **23**, 133.
Wani, M. C., Kepler, J. A., Thompson, J. B., Wall, M. E., and Levine, S. G. (1970). *Chem. Commun.* p. 404.
Wendler, N. L., Slates, H. L., and Tishler, M. (1951). *J. Amer. Chem. Soc.* **73**, 3816.
Wendler, N. L., Taub, D., and Kuo, C. H. (1969a). U.S. Patent 3,435,044.
Wendler, N. L., Taub, D., and Kuo, C. H. (1969b). U.S. Patent 3,450,706.
Wenkert, E., and Roychaudhuri, D. K. (1958). *J. Amer. Chem. Soc.* **80**, 1613.
White, E. H., and Dunathan, H. C. (1956). *J. Amer. Chem. Soc.* **78**, 6055.
Williamson, J. B. (1970). British Patent 1,209,378.
Woods, G. F., and Scotti, F. (1961). *J. Org. Chem.* **26**, 312.
Woods, G. F., and Viola, A. (1956). *J. Amer. Chem. Soc.* **78**, 4380.
Yakhontov, L. N., Azimov, V. A., Lapan, E. I., and Ordzhonikidze, S. (1969). *Tetrahedron Lett.* p. 1909.
Yates, P., and Field, G. F. (1970). *Tetrahedron* **26**, 3135.
Zellner, R. J. (1961). U.S. Patent 3,008,965.

CHAPTER 2

Homogeneous Hydrogenation

Noble metal, homogeneous hydrogenation catalysts have been in recent years the subject of intensive study, especially from a theoretical viewpoint (Lyons *et al.*, 1970). Synthetic applications of these catalysts have lagged behind the theoretical development for a single important reason; noble metal, heterogeneous hydrogenation catalysts are excellent, versatile, efficient catalysts (Rylander, 1967), which limits the need for the use of a homogeneous catalyst to those applications where heterogeneous catalysts are in some respect deficient. So far, this need has been limited to the hydrogenation of olefins where some aspect of selectivity is involved, to incipient aromatic systems where disproportionation is possible, to systems that poison heterogeneous catalysts, to selective labeling reactions, and to asymmetric hydrogenations. These applications are reviewed here; those papers, some of which report elegant work, dealing with mechanisms and kinetics of homogeneous hydrogenation will not be discussed.

CATALYSTS

New homogeneous hydrogenation catalysts are being discovered at a rapid rate. All six noble metals, palladium (Maxted and Ismail, 1964; Tayim and Bailar, 1967b), platinum (Cramer *et al.*, 1963; Davis *et al.*, 1963), rhodium (Vaska and Rhodes, 1965; Young *et al.*, 1965; James and Rempel, 1966;

Mague and Wilkinson, 1966; Osborn *et al.*, 1966; Jardine *et al.*, 1967; Montelatici *et al.*, 1968; O'Connor and Wilkinson, 1968, 1969; Chevallier *et al.*, 1969; Piers and Cheng, 1969; Shapley *et al.*, 1969; Lehman *et al.*, 1970; Dewhirst, 1970; Abley *et al.*, 1971; Masters *et al.*, 1971), ruthenium (Halpren *et al.*, 1961; Harrod *et al.*, 1961; Halpren, 1965; Jardine and McQuillin, 1968; Hui and James, 1969; Ogata *et al.*, 1970; Khan *et al.*, 1971), iridium (Haddad *et al.*, 1964; Eberhardt and Vaska, 1967; Yamaguchi, 1967; Baddley and Fraser, 1969; Guistiniani *et al.*, 1969; Jardine *et al.*, 1969; van Gaal *et al.*, 1970; White *et al.*, 1971), and osmium (Vaska, 1965; Fotis and McCollum, 1967; Dewhirst, 1969) make active catalysts and these have appeared already in great variety. It is much too early to judge what type of catalyst will ultimately show the greatest synthetic utility.

There are, at present, practical limitations to homogeneous hydrogenation catalysts, but these may be overcome with further work. Homogeneous catalysts, with some exceptions, are not as active, on a weight of metal basis, as heterogeneous catalysts, despite every metal atom in the system being available for reaction. Homogeneous catalysts are more difficult to separate from the product. Unless the product crystallizes from the reaction mixture, simple filtration will not suffice and extraction or distillation processes must be used. Some effort has been made recently to overcome this difficulty by supporting the catalysts (Rony, 1969; Acres *et al.*, 1966). Grubbs and Kroll (1971) used polystyrene beads (200–400 mesh) and anchored the catalysts through a sequence of reactions involving chloromethylation of the aromatic ring, followed by displacement of the halogen with lithodiphenylphosphine. The beads were then equilibrated with a twofold excess of tris(triphenylphosphine)-chlororhodium for 2 to 4 weeks. The catalysts are recovered easily by filtration and can be reused many times. The technique of supporting catalysts removes the limitation imposed on rate by solubility requirements of homogeneous catalysts; any amount of supported catalysts can be effectively used provided the system does not become diffusion-controlled.

$$\text{poly}-C_6H_5 \xrightarrow[SnCl_4]{CH_3CH_2OCH_2Cl} \text{poly}-C_6H_4-CH_2Cl \xrightarrow{LiPPh_2}$$

$$\text{poly}-C_6H_4-CH_2PPh_2 \xrightarrow{ClRh(Ph_3P)_3} \text{poly}-C_6H_4-CH_2P(Ph)_2RhL_n$$

$$L = Ph_3P$$

OLEFINS

Hydrogenation of olefins, which proceeds readily over certain homogeneous catalysts, has been studied more than any other reduction. Some catalysts, such as tris(triphenylphosphine)hydridochlororuthenium(II), reduce olefins so rapidly in benzene–ethanol that even with vigorous stirring, the reaction is apt to be diffusion-controlled (Hallman *et al.*, 1968). Ethanol was once thought to be actively involved in the hydrogenation, for in the absence of such a solvent the hydrogenation was extremely slow (Evans *et al.*, 1965). Later work revealed ethanol was acting merely as a base to promote formation of the active catalyst (Hallman *et al.*, 1968). Solvents also control the degree of dissociation of the catalytic species (Hartwell and Clark, 1970). Interaction of the olefin and catalyst is complex and the rate of hydrogenation depends on the concentration of the substrate and catalyst; with certain substrates a plot of rate versus substrate concentration shows both a maximum and minimum (Jardine *et al.*, 1969).

The rates of hydrogenation of olefins are very sensitive to olefin structure and this fact has led to interesting synthetic applications. The catalyst, tris-(triphenylphosphine)hydridochlororuthenium(II) (Skapski and Troughton, 1968), has been shown to be remarkably selective for the hydrogenation of terminal olefins in preference to internal olefins (Hallman *et al.*, 1967). Terminal olefins are reduced at extremely high rates, whereas internal olefins are reduced very slowly. Brown and Piszkiewicz (1967) using tris(triphenylphosphine)-chlororhodium(I) reported that selective reduction of a terminal double bond in the presence of other more-substituted double bonds is a general reaction. In heterogeneous systems, the same type of result may be achieved over ruthenium-on-carbon (Berkowitz and Rylander, 1959).

Substituted Olefins

Homogeneous catalysts are able to reduce olefinic unsaturation in a variety of compounds including α,β-unsaturated carboxylic acids, esters, aldehydes, ketones, nitriles, and nitro compounds. The general scope of the reaction was demonstrated with tris(triphenylphosphine)chlororhodium(I) as catalyst, but undoubtedly other homogeneous catalysts can be used also. Hydrogenations were carried out in deoxygenated benzene or absolute ethanol at 40°–60°C and 60–100 psig hydrogen pressure for 12 to 18 hours (Harmon *et al.*, 1969). Several investigators have used this catalyst for the hydrogenation of α,β-unsaturated aldehydes to saturated aldehydes (Osborn *et al.*, 1966; Birch and Walker, 1966b; Djerassi and Gutzwiller, 1966; Jardine *et al.*, 1967), but unwanted decarbonylation frequently accompanies the reduction (Jardine and Wilkinson, 1967). This difficulty can be overcome to a large extent by carrying out the reduction in absolute ethanol (Harmon *et al.*, 1969).

The general applicability of homogeneous catalysts in reductions of substituted olefins is established, but it is debatable whether or not, at present, they are useful inasmuch as all reductions reported so far can be done much more simply and with better yield over appropriate heterogeneous catalysts.

Diminished Isomerization

Homogeneous catalysts have proved useful where heterogeneous catalysts, because of their greater isomerizing tendencies (Rylander, 1971), have led to migration of the double bond into an inaccessible position or to mixtures of products. For example, hydrogenation of the diene, coronopilin (**I**) over $(Ph_3P)_3RhCl$ affords a low-melting form of dihydrocoronopilin (**II**), whereas platinum oxide (a catalyst among heterogeneous species of relatively low isomerizing activity) affords the isomerized product, isocoronopilin (**III**),

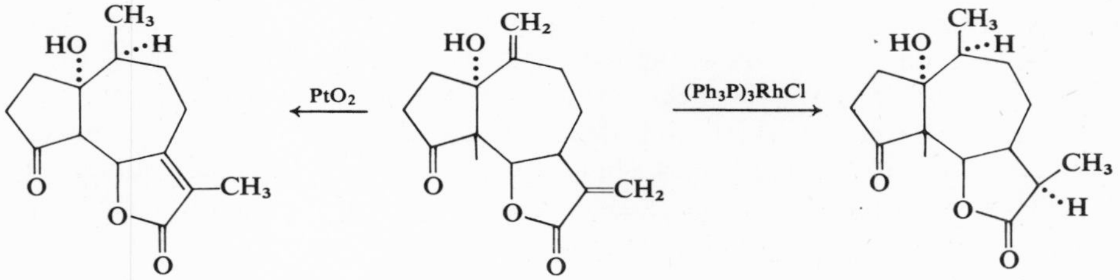

(III) ← PtO_2 — (I) — $(Ph_3P)_3RhCl$ → (II)

in 63 to 100% yield, depending on whether or not the catalyst was prehydrogenated (Ruesch and Mabry, 1969).

Similarly, the methylene groups of psilostachyine and confertiflorin are reduced smoothly over tris(triphenylphosphine)chlororhodium(I), but are mainly isomerized by heterogeneous catalysts (Biellmann and Jung, 1968). The rhodium catalyst also proved effective in a methylene reduction in the stereoselective total synthesis of seychellene (Piers *et al.*, 1969). This catalyst proved to be completely stereoselective in hydrogenation of the complex olefin, **IV**, affording a single isomer, **V**, whereas over platinum oxide a mixture of epimers is obtained (Piers and Cheng, 1969).

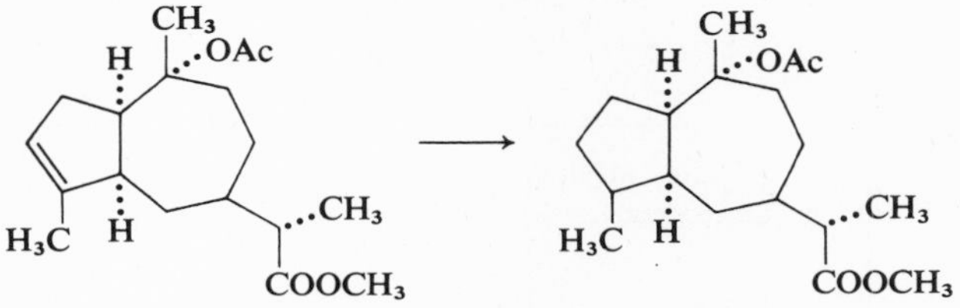

(IV) → (V)

In certain compounds, isomerization could not be avoided even over the homogeneous rhodium catalyst. Damsin, with a structure closely related to coronopilin, affords a 3:2 mixture of isodamsin and dihydrodamsin (Ruesch and Mabry, 1969; Biellmann and Jung, 1968).

An interesting catalyst that has so far seen very little use is $py_2(dmf)RuCl_2(BH_4)$ (Jardin and McQuillin, 1969b). This catalyst reduces 3-oxo-$\Delta^{4,5}$-steroids readily, whereas $(Ph_3P)_3RhCl$ does not, and in certain cases, the reduction was more stereospecific than obtained with heterogeneous catalysts.

Stereochemistry

The stereochemical outcome of an olefin hydrogenation depends in large measure on the extent of prior isomerization of the double bond into a new position before saturation and on the stereochemical requirements of the catalyst (Senda *et al.*, 1972). Since homogeneous and heterogeneous catalysts are known to differ widely in both their isomerizing tendencies and steric requirements, it might be anticipated that the use of homogeneous catalysts could provide stereochemical results not obtainable with heterogeneous catalysts. Several workers have demonstrated the validity of this premise. Mitchell (1970) hydrogenated 4-*t*-butylmethylenecyclohexane with a number of catalysts and established that large differences in isomer distribution could be obtained by appropriate choice of catalyst (see Table I).

$$(CH_3)_3C\text{-}C_6H_{10}{=}CH_2 \xrightarrow{H_2} (CH_3)_3C\text{-}C_6H_{10}\text{-}CH_3 + (CH_3)_3C\text{-}C_6H_{10}\cdots CH_3$$

TABLE I
HYDROGENATION OF 4-*t*-BUTYLMETHYLENECYCLOHEXANE

Solvent	Catalyst	% *cis*[a]	% *trans*[a]
MeOH	$H_2PtCl_6 \cdot SnCl_2$	7	93
C_6H_6	$OsHCl(CO)(Ph_3P)_3$	64	36
C_6H_6	$(Ph_3)_3RhCl$	68	32
C_6H_6	5% Rh-on-C	80	20
HOAc	PtO_2	83[b]	17

[a] % of saturated product.
[b] Siegel and Dmuchovsky (1962).

Small amounts of sodium hydroxide have a remarkable effect on this reaction when it is carried out with tris(tri-*o*-tolylphosphine)chlororhodium in benzene–ethanol. Without sodium hydroxide, the product is 85% *cis*-4-*t*-butylmethylcyclohexane, whereas when sodium hydroxide is present, the product contains 82% *trans* isomer. In contrast, the use of sodium hydroxide has no effect on stereochemistry of hydrogenation over heterogeneous rhodium-on-carbon (Mitchell, 1970).

Tyman and Willis (1970) made a comparison of catalysts using 2-butyl-4-methylenepyran as a substrate (see Table II). It is instructive to note the large range in isomer distribution that can be obtained by appropriate choice of catalyst and to contrast the results obtained here with those given above in hydrogenation of 4-*t*-butylmethylenecyclohexane. Without experience with a system, it seems difficult to predict the effect of various catalysts.

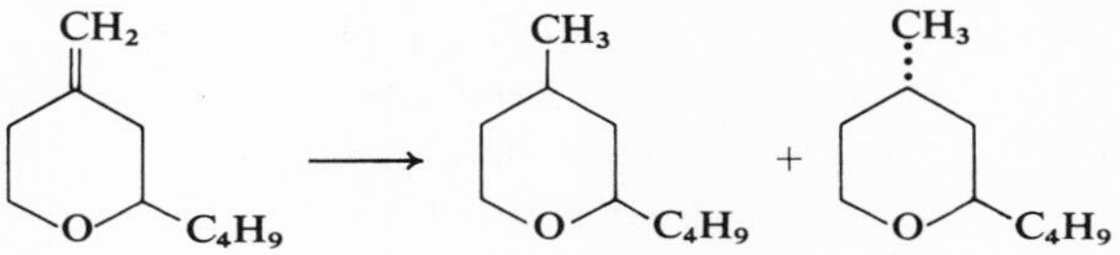

CH_2 CH_3 CH_3

O C_4H_9 → O C_4H_9 + O C_4H_9

TABLE II

HYDROGENATION OF 2-BUTYL-4-METHYLENEPYRAN

Catalyst	% *cis* isomer	% *trans* isomer
Pd/C	83	17
PtO_2	53	47
$H_2PtCl_6 \cdot SnCl_2$	90	10
$(Ph_3)_3RhCl$	68	32

Vinylcyclopropanes

Facile cleavage of vinylcyclopropanes by heterogeneous catalysts under hydrogenation conditions is well documented (Rylander, 1967) and maintenance of the intact ring sometimes is difficult. Interesting use was made of tris-(triphenylphosphine)chlororhodium(I) in this connection in the selective hydrogenation of vinylcyclopropanes in the steroid series. 3α,5-Cyclo-5α-cholest-6-ene was reduced to 3α,5-cyclo-5α-cholestane and 1α,5-cyclo-5α-cholest-2-ene was reduced to 1α,5-cyclo-5α-cholestane with the cyclopropane ring remaining intact. The latter substrate, when reduced over 10% palladium-on-carbon, rapidly absorbed 2 moles of hydrogen, affording 5α-cholestane (Laing and Sykes, 1968). Other workers in a study of the mechanism of homogeneous hydrogenation reduced seven simple vinylcyclopropanes over tris-(triphenylphosphine)chlororhodium(I) and obtained in each case the saturated cyclopropane as the major product with an average yield of 89% (Heathcock and Poulter, 1969).

Dienes

Homogeneous catalysts of the type $MX_2(QPh_n)_2$, where M = Pt or Pd, X = halogen, and Q = P, As, S, or Se, when used in the presence of a co-catalyst, $SnCl_2$, have been found to be effective for the selective hydrogenation of dienes to monoenes (Bailar and Itatani, 1966; (Itatani and Bailar, 1967). Platinum–tin complexes have received the most attention (Frankel, 1970). Little or no saturated products are found provided the double bonds are not terminal (Adams *et al.*, 1968a, b). Hydrogenation of nonconjugated, nonterminal diolefins is preceded by isomerization into conjugation, a reaction that is fast relative to hydrogenation. The reaction conditions required depend on the nature of the catalyst complex and on the solvent. With a catalyst composed of 0.5 mmole of platinum and 5.0 mmole of stannous chloride, convenient reaction conditions are a pressure of 500–600 psig, a temperature of 50°–110°C, with 50 ml of solvent and 30 moles of substrate such as methyl linolenate (Tayim and Bailar, 1967b). The source of hydrogen may be either hydrogen gas or methanol; both sources give comparable results, but hydrogen gas is preferred. Tin is essential to the process and is thought to activate the catalyst by being coordinated to it through the ligand $SnCl_3$ (Tayim and Bailar, 1967a).

Other workers found monoenes, whether terminal or internal, could be reduced readily under mild conditions with a platinum to tin ratio of 1:5, with extra chloride or bromide ions, and with solvents other than methanol (van Bekkum *et al.*, 1967).

Dienes may arise in good yield as the reduction products of trienes over homogeneous platinum–tin catalysts. Formation of dienes in high selectivity from methyl linolenate results from the remaining double bonds being separated by several methylene groups, a consequence of preferential attack at the middle 12 double bond (Frankel *et al.*, 1967). Very little linoleate is hydrogenated as long as linolenate is present (van't Hof and Linsen, 1967).

Iridium trichloride–stannous chloride catalysts are effective in selectively hydrogenating 1,5,9-cyclododecatriene to cyclododecene in alcohol solvent at 100°C. At least 1 mole of tin should be present to solubilize the iridium and to prevent formation of insoluble materials. More than 5 moles of tin suppresses the rate of hydrogenation (Gosser, 1970). The homogeneous catalyst, $Ir(COD)^+$ (COD = cyclooctadiene), is very effective for the rapid reduction of 1,5-cyclooctadiene solely to cyclooctene. In contrast, hydrogenation of norbornadiene catalyzed by $Rh(NBD)^+$ affords a dimeric species, $C_{14}H_{18}$, in 80% yield (Schrock and Osborn, 1971).

An interesting contrast between homogeneous and heterogeneous catalysts is provided by hydrogenation of the complex diene, eremophilone. Palladium on a variety of supports and in various solvents always led to a more rapid reduction of the conjugated double bonds, whereas reduction with tris-

(triphenylphosphine)chlororhodium(I) afforded 13,14-dihydroeremophilone (Brown and Piszkiewicz, 1967).

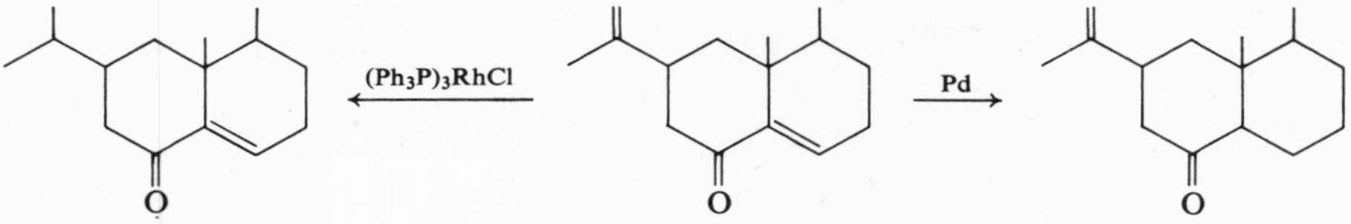

$(Ph_3P)_3RhCl$ Pd

Birch and Walker (1967b), considering the catalyst $(Ph_3P)_3RhCl$ as a pure hydrogen donor, investigated the possibility of reducing quinones to endiones instead of the usual aromatic products. Their efforts met with partial success and selected quinones, 1,4-naphthoquinone, juglone, and 2,3-dimethoxybenzoquinone were converted to the corresponding endiones. Benzoquinone itself was reduced to quinhydrone or quinol, and other quinones of high oxidation potential, diphenoquinone β-naphthoquinone, and 2,6-naphthoquinone, appeared to cause destruction of the catalyst.

Steroidal Dienes

Homogeneous catalysts seem particularly suitable in hydrogenation of multiple unsaturated steroids where achievement of correct selectivity and stereochemistry may pose a problem. Triphenylphosphine complexes of both rhodium and ruthenium have been used successfully in these reductions. The ruthenium complex apparently requires less control than rhodium catalysts for the reduction stops spontaneously at the monoene. Over ruthenium catalysts at 162 atm pressure, 1,4-androstadiene-3,17-dione was reduced in high yield of 92% to 4-androstene-3,17-dione accompanied by 8% androstane-3,17-dione.

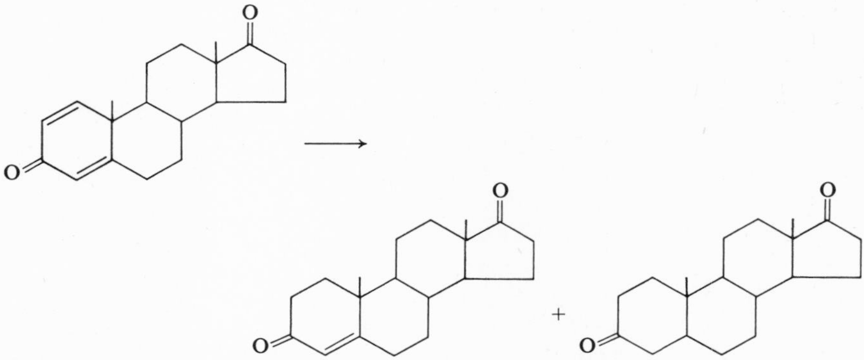

The yield of monoene increases with pressure; it is only 66% at 10 atm. Androstane-3,17-dione, formed in constant ratio throughout the hydrogenation, does not arise from 4-androstene-3,17-dione, but probably from 5α-androst-1-ene-3,17-dione which is readily reduced under the conditions of the reaction. The 17β-hydroxy-, acetoxy-, and acetyl derivatives of 1,4-androstadiene-3-one were also hydrogenated to the corresponding 4-en-3-ones with high selectivity (Nishimura and Tsuneda, 1969).

Complex triphenylphosphine rhodium catalysts do not require as vigorous conditions as ruthenium catalysts, and reductions may be carried out smoothly at room temperature and pressure. Reduction in the presence of tris(triphenylphosphine)chlororhodium(I) of either 1,4-androstadiene-3,17-dione or 4,6-androstadiene-3,17-dione afforded the 4-ene-3,17-dione in 75–80% yield, the remainder being unchanged substrate. These results contrast sharply with those from reductions by heterogeneous catalysts, which frequently produce a multiplicity of products (Djerassi and Gutzwiller, 1966). Tris(triphenylphosphine)chlororhodium(I) has been used successfully also in the selective hydrogenation of 3-oxo-11α-acetoxy-20-ethylenedioxy-Δ^4-pregnen (Wieland and Anner, 1968).

SULFUR COMPOUNDS

Traces of divalent sulfur compounds are frequently severe poisons in heterogeneous catalytic hydrogenation over platinum metals, and reduction of molecules containing an exposed divalent sulfur is a formidable problem. By the use of homogeneous catalysts, both of these difficulties may be circumvented. Birch and Walker (1967a) found only a moderate decrease in the rate of hydrogenation of 1-octene, dehydrolinalool, or ergosterol when 2.5 moles of thiophenol was added per mole of tris(triphenylphosphine)chlororhodium(I). Very large excesses of thiophenol cause severe inhibition, perhaps due to a shift in the equilibrium between a normal catalyst species and a catalyst-thiophene complex. Sulfides exert a much smaller inhibiting effect on the catalyst than does thiophene. Earlier workers (Rylander *et al.*, 1962), in testing whether or not a palladium system was truly homogeneous, found a substantial increase in the rate of hydrogenation when thiophene was added to the system, probably due to a change in the active catalyst through complexing with thiophene. Rhodium complexes with sulfur-containing ligands, $RhCl_3(SEt_2)_3$, have been shown to be effective catalysts for olefin reduction in dimethylacetamide solutions (James *et al.*, 1968).

Unsaturated thiophene derivatives are reduced smoothly to the saturated thiophene by $(Ph_3P)_3RhCl$ at room temperature and 3 to 4 atm pressure in degassed benzene–hexane. This contrasts sharply with the use of heterogeneous catalysts, which require either a very high catalyst loading or severe conditions. Examples of compounds that are smoothly reduced with the homogeneous catalyst include 4-(2-thenoyl)-1-butene, 5(-2-thienyl)-1-pentene, 2-crotonylthiophene, and 2-propylidenethiophene. With 300 mg of catalyst per 0.05 mole of sulfur compound, reduction was complete in 2 to 3 hours (Hornfeldt *et al.*, 1968). Several complex rhodium catalysts proved effective in hydrogenation of unsaturated polysulfones at elevated temperatures and pressures, a reaction that proved difficult to achieve with heterogeneous catalysts (Youngman *et al.*, 1969).

LABELING

Specific labeling of olefins by deuteration over heterogeneous catalysts is often difficult due to double-bond migration, allylic interchanges, excessive deuteration, and, in dienes, to nonspecific attack. These difficulties may be overcome to a large extent by use of certain homogeneous hydrogenation catalysts. Treatment of methyl oleate and methyl linoleate with deuterium in the presence of tris(triphenylphosphine)chlororhodium(I) affords only dideutero and tetradeutero derivatives, respectively (Birch and Walker, 1966a). Trisubstituted alkenes are reduced very slowly but with high specificity to products with only two deuterium atoms (Hussey and Takeuchi, 1969). Deuteration of 1,4-dimethyl- and 1-methyl-4-isopropylcyclohexenes affords dideutero products ranging from 84% to 98%, with the resulting *trans*-cyclohexanes having a higher percentage of dideutero compounds than the corresponding *cis* isomers (Hussey and Takeuchi, 1969). Morandi and Jensen (1969) selectively dideuterated 17 monoolefins. Ergosterol on deuteration affords the 5,6-dideutero compound with the 7 and 22 double bonds unchanged (Birch and Walker, 1966a). Deuteration of 22-dihydroergosteryl acetate affords ergost-7-en-3β-ol-5α,6-d_2-3β-acetate (Birch and Walker, 1966b).

Differences between homogeneous and heterogeneous catalysts are accentuated by certain types of substrate structure. Catalytic deuteration of Δ^1-cholesten-3-one over either heterogeneous palladium or homogeneous tris-(triphenylphosphine)chlororhodium(I) proceeds similarly with attack from the α-face; on the other hand, products of deuteration of 1,4-androstadiene-3,17-dione depend markedly on the catalyst. In this diene, heterogeneous palladium promotes attack from the β-face, whereas the homogeneous catalyst promotes attack at the α-side, affording 85% Δ^4-androstene-3,17-dione-1α,2α-d_2 (Djerassi and Gutzwiller, 1966), with a small amount of 2β-d isomer (Brodie *et al.*, 1969).

Voelter and Djerassi (1968) examined in detail the effect of reaction time, catalyst concentration, and reaction media on the isotopic distribution in deuteration of Δ^2-cholestenone and other steroids. The use of protoic solvents, methanol and ethanol, led to appreciable amounts, up to 39%, of d_1 compounds through exchange with the solvent, whereas with benzene, acetone, or tetrahydrofuran, monodeuteration was kept to within a few percent and yields of 95% d_2 compounds were obtained. Longer reaction times or higher catalyst loadings favor more extensive labeling. Other deuterations of steroids with $(Ph_3P)_3RhCl$ have been reported by Biellmann and Liesenfelt (1966).

The homogeneous catalyst, tris(triphenylphosphine)nitrosylrhodium was used to deuterate cyclohexene affording $C_6H_{10}D_2$ in >99% isotopic purity. The authors (Collman *et al.*, 1969) pointed out that only one other catalyst $(Ph_3P)_3RhCl$, has been reported to catalyze deuteration without HD scrambling.

Ruthenium(II) chloride catalyzes the homogeneous hydrogenation of certain activated double bonds such as those in maleic, fumaric, acrylic, or crotonic acids. Deuteration of fumaric acid in water was shown to be stereospecifically *cis*, affording predominately DL-2,3-dideuterosuccinic acid, with entering deuterium coming from D_2O and not from D_2 (Chalk *et al.*, 1959).

Diminished Disproportionation

Hydrogenation of olefins contained in an incipient aromatic system may sometimes be troublesome due to disproportionation of the olefin into a saturated compound and an aromatic (Birch and Walker, 1966a). This particular difficulty may be largely circumvented by the use of a homogeneous catalyst. For instance, hydrogenation of methyl 1,4,5,8-tetrahydro-1-naphthoate with tris(triphenylphosphine)chlororhodium(I), affords the 9,10-octalin ester in 96% yield along with only 4% of the tetralin derivative.

$COOCH_3$ $\xrightarrow[H_2]{(Ph_3P)_3RhCl}$ $COOCH_3$

In contrast, hydrogenation over platinum oxide in ether affords the 9,10-octalin ester in only 56% yield with 26% of the tetralin derivative plus two unidentified products; over 5% palladium-on-barium sulfate the yield of the 9,10-octalin ester was just 49%, with 26% of the tetralin and three unidentified products. A similar lack of disproportionation was found in reduction of either isotetralin or 1,4-dihydrotetralin to 9,10-octalin with tris(triphenylphosphine)chlororhodium(I) in benzene–ethanol. However, in this case, an unexpected isomerization occurred and both substrates afforded in addition to 9,10-octalin, 20% of 1,9-octalin (Sims *et al.*, 1969). Hydrogenation of 2,5-dihydrobenzylamine toluene-*p*-sulfonate over tris(triphenylphosphine)chlororhodium(I) affords cyclohexylmethylamine toluene-*p*-sulfonate on absorption of 2 moles of hydrogen (Birch and Walker, 1966a).

$$CH_2NHSO_2C_6H_4CH_3 \xrightarrow[(Ph_3P)_3RhCl]{2H_2} CH_2NHSO_2C_6H_4CH_3$$

Disproportionation may not be avoided if homogeneous catalysts are used indiscriminately. (Cyclohexa-1,3-diene)(pentamethylcyclopentadienyl)-rhodium is a very active catalyst for disproportionation of cyclohexa-1,3-diene to cyclohexene and benzene (Moseley and Maitlis, 1969); cyclohexa-1,4-diene is disproportionated by several iridium complexes (Lyons, 1969).

Asymmetric Hydrogenations

The principle that optically inactive compounds can be converted to chiral compounds by hydrogenation over optically active heterogeneous catalysts has been known for many years, but despite work by many investigators, the optical yield was never raised beyond a few percent. Very much more rapid progress has been made in asymmetric hydrogenations employing chiral homogeneous catalysts (Horner *et al.*, 1968a), and in only a few years optical yield has been increased from several percent to nearly quantitative. In homogeneous catalysts, chirality may lie either at the metal bonding or at a site removed from the metal, and good optical yields have been obtained with both types of dissymmetry. Naturally occurring optically active compounds may be used conveniently as precursors for the chiral ligand, obviating the need for a classical resolution step in preparing the catalyst.

Knowles and Sebacky (1968) using rhodium catalyst prepared from optically active (−)-methylpropylphenylphosphine and rhodium chloride reduced α-phenylacrylic acid to hydratropic acid with 15% optical excess. Later it was found easier to generate the catalyst *in situ* from $[Rh(1,5\text{-hexadiene})Cl]_2$ (Knowles *et al.*, 1970). Other workers found that α-ethylstyrene and α-methoxystyrene can be hydrogenated to (S)-(+)-2-phenylbutane (7 to 8% optical yield) and to (R)-(+)-1-methoxy-1-phenylethane (3 to 4% optical yield) by use of a phosphine–rhodium complex formed *in situ* from $Rh(1,5\text{-hexadiene})Cl_2$ and (S)-(+)-methylphenyl-*n*-propylphosphine in benzene (Horner *et al.*, 1968b). Abley and McQuillin (1969) reduced methyl 3-phenylbut-2-en-oate to (+)- or (−)-methyl-3-phenylbutanoate in better than 50% optical yield by employing a catalyst prepared by reducing py_3RhCl_3 with sodium borohydride in (+)- or

(—)-1-phenylethylformamide as solvent. If the amide is coordinated to the metal through the carbonyl oxygen, the optically active center is five atoms away from the center of induced asymmetry. Catalysts of this type are highly active (Jardine and McQuillin, 1969a), and a number of chiral amides may be used (Abley and McQuillin, 1971).

$$\underset{H_3C}{\overset{Ph}{>}}C{=}CHCOCH_3 \ (C{=}O) \xrightarrow[RhCl_3,\ NaBH_4]{Ph\overset{*}{C}HNHCHO\ (CH_3)} \underset{H_3C}{\overset{Ph}{>}}\overset{*}{C}HCH_2COCH_3 \ (C{=}O)$$

Several workers have used naturally occurring optically active compounds as precursors for chiral hydrogenation catalysts. Dang and Kagan (1971) achieved optical yields as high as 72% by means of an optically active diphosphine–rhodium(I) complex prepared from (+)-ethyl tartrate via the ditosylate, and subsequent interaction of 2 moles of the diphosphine with $[Rh(cyclooctene)_2Cl]_2$.

$$\text{(HO)(H)C(COOEt)–C(COOEt)(H)(HO)} \longrightarrow (HC_3)_2C\text{(O–CH(CH}_2\text{OTs)–CH(CH}_2\text{OTs)–O)} \longrightarrow (HC_3)_2C\text{(O–CH(CH}_2\text{PPh}_2\text{)–CH(CH}_2\text{PPh}_2\text{)–O)}$$

Hydrogenation of atropic acid in the presence of this complex and triethylamine in benzene–ethanol affords quantitatively yields of (S)-hydratropic acid with an optical purity of 63%. The high optical yield was obtained only with the free acid; hydrogenation of the methyl ester of atropic acid produces methyl hydratropate of low optical purity (7%) and having the (R)-configuration.

$$\underset{H}{\overset{H}{>}}C{=}C\underset{COOH}{\overset{Ph}{<}} \xrightarrow[H_2]{*\text{diphosphine-rhodium}} H_3C\overset{*}{C}HCOOH \ (Ph)$$

α-Acetamidocinnamic acid is reduced rapidly and quantitatively with this complex to (R)-*N*-acetylphenylalanine with an optical yield of 72%. Similarly, α-phenylacetamidoacrylic acid is reduced to (R)-*N*-phenylacetylalanine with an optical purity of 68%. The authors (Dang and Kagan, 1971) ascribe the high stereoselectivity to the conformational rigidity of the catalyst together with participation of the acid function of the substrates.

Another chiral catalyst based on natural products is prepared by interaction of menthyl chloride with lithium diphenylphosphide to afford neomenthyldiphenylphosphine, which in turn is treated with an olefin–rhodium(I) complex and hydrogenated. Reduction of (E)-β-methylcinnamic acid with this catalyst at 300 psig of hydrogen pressure and 60°C in the presence of six equivalents of triethylamine affords (S)-3-phenylbutanoic acid in 61 % optical excess (Morrison *et al.*, 1971).

Exceptionally high optical yields are obtained in the hydrogenation of α-acylaminoacrylic acids in the presence of certain optically active phosphine–rhodium catalysts where the chirality resides in the phosphorus atom. Optical yields were found to vary greatly with relatively minor change in ligand structure, showing clearly the need for carefully tailored catalysts. The most successful of a number of catalysts tested was derived from chiral *o*-anisylmethylcyclohexylphosphine and rhodium–olefin complexes or rhodium chloride. Optical yields of 85 to 90 % were obtained, which, coupled with the probability that the phosphine may have been only 95 % optically pure, indicates an almost complete stereospecific hydrogenation. The authors (Knowles *et al.*, 1972) believe that optimum optical yields are obtained with exactly two ligands. A catalyst containing one chiral ligand gave a slow hydrogenation and a low optical bias; addition of another chiral ligand *in situ* gave fast rates and a high optical purity.

$$HO\text{-}C_6H_3(OCH_3)\text{-}CH{=}C(NHCOPh)COOH \longrightarrow HO\text{-}C_6H_3(OCH_3)\text{-}CH_2\overset{*}{C}H(NHCOPh)COOH$$

MISCELLANEOUS

A number of functional groups including aldehydes (Coffey, 1967; Heil and Marko, 1968), acetylenes (Trocha-Grimshaw and Henbest, 1968; Shapley *et al.*, 1969; Legzdins *et al.*, 1970; Schrock and Osborn, 1971), ketones (Schrock and Osborn, 1970), imines, nitro and azo compounds, and carbocyclic and heterocyclic aromatics (Jardine and McQuillin, 1970, 1972) have been hydrogenated over homogeneous catalysts. The reductions are of much theoretical interest, but at present lack practical importance inasmuch as the reductions can all be done better over heterogeneous catalysts. A useful synthesis of hydrazobenzene uncontaminated by any trace of aniline involves the $[Rh(CO)_2Cl]_2$-catalyzed reduction of azobenzene by lithium aluminum hydride. Substituents have a profound effect on the reaction. Azophenetole gives 4-ethoxyaniline, whereas azoxyanisole affords hydrazoanisole, both quantitatively (Bruce *et al.*, 1971). Various homogeneous rhodium catalysts have been used for hydrogenation under high pressure of the remaining unsaturation in partially epoxidized polybutadiene (Frew, 1969).

REFERENCES

Abley, P., and McQuillin, F. J. (1969). *Chem. Commun.* p. 477.
Abley, P., and McQuillin, F. J. (1971). *J. Chem. Soc., C* p. 844.
Abley, P., Jardine, I., and McQuillin, F. J. (1971). *J. Chem. Soc., C* p. 840.
Acres, G. J. K., Bond, G. C., Cooper, B. J., and Dawson, J. A. (1966). *J. Catal.* **6**, 139.
Adams, R. W., Batley, G. E., and Bailar, J. C., Jr. (1968a). *Inorg. Nucl. Chem. Lett.* **4**, 453.
Adams, R. W., Batley, G. E., and Bailar, J. C., Jr. (1968b). *J. Amer. Chem. Soc.* **90**, 6051.
Baddley, W. H., and Fraser, M. S. (1969). *J. Amer. Chem. Soc.* **91**, 3661.
Bailar, J. C., Jr., and Itatani, H. (1966). *J. Amer. Oil Chem. Soc.* **43**, 337.
Berkowitz, L. M., and Rylander, P. N. (1959). *J. Org. Chem.* **24**, 708.
Biellmann, J. F., and Jung, M. J. (1968). *J. Amer. Chem. Soc.* **90**, 1673.
Biellmann, J. F., and Liesenfelt, H. (1966). *Bull. Soc. Chim. Fr.* [5] p. 4029.
Birch, A. J., and Walker, K. A. M. (1966a). *J. Chem. Soc., C* p. 1894.
Birch, A. J., and Walker, K. A. M. (1966b). *Tetrahedron Lett.* No. 41, p. 4939.
Birch, A. J., and Walker, K. A. M. (1967a). *Tetrahedron Lett.* No. 20, p. 1935.
Birch, A. J., and Walker, K. A. M. (1967b). *Tetrahedron Lett.* No. 36, p. 3457.
Brodie, H. J., Kripalani, K. J., and Possanza, G. (1969). *J. Amer. Chem. Soc.* **91**, 1241.
Brown, M., and Piszkiewicz, L. W. (1967). *J. Org. Chem.* **32**, 2013.
Bruce, M. I., Goodall, B. L., Igbal, M. Z., and Stone, F. G. A. (1971). *Chem. Commun.* p. 661.
Chalk, A. J., Halpren, J., and Harkness, A. C. (1959). *J. Amer. Chem. Soc.* **81**, 5854.
Chevallier, Y., Stern, R., and Sajus, L. (1969). *Tetrahedron Lett.* No. 15, p. 1197.
Coffey, R. S. (1967). *Chem. Commun.* p. 923.
Collman, J. P., Hoffman, N. W., and Morris, D. E. (1969). *J. Amer. Chem. Soc.* **91**, 5659.
Cramer, R. D., Jenner, E. L., Lindsey, R. V., Jr., and Stolberg, U. G. (1963). *J. Amer. Chem. Soc.* **85**, 1691.
Dang, T. P., and Kagan, H. B. (1971). *Chem. Commun.* p. 481.
Davis, A. G., Wilkinson, G., and Young, J. F. (1963). *J. Amer. Chem. Soc.* **85**, 1692.
Dewhirst, K. C. (1969). U.S. Patent 3,454,644.
Dewhirst, K. C. (1970). U.S. Patent 3,489,786.
Djerassi, C., and Gutzwiller, J. (1966). *J. Amer. Chem. Soc.* **88**, 4537.
Eberhardt, G. G., and Vaska, L. (1967). *J. Catal.* **8**, 183.
Evans, D., Osborn, J. A., Jardine, F. H., and Wilkinson, G. (1965). *Nature (London)* **208**, 5016.
Fotis, P., Jr., and McCollum, J. D. (1967). U.S. Patent 3,324,018.
Frankel, E. N. (1970). *In* "Topics in Lipid Chemistry" (F. D. Gunstone, ed.), p. 161. Logos Press, London.
Frankel, E. N., Emken, E. A., Itatani, H., and Bailar, J. C., Jr. (1967). *J. Org. Chem.* **32**, 1447.
Frew, D. W., Jr. (1969). U.S. Patent 3,480,609.
Gosser, L. W. (1970). U.S. Patent 3,513,210.
Grubbs, R. H., and Kroll, L. C. (1971). *J. Amer. Chem. Soc.* **93**, 3062.
Guistiniani, M., Dolcetti, G., Nicolini, M., and Belluco, U. (1969). *J. Chem. Soc., A* p. 1961.
Haddad, Y. M. Y., Henbest, H. B., Husbands, J., and Mitchell, T. R. B. (1964). *Proc. Chem. Soc., London* p. 361.
Hallman, P. S., Evans, D., Osborn, J. A., and Wilkinson, G. (1967). *Chem. Commun.* p. 305.
Hallman, P. S., McGarvey, B. R., and Wilkinson, G. (1968). *J. Chem. Soc., A* p. 3143.
Halpren, J. (1965). *Proc. Int. Congr. Catal. 3rd, 1964* Vol. I, p. 146.
Halpren, J., James, B. R., and Kemp, A. L. W. (1961). *J. Amer. Chem. Soc.* **83**, 4097.

Harmon, R. E., Parsons, J. L., Cooke, D. W., Gupta, S. K., and Schoolenberg, J. (1969). *J. Org. Chem.* **34**, 3684.
Harrod, J. F., Ciccone, S., and Halpren, J. (1961). *Can. J. Chem.* **39**, 1372.
Hartwell, G. E., and Clark, P. W. (1970). *Chem. Commun.* p. 1115.
Heathcock, C. H., and Poulter, S. R. (1969). *Tetrahedron Lett.* No. 32, p. 2755.
Heil, B., and Marko, L. (1968). *Acta Chim. (Budapest)* **55**, 107.
Horner, L., Büthe, H., and Siegel, H. (1968a). *Tetrahedron Lett.* No. 37, p. 4023.
Horner, L., Siegel, H., and Büthe, H. (1968b). *Angew Chem., Int. Ed. Engl.* **7**, 942.
Hornfeldt, A.-B., Gronowitz, J. S., and Gronowitz, S. (1968). *Acta Chem. Scand.* **22**, 2725.
Hui, B., and James, B. R. (1969). *Chem. Commun.* p. 198.
Hussey, A. S., and Takeuchi, Y. (1969). *J. Amer. Chem. Soc.* **91**, 672.
Itatani, H., and Bailar, J. C., Jr. (1967). *J. Amer. Oil Chem. Soc.* **44**, 147.
James, B. R., and Rempel, G. L. (1966). *Can. J. Chem.* **44**, 233.
James, B. R., Ng, F. T. T., and Rempel, G. L. (1968). *Inorg. Nucl. Chem. Lett.* **4**, 197.
Jardine, F. H., and Wilkinson, G. (1967). *J. Chem. Soc., C* p. 270.
Jardine, F. H., Osborn, J. A., and Wilkinson, G. (1967). *J. Chem. Soc., A* p. 1574.
Jardine, I., and McQuillin, F. J. (1968). *Tetrahedron Lett.* No. 50, p. 5189.
Jardine, I., and McQuillin, F. J. (1969a). *Chem. Commun.* p. 477.
Jardine, I., and McQuillin, F. J. (1969b). *Chem. Commun.* p. 502.
Jardine, I., and McQuillin, F. J. (1970) *Chem. Commun.* p. 626.
Jardine, I., and McQuillin, F. J. (1972). *Tetrahedron Lett.* No. 2, p. 173.
Jardine, I., Howsam, R. W., and McQuillin, F. J. (1969). *J. Chem. Soc., C* p. 260.
Khan, M. M. T., Andal, R. K., and Manoharan, P. T. (1971). *Chem. Commun.* p. 561.
Knowles, W. S., and Sabacky, M. J. (1968). *Chem. Commun.* p. 1445.
Knowles, W. S., Sabacky, M. J., and Vineyard, B. D. (1970). *Ann. N.Y. Acad. Sci.* **172**, p. 232.
Knowles, W. S., Sabacky, M. J., and Vineyard, B. D. (1972). *Chem. Commun.* p. 10.
Laing, S. B., and Sykes, P. J. (1968). *J. Chem. Soc., C* p. 421.
Legzdins, P., Mitchell, R. W., Rempel, G. L., Ruddick, J. D., and Wilkinson, G. (1970). *J. Chem. Soc., A* p. 3322.
Lehman, D. D., Shriver, D. F., and Wharf, I. (1970). *Chem. Commun.* p. 1486.
Lyons, J. E. (1969). *Chem. Commun.* p. 564.
Lyons, J. E., Rennick, L. E., and Burmeister, J. L. (1970). *Ind. Eng. Chem., Prod. Res. Develop.* **9**, 2.
Mague, J. T., and Wilkinson, G. (1966). *J. Chem. Soc., A* p. 1736.
Masters, C., McDonald, W. S., Raper, G., and Shaw, B. L. (1971). *Chem. Commun.* p. 210.
Maxted, E. B., and Ismail, S. M. (1964). *J. Chem. Soc., London* p. 1750.
Mitchell, T. R. B. (1970). *J. Chem. Soc., B* p. 823.
Montelatici, S., van der Ent, A., Osborn, J. A., and Wilkinson, G. (1968). *J. Chem. Soc., A* p. 1054.
Morandi, J. R., and Jensen, H. B. (1969). *J. Org. Chem.* **34**, 1889.
Morrison, J. D., Burnett, R. E., Aguiar, A. M., Morrow, C. J., and Phillips, C. (1971). *J. Amer. Chem. Soc.* **93**, 1303.
Moseley, K., and Maitlis, P. M. (1969). *Chem. Commun.* p. 1156.
Nishimura, S., and Tsuneda, K. (1969). *Bull. Chem. Soc. Jap.* **42**, 852.
O'Connor, C., and Wilkinson, G. (1968). *J. Chem. Soc., A* p. 2665.
O'Connor, C., and Wilkinson, G. (1969). *Tetrahedron Lett.* No. 18, p. 1375.
Ogata, I., Iwata, R., and Ikeda, Y. (1970). *Tetrahedron Lett.* No. 34, p. 3011.
Osborn, J. A., Jardine, F. H., Young, J. F., and Wilkinson, G. (1966). *J. Chem. Soc., A* p. 1711.
Piers, E., and Cheng, K. F. (1969). *Chem. Commun.* p. 562.
Piers, E., Britton, R. W., and de Waal, W. (1969). *Chem. Commun.* p. 1069.

Rony, P. R. (1969). *J. Catal.* **14**, 142.
Rüesch, H., and Mabry, T. J. (1969). *Tetrahedron* **25**, 805.
Rylander, P. N. (1967). "Catalytic Hydrogenation over Platinum Metals," p. 469. Academic Press, New York.
Rylander, P. N. (1971). *Advan. Chem. Ser.* **98**, 150.
Rylander, P. N., Himelstein, N., Steele, D. R., and Kreidl, J. (1962). *Engelhard Ind., Tech. Bull.* **3**, 61.
Schrock, R. R., and Osborn, J. A. (1970). *Chem. Commun.* p. 567.
Schrock, R. R., and Osborn, J. A. (1971). *J. Amer. Chem. Soc.* **93**, 3089.
Senda, Y., Iwaski, T., and Mitsui, S. (1972). *Tetrahedron* **28**, 4059.
Shapley, J. R., Schrock, R. R., and Osborn, J. A. (1969). *J. Amer. Chem. Soc.* **91**, 2816.
Siegel, S., and Dmuchovsky, B. (1962). *J. Amer. Chem. Soc.* **84**, 3132.
Sims, J. J., Honwad, V. K., and Selman, L. H. (1969). *Tetrahedron Lett.* No. 2, p. 87.
Skapski, A. C., and Troughton, P. G. H. (1968). *Chem. Commun.* p. 1230.
Tayim, H. A., and Bailar, J. C., Jr. (1967a). *J. Amer. Chem. Soc.* **89**, 3420.
Tayim, H. A., and Bailar, J. C., Jr. (1967b). *J. Amer. Chem. Soc.* **89**, 4330.
Trocha-Grimshaw, J., and Henbest, H. B. (1968). *Chem. Commun.* p. 757.
Tyman, J. H. P., and Willis, B. J. (1970). *Tetrahedron Lett.* p. 4507.
van Bekkum, H., van Gogh, J., and van Minnen-Pathuis, G. (1967). *J. Catal.* **7**, 292.
van Gaal, H., Cuppers, H. G. A. M., and van der Ent, A. (1970). *Chem. Commun.* p. 1694.
van't Hof, L. P., and Linsen, B. G. (1967). *J. Catal.* **7**, 295.
Vaska, L. (1965). *Inorg. Nucl. Chem. Lett.* **1**, 89.
Vaska, L., and Rhodes, R. E. (1965). *J. Amer. Chem. Soc.* **87**, 4970.
Voelter, W., and Djerassi, C. (1968). *Chem. Ber.* **101**, 58.
White, C., Gill, D. S., Kang, J. W., Lee, H. B., and Maitlis, P. M. (1971). *Chem. Commun.* p. 734.
Wieland, von P., and Anner, G. (1968). *Helv. Chim. Acta* **51**, 1698.
Yamaguchi, M. (1967). *Kogyo Kagaku Zasshi* **70**, 675.
Young, J. F., Osborn, J. A., Jardine, F. H., and Wilkinson, G. (1965). *Chem. Commun.* p. 131.
Youngman, E. A., Bauer, R. S., Dewhirst, K. C., Holler, H. V., and Lunk, H. E. (1969). U.S. Patent 3,444,145.

CHAPTER 3

Oxidation

Oxidation of organic compounds in the presence of noble metal catalysts may effect a variety of changes including introduction of alcohol, carbonyl, epoxide, and halogen functions into the molecule, as well as inducing various coupling and degradative reactions. For organizational purposes, the oxidative reactions of various functions are considered individually in the following discussion.

Oxidation of Olefins in Aqueous Systems

Liquid-phase olefin oxidation by noble metal catalysis has received much attention in the last decade due largely to the successful development of the Wacker process for production of acetaldehyde from ethylene (Smidt *et al.*, 1959a). The process is based on an elegant method of converting the stoichiometric reaction of ethylene and palladium chloride, known since 1894 (Phillips, 1894), to a catalytic system.

$$CH_2{=}CH_2 + H_2O + PdCl_2 \rightarrow CH_3CHO + Pd + 2\ HCl$$

Finely divided palladium can be reoxidized with air, but the oxidation is faster if palladium is reoxidized by cupric ion and the cuprous ion thus formed air-oxidized to cupric. The last step is rate-determining (Teramoto *et al.*, 1963; Bryant and McKeon, 1969).

$$Pd + 2\,CuCl_2 \rightarrow PdCl_2 + 2\,CuCl$$

$$2\,CuCl + \tfrac{1}{2}\,O_2 + 2\,HCl \rightarrow 2\,CuCl_2 + H_2O$$

The overall equation for the reaction becomes

$$CH_2{=}CH_2 + \tfrac{1}{2}\,O_2 \rightarrow CH_3CHO$$

These simple equations belie the complexity of the system, which has been the subject of intensive investigation and several reviews (Stern, 1967; Aguilo, 1967; Henry, 1968; Moiseev, 1969; Maitlis, 1971). Suffice it to note here that the equations illustrate a general phenomenon encountered repeatedly, i.e., a stoichiometric reaction and reduction of a noble metal becomes catalytic with respect to the metal when it is reoxidized and the cycle repeated. All noble metals may be so used in conjunction with a variety of oxidants. Ruthenium- and osmium-catalyzed oxidations are discussed in another chapter.

The oxidation of olefins in aqueous systems catalyzed by palladium is general. It can be carried out both homogeneously and heterogeneously (Smidt *et al.*, 1964; Hargis and Young, 1968). Higher homologs of ethylene afford predominately ketones (Smidt *et al.*, 1959a; Smidt and Krekeler, 1963; Hafner *et al.*, 1962). The ratio of aldehyde to ketone arising from terminal olefins can be affected by the pH, temperature, and the nature of the palladium complex, but the influence of these variables is of more interest in mechanistic studies than in synthetic applications; relatively little can be done to change the product distribution substantially for most olefins. Styrene, on the other hand, can be made to form predominately either acetophenone or phenylacetaldehyde (Hafner *et al.*, 1962). Electron-withdrawing substituents on the aromatic ring of styrene favor substituted aldehydes, electron-releasing substituents favor acetophenones (Okada *et al.*, 1968). Halides on either the α- or β-carbon are lost during the reaction (Smidt *et al.*, 1959a). Oxidation of olefins to carbonyl compounds having the carbon skeleton intact is limited essentially to those olefins having at least one hydrogen on each olefinic carbon; more substituted olefins tend to undergo oxidative cleavage of the double bond, rearrangements, or hydration (Hüttel and Bechter, 1959; Hüttel *et al.*, 1961; Smidt *et al.*, 1959a).

1-Methylcyclobutene undergoes an oxidative rearrangement in the presence of thallium(II), palladium(II), or mercury(II) to afford quantitative yields of cyclopropyl methyl ketone. These metals were used stoichiometrically in this reaction, but one assumes a suitable redox system could be found to permit

$$\text{(1-methylcyclobutene)} + Pd^{2+} + H_2O \longrightarrow \text{(cyclopropyl)}\!-\!\underset{\underset{O}{\|}}{C}CH_3 + Pd^0 + 2H^+$$

catalytic use of the metal (Byrd *et al.*, 1971). A cyclopropane ring itself shows some olefinic character in this type reaction; treatment of phenylcyclopropane with palladium chloride in water affords propiophenone in 60% yield and benzyl methyl ketone in 35% yield (Ouellette and Levin, 1968, 1971).

$$C_6H_5\text{-}C_3H_5 \xrightarrow[H_2O]{PdCl_2} C_6H_5\text{-}C(=O)CH_2CH_3 + C_6H_5\text{-}CH_2C(=O)CH_3$$

Higher olefins are oxidized with more difficulty due to solutility limitations. Clement and Selwitz (1964, 1968) described an improved procedure for converting α-olefins to methyl ketones with palladium chloride as a catalyst, a procedure especially useful for olefins of higher molecular weight (hexene and larger) that do not react readily with an entirely aqueous solution of palladium chloride as do lower homologs. With this method that employs aqueous dimethylformamide as solvent 1-dodecene is converted to 2-dodecanone in yields up to 85%; the ketonic material is 96–99% methyl ketone. Best yields were obtained with solvent systems containing 12–17% water by volume and with a regulated addition of olefin to prevent accumulation of unchanged olefin. Typically, a flask containing a gas-dispersing tube and a dropping funnel is charged with 50 ml of dimethylformamide, 7 ml of water, 0.020 mole of palladium chloride, and 0.020 mole cupric chloride. The stirred solution is heated to 60°C and oxygen passed through the reaction mixture at 3.3 liters/hour while 0.20 mole of olefin is added dropwise over a 2.5 hour period. The mixture is heated to 60°–70°C for an additional half-hour, then the liquid phase is removed, washed with water, dried, and fractionally distilled. Other water-miscible solvents such as dimethyl sulfoxide, acetone, acetic acid, tetrahydrofuran, dioxane, and acetonitrile give much inferior results. Dimethylformamide may accelerate the reoxidation of cuprous chloride (Tamura and Yasui, 1969a). The presence of nitrate ions is said to be advantageous in oxidations of this type (MacLean and Stautzenberger, 1968).

With slight modification of the procedure *p*-benzoquinone may be used as an oxidant instead of oxygen and copper salts. For example, to a solution of 0.01 mole of palladium chloride, 0.05 mole of *p*-benzoquinone, 0.14 mole of water, and 25 ml of dimethylformamide is added 0.10 mole of 1-hexene. The temperature is initially controlled between 25° to 50°C by an ice bath. After 20 hours at 25°C, a 69% yield of ketone, 98% hexanone-2, is obtained. Acetic acid,

$$CH_3CH_2CH_2CH_2CH{=}CH_3 + O{=}C_6H_4{=}O \xrightarrow[H_2O]{PdCl_2} CH_3CH_2CH_2CH_2C(=O)CH_3 + HO\text{-}C_6H_4\text{-}OH$$

acetone, dimethyl sulfoxide, and tetrahydrofuran as solvents give inferior results. These techniques are applicable also to olefinic compounds other than hydrocarbons; 10-undecenoic acid was converted to 10-ketoundecanoic acid in 83% yield (Clement and Selwitz, 1968).

Ferric sulfate has been used successfully as an oxidant to reoxidize reduced palladium (Smidt *et al.*, 1959b). Oxidation of 4-methoxy-1-propenylbenzene with aqueous palladium chloride at 70°C affords *p*-methoxyphenylacetone in 91% yield.

$$H_2O + CH_3O{-}C_6H_4{-}CH{=}CHCH_3 + 2Fe^{3+} \xrightarrow{Pd} CH_3O{-}C_6H_4{-}CH_2\underset{\underset{O}{\parallel}}{C}CH_3 + 2Fe^{2+} + 2H^+$$

Similarly, crotonaldehyde at 25°C was converted to acetylacetaldehyde, which cyclizes to 1,3,5-triacetylbenzene in 85% yield.

$$H_2O + CH_3CH{=}CHCHO \xrightarrow[Fe^{3+}]{Pd} CH_3\underset{\underset{O}{\parallel}}{C}CH_2CHO \longrightarrow 1,3,5\text{-}C_6H_3(COCH_3)_3$$

Oxidation of Olefins in Nonaqueous Systems

Oxidation of olefins in nonaqueous systems is complex. The products may be monosubstituted vinyl or allylic compounds or disubstituted saturated materials as well as compounds arising from isomerization of the double bond and from oligomerizations. The product distribution is affected by a number of variables, including olefin structure, the nucleophile, temperature, medium acidity, the nature and concentration of ions, as well as the nature of the catalyst. Oxygen may have an important influence on the products, especially in the absence of chloride (Brown *et al.*, 1969).

Oxidation of Ethylene

Olefins interact with carboxylic acid in the presence of palladium salts to afford vinyl esters and palladium metal (Moiseev *et al.*, 1960; Stern and Spector, 1961; Stern, 1963). In the presence of an oxidizing agent, the reaction becomes catalytic with respect to palladium and it has been made the basis of both homogeneous and heterogeneous syntheses of vinyl acetate (Szonyi, 1968;

Schwerdtel, 1968). The processes appear to proceed by radically different mechanisms (Nakamura and Yasui, 1970, 1971).

$$CH_2{=}CH_2 + HOAc \xrightarrow[Pd]{O_2} CH_2{=}CHOAc + H_2O$$

Homogeneous palladium-catalyzed oxidations of ethylene are complicated and by proper choice of conditions, the major products can be vinyl acetate, acetaldehyde, glyoxal, ethylidene diacetate, 1,2-diacetoxyethane, 2-acetoxyethanol, and 1-acetoxy-2-chloroethane. The ratio of these products depends on the solvent, nature and concentration of ligands, on the oxidant present, and on the agitation (Henry, 1964). With a copper redox system, the yield of vinyl acetate depends markedly on the concentration of chloride; increasing chloride (as lithium chloride) first diminishes the yield of vinyl acetate by increasing the yield of acetaldehyde and ethylidene diacetate and then as the concentration is increased still further to 2 *M*, these are supplanted by esters of ethylene glycol (55% yield) (Clark *et al.*, 1969; Henry, 1967). Palladium-catalyzed oxidation of ethylene in the presence of copper chloride and alkali or alkaline earth chlorides affords chloroethanol as the major product (Stangl and Jira, 1970). Chloride also catalyzes the reoxidation of palladium metal; with acetate as the only ligand palladium metal is not oxidized (Tamura and Yasui, 1968a). The yield of vinyl acetate is increased by the presence of dimethylacetamide (Clark *et al.*, 1968), which may inhibit decomposition of vinyl acetate to acetaldehyde and acetic anhydride (Schultz and Rony, 1970). A variation of the vinyl acetate process is to carry out the oxidation in the presence of sufficient water to hydrolyze part of the vinyl acetate formed into acetaldehyde and acetic acid. Under the conditions of the reaction, acetaldehyde is oxidized to acetic acid (British Patent 1,194,083). The overall reaction then becomes

$$2\,C_2H_4 + 1.5\,O_2 \rightarrow CH_2{=}CHOAc$$

with ethylene the source of the acetic acid.

In the presence of palladium acetate, vinyl acetate undergoes oxidative coupling to form 1,4-diacetoxy-1,3-butadiene in moderate yields (Kohll and van Helden, 1967). This reaction does not occur with palladium chloride; instead acetaldehyde and acetic anhydride are formed (Clement and Selwitz, 1962). The coupling reaction can be made catalytic with respect to palladium by addition of cupric acetate.

$$CH_3CHO + (CH_3CO)_2O \xleftarrow{PdCl_2} CH_2{=}CHOAc \xrightarrow{Pd(OAc)_2} AcOCH{=}CHCH{=}CHOAc$$

Ethylidene diacetate can be obtained as the major product of ethylene oxidation by carrying out the reaction at relatively high temperature, 95°C, in the presence of palladium and iron acetates, lithium chloride, and nitric and acetic acids (van Helden *et al.*, 1968). Ethylidene diacetate evidently does not arise from intermediate vinyl acetate, for the reaction carried out in the presence of CH_3COOD affords ethylidene diacetate containing very little deuterium (Moiseev and Vargaftik, 1965). Higher nitrate ion concentrations in the above reaction favor 1,2-diacetoxyethane and 2-acetoxyethanol (Tamura and Yasui, 1968b).

Excellent yields of ethylene glycol diacetate can be obtained by the oxidation of ethylene in acetic acid using benzoquinone as an oxidant and palladium chloride as catalyst (British Patent 1,058,995).

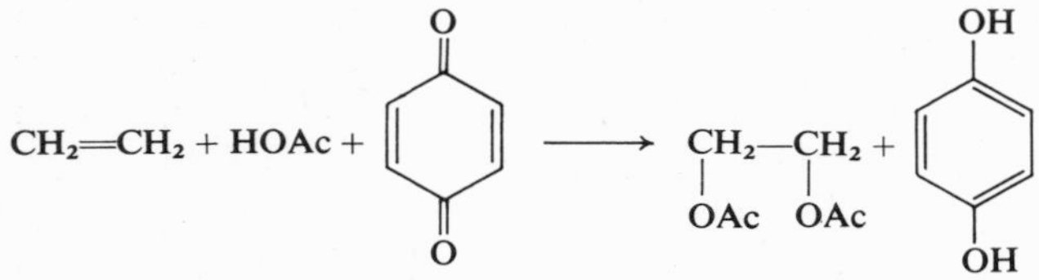

$$CH_2{=}CH_2 + HOAc + \text{(p-benzoquinone)} \longrightarrow \underset{\text{OAc}}{CH_2}{-}\underset{\text{OAc}}{CH_2} + \text{(hydroquinone)}$$

With nitric acid as an oxidant, glyoxal is the major product. Ethylene bubbled through a solution of about 20% nitric acid containing catalytic amounts of a palladium salt and lithium chloride at 35°–50°C is converted to glyoxal in 80–90% yield (Platz and Fuchs, 1967). Lithium chloride is believed to form a complex with palladium, making it more soluble and also redissolving any palladium metal that precipitates during the reaction (Gourlay, 1969). The consumption of nitric acid can be decreased by blowing air through the mixture (Platz and Nohe, 1966).

$$CH_2{=}CH_2 + HNO_3 \xrightarrow{Pd} OHCCHO$$

With palladium chloride as catalyst and lithium nitrate as oxidant, ethylene and acetic acid are converted to ethylene glycol monoacetate in 90% yield (Tamura and Yasui, 1969b). Lithium nitrate is more effective than either ferric or cupric nitrates.

$$CH_2{=}CH_2 + HOAc \rightarrow HOCH_2CH_2OAc$$

Oxidation of Higher Olefins

The same type of products that are formed in oxidation of ethylene also occur with higher olefins plus products derived from allylic attack (Kitching *et al.*, 1966; Collman *et al.*, 1967; James and Ochiai, 1971) and migration of the double bond as well as by epoxidation (Takao *et al.*, 1970a). Additionally, the product may contain both positional and geometric isomers. The situation is further complicated by a number of conflicting reports (Maitlis, 1971). Because of this complexity and despite a fairly large amount of work in the area, only a few synthetically useful reactions have emerged.

Under conditions similar to that used for heterogeneously catalyzed production of vinyl acetate, propylene is converted to allyl acetate. A mixture of propylene, acetic acid, and oxygen in 15.6 to 13.7 to 3 volume percent ratios passed over a 4% palladium-on-titanium dioxide catalyst at 9.69 gaseous volume hourly space velocity affords allyl acetate in high yield at the rate of 1.66 moles/liter of catalyst per hour (Ketley, 1970).

$$H_2C{=}CHCH_3 + HOAc + \tfrac{1}{2}\,O_2 \rightarrow H_2C{=}CHCH_2OAc + H_2O$$

Acetoxylation of propylene in the liquid phase is considerably more complex and affords varying amounts of *cis*- and *trans*-1-acetoxypropene, 2-acetoxypropene, and allyl acetate depending on conditions, as well as acetone, propylidene diacetate, and 1,3-diacetoxypropane. In a chloride-free media in the presence of oxygen, allyl acetate is the major product (Neth. Patent 6,508,238). The reaction is sensitive to the solvent (Clark *et al.*, 1968). With palladium chloride *cis*- and *trans*-1-acetoxypropene predominate (Belov *et al.*, 1965), whereas under slightly different conditions palladium chloride and disodium hydrogen phosphate in acetic acid afford 64% 2-acetoxypropene (Stern, 1963). Monoesters of propylene glycol are obtained with palladium chloride and nitrate ions in acetic acid (Tamura and Yasui, 1968b). Similar complex mixtures are obtained in acetoxylations of butenes (Belov and Moiseev, 1966; Kitching *et al.*, 1966; Henry, 1967).

The homogeneous vinylation reaction of higher olefins was examined thoroughly by Schultz and Gross (1968) and the effect of various reaction parameters on product distribution determined. With α-olefins primary esters could be produced in greater than 75% specificity. An important variable in determining the product distribution is the concentration of acetate ion as illustrated by comparison of the following two experiments using 1-octene. Oxygen was bubbled for 3 hours at 100°C through a solution with the composition 0.45 gm of palladous acetate, 7.20 gm of cupric acetate, 26.4 gm of lithium acetate, 150 ml of acetic acid, and 30 ml of 1-octene. At the end of this time, 14.0 gm of ester was isolated containing greater than 90% by weight of primary octenyl acetate. When the experiment was repeated except that the lithium acetate was reduced tenfold to 2.6 gm, the ester isolated was more than 60% 2-acetoxy-1-octene (Wright, 1970).

$$\underset{\displaystyle OCOCH_3}{CH_3(CH_2)_5\overset{|}{C}{=}CH_2} \xleftarrow[O_2]{\text{low } OAc^-} CH_3(CH_2)_5CH{=}CH_2 \xrightarrow[O_2]{\text{high } OAc^-} CH_3(CH_2)_5CH{=}CHOCOCH_3$$

With chloride ion present, the percentage of 1-acetate produced in the vinylation reaction of 1-hexene depends on the acetate–chloride ratio and reaches a minimum when this ratio is 1. The overall yield of total acetates is favored, however, by higher acetate–chloride ratios. Sodium acetate gives the highest yields of a group which includes sodium, lithium, magnesium, calcium, strontium, barium, zinc, cadmium, mercury, and aluminum acetates. Whether or not temperature is an important parameter affecting the yield of primary acetate depends on the acetate–chloride ratio; at a ratio of 2.25 variations in temperature between 80° and 150°C had little effect on the yield of 1-acetate, whereas at a ratio of 0.62 the temperature had a marked effect on both total yield of acetate and yield of 1-acetate. With increasing temperature between 80° and 150°C, the yield of 1-acetate in the hexenyl acetate fraction falls from 84 to 52%, whereas based on total product the yield increases from 14 to 41%. The variation occurs because of the marked effect of temperature on the amount of high-boilers; at 80°C high-boilers constituted 84% of the product, whereas at 150°C, only 21%. This paper presents an interesting reaction mechanism which is at variance with other frequently seen vinylation mechanisms (Schultz and Gross, 1968). Levanda and Moiseev (1968) obtained mainly secondary acetates on acetoxylation of 1-hexene by palladium chloride in the absence of copper.

Olefins are air-oxidized in the presence of iridium (Collman *et al.*, 1967; Takao *et al.*, 1970a, b) or rhodium complexes (Takao *et al.*, 1970c), affording carbonyl derivatives. Oxidation over iridium chloride or iridium–tin complexes is particularly interesting for additionally epoxides are formed in yields up to 42%. Homogeneous epoxidation by air is a potentially attractive industrial synthesis if it could be developed considerably beyond the present art. Some remarkable solvent effects were found with rhodium catalysts. Oxidation of styrene with either $RhCl_3$ or $RhCl(Ph_3P)_3$ in the polar solvent, dioxane, affords predominantly benzaldehyde, whereas in the protic solvent, ethanol, acetophenone is obtained in excellent yield. In nonpolar toluene, a mixture of acetophenone and benzaldehyde results. Appreciable styrene oxide is formed in the presence of $RuCl_2(Ph_3P)_3$ (Lyons and Turner, 1972). Oxygenation of (+)-carvomenthene (**I**) in benzene containing $RhCl(Ph_3P)_3$ affords a mixture of carvotanacetone (**II**) and piperitone (**III**) plus several alcohols. The carvotanacetone obtained in this oxidation is completely racemized, which led the authors (Baldwin and Swallow, 1969) to suggest that the reaction proceeds through a symmetrized intermediate, which is probably a radical.

H O + O

(**I**) (**II**) (**III**)

Cyclohexene and cyclopentene have been oxidized with a number of noble metal complex catalysts at 14–36 psig and 25°–60°C in benzene or methylene dichloride.

$+ O_2$ — $RhCl(CO)(Ph_3P)_2$, $PtO_2(Ph_3P)_2$, $IrI(CO)(Ph_3P)_2$, $IrCl(N_2)(Ph_3P)_2$, $RhCl(Ph_3P)_3$ → O + O

Oxidation of cyclohexanone affords 2-cyclohexenone as the major product together with smaller amounts of cyclohexene oxide and several other minor unidentified products. The iridium chloride complex, $IrCl(CO)(Ph_3P)_2$, is a poor catalyst for oxidation in benzene solution, whereas the analogous iodide complex, $IrI(CO)(Ph_3P)_2$, is an effective catalyst. The iodide is known to form a much more stable oxygen adduct than the chloride, and probably a prerequisite for catalysis is prior coordination of the complex with molecular oxygen (Collman *et al.*, 1967). Other workers (Kurkov *et al.*, 1968) oxidized cyclohexene in the presence of a number of rhodium catalysts including tris-(triphenylphosphine)rhodium chloride, rhodium acetylacetonate, and rhodium 2-ethylhexanoate, and concluded the oxidations are free-radical in nature. The product distribution was similar with all catalysts. The oxidations were carried out with 4.8 *M* cyclohexene in benzene at 60°C for 10 hours. Solutions were $1.4–2.9 \times 10^{-3}$ *M* in catalysts.

→ =O (35–57%) + —OH (20–28%) + —OOH (8–14%) + unidentified

OXIDATION OF COMPLEX OLEFINS

Catalytic oxidation has proved a useful synthetic tool in the tetracycline series (Muxfeldt *et al.*, 1962, 1968; Muxfeldt and Rogalski, 1965).

H H $N(CH_3)_2$ OH NH_2 OH O OH O O — DMF, Pt, O_2 → H H $N(CH_3)_2$ OH NH_2 OH OH O HO O O

A convenient synthesis of acetates of C_{10}-terpene alcohols involves interaction of acetic acid and a terpene catalyzed by palladium–copper. A mixture of palladium acetate, copper diacetate, sodium acetate, acetic acid, and myrcene heated at reflux under a stream of oxygen for 15 hours affords 5% linalyl acetate, 11% 2-acetoxy-3-methylene-7-methyl-oct-6-ene, 5% 3-acetoxymethyl-7-methyl-octa-2,6-diene, 31% neryl acetate, and 36% geranyl acetate (Suga *et al.*, 1971). Other workers have reported obtaining mixtures of 1-acetoxy-1-alkenes and 1-acetoxy-2-alkenes from acetoxylation of long-chain α-olefins (Vargaftik *et al.*, 1962; Clark *et al.*, 1964).

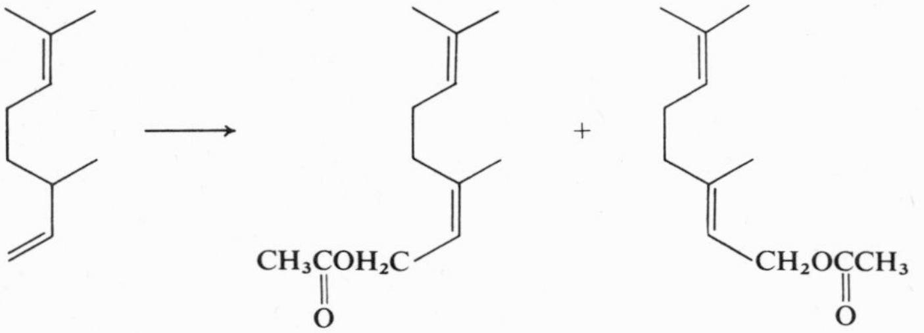

Myrcene Neryl acetate Geranyl acetate

An interesting synthesis of *syn*-7-norbornenol from norbornene involves the use of cupric chloride as both oxidant for regeneration of reduced palladium and reagent to introduce chloride. The intermediate *exo*-2-chloro-*syn*-7-acetoxynorbornane is dehydrochlorinated by potassium *t*-butoxide in dimethyl sulfoxide to afford *syn*-7-norbornenol. Only 5 to 10% of diacetate is formed due to the relatively high chloride concentration. The author (Baird, 1966) pictured formation of *exo*-2-chloro-*syn*-acetoxynorbornane as involving rearrangement of a norbornyl cation.

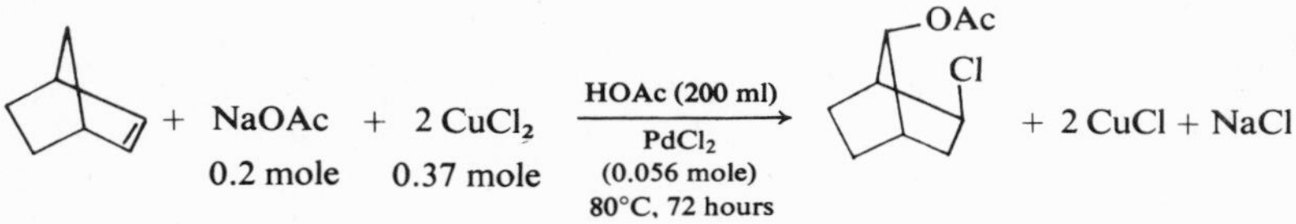

This reaction has been extended to more complex norbornene systems (Baker *et al.*, 1970). Reaction of **IV** with the benzene ring in the *endo* configuration affords two major products, **V** and **VI**. The corresponding compound with *exo* configuration affords a mixture of four products.

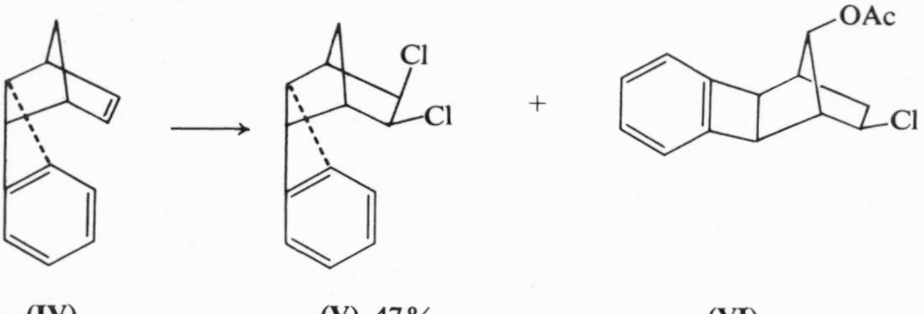

(IV) (V) 47% (VI)

A similar reaction was applied by Battiste and Nebzydoski (1969) to synthesis in the *exo*-tricyclo[4.2.1.$0^{2,5}$]non-3-ene series. The mechanism of these reactions has been discussed (Adderley *et al.*, 1971).

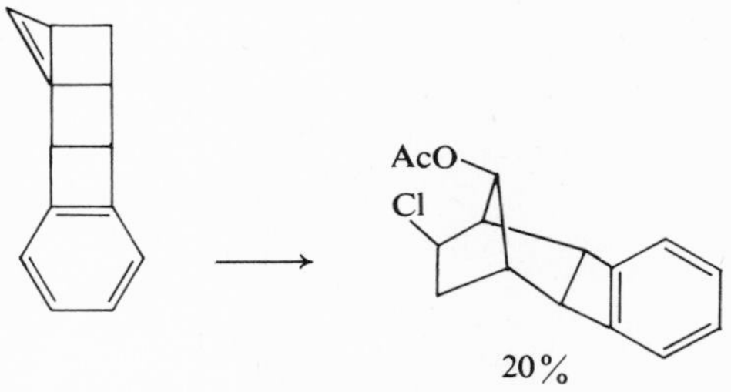

AcO
Cl
20%

Oxidative Addition of Alcohols to Olefins

Oxidative addition of alcohols to olefins provides in favorable circumstances a convenient route to acetals or vinyl ethers (Clark *et al.*, 1970). The reaction discovered independently by Stern and Spector (1961) and Moiseev *et al.* (1960) has since been the subject of a number of patents covering catalytic aspects. It is tempting to assume that the acetal is derived from a vinyl ether by addition of alcohol, but this is not the case with monohydroxy compounds since no deuterium appears in either the methyl vinyl ether (François, 1969) or the dimethylacetal (Moiseev and Vargaftik, 1965) derived by oxidative addition of CH_3OD to ethylene.

$$CH_2{=}CH_2 + \xrightarrow{CH_3OD} CH_2{=}CHOCH_3 \xrightarrow{CH_3OD} CH_3CH(OCH_3)_2$$

The products obtained in palladium–copper-catalyzed air oxidation of olefins in alcohols depend markedly on the conditions. Acetals are formed most readily with lower alcohols (Ketley and Fisher, 1968; Ketley and Braatz, 1968) or from 1,2-diols. The oxidations are much faster than those run under comparable conditions in aqueous media. Even small amounts of water, 10–20%, have a strong adverse effect on rate. Isomerization of the double bond may precede oxidation, leading to mixtures of isomers, but the isomer distribution can be controlled to a large extent by the temperature; lower temperatures favor a single product (Lloyd and Luberoff, 1969).

The tendency to acetal formation is greater when oxidations are carried out in diol solvents. Oxidation of cyclohexene in ethanol affords cyclohexanone

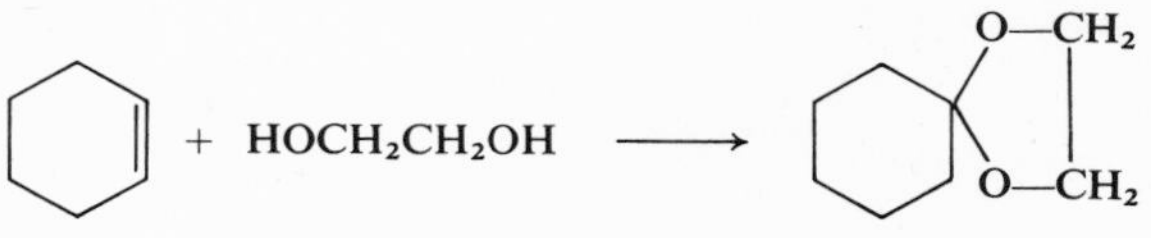

$$+ \ HOCH_2CH_2OH \longrightarrow$$

O—CH_2
O—CH_2

in 95% yield, whereas in ethanediol a high yield of 1,4-dioxospiro[4,5]decane is obtained.

Oxidation of ethylene in ethanediol gives 2-methyl-1,3-dioxolane in 91% yield; acrylonitrile and ethanediol afford 1,3-dioxolane-2-acetonitrile in high yield. Oxidation of 1-hexene in glycerol affords 2-*n*-butyl-2-methyl-1,3-dioxolane-4-methanol in 70% yield together with 26% hexanone. Vinyl ethers tend to form acetals readily. Oxidation of 2,5-dihydrofuran in ethanol affords 3,3-diethoxytetrahydrofuran in 94% yield (Lloyd and Luberoff, 1969).

$$\text{2,5-dihydrofuran} + 2C_2H_5OH \xrightarrow[\substack{120\ \text{minutes} \\ 50°C \\ 3\ \text{atm}\ O_2}]{\substack{0.02\ M\ PdCl_2 \\ 0.10\ M\ CuCl_2}} \text{3,3-}(OC_2H_5)_2\text{-tetrahydrofuran}$$

Vinyl ethers in the presence of palladium catalysts undergo a facile stereospecific interchange with alcohols that proceeds with inversion of configuration (McKeon *et al.*, 1972). With palladium chloride, the interchange reaction ceases due to precipitation of palladium metal and formation of acetals. This difficulty can be overcome by the use of *cis* complexes of palladium acetate with bidentate ligands; these complex catalysts that do not decompose under reaction conditions permit synthesis routes to a variety of vinyl ethers and esters through interchange reactions (McKeon and Fitton, 1972).

$$H_2C{=}CHOR + R'OH \rightleftharpoons H_2C{=}CHOR' + ROH$$

Oxidative Coupling

In the presence of noble metal catalysts, a number of compounds undergo oxidative coupling reactions with the formation of new carbon–carbon, carbon–oxygen, or carbon–nitrogen bonds. During the coupling reaction, the catalyst frequently is reduced, but becomes reoxidized to an active catalytic species in a cyclic system.

Aromatics

Benzene, in acetic acid solution containing palladium chloride or palladium acetate and sodium acetate (van Helden and Verberg, 1965) or palladium chloride and silver nitrate (Fujiwara *et al.*, 1970) or in 20% sulfuric acid solution containing palladium sulfate (Davidson and Triggs, 1966) is dimerized to biphenyl. Palladium(II) is reduced to the metal.

$$2\,C_6H_6 + Pd(II) + \xrightarrow{90\text{–}110°C} C_6H_5\text{–}C_6H_5 + Pd^0 + 2H^+$$

The corresponding reactions of toluene are complicated by side-chain attack; in acetic acid at 100°C benzyl acetate is formed in 45% yield and in sulfuric acid toluene is oxidized to benzoic acid. Acetoxylation can be completely suppressed by addition of perchloric acid (Davidson and Triggs, 1966) or mercuric salts (Unger and Fouty, 1969). Perchloric acid is also a catalyst for the coupling reaction, presumably operating through increasing the electrophilicity of palladium (Davidson and Triggs, 1968b). Dimerization of toluene affords all six possible nuclear-coupled products with the isomer distribution being $p \simeq m \gg o$.

Dimerizations of aromatics may be made catalytic with respect to palladium by carrying out the reaction under 50 atm pressure of oxygen. Acetoxylation is almost completely suppressed (Davidson and Triggs, 1967, 1968a). Later workers found the use of acetic acid and other solvents results in decreased yields (Itatani and Yoshimoto, 1971). In addition, catalytic activity decreases with an increasing concentration of palladium acetate. Extreme caution must be exercised in carrying out dimerizations in the presence of high pressures of oxygen; cumene, anisole, and tetraline explode under the conditions of the reaction. Diphenyl ether undergoes a smooth intramolecular coupling to afford phenylene oxide in excellent yields.

$$C_6H_5{-}O{-}C_6H_5 + \tfrac{1}{2}O_2 \xrightarrow{Pd(OAc)_2} (C_6H_4)_2O + H_2O$$

High yields of biaryls are obtained also by oxidative coupling of substituted benzenes in the presence of ethylene–palladium chloride and silver nitrate as oxidant (Fujiwara, 1970). Ferric sulfate has also been used as an oxidant (Frevel *et al.*, 1970). Aromatic and mercurated aromatics undergo substitution reactions by nucleophiles such as OAc^-, N_3^-, Cl^-, NO_2^-, Br^-, CN^-, and SCN^- in the presence of oxidants such as Cr(VI), $Pb(OAc)_4$, $NaClO_3$, $KMnO_4$, $NaNO_3$, and $NaNO_2$ with Pd(II) as a catalyst, in patterns characteristic of an electrophilic substitution reaction. Without the oxidant present, coupled aromatics are the main products. The reaction has little utility for simple compounds because of the expense of the oxidizing agent and the slow rate, but it might prove quite useful for aromatics not easily available otherwise (Henry, 1971).

$$R{-}C_6H_5 + X^- \xrightarrow[Pd(II)]{[Oxid.]} R{-}C_6H_4{-}X$$

Toluene in acetic acid–potassium acetate may be oxidized by palladium salts to benzyl acetate (Bryant *et al.*, 1968a, b) or to bitolyls (van Helden and

Balder, 1964; van Helden and Verberg, 1965), the major product being determined by the palladium salt and by the palladium salt–potassium (or sodium) acetate ratio. With palladium acetate, benzylic oxidation occurs predominately regardless of the reactant ratios, but with palladium chloride as an oxidant, the major course of the reaction shifts from 64.3% bitolyls with a sodium acetate–palladium chloride ratio of 5 to 68% benzylic oxidation when the ratio is 20. At a ratio of 10, the overall yield of both reactions is sharply and reproducibly lower. These phenomena have been interpreted as meaning that the palladium species responsible for coupling is altered as the acetate concentration is increased (Bryant *et al.*, 1968b). The coupling reaction was thought to require an aggregate of two or more palladium ions connected by bridging ligands, whereas a palladium species of lower aggregation is responsible for benzylic oxidation. Bridged structures are more easily broken down in palladium acetate than in palladium chloride, since chloride is a more effective bridging ligand than acetate. Davidson and Triggs (1966) proposed that dimerization proceeds via a σ-phenylpalladium(II) complex, since other metals yield stable phenyl derivatives under identical conditions.

Vanderwerff (1970) considered in detail the types of products to be derived from oxidative dimerization of aromatic hydrocarbons and was able to predict from equilibrium considerations whether methyl-substituted aromatics would undergo benzylation or nuclear coupling. The oxidation system consisted of the substrate in trifluoroacetic acid, a platinum- or palladium-on-carbon catalyst, and dispersed excess oxygen. The acidity can be increased by addition of boron trifluoride hydrate. Vanderwerff pointed out that a variety of systems combining a strong Bronsted acid and an oxidant may effect oxidative dimerization, although many of these systems are impractical for preparative work. Among such systems are nitrogen dioxide or cerium(IV) trifluoroacetate in trifluoroacetic acid, anhydrous iron(III) chloride suspended in *n*-hexane or carbon tetrachloride, hydrated copper(II) sulfate in concentration sulfuric acid, and oxygen with platinum-on-carbon in 85% phosphoric acid saturated with boron trifluoride.

Aromatics and Olefins

A convenient route to a variety of aromatic olefins is the direct substitution of an aromatic nucleus for a hydrogen of the olefinic function in the presence of stoichiometric amounts of palladium acetate in acetic acid (Moritani *et al.*, 1971b). The reaction can be made catalytic with respect to palladium by reoxidation under oxygen pressure (Shue, 1971) or with air or oxygen in the presence of catalytic amounts of cupric acetate or silver acetate. Styrene and a number of substituted styrene and stilbene derivatives have been made by this reaction, which affords the more thermodynamically stable olefin isomer.

Olefins, such as butadiene, that easily form a π-allyl complex give acetates as major products (Danno *et al.*, 1969).

Styrene on interaction with substituted aromatics affords unsymmetrical *trans*-stilbenes (Fujiwara *et al.*, 1968c). The isomers obtained are controlled by the directing influence of the substituent; an alkyl group affords *ortho–para* derivatives, nitro groups *meta* derivatives. Chlorobenzene affords considerable amounts of all isomers, suggesting loss of directing effect through coordination with palladium (Fujiwara *et al.*, 1969b). Coupling of *p*-bromostyrene with benzene is accompanied by some hydrogenolysis of the halogen (Moritani *et al.*, 1969).

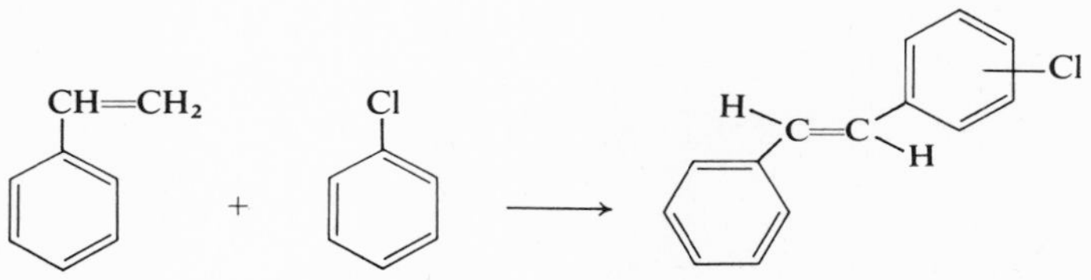

Stilbenes are formed in this reaction only if acetic acid or chloroacetic acid are present; no stilbene is formed when hydrogen chloride, acetone, ethanol, or ethyl ether is substituted for acetic acid (Fujiwara *et al.*, 1968a). The yields also depend on the amount of sodium acetate present increasing with increasing amounts of acetate (Fujiwara *et al.*, 1968b). Condensation of deuterated styrene with benzene established that stilbene is formed without a hydride shift (Danno *et al.*, 1970a). Support for a σ-bonded olefin–palladium(II) intermediate has been obtained (Danno *et al.*, 1970b; Moritani *et al.*, 1971a, b).

Benzene couples with ethylene to afford styrene (Shue, 1971). The reaction provides a potentially attractive industrial synthesis of styrene, but great advances over the present art are required to make the reaction economical. 1-Butene couples with benzene at the terminal carbon (Moritani *et al.*, 1969). Benzene and either *cis*- or *trans*-butenes afford 2-phenyl-*cis*-butene as the major product (Fujiwara *et al.*, 1969a).

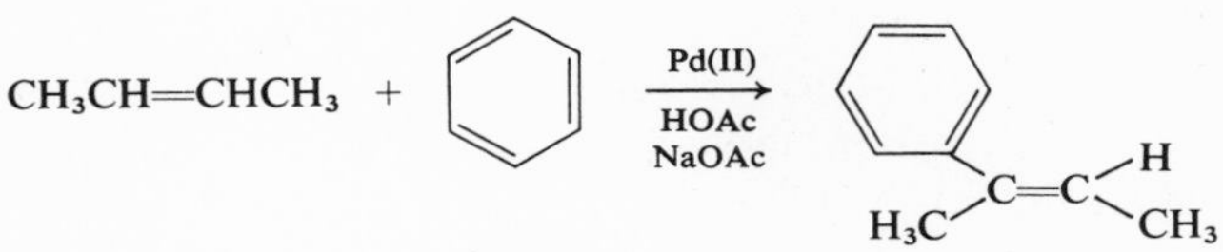

$CH_3CH{=}CHCH_3$ + (benzene) $\xrightarrow[\text{HOAc, NaOAc}]{\text{Pd(II)}}$

Phenylbutenes are also formed by interaction of styrene and ethylene in the presence of the palladium–styrene complex, di-μ-chloro-dichlorobis(styrene) dipalladium(II). The major isomer is *trans*-1-phenyl-1-butene. In the absence of ethylene, styrene is converted to *trans*-1,3-diphenyl-1-butene (Kawamoto *et al.*, 1970).

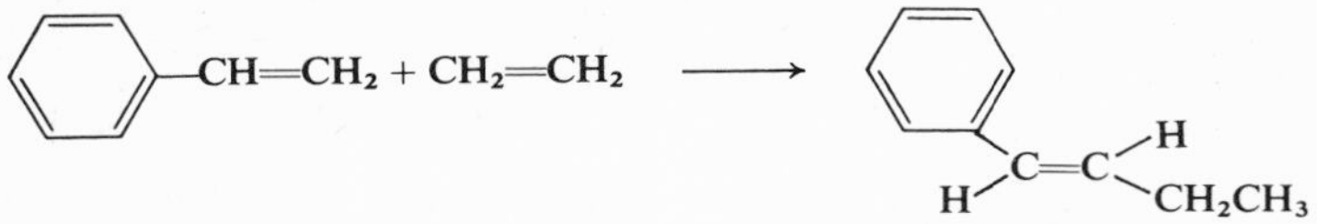

$$C_6H_5CH{=}CH_2 + CH_2{=}CH_2 \longrightarrow C_6H_5(H)C{=}C(H)CH_2CH_3$$

Coupling of ethylene with dehydroabietic acid in the presence of palladium and silver acetates affords dimeric dibasic acids (Schuller, 1971).

A modification of the above reactions is the intramolecular oxidative cyclization of the olefins, $PhR_2CCH{=}CH_2$ ($R = Ph$ or CH_3), to indenes (Bingham *et al.*, 1970). The reaction as described was not catalytic but presumably could be easily made so.

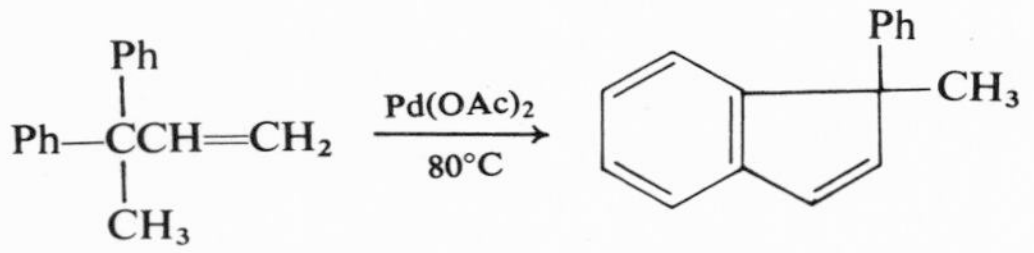

$$Ph{-}C(Ph)(CH_3)CH{=}CH_2 \xrightarrow[80°C]{Pd(OAc)_2} \text{1-Ph-1-CH}_3\text{-indene}$$

Ferrocene interacts with styrene in the presence of palladium acetate to produce *trans*-α-styrylferrocene. The reaction provides a convenient method of introducing alkenyl groups into ferrocene, a normally troublesome reaction. The reaction as described was stoichiometric with respect to palladium (Asano *et al.*, 1970). It was later extended to acrylonitrile, methyl acrylate, and acrylaldehyde (Asano *et al.*, 1971).

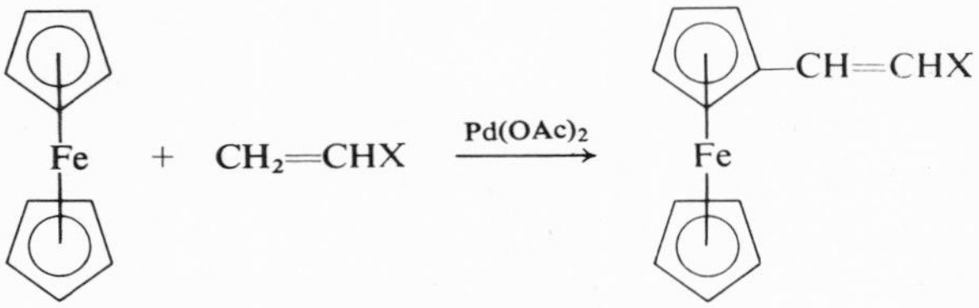

$$\text{Fe(C}_5\text{H}_5)_2 + CH_2{=}CHX \xrightarrow{Pd(OAc)_2} \text{C}_5\text{H}_5\text{FeC}_5\text{H}_4{-}CH{=}CHX$$

$X = Ph, CN, CO_2CH_3, CHO$

Ferrocenylation of olefins may be carried out also by interaction of chloromercuriferrocene and olefins in the presence of cupric chloride and catalytic amounts of palladium (Kasahara *et al.*, 1972a). Without olefin present, chloromercuriferrocenes couple in the presence of palladium to afford biferrocenyls. In the presence of carbon monoxide, carbonyl derivatives are formed (Kasahara *et al.*, 1972b).

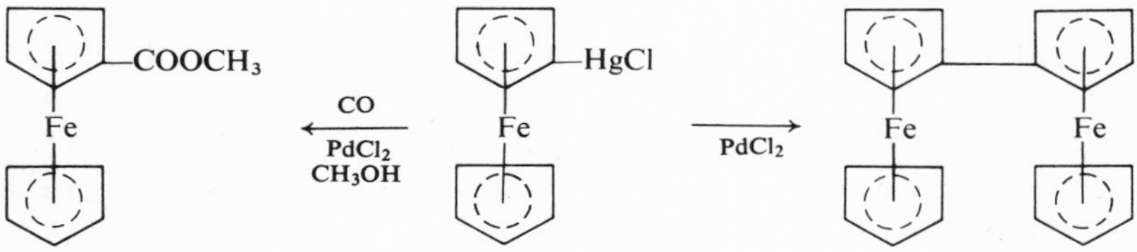

$$\text{FcCOOCH}_3 \xleftarrow[PdCl_2,\ CH_3OH]{CO} \text{FcHgCl} \xrightarrow{PdCl_2} \text{Fc{-}Fc}$$

Heck has made extensive use of arylmercuric salts to achieve synthesis of a variety of substituted aromatic compounds by coupling reactions in the presence of a palladium salt and an oxidant (Heck, 1968a–f). Arylbutenyl acetates may be prepared by this technique from interaction of a conjugated diene, an arylmercuric salt, and lead tetraacetate with catalytic amounts of palladium acetate. For example, 16.8 gm (50 mmole) of phenylmercuric acetate, 0.11 gm (0.5 mmole) of palladium acetate, and 24.35 gm (55) mmole of lead tetraacetate in 50 ml of acetonitrile, and 10 ml of butadiene contained in a heavy-walled Pyrex bottle affords 1-phenyl-3-buten-2-yl acetate in 78% yield. None of the allylic isomer, 1-phenyl-2-buten-4-yl acetate, is found. Under the same conditions, isoprene reacts only with lead tetraacetate, producing 1,2-diacetoxy-2-methyl-3-butene (Heck, 1968f).

$$C_6H_5HgOAc + CH_2{=}CHCH{=}CH_2 + Pb(OAc)_4 \xrightarrow{Pd(OAc)_2} C_6H_5CH_2CH(OAc)CH{=}CH_2 + Pb(OAc)_2 + Hg(OAc)_2$$

A useful method of introducing 2-haloethyl groups into aromatic systems is by interaction of arylmercuric halides with olefins in the presence of cupric halides and catalytic amounts of palladium salts. Acetic acid or aqueous acetic acid is the best solvent and lithium chloride in the reaction mixture has a beneficial effect, probably because of its solubilizing action on cupric chloride. Treatment of a solution of 15.7 gm (50 mmole) of phenylmercuric chloride, 2.1 gm (50 mmole) of lithium chloride, 13.4 gm (100 mmole) of anhydrous cupric chloride, 40 ml of acetic acid, 5 ml of water, and 5 ml of 0.1 *M* Li_2PdCl_4 in acetic acid with ethylene at 30 psig overnight affords 2-phenylethyl chloride in 76% yield. Aryl ethyl bromides are obtained by the same procedure but in lower yield (Heck, 1968c).

$$CH_2{=}CH_2 + C_6H_5HgCl + CuCl_2 + LiCl \xrightarrow[HOAc]{Li_2PdCl_4} C_6H_5CH_2CH_2Cl$$

Allylaromatic derivatives are obtained by interaction of arylmercuric salts and allylic halides in the presence of palladium salts as catalysts. Yields range from 31 to 87%. Isomerization of the initially formed allylaromatic compound into a propenyl derivative usually occurs only if there is insufficient catalyst or reoxidant present or if the allylic halide concentration is below 0.1 *M*. Phenylmercuric chloride and 3-chloro-1-butene with lithium palladium chloride and cupric chloride as catalyst, produces the expected 1-phenyl-2-butene in 50% yield with little of the isomeric phenylbutenes being formed. However, under the same conditions *trans*-crotyl chloride and phenylmercuric chloride forms a mixture of at least seven products including 34% 2-phenyl-2-butene, 29% 1-phenyl-1-butene, and 15% 1-phenyl-2-butene (Heck, 1968c).

$$C_6H_5HgCl + CH_2{=}CHCH(Cl)CH_3 \xrightarrow[CuCl_2]{Li_2PdCl_4} C_6H_5CH_2CH{=}CHCH_3 + HgCl_2$$

Heck's synthesis can be adapted also to the preparation of numerous 2- and 3-arylaldehydes and ketones by employing enol esters of aldehydes and ketones (Heck, 1968d) or primary or secondary alcohols, respectively (Heck, 1968b). In the reaction of allylic alcohols with phenylmercuric chloride, the presence of dicyclohexylethylamine produces improved yields apparently by preventing formation of halides from the allylic alcohol and by-product hydrogen chloride. The reaction appears to be quite general and useful, but the yields are widely variable and depend on the substrate structures. Tertiary allylic alcohols react as simple olefins, giving arylated derivatives.

$$PhHgX + CH_2{=}CHCH(R)OH \xrightarrow[CuCl_2]{Li_2PdCl_2} PhCH_2CH_2C(=O)R$$

An unusual variation of Heck's arylation procedure resulted in simultaneous addition to the double bond of indene or 1,2-dihydronaphthalene of a phenyl group and an anion part of a protic solvent (Horino and Inque, 1971).

$$\text{indene} + C_6H_5HgCl \xrightarrow[ROH]{Li_2PdCl_2} \text{2-phenylindene} + \text{1-OR-2-phenylindan}$$

$R = CCH_3$ (C=O), CH_3

ACETOXYLATION

Acetoxylation of aromatics and alkylaromatics is a reaction of considerable synthetic utility with promise of commercial adaption. A potentially attractive commercial synthesis of phenol from benzene involves acetoxylation as an initial step.

$$C_6H_6 + HOAc + \tfrac{1}{2}O_2 \longrightarrow C_6H_5\text{—}OAc + H_2O$$

Phenyl acetate, formed in the process, is converted to phenol either through hydrolysis

$$C_6H_5\text{—}OAc + H_2O \longrightarrow C_6H_5\text{—}OH + HOAc$$

or through pyrolysis affording phenol and ketene.

$$C_6H_5\text{—}OAc \longrightarrow C_6H_5\text{—}OH + CH_2{=}C{=}O$$

In either case, acetic acid is recycled so that the overall reaction consumes only benzene and oxygen. Major difficulties in the reaction are overoxidation to carbon dioxide and insufficient catalyst life. Neither of these difficulties appears insurmountable. At present, preferred catalysts are base-activated palladium modified by an additional metal. In a typical experiment, acetic acid, benzene, oxygen, and nitrogen in mole ratio of 4:1:0.5:1.5 are passed over a palladium–gold on silica gel catalyst modified by cadmium and potassium acetates at 155°C. Conversions to phenylacetate of 70–75% are obtained (Arpe and Hornig, 1970). Other catalysts for this purpose are supported palladium promoted by bismuth or selenium (British Patent 1,200,708) alternatively or additionally activated by a salt of a carboxylic acid (British Patent 1,200,392). Most acetoxylations have used palladium, but apparently equally good results may be obtained with iridium, ruthenium, or rhodium salts; platinum and nickel are definitely inferior as catalysts. Nitric acid instead of oxygen can serve as the oxidant to reoxidize reduced palladium (Selwitz, 1970).

Acetoxylation proceeds much more readily with trifluoroacetic acid than acetic acid as well as producing a different isomer distribution. Naphthalene reacts with $Pd(OOCF_3)_2$ in CF_3COOH at room temperature to afford α-naphthyltrifluoroacetate, whereas acetic acid requires elevated temperature and produces about equal amounts of α- and β-naphthylacetates (Arzoumanidis *et al.*, 1972).

Alkylbenzenes undergo predominately either nuclear or side-chain acetoxylation depending on the catalyst. Interaction of carboxylic acids and methyl-benzenes proceeds smoothly in the presence of air and certain palladium catalyst systems to afford benzyl esters. The process is sufficiently selective and efficient to make it the method of choice for preparation of many benzyl esters. At high conversions, some diacetates are formed (Fitton *et al.*, 1969). Stannous acetate and charcoal are powerful promoters for the oxidation. An example of the reaction is the oxidation of toluene in acetic acid. A mixture of 8 moles of acetic acid, 1.1 moles of potassium acetate, 1.0 mole of toluene, 0.06 mole of stannous acetate, 0.016 mole of palladium(II) acetate, and 33.6 gm of charcoal was stirred at 100°C for 9 hours while air was blown over the surface. After filtration, extraction, and distillation, benzyl acetate was obtained in 41% yield (Bryant *et al.*, 1968a).

$$C_6H_5CH_3 + HOAc + \tfrac{1}{2}O_2 \xrightarrow[Sn]{Pd} C_6H_5CH_2OAc + H_2O$$

The same reaction provides an excellent synthesis of xylyl acetates and xylylene diacetates. The ratio of xylyl acetate to xylylene diacetate is independent of the degree of completion as long as xylene is present; after xylene is gone, the percentage of xylylene diacetate increases.

$$CH_3C_6H_4CH_3 + HOAc + \tfrac{1}{2}O_2 \xrightarrow[Sn]{Pd} CH_3C_6H_4CH_2OAc + AcOCH_2C_6H_4CH_2OAc$$

In a typical experiment, 482 gm (8.04 moles) of acetic acid, 108 gm (1.1 moles) of potassium acetate, 106 gm (1.0 mole) of *p*-xylene, 14.2 gm of stannous acetate, 33.6 gm of charcoal, and 3.6 gm (0.016 mole) of palladium acetate are stirred at 100°C for 8 hours. The product consists of 72.5 gm of *p*-xylyl acetate, 40.9 gm of α,α′-*p*-xylylene diacetate, 1.3 gm of α,α-*p*-xylylene diacetate, 0.6 gm of *p*-acetoxymethylbenzylidene diacetate, and 3.2 gm of α-acetoxy-*p*-tolualdehyde. *ortho*- and *meta*-Xylenes are similarly oxidized (Bryant *et al.*, 1969). Acetoxylation of methyl groups is enhanced by electron-donating substituents such as alkoxy groups, but some nuclear acetoxylation may also occur. Electron-withdrawing ring substituents greatly decrease the rate of acetoxylation (Bryant *et al.*, 1968a; Bushweller, 1968).

In contrast to the above reactions, nuclear acetoxylation comes the preferred course if excess of acetate is avoided and the reaction is carried out under oxygen. The presence of even a small amount of sodium acetate markedly lowers the yield of the nuclear acetoxylation with a concomitant increase in the yield of side-chain acetates. A mixture of *p*-xylene in glacial acetic acid containing palladium acetate heated to 110°C with a slow stream of oxygen bubbling through the solution affords a mixture of 70% 2,5-dimethylphenyl acetate and 30% *p*-methylbenzyl acetate after 150 hours. The yield of products is 450% based on palladium acetate.

CH_3 ... CH_3 $\xrightarrow[\text{Pd(OAc)}_2,\ 110^\circ\text{C}]{\text{HOAc}}$ CH_3, OAc ... CH_3 (70%) + CH_2OAc ... CH_3 (30%)

A number of substituted aromatics were subjected to this reaction. The orientation of the nuclear acetoxylation products suggested that, at least for the *ortho,para*-directing substituents, acetoxylation by palladium acetate takes place as an electrophilic addition process with oxypalladation products being formed as intermediates (Eberson and Gomez-Gonzales, 1971).

OCH_3 → OCH_3, PdOAc, OAc + OCH_3, OAc, PdOAc → OCH_3, OAc (97%)

An unusual carboxylation reaction has been reported to occur under conditions similar to those usually used for acetoxylation (Sakakibara *et al.*, 1969). The yields of acid are small and the origin of the carboxyl group obscure. The reaction would be useful if it could be further developed.

CH_3O–C_6H_5 $\xrightarrow[\text{PdCl}_2,\ (\text{AcO})_2\text{O},\ 100^\circ\text{C, 5 hours}]{\text{CH}_3\text{COOH, NaOAc}}$ CH_3O–C_6H_4–COOH

Oxidation of Alkylaromatics

Air oxidation of alkylaromatics affords ketones, oxidation occurring at the α-methylene group. Air oxidation at 130°C of 20 gm of the alkylaromatics acenaphthene, ethylbenzene, fluorene, and *n*-propylbenzene in the presence of 100 mg of tris(triphenylphosphine)rhodium chloride affords the corresponding ketones, acenaphthenone, acetophenone, fluorenone, and propiophenone, each in better than 90% yield. However, the conversions were only 10–23%. A more facile oxidation occurs with tetralin and conversions are 48–60% at 55°C. No ketones were formed in the absence of the rhodium complex. The use of oxygen instead of air about doubles the conversion (Blum *et al.*, 1967). Improved results by the use of biphyllic ligands are claimed (Fenton, 1969). Other potential catalysts, palladous chloride, rhodium trichloride (trihydrate), chlorocarbonylbis(triphenylphosphine)rhodium, chlorocarbonylbis(triphenylphosphine)iridium, trichlorotris(triphenylarsine)rhodium, and trichlorotris-(triphenylstilbene)rhodium were ineffective in these oxidations (Blum *et al.*, 1967). Certain palladium(0) complexes are effective catalysts (Stern, 1970).

The course of this reaction is influenced apparently by the solvent. Oxidation of VII by air in refluxing benzene affords the alcohol, **VIII**, in 48% yield, whereas without solvent on a steambath, the ketone, **IX**, is obtained in good yield at 40% conversion (Birch and Subba Rao, 1968).

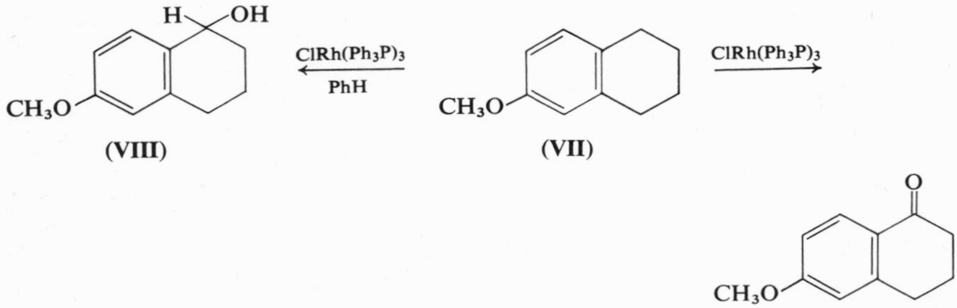

H OH

CH_3O **(VIII)** $\xleftarrow[PhH]{ClRh(Ph_3P)_3}$ CH_3O **(VII)** $\xrightarrow{ClRh(Ph_3P)_3}$ CH_3O O **(IX)**

There are experimental indications that oxidations of alkylbenzenes over rhodium catalysts such as tris(triphenylphosphine)chlororhodium, rhodium acetylacetonate, or rhodium 2-ethylhexanoate are free-radical in nature (Kurkov *et al.*, 1968). Some hydroperoxides form and the reaction is completely inhibited by 2% hydroquinone.

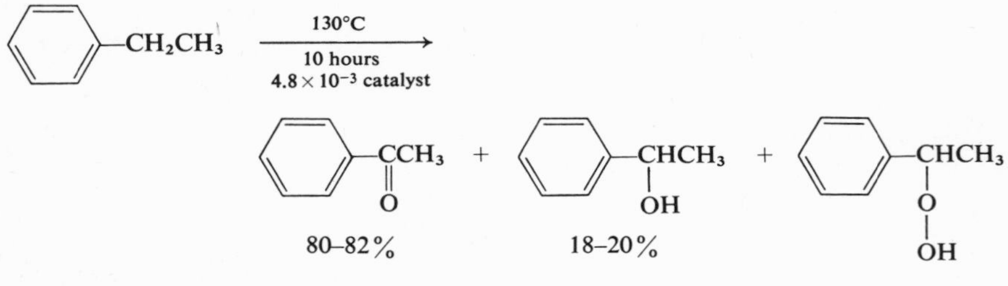

$C_6H_5CH_2CH_3 \xrightarrow[\substack{10\ \text{hours} \\ 4.8\times10^{-3}\ \text{catalyst}}]{130°C}$

$C_6H_5CCH_3$ (=O) 80–82% + $C_6H_5CHCH_3$ (OH) 18–20% + $C_6H_5CHCH_3$ (O–OH)

Oxidation of 2,3-cyclopentenoindole (**X**) and tetrahydrocarbazole (**XI**) over reduced platinum oxide in ethyl acetate affords lactams, **XII**, via an intermediate hydroperoxide.

Catalytic oxidation of 2,3-cycloheptenoindole, $n = 3$, proceeds similarly, but the ten-membered lactam was converted during isolation to tetrahydro-

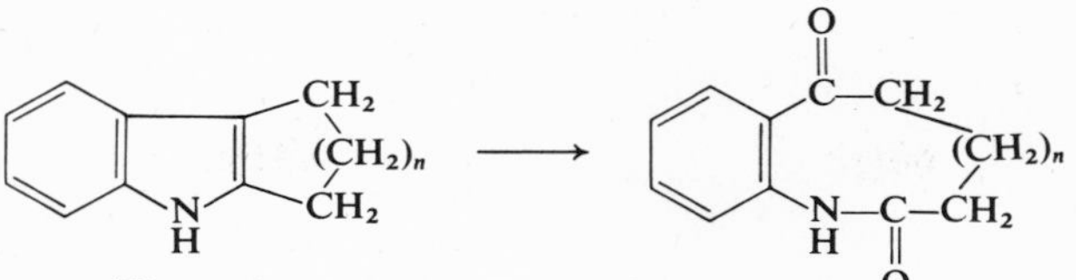

(X) $n = 1$
(XI) $n = 2$

(XII)

phenanthridone. On the other hand, oxidation of 2,3-cyclooctenoindole (**XIII**), $n = 4$, proceeds by attack at the 2-position and affords **XIV**. Compounds in this series also undergo autooxidation, affording the same products (Witkop *et al.*, 1951).

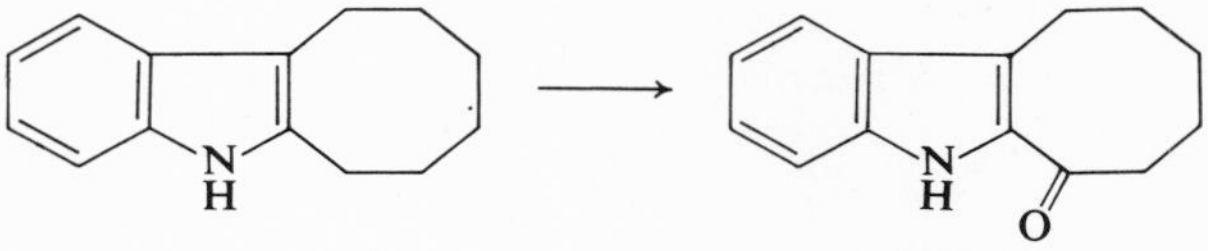

(XIII) (XIV)

The diquaternary salt, **XV**, was oxidized to **XVI** in 83% yield by refluxing 1 gm of **XV** in 100 ml of nitromethane with 10% palladium-on-carbon. The dibromide salt could not be similarly oxidized, possibly due to insolubility. Presumably nitromethane functions as the oxidant (Glover and Morris, 1964).

(XV) $2X^-$
X = picrate

(XVI)

Oxidation of Alcohols

Primary alcohols in the presence of oxygen and a noble metal catalyst can be converted to aldehydes or acids and secondary alcohols afford ketones. The reaction is useful mainly because of the frequently obtained high selectivity with polyfunctional molecules and because of its applicability to sensitive compounds. Various workers consider the oxidation to be in fact a dehydrogenation in which oxygen simply serves as a hydrogen acceptor, removing it from the equilibrium by formation of water (Wieland, 1912, 1913, 1921; Muller and Schwabe, 1928; Rottenberg and Baertschi, 1956), but other views have also been expressed (Macrae, 1933; Rottenberg and Thurkauf, 1959). Regardless of the mechanism, it is useful to distinguish this type of conversion from dehydrogenation without an acceptor, since the latter only takes place at elevated temperatures and is not applicable to heat sensitive molecules.

Catalysts

The vast majority of workers in this area use some form of platinum catalyst, although success has been reported also with other metals. A comparative study of platinum, palladium, rhodium, and ruthenium catalysts for oxidation of alcohols failed to reveal a cause for this overwhelming preference for platinum catalysts. Other metals at times functioned as well or better, but the results were highly variable and depended critically on the substrate and solvent. The conclusion was reached that platinum was probably the best first choice because of substantial precedent and a history of success, but in case of failure it was decidedly worthwhile to investigate other noble metals (Rylander and Kilroy, 1962).

Both supported and unsupported platinum catalysts are used in this type of reaction. Powdered carbon is a favorite support for batch work and in continuous processing carbon or alumina granules may be used.

Williams and Lutz (1968) examined the effect of metal, metal concentration, carrier, temperature, and oxygen pressure on the rate of oxidation of long-chain primary alcohols to the corresponding acids. Platinum, preferably supported on carbon, proved to be by far the most effective noble metal catalyst. The rate is roughly proportional to the metal concentration on the carrier (based on constant weight of catalyst) and to the oxygen pressure. Temperatures around 40° to 60°C give good results. Platinum oxide proved inactive in these oxidations unless first reduced with hydrogen. Solvents such as heptane or benzene substantially increased the rate of oxidation by lowering the viscosity of the media. Other workers set circumscribed limits to the effective range of catalyst, asserting that too high a catalyst loading has an adverse effect on yield (Trenner, 1947). Platinum oxide prepared according to the procedure of Adams and Shriner (1923) is used frequently as a catalyst; its functioning was examined in some detail by Glattfeld and Gershon (1938) in connection with oxidation of sugar alcohols. The oxide itself is inactive and is first converted to catalytically active platinum black by a stoichiometric oxidation–reduction with the sugar. The time required varies inversely with the temperature; at 30°C the reaction is very slow, whereas at 75°C, it is much faster. The effectiveness of platinum oxide catalysts was not changed by varying the temperature of preparation from 400° to 600°C in marked contrast to the sensitivity to preparation temperature found when these catalysts are used for hydrogenation. Air was found to be a more efficient oxidizing agent than oxygen inasmuch as oxygen retards reduction of the catalyst to an active form. If, however, the reaction is begun with platinum black, oxygen is more efficient than air. Later workers found air could be used instead of oxygen without an adverse effect on either the reaction rate or yield of product (Sneeden and Turner, 1955b). Tsou and Seligman (1952) distinguished between hydrogen-reduced platinum oxide and platinum black as oxidation catalysts. There were able to prepare phenyl β-D-glucopyruronoside in 32% yield by oxidation of phenyl β-D-glucopyranoside over platinum black, whereas this reaction could not be achieved over fresh hydrogen-reduced platinum oxide (Marsh, 1952). The authors cautioned that all adsorbed hydrogen on reduced platinum oxide must be removed before use if success is to be obtained.

Homogeneous noble metal catalysts will also convert alcohols to ketones, aldehydes, or acids (Vaska and DiLuzio, 1961; Nikiforova *et al.*, 1963; Charman, 1966); the reaction has found little synthetic use. The metal is reduced in the process, but the system can be made catalytic by an *in situ* regeneration of the reduced metal (Lloyd, 1967; Brown *et al.*, 1969).

Carbohydrates

Catalytic oxidation of carbohydrates over platinum metals has proved a very powerful tool for controlled transformations of these substances. The oxidations are characterized by extremely high selectivity which has been likened to that obtained in natural processes. Selective oxidations of mono- and oligosaccharides with platinum catalysts and air have led to efficient preparations of uronic acids, aminouronic acids, uronosides, aldonic acids, and intermediates in ascorbic acid synthesis (Heyns, 1970). These reactions have proved a useful means of constitutional and conformational analyses (Heyns, 1963). Axial hydroxyls are dehydrogenated specifically; equatorial groups are not attacked. Selective oxidation of axial hydroxyls has been rationalized by assuming a dehydrogenation mechanism with preferential attack at the more reactive equatorial hydrogen (Brimacombe *et al.*, 1965). Additionally, there is strong tendency for the preferential oxidation of primary hydroxyl over secondary hydroxyl, a fact that permits practical syntheses of uronic acids. An excellent review of carbohydrate oxidation over noble metals covering the literature through 1959 has been written by Heyns and Paulsen (1963). A few selected examples of carbohydrate oxidations are given below to illustrate the general procedures and the types of selectivity involved.

Reaction Conditions and Procedure

The equipment needed for oxidation is similar to that required for catalytic hydrogenation. Oxygen must be brought into intimate contact with the substrate, preferably in solvent. Provision may be made for measuring the amount of gas absorbed either by pressure drop or displacement. Agitation should be vigorous despite the relatively slow reaction rate. One procedure is illustrated by the conversion of 1,2-isopropylidene-D-glucose to D-glucuronic acid. Oxidation is carried out using a 13% platinum-on-Darco-G-60 catalyst prepared by formaldehyde reduction of chloroplatinic acid. Sixty-six grams (0.3 mole) 1,2-isopropylidene-D-glucose and 6.3 gm (0.075 mole) of sodium bicarbonate are dissolved in 900 ml of water in a 3-liter, three-neck, creased, round-bottom flask. After solution 6.8 gm of 13% platinum-on-carbon is introduced and the mixture is vigorously stirred at 3000 rpm while compressed air that has been cleaned by passage through concentrated sulfuric acid is passed into the solution at 112 liters/hour. Reaction temperature is maintained at 50°C by means of a water bath. After 1.25 hours, the pH of the solution drops to 7.5 and three more portions of 6.3 gm of sodium bicarbonate are added over the next 7 hours. After 11.5 hours, the reaction is stopped, filtered, concentrated, and 1,2-isopropylidene-D-glucuronic acid is isolated in 53.5% yield.

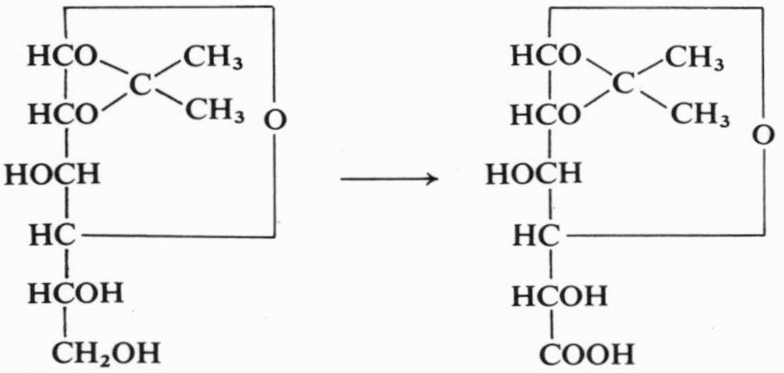

```
 ┌──────────────────┐            ┌──────────────────┐
HCO         CH3      │           HCO         CH3      │
 |     C             │            |     C             │
HCO         CH3      O    ──→    HCO         CH3      O
 |                   │            |                   │
HOCH                 │           HOCH                 │
 |                   │            |                   │
HC───────────────────┘           HC───────────────────┘
 |                                |
HCOH                             HCOH
 |                                |
CH2OH                            COOH
```

Both vigorous stirring and a high gas flow contribute to a fast oxidation, as shown in Table I (Mehltretter *et al.*, 1951); the yield in the example is unaffected by changes in these variables. Other workers used a very high catalyst loading in this oxidation (20.6 gm 1,2-isopropylidene glucose with 75 gm 0.5% platinum-on-alumina) and achieved a 74% yield (Reiners, 1958).

TABLE I
OXIDATION OF 1,2-ISOPROPYLIDENE-D-GLUCOSE

Air flow (liters/hour)	Stirring (rpm)	Reaction time (hours)	Yield (%)
112	3500	4.5	57
30	3500	10.5	57
112	1000	22.5	57

Oxidations of this type may also be carried out in continuous processing. Reiners (1958) oxidized a 5.8% solution of methyl-α-D-glucoside containing 1.3 equivalents of sodium bicarbonate over 0.5% platinum-on-alumina pellets at 70°C and obtained the corresponding uronic acid in high yield at 25% conversion. The conversion and yield remained substantially constant when the space velocity was increased by 70%.

Improved yields of uronic acids may be obtained by the use of secondary alcohols as antifoaming agents and by oxidation modifiers. Benzyl 2-acetamino-2-deoxy-α-D-glucopyranoside (25 gm), 15 gm of 10% platinum-on-carbon, and 500 ml of water are charged to a four-necked flask equipped with stirrer, reflux condenser, dropping funnel, and inlet capillary. Air or oxygen is bubbled into the violently stirred solution maintained at 95°C with the pH adjusted to 6.8–7.7 by addition of sodium bicarbonate solution. A secondary alcohol such as isopropanol or isobutanol is added at intervals to prevent excessive foaming and overoxidation. The reaction is continued until the starting material, as determined by thin-layer chromatography, disappears. The yield of benzyl 2-acetamino-2-deoxy-α-D-glucopyranosiduronic acid is 77% (Yoshimura *et al.*, 1969).

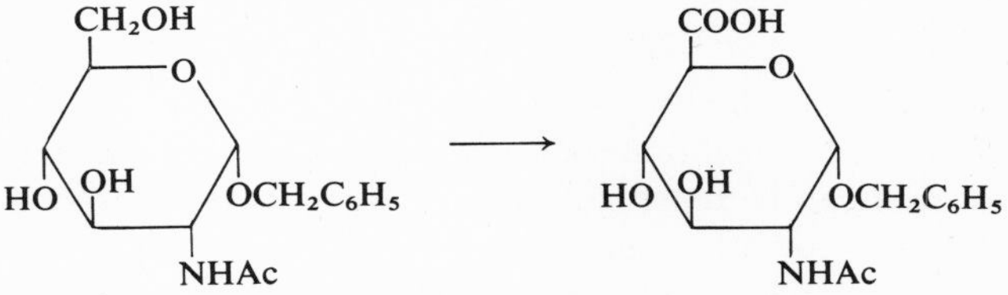

CH_2OH, O, HO, OH, $OCH_2C_6H_5$, NHAc → COOH, O, HO, OH, $OCH_2C_6H_5$, NHAc

An efficient apparatus for catalytic oxidation has been described by von Schuching and Frye (1965). A standard chromatographic column, 45 × 600 mm, containing a fused coarse sinter is cut off 24 cm above the disk. The top of the tube is left open and a mechanical stirrer, maximum speed 1400 rpm is centered above it. A 77 mm diameter stirring rod, drawn into a butterfly shape at the bottom, extends to a few millimeters above the disk. The stirring rod is held steady at high speeds by a glass sleeve. Heating is effected by an electric heating tape regulated by a rheostat.

As an example of the apparatus in use, 2.3 gm of 1,2-mono-*O*-cyclohexylidene-L-xylofuranose, 300 mg of sodium bicarbonate, and 1.7 gm of 4.5% platinum-on-carbon in 100 ml of water is placed in the apparatus. The reaction is stirred at 1400 rpm and held at 50°C while oxygen is blown in the suspension

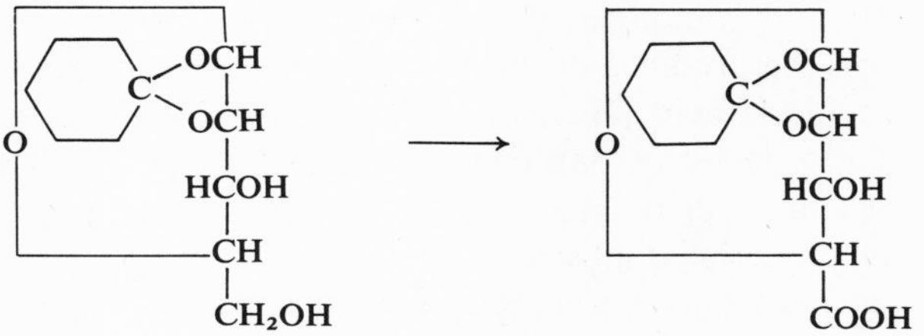

C, OCH, OCH, O, HCOH, CH, CH_2OH → C, OCH, OCH, O, HCOH, CH, COOH

at 50 ml/minute. After 30 minutes, another 1.7 gm of catalyst and another 300 mg of sodium bicarbonate is added. After 2 hours, the reaction is stopped and 1,2-mono-*O*-cyclohexylidene-L-xyluronic acid is isolated in 70% yield (von Schuching and Frye, 1965).

D-Glucosaccharic acid is prepared in 54% yield by air oxidation at 50°C of a vigorously agitated solution of glucose in the presence of 10% platinum-on-carbon. The solution is kept above pH 5.0 by addition of potassium bicarbonate as needed (Mehltretter *et al.*, 1949).

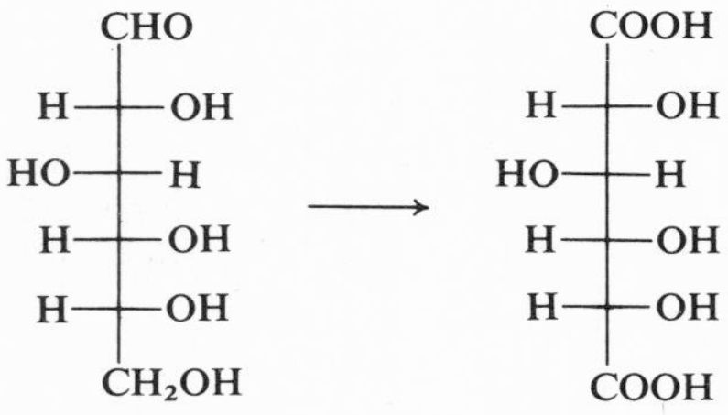

$$\begin{array}{c} CHO \\ H-C-OH \\ HO-C-H \\ H-C-OH \\ H-C-OH \\ CH_2OH \end{array} \longrightarrow \begin{array}{c} COOH \\ H-C-OH \\ HO-C-H \\ H-C-OH \\ H-C-OH \\ COOH \end{array}$$

In a similar manner Trenner (1947) oxidized 4,5-acetone-2,5-furanose-*d*-gluconic acid over 13% platinum-on-carbon to the corresponding saccharic acid. The catalyst can be reused repeatedly if care is taken to avoid catalyst contact with nitrogen-containing compounds.

Amino Sugars

Oxidation of amino sugars may proceed with more difficulty than other carbohydrates due to the inhibiting effect of the nitrogen atom on the catalysts, inhibition probably arising through overly strong adsorption of the nitrogen atom. Oxidation of carbohydrates over platinum catalysts has been shown to be inhibited by nitrogenous compounds. Methyl-α-D-glucopyranoside failed to undergo oxidation over platinum-on-carbon when ammonium carbonate or glycine was added to the reaction mixture. This inhibition by nitrogen compounds was cited as the reason for the failure to oxidize α-D-glucosylamine, methyl *N*-benzyloxycarbonyl-α-D-glucosaminide, and methyl *N*-acetyl-α-D-glucosaminide (Barker *et al.*, 1958). Nonetheless, with proper reaction conditions, amino sugars will successfully undergo catalytic oxidation. D-Glucosaminic acid is readily obtained by the oxidation at 30°C of D-glucosamine hydrochloride in the presence of enough potassium bicarbonate to interact with the hydrochloric acid (Heyns and Koch, 1953); in the same manner L-glucosamine affords L-glucosaminic acid (Hardegger and Lohse, 1957).

Oxidation of amino sugars may be facilitated by suitable protection of the amino function, rendering it more stable to oxidation and less prone to inhibit the catalyst. The *N*-carbobenzoxy group proved a quite suitable derivative for blocking α-methyl-α-D-glucosaminide in preparation of methyl-α-D-glucosamineuronide. The *N*-carbobenzoxy function provides good crystallizability to the products and it can be easily removed by hydrogenation (Heyns and Paulsen, 1955; Kiss and Wyss, 1972). Other workers used an acetyl group as a protecting function in oxidation of D-glucosaminide derivatives (Marsh and Lewy, 1958) and aminocyclitols (Suami and Ogawa, 1970).

A procedure for degrading ribopolynucleotides requires conversion of a terminal hydroxyl group into a suitable carbonyl function that will facilitate elimination of its β-phosphate. Success of the method depends on quantitative oxidation of a terminal hydroxymethyl group of deoxyribopolynucleotide to the corresponding uronic acid. Oxidation is carried out with a deactivated platinum oxide catalyst (one used in a previous experiment) and hydrogen peroxide. The deactivated catalyst is used to prevent irreversible adsorption of the polynucleotide. Hydrogen peroxide gives a rapid oxidation; it could not be replaced by oxygen. In a model experiment, 10 ml of 0.6% hydrogen peroxide was added dropwise with stirring over 1 hour to a mixture of 1.06 micromoles of thymidylyl-(3′,5′)-thymidine in 1 ml of 0.01 *N* sodium acetate buffer, pH 6.0, at 90°C, and 200 mg of freshly reduced, previously deactivated platinum oxide catalyst. The resulting uronic acid was isolated as its ammonium salt in 90% yield (Vizsolyi and Tener, 1962). Later workers found oxygen and platinum oxide quite suitable for selective oxidation of nucleosides and nucleotides (Moss *et al.*, 1963).

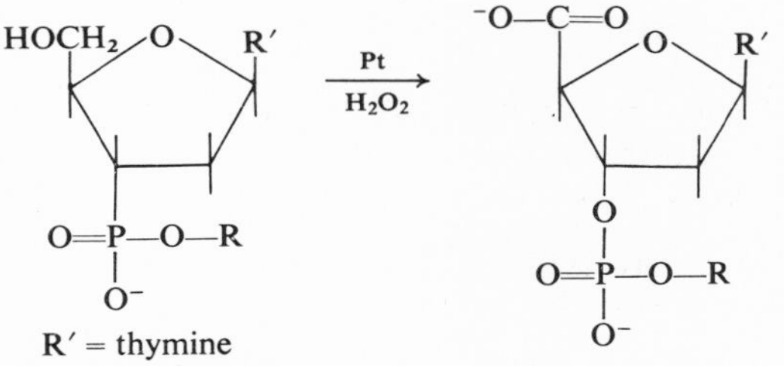

$\xrightarrow[H_2O_2]{Pt}$

R′ = thymine
R = thymidine

Cyclitols

Oxidation of cyclitols is characterized by both preferential and limited attack. Axial alcohols are oxidized preferentially and, where two or more axial hydroxyls are present, only one is attacked. Scyllitol, having no axial hydroxyls, is not oxidizable catalytically under the usual conditions. In *myo*-inositol only the single axial hydroxyl is attacked affording *myo*-inosose-2 over platinum-on-carbon at 60°C (Heyns and Paulsen, 1953).

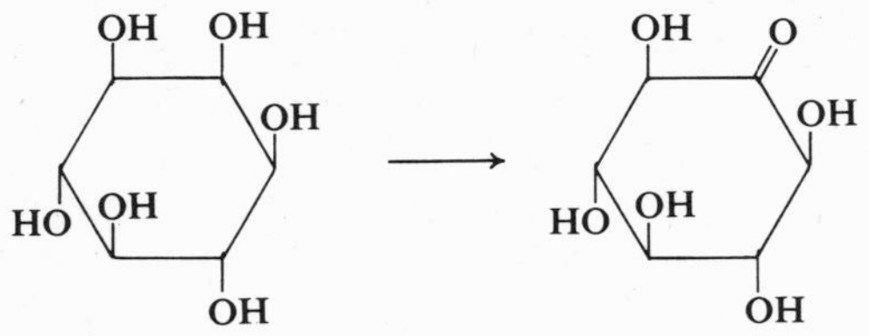

In cyclitols having two axial hydroxyls only one is oxidized by platinum and oxygen, and the reaction nicely complements bacterial oxidations that yield diketones. Selective oxidation of axial hydroxyls has been rationalized by assuming a dehydrogenation mechanism with preferential attack at the more reactive equatorial hydrogen (Brimacombe *et al.*, 1965). Catalytic oxidation of 2.85 gm of *neo*-inositol in 700 ml of water over 1.86 gm of reduced platinum oxide at 70°C for 30 minutes affords *myo*-inosose-5, isolated as its phenylhydrazone, in 23–47% yield. Only one of the two symmetrical axial hydroxyls is oxidized and the reaction stops at the monoketone stage (Allen, 1956, 1962). Catalytic hydrogenation of the ketone reforms *neo*-inositol.

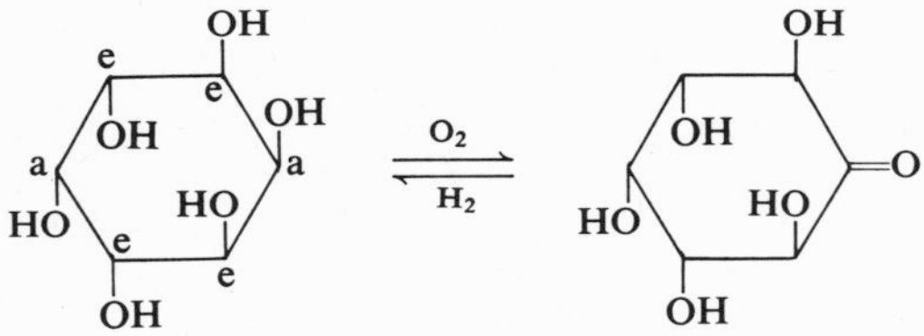

$\xrightleftharpoons[H_2]{O_2}$

Where the two axial hydroxyls are unlike, as in pinitol and quebrachitol, one of these is oxidized to the exclusion of the other (Anderson *et al.*, 1957; Post and Anderson, 1962).

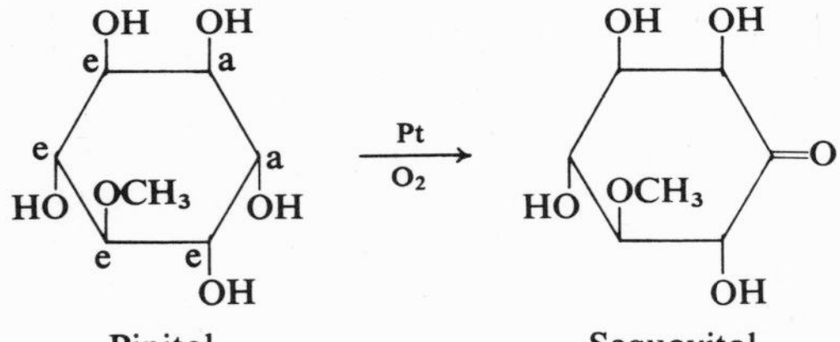

$\xrightarrow[O_2]{Pt}$

Pinitol Sequoyitol

Oxidations are also selective when three axial hydroxyls are present. The fully symmetrical *cis*-inositol is oxidized only to the symmetrical *cis*-inosose. *Muco*-inositol and *allo*-inositol afford *muco*-inosose-1 and *allo*-inosose-1, respectively (Angyal and Anderson, 1959). The conclusion was reached that the axial hydroxyl is oxidized which has an axial and an equatorial hydroxyl in its vicinity.

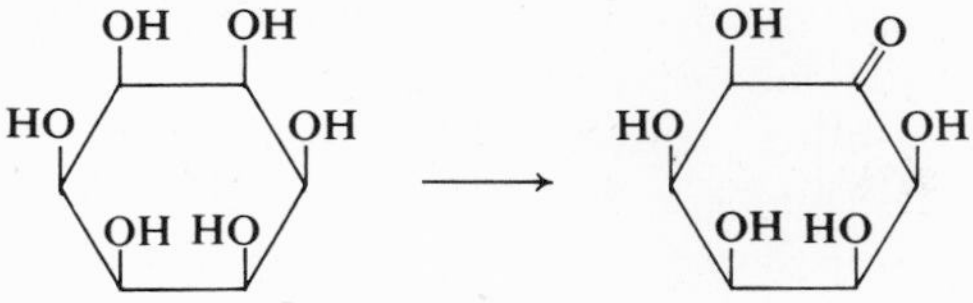

cis-Inositol → *cis*-Inosose

Aminocyclitols are also preferentially oxidized at the axial hydroxyl group. The sensitive amino group must be protected; Heyns and Paulsen (1956) favor the carbobenzoxy group, but the acetyl group has also been used. Selective oxidation of di-*N*-acetyl-*myo*-inosodiamine-4,6 in water over platinum oxide at 40°C under 10 psig oxygen pressure affords di-*N*-acetyl-2-oxo-*myo*-inosodiamine-4,6 (Suami and Ogawa, 1970).

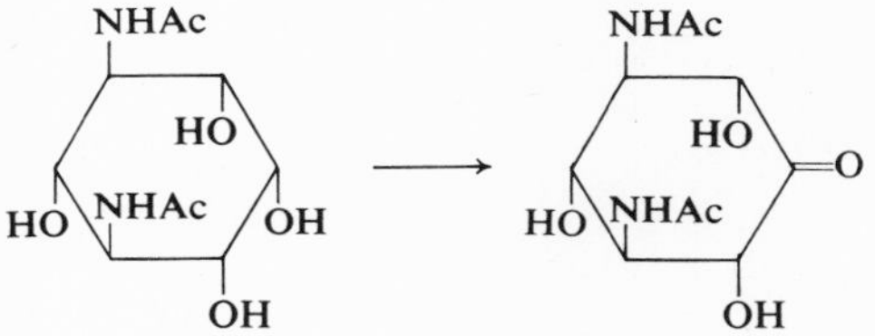

Steroids

Hydroxyl functions in steroids are oxidized selectively by oxygen and noble metal catalysts but the rules governing the selectivity are not nearly so clear-cut as with carbohydrates. Probably the steric requirements imposed by the steroid overshadow distinctions between primary and secondary hydroxyls and between axial and equatorial configurations. Both cholestan-3-α-ol and cholestan-3-β-ol were oxidized with equal ease to 3-cholestanone, but all attempts at oxidation of cholesterol met with failure (Sneeden and Turner, 1955a, b). Oxidation of dihydroouabagenin resulted in preferential oxidation of the secondary 3-hydroxyl instead of the primary hydroxyl. A solution of 2.4 gm of dihydroouabagenin (**XVII**) in 100 ml of water was stirred under an oxygen atmosphere for 48 hours with platinum black (prepared from 300 mg of platinum oxide). By the end of this time, one equivalent of oxygen had been adsorbed and 2.0 gm of **XVIII** was isolated as a result of preferential attack at the 3-position (Sneeden and Turner, 1955a; Turner and Meschino, 1958). Similar oxidations of derivatives of strophanthidin over platinum black were reported by others (Sih *et al.*, 1963; Kupchan *et al.*, 1967).

A number of other 3-hydroxy steroids have been oxidized to the corresponding ketones in 50–75% yield over reduced platinum. The yields obtained were superior to those of partial Oppenauer oxidation. In the bile esters methyl

3α,6α-dihydroxycholanate, methyl 3α,12α-dihydroxycholanate, and methyl 3α,7α,12α-trihydroxycholanate attack was specific at C-3 in contrast with preferential attack at other positions when chromic acid or *N*-bromosuccinimide are used (Sneeden and Turner, 1955b). Variable results were obtained in selective oxidation over platinum oxide of the 3α-hydroxy group of 3α,17β-dihydroxy-5β-androst-9(11)-en-12-one. The specificity of the catalyst changed from batch to batch affording differing amounts of the 3,12-17-trione as by-product. Oxidation was carried out by bubbling oxygen through a solution of 585 mg of substrate in 100 ml of acetone and 10 ml of water at room temperature containing platinum prepared by a 2 hour previous hydrogenation of 176 mg of platinum oxide. Further quantities of catalyst were added after 1, 2, and 3 days. The best yield was 80% (Coombs and Roderick, 1967).

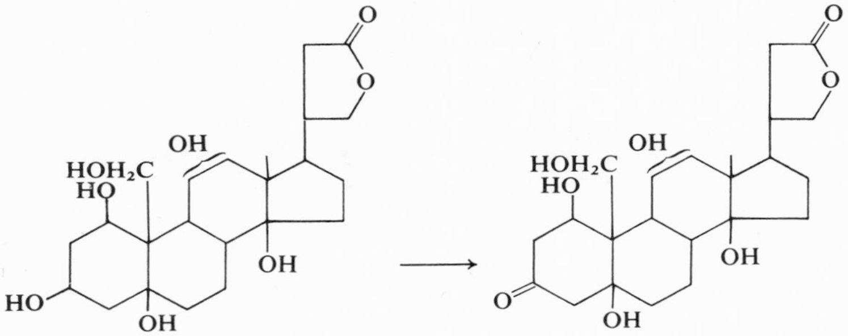

(XVII) → (XVIII)

Simple Alcohols

Simple primary alcohols are converted by oxygen and a noble metal catalyst to either the corresponding aldehyde or acid. Heyns and Blazejewicz (1960) in a systematic investigation delimited the scope of the reaction and came to a number of generalities. Water is an excellent solvent to use for oxidation of water-soluble alcohols, whereas water-insoluble alcohols are best oxidized in a paraffin or benzene. Acetone, methyl ethyl ketone, and dioxane may also be used as solvents. An important factor in successful oxidation is prevention of catalyst agglomeration. With water-insoluble substrates agglomeration may be controlled by the substrate concentration, which should be neither too high nor too low. Recommended concentrations for maximum rate are 2–7% of substrate in the solvent; the rate falls off at both higher and lower concentration.

Aldehydes are formed in neutral solution, but with lower molecular weight compounds, the yields are only fair due to inhibition of the catalyst by acid formed in the reaction. Higher molecular weight alcohols oxidized in heptane solvent afford the corresponding aldehydes in excellent yield; myristyl alcohol and cetyl alcohol are converted to myristaldehyde and palmitaldehyde in 91 and 95% yields, respectively. Ethylene glycol and its aqueous solutions are air-oxidized to glyoxal over palladium–platinum or platinum-on-carbon or -alumina. By-products of the reaction are glycolic acid, formic acid, formaldehyde, and glycolaldehyde. High ethylene glycol concentrations promote glyoxal formation, whereas increasing water concentrations promote acidic products. The reaction rate is enhanced by an increase in pressure (Ioffe *et al.*, 1962).

Benzaldehyde has been obtained in good yield by oxidation of benzyl alcohol over reduced platinum oxide in *n*-heptane at 60°C (Heyns and Blazejewicz, 1960) and in ethyl acetate at room temperature (Sneeden and Turner, 1955b). Formation of benzaldehyde in good yield is surprising inasmuch as benzaldehyde is oxidized much more easily than benzyl alcohol. Probably the alcohol is adsorbed selectively on the catalyst. Oxidation of benzaldehyde over 5% palladium-on-silica proceeds about 10 times more rapidly than oxidation of benzyl alcohol; however, when only 9 mole percent of benzyl alcohol is added to an oxidation of benzaldehyde the rate immediately drops to the lower value obtained with benzyl alcohol (Rylander and Kilroy, 1958). It might be generally good practice in synthesis of an aromatic aldehyde by catalytic oxidation of an alcohol to curtail the reaction while some alcohol still remains.

Hydroxybenzaldehydes can be obtained in excellent yield by interaction of oxygen and hydroxymethylphenols in the presence of a catalyst; little or no carboxylic acid is formed. Preferred conditions for the reaction are an alkaline medium containing borates with pure oxygen gas as oxidant. For example, in a water-cooled 500 ml flask is placed 1.24 gm of 10% palladium-on-carbon, 200 ml of 1 *N* potassium hydroxide solution, 12.4 gm of *o*-hydroxybenzyl alcohol, and 6.3 gm of boric acid. The flask is connected to a gas-measuring buret, filled with oxygen, and stirred at 25°C until oxygen absorption ceases (about 11 hours). Salicyclic aldehyde is obtained in 83.5% yield after acidification of the reaction mixture with sulfuric acid followed by steam distillation. Without added boric acid, the yield is the same, but the reaction requires 45 hours. Use of air instead of oxygen affords the product in 74% yield, and use of platinum-on-carbon instead of palladium gives only a 12% yield. Palladium-catalyzed oxidation of 2-hydroxymethyl-4-methylphenol, 2-hydroxymethyl-6-methylphenol, 2-hydroxymethyl-6-ethoxyphenol, 2-hydroxymethyl-6-chlorophenol, and 2-methoxy-4-hydroxymethylphenol affords the corresponding aldehyde in 84, 96, 85, 83, and 62% yields, respectively (British Patent 987, 947). These examples suggest the yield is better when the hydroxymethyl group is *ortho* to the phenol.

Lutz and Williams (1969) reported an interesting metal-catalyzed oxidation of a hindered 4-hydroxymethylphenol; the product was either the hydroxybenzaldehyde or the diphenoquinone, depending on the conditions of the

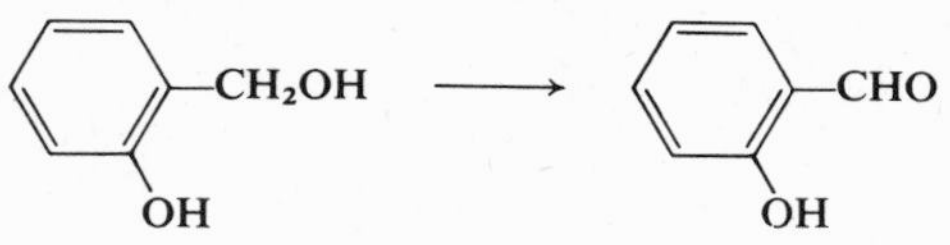

CH₂OH → CHO

reaction. In aqueous solution buffered at pH 8, 3,5-di-*t*-butyl-4-hydroxybenzyl alcohol is converted by platinum-catalyzed air oxidation to 3,5-di-*t*-butyl-4-hydroxybenzaldehyde, isolated in 46% yield, whereas in strong aqueous base, where phenoxide ions are present, nearly quantitative yields of 3,3′,5,5′-tetra-*t*-butyldiphenoquinone are obtained. Without platinum present, the major product (38%) is 2,6-di-*t*-butyl-1,4-benzoquinone. Either 10% platinum-on-carbon or platinum oxide (prereduced by hydrogen at 25°C, 225 psig) can be used as catalysts.

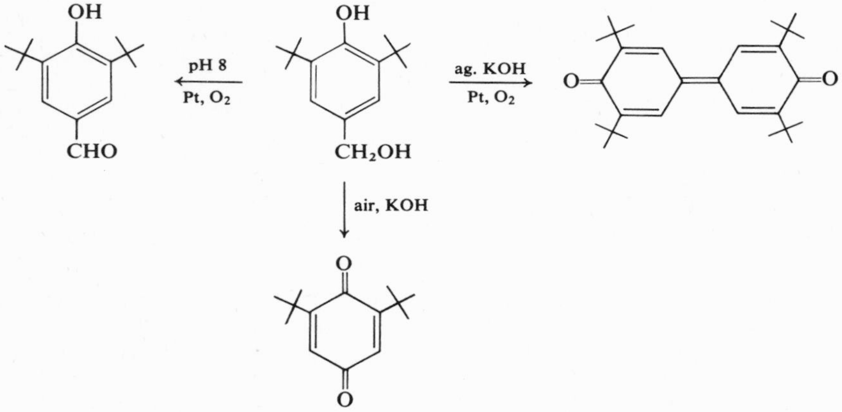

The solvent may have an effect on the extent of oxidation. The major product from the oxidation of 5-hydroxymethyluracil over prereduced platinum oxide in 50% aqueous acetic acid is 5-formyluracil, whereas in water 5-carboxyuracil is formed (Cline *et al.*, 1959).

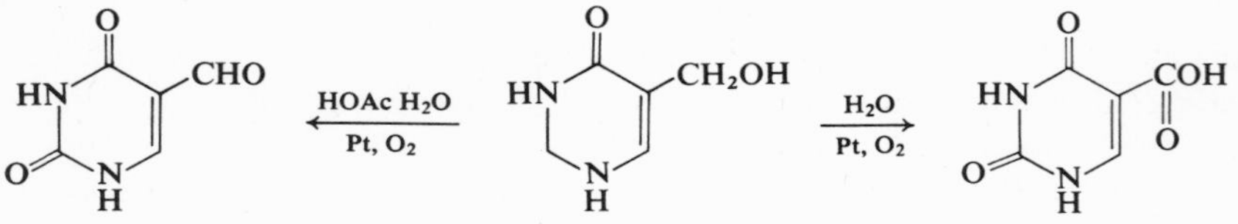

Secondary alcohols are converted smoothly to the corresponding ketones. Yields are usually very high. The rates of oxidation depend on the structure and decrease with increased branching and increased molecular weight. Heyns (1970) has presented a number of generalities governing oxidations in the bicyclic [3.3.0], [3.2.1], and [2.2.1] ring systems. *Endo* hydroxyls are much more susceptible to dehydrogenation than are *exo* hydroxyls. In norbornanols, the order of decreasing reactivity is *endo* OH > *exo* OH > 7-OH. An *endo* hydroxyl in the presence of a vicinal *exo* hydroxyl is dehydrogenated at a slower rate than other *endo* hydroxyls; *exo* hydroxyls are dehydrogenated much faster in molecules containing a 7-*syn* hydroxyl than in those with a 7-*anti* hydroxyl. Intramolecular hydrogen bonds considerably enhance the rate of catalytic oxidative dehydrogenation.

Rhodium-on-carbon prereduced by hydrogen is said to be a particularly useful catalyst for oxidation of hydroquinone to quinone. The process was developed to regenerate quinone used with palladium salts in synthesis of vinyl esters. Oxidation is carried out in acetic acid solution at 50°C under 5 atm of oxygen. Benzoquinone is regenerated quantitatively (Achard and Perras, 1968).

The double bonds of unsaturated alcohols, in general, remain unchanged during catalytic oxidation of the hydroxyl function and *cis–trans* isomerization does not occur. Oxidation of 2-methyl-2-buten-1-ol over platinum oxide in heptane at 60°C affords the unsaturated aldehyde in 77% yield and similarly geraniol is converted to citral in 63% yield (Heyns and Blazejewicz, 1960). A synthesis of maltol, 2-methyl-3-hydroxy-4-pyrone, involves air oxidation of suitable precursors at pH 9 in the presence of a platinum catalyst (Schleppnik and Oftedahl, 1969, 1970).

O_2, Pt →　　← O_2, Pt, 50°C

A novel synthesis of cinnamolide (**XX**) involves the oxidation of a hydroxyl group and an allyl methyl group in bicyclofarnesol (**XIX**) with palladium chloride (Yanagawa *et al.*, 1970). This reaction was stoichiometric with respect to palladium chloride, but other oxidations in this series used catalytic palladium with cupric chloride as oxidant.

(XIX) 200 mg → 100 mg $PdCl_2$, 1 ml isopropyl ether, 5 ml H_2O, reflux, 48 hours → **(XX)**

Oxidative Dehydrogenation

Oxidative dehydrogenation combines dehydrogenation of a substrate with oxidation of liberated hydrogen. The two reactions permit dehydrogenation to occur at lower temperatures than are required if the liberated hydrogen is not removed from the equilibrium. Additionally, oxidation of hydrogen provides heat for the endothermic dehydrogenation. Oxygen, sulfur, sulfur dioxide, chlorine, bromine, and iodine are among suitable oxidants (Pasternak and Vadekar, 1970). Noble metals, apart from their use with alcohols, have not figured prominently in this reaction, perhaps due to their tendency to promote excessive oxidation (Farkas, 1970); palladium is one of the most active catalysts for oxidation of hydrocarbons (Moro-oka *et al.*, 1969). Overoxidation has been inhibited by incorporating sulfuric or phosphoric acid in the catalyst (British Patent, 1,098,697).

A thorough study of oxidative dehydrogenation of tetralin and decalins over 1% platinum-on-alumina in the range 350° to 650°C and 10–30 atm pressure established suitable operating conditions for the reaction. The catalyst shows an instability that does not lessen with use. Deactivated catalysts can be temporarily reactivated by treatment with hydrogen, suggesting that the catalyst might be stabilized by addition of hydrogen to the feed or perhaps by operating the reactor at high conversions, which would increase the partial pressure of hydrogen over the catalyst during operation (Uchida *et al.*, 1971). Earlier workers examined oxidative dehydrogenation of cyclohexane to benzene over 0.5% platinum-on-alumina (Jouy and Balaceanu, 1961).

A unique method for preparation of α,β-unsaturated carbonyl compounds involves oxidative dehydrogenation by air or oxygen in the presence of catalytic amounts of Pd(II) and a co-catalyst. Of 45 catalysts tested for this reaction, palladium compounds were the most active and selective, with rhodium, osmium, iridium, and platinum showing decreasing catalytic activity. The best complexes are soluble ones such as dichlorobis(triphenylphosphine)palladium-(II) and palladium(II) acetylacetonate. Palladium chloride has modest activity but the Lewis acid character of this salt promotes aldol condensation. Either copper(II) or quinone are effective co-catalysts. The reaction may be carried out neat or with acetic acid or benzoic acid solvents. In unsymmetrical compounds, most dehydrogenation occurs at the least-substituted carbons (Theissen, 1970, 1971).

Some catalytic oxidative dehydrogenations occur so readily they pose a problem in isolation of the product. On catalytic hydrogenation of 1,6-dimesitoyl-1-cyclohexene over platinum oxide 1 mole of hydrogen is absorbed, but when the reaction mixture is exposed to air with the catalyst still present, an immediate quantitative oxidation occurs reforming the starting material

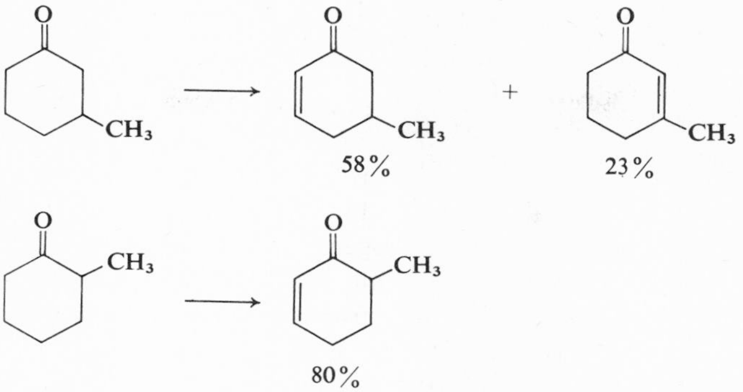

O, CH_3 → O, CH_3 (58%) + O, CH_3 (23%)

O, CH_3 → O, CH_3 (80%)

(Fuson *et al.*, 1960). A similar example was reported earlier by Fuson and Foster (1943).

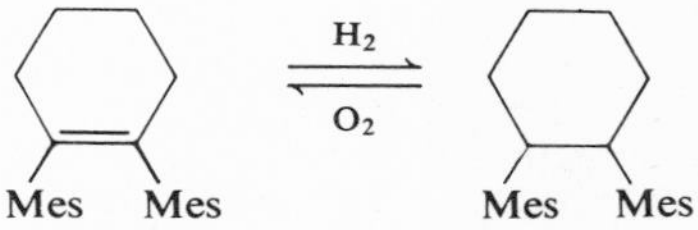

Mes Mes ⇌ (H_2 / O_2) Mes Mes

Oxidative coupling of phenylacylaniline presumably arises through an oxidative dehydrogenation. Phenylacylaniline, when shaken with ethanol and 5% palladium-on-carbon at 20°C for 16 hours, is converted to an amino ketone, **XXI**, in 90% yield. The product is assumed to arise by a Michael-type reaction between phenylacylaniline and intermediate phenylglyoxal anil. Without catalyst only 2% of **XXI** was formed after agitation in oxygen for 6 days (Fraser *et al.*, 1963).

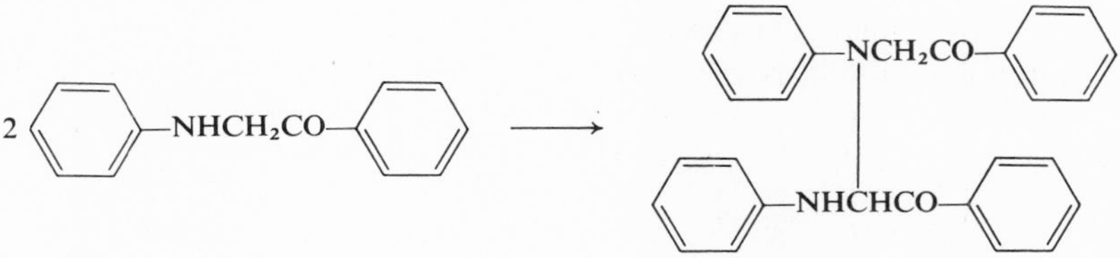

2 C_6H_5–$NHCH_2CO$–C_6H_5 → C_6H_5–NCH_2CO–C_6H_5 / C_6H_5–$NHCHCO$–C_6H_5

(XXI)

Oxidative dehydrogenation of propyl amine over palladium-on-alumina at 90°–175°C affords a mixture of acrylonitrile and propionitrile (McClain and Mador, 1968).

N-DEALKYLATION

Hess and Boekelheide (1969) described an oxidative procedure that might have general utility in the demethylation of tertiary amines. Demethylation of **XXII** in boiling ethanol containing 30% palladium-on-carbon is smoothly accomplished on bubbling oxygen through the mixture from a fritted glass inlet tube. The product, **XXIII**, is obtained in 75% yield after chromatography over silica gel. The authors suggested the reaction proceeds through an intermediate methylene immonium ion.

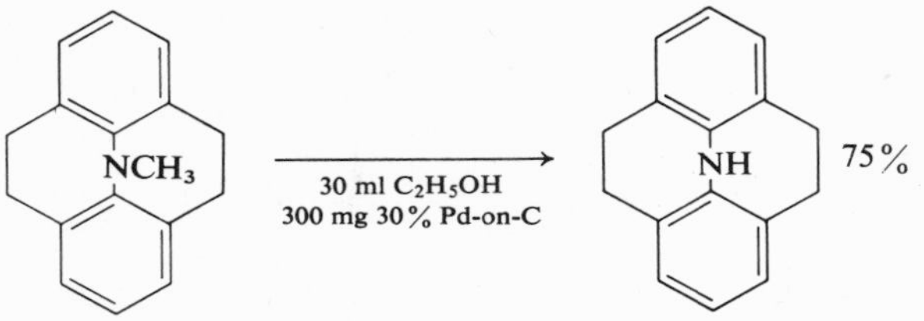

NCH_3 → (30 ml C_2H_5OH, 300 mg 30% Pd-on-C) NH 75%

(XXII) 300 mg **(XXIII)**

A similar demethylation was successfully applied to clindamycin, using prereduced platinum oxide catalyst. In a typical experiment, 50 gm of clindamycin hydrochloride is dissolved in 800 ml of water and stirred vigorously at 25°C with 2 to 3 times its weight of prereduced platinum oxide while air or oxygen is bubbled into the reaction mixture for several days. Recrystallization of the product from ethanol affords 1′-demethylclindamycin hydrochloride in 52% yield. *N*-Dealkylation is not restricted to methyl groups in this series since the *N*-ethyl and *N*-butyl groups are similarly converted to 1′-demethylclindamycin in 30% and 50% yields, respectively (Birkenmeyer and Dolak, 1970). No reaction occurs in this series when the substrates are subjected to oxidation in nonpolar solvents such as benzene, conditions suggested earlier for the conversion of *N*-methyl tertiary amines to *N*-formyl secondary amines (Davis and Rosenblatt, 1968). No reaction occurs if the platinum oxide is not first prereduced (Birkenmeyer, 1971).

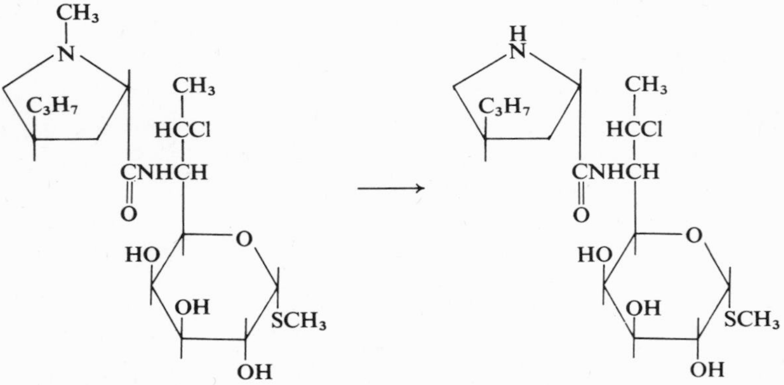

In nonpolar solvent *N*-methyl groups in tertiary amines have been converted to *N*-formyl groups by oxidation in the presence of platinum black. *N*-Methylpiperidine is converted quantitatively to *N*-formylpiperidine by shaking the amine with platinum black for 20 hours under an oxygen atmosphere at room temperature. Trimethylamine gives dimethylformamide in 74% yield after 48 hours. The oxidation is selective inasmuch as *N*-benzyl or *N*-ethyl groups in tertiary amines are not similarly attacked (Davis and Rosenblatt, 1968).

$$\text{C}_5\text{H}_{10}\text{N—CH}_3 \longrightarrow \text{C}_5\text{H}_{10}\text{NCHO}$$

Nitrilotriacetic acid is oxidized by air in the presence of 5% palladium-on-carbon at pH 8.5 and 90°C to a mixture of imidodiacetic acid and oxalic acid; the latter undergoes further oxidation to carbon dioxide and water (Tetenbaum and Stone, 1970).

$$N(CH_2COOH)_3 \rightarrow HN(CH_2COOH)_2 + (CO_2H)_2$$

REFERENCES

Achard, R., and Perras, P. (1968). U.S. Patent 3,383,406.
Adams, R., and Shriner, R. L. (1923). *J. Amer. Chem. Soc.* **45**, 2171.
Adderley, C. J. R., Nebzydoski, J. W., and Battiste, M. A. (1971). *Tetrahedron Lett.* p, 3545.
Aguilo, A. (1967). *Advan. Organometal. Chem.* **5**, 321.
Allen, G. R., Jr. (1956). *J. Amer. Chem. Soc.* **78**, 5691.
Allen, G. R., Jr. (1962). *J. Amer. Chem. Soc.* **84**, 3128.
Anderson, L., DeLuca, E. S., Bieder, A., and Post, G. G. (1957). *J. Amer. Chem. Soc.* **79**, 1171.
Angyal, S. J., and Anderson, L. (1959). *Advan. Carbohyd. Chem.* **14**, 135.
Arpe, von H.-J., and Hornig, L. (1970). *Erdoel Kohle, Erdgas, Petrochem.* **23**, 79.
Arzoumanidis, G. G., Rauch, F. C., and Blank, G. (1972). *Abstr. 163rd Meet., Amer. Chem. Soc. Inorganic Sec.*, paper 73.
Asano, R., Moritani, I., Fujiwara, Y., and Teranishi, S. (1970). *Chem. Commun.* p. 2932.
Asano, R., Moritani, I., Sonoda, A., Fujiwara, Y., and Teranishi, S. (1971). *J. Chem. Soc., C* p. 3691.
Baird, W. C., Jr. (1966). *J. Org. Chem.* **31**, 2411.
Baker, R., Halliday, D. E., and Mason, T. J. (1970). *Tetrahedron Lett.* p. 591.
Baldwin, J. E., and Swallow, J. C. (1969). *Angew. Chem., Int. Ed. Engl.* **8**, p. 601.
Barker, S. A., Bourne, E. J., Fleetwood, J. G., and Stacey, M. (1958). *J. Chem. Soc., London* p. 4128.
Battiste, M. A., and Nebzydoski, J. W. (1969). *J. Amer. Chem. Soc.* **91**, 6887.
Belov, A. P., and Moiseev, I. I. (1966). *Izv. Adad. Nauk SSSR Ser. Khim.* p. 114.
Belov, A. P., Pek, Yu. G., and Moiseev, I. I. (1965). *Izv. Adad. Nauk SSSR Ser. Khim.* p. 2170.

Bingham, A. J., Dyall, L. K., Norman, R. O. C., and Thomas, C. B. (1970). *J. Chem. Soc., C* p. 1879.
Birch, A. J., and Subba Rao, G. S. R. (1968). *Tetrahedron Lett.* p. 2917.
Birkenmeyer, R. D. (1971). Personal communication.
Birkenmeyer, R. D., and Dolak, L. A. (1970). *Tetrahedron Lett.* p. 5049.
Blum, J., Rosenman, H., and Bergmann, E. D. (1967). *Tetrahedron Lett.* p. 3665.
Brimacombe, J. S., Cook, M. C., and Tucker, L. C. N. (1965). *J. Chem. Soc., London* p. 2292.
Brown, R. G., Davidson, J. M., and Triggs, C. (1969). *Amer. Chem. Soc., Div. Petrol. Chem., Prepr.* **14**, B23.
Bryant, D. R., and McKeon, J. E. (1969). *Amer. Chem. Soc., Div. Petrol. Chem., Prepr.* **14**, No. 2, B1A.
Bryant, D. R., McKeon, J. E., and Ream, B. C. (1968a). *J. Org. Chem.* **33**, 4123.
Bryant, D. R., McKeon, J. E., and Ream, B. C. (1968b). *Tetrahedron Lett.* No. 30, p. 3371.
Bryant, D. R., McKeon, J. E., and Ream, B. C. (1969). *J. Org. Chem.* **34**, 1106.
Bushweller, C. H. (1968). *Tetrahedron Lett.* p. 6123.
Byrd, J. E., Cassar, L., Eaton, P. E., and Halpren, J. (1971). *Chem. Commun.* p. 40.
Charman, H. B. (1966). *Nature* (*London*) **212**, 278.
Clark, D., Hayden, P., Walsh, W. D., and Jones, E. W. (1964). British Patent 964,001.
Clark, D., Hayden, P., and Smith, R. D. (1968). *Discuss. Faraday Soc.* **46**, 98.
Clark, D., Hayden, P., and Smith, R. D. (1969). *Amer. Chem. Soc., Div. Petrol. Chem., Prepr.* **14**, No. 2, B10.
Clark, D., Hayden, P., and Charlton, J. (1970). U.S. Patent 3,515,758.
Clement, W. H., and Selwitz, C. M. (1962). *Tetrahedron Lett.* No. 23, p. 1081.
Clement, W. H., and Selwitz, C. M. (1964). *J. Org. Chem.* **29**, 241.
Clement, W. H., and Selwitz, C. M. (1968). U.S. Patent 3,370,073.
Cline, R. E., Fink, R. M., and Fink, K. (1959). *J. Amer. Chem. Soc.* **81**, 2521.
Collman, J. P., Kubota, M., and Hosking, J. W. (1967). *J. Amer. Chem. Soc.* **89**, 4809.
Coombs, M. M., and Roderick, H. R. (1967). *J. Chem. Soc., C* p. 1819.
Danno, S., Moritani, I., and Fujiwara, Y. (1969). *Tetrahedron* **25**, 4809.
Danno, S., Moritani, I., and Fujiwara, Y. (1970a). *Chem. Commun.* p. 610.
Danno, S., Moritani, I., Fujiwara, Y., and Teranishi, S. (1970b). *Bull. Chem. Soc. Jap.* **43**, 3966.
Davidson, J. M., and Triggs, C. (1966). *Chem. Ind.* (*London*) p. 457.
Davidson, J. M., and Triggs, C. (1967). *Chem. Ind.* (*London*) p. 1361.
Davidson, J. M., and Triggs, C. (1968a). *J. Chem. Soc., A* p. 1324.
Davidson, J. M., and Triggs, C. (1968b). *J. Chem. Soc., A* p. 1331.
Davis, G. T., and Rosenblatt, D. H. (1968). *Tetrahedron Lett.* No. 38, p. 4085.
Eberson, L., and Gomez-Gonzales, L., J.C.S. (1971). *Chem. Commun.* p. 263.
Farkas, A. (1970). *Hydrocarbon Process.* No. 7, 121.
Fenton, D. M. (1969). U.S. Patent 3,422,147.
Fitton, P., McKeon, J. E., and Ream, B. C. (1969). *Chem. Commun.* p. 370.
François, P. (1969). *Ann. Chim.* (*Paris*) **4**, 371.
Fraser, E., Paterson, W., and Proctor, G. R. (1963). *J. Chem. Soc., London* p. 5107.
Frevel, L. K., Kressley, L. J., and Strojny, E. J. (1970). U.S. Patent 3,494,877.
Fujiwara, Y. (1970). *Bull. Chem. Soc. Jap.* **43**, 863.
Fujiwara, Y., Moritani, I., and Matsuda, M. (1968a). *Tetrahedron* **24**, 4819.
Fujiwara, Y., Moritani, I., Matsuda, M., and Teranishi, S. (1968b). *Tetrahedron Lett.* No. 5, p. 633.
Fujiwara, Y., Moritani, I., Matsuda, M., and Teranishi, S. (1968c). *Tetrahedron Lett.* p. 3863.
Fujiwara, Y., Moritani, I., Danno, S., Asano, R., and Teranishi, S. (1969a). *J. Amer. Chem. Soc.* **91**, 7166.

Fujiwara, Y., Moritani, I., Asano, R., Tanaka, H., and Teranishi, S. (1969b). *Tetrahedron* **25**, 4815.
Fujiwara, Y., Moritani, I., Ikegami, K., Tanaka, R., and Teranishi, S. (1970). *Bull. Chem. Soc. Jap.* **43**, 863.
Fuson, R. C., and Foster, R. E. (1943). *J. Amer. Chem. Soc.* **65**, 913.
Fuson, R. C., Hatchard, W. R., Kottke, R. H., and Fedrick, J. L. (1960). *J. Amer. Chem. Soc.* **82**, 4330.
Glattfeld, J. W. E., and Gershon, R. B. (1938). *J. Amer. Chem. Soc.* **60**, 2013.
Glover, E. E., and Morris, G. H. (1964). *J. Chem. Soc.*, London p. 3366.
Gourlay, G. (1969). U.S. Patent 3,471,567.
Hafner, W., Jira, R., Sedlmeier, J., and Smidt, J. (1962). *Chem. Ber.* **95**, 1575.
Hardegger, E., and Lohse, F. (1957). *Helv. Chim. Acta* **40**, 2383.
Hargis, C. W., and Young, H. S. (1968). U.S. Patent 3,379,651.
Heck, R. F. (1968a). *J. Amer. Chem. Soc.* **90**, 5518.
Heck, R. F. (1968b). *J. Amer. Chem. Soc.* **90**, 5526.
Heck, R. F. (1968c). *J. Amer. Chem. Soc.* **90**, 5531.
Heck, R. F. (1968d). *J. Amer. Chem. Soc.* **90**, 5535.
Heck, R. F. (1968e). *J. Amer. Chem. Soc.* **90**, 5538.
Heck, R. F. (1968f). *J. Amer. Chem. Soc.* **90**, 5542.
Henry, P. M. (1964). *J. Amer. Chem. Soc.* **86**, 3246.
Henry, P. M. (1967). *J. Org. Chem.* **32**, 2575.
Henry, P. M. (1968). *Advan. Chem. Ser.* **70**, 126.
Henry, P. M. (1971). *J. Org. Chem.* **36**, 1886.
Hess, B. A., Jr., and Boekelheide, V. (1969). *J. Amer. Chem. Soc.* **91**, 1672.
Heyns, K. (1963). *Angew. Chem., Int. Engl. Ed.* **2**, 402.
Heyns, K. (1970). *Angew. Chem., Int. Engl. Ed.* **9**, 383.
Heyns, K., and Blazejewicz, L. (1960). *Tetrahedron* **9**, 67.
Heyns, K., and Koch, W. (1953). *Chem. Ber.* **86**, 110.
Heyns, K., and Paulsen, H. (1953). *Chem. Ber.* **86**, 833.
Heyns, K., and Paulsen, H. (1955). *Chem. Ber.* **88**, 188.
Heyns, K., and Paulsen, H. (1956). *Chem. Ber.* **89**, 1152.
Heyns, K., and Paulsen, H. (1963). *In* "Newer Methods of Preparative Organic Chemistry" (W. Foerst, ed.), Vol. 2, p. 303. Academic Press, New York.
Horino, H., and Inque, N. (1971). *Bull. Chem. Soc. Jap.* **44**, 3210.
Hüttel, R., and Bechter, M. (1959). *Angew. Chem.* **71**, 456.
Hüttel, R., Kratzer, J., and Bechter, M. (1961). *Chem. Ber.* **94**, 766.
Ioffe, I. I., Klimova, N. V., and Makeev, A. G. (1962). *Kinet. Katal.* **3**, 107; *Chem. Abstr.* **57**, 15865 (1962).
Itatani, H., and Yoshimoto, H. (1971). *Chem.* (*London*) p. 674.
James, B. R., and Ochiai, E. (1971). *Can. J. Chem.* **49**, 975.
Jouy, M., and Balaceanu, J. C. (1961). *Actes Congr. Int. Catal. 2nd, 1960* Vol. I, p. 645.
Kasahara, A., Izumi, T., Saito, G., Yodona, M., Saito, R., and Goto, Y. (1972a). *Bull. Chem. Soc. Jap.* **45**, 894.
Kasahara, A., Izumi, T., and Ohnishi, S. (1972b). *Bull. Chem. Soc. Jap.* **45**, 951.
Kawamoto, K., Imanaka, T., and Teranishi, S. (1970). *Bull. Chem. Soc. Jap.* **43**, 2512.
Ketley, A. D. (1970). U.S. Patent 3,517,054.
Ketley, A. D., and Braatz, J. A. (1968). *Chem. Commun.* p. 169.
Ketley, A. D., and Fisher, L. P. (1968). *J. Organometal. Chem.* **13**, 243.
Kiss, J., and Wyss, P. C. (1972). *Tetrahedron Lett.* p. 3055.
Kitching, W., Rappoport, Z., Winstein, S., and Young, W. G. (1966). *J. Amer. Chem. Soc.* 88, 2055.

Kohll, C. F., and van Helden, R. (1967). *Rec. Trav. Chim. Pays-Bas* **86**, 193.
Kupchan, S. M., Mokotoff, M., Sandhu, R. S., and Hokin, L. E. (1967). *J. Med. Chem.* **10**, 1025.
Kurkov, V. P., Pasky, J. Z., and Lavigne, J. B. (1968). *J. Amer. Chem. Soc.* **90**, 4743.
Levanda, O. G., and Moiseev, I. I. (1968). *Zh. Org, Khim.* **4**, 1533.
Lloyd, W. G. (1967). *J. Org. Chem.* **32**, 2816.
Lloyd, W. G., and Luberoff, B. J. (1969). *J. Org. Chem.* **34**, 3949.
Lutz, E. F., and Williams, P. H. (1969). *J. Org. Chem.* **34**, 3656.
Lyons, J. E., and Turner, J. O. (1972). *Tetrahedron Lett.* p. 2903.
McClain, D. M., and Mador, I. L. (1968). U.S. Patent 3,396,190.
McKeon, J. E., and Fitton, P. (1972). *Tetrahedon* **28**, 233.
McKeon, J. E., Fitton, P., and Griswold, A. A. (1972). *Tetrahedron* **28**, 227.
MacLean, A. F., and Stautzenberger, A. L. (1968). U.S. Patent 3,384,669.
Macrae, T. F. (1933). *Biochem. J.* **27**, 1248.
Maitlis, P. M. (1971). "The Organic Chemistry of Palladium," Vol. 1. Academic Press, New York.
Marsh, C. A. (1952). *J. Chem. Soc., London* p. 1578.
Marsh, C. A., and Lewy, G. A. (1958). *Biochem. J.* **68**, 617.
Mehltretter, C. L., Rist, C. E., and Alexander, B. H. (1949). U.S. Patent 2,472,168.
Mehltretter, C. L., Alexander, B. H., Mellies, R. L., and Rist, C. E. (1951). *J. Amer. Chem. Soc.* **73**, 2424.
Moiseev, I. I. (1969). *Amer. Chem. Soc., Div. Petrol. Chem., Prepr.* **14**, B49.
Moiseev, I. I., and Vargaftik, M. N. (1965). *Izv. Akad. Nauk SSSR Ser. Khim.* p. 744.
Moiseev, I. I., Vargaftik, M. N., and Syrkin, Ya. K. (1960). *Dokl. Akad. Nauk SSSR* **133**, 801.
Moritani, I., Fujiwara, Y., and Teranishi, S. (1969). *Amer. Chem. Soc., Div. Petrol. Chem., Prepr.* **14**, B172.
Moritani, I., Fujiwara, Y., and Danno, S. (1971a). *J. Organometal. Chem.* **27**, 279.
Moritani, I., Danno, S., Fujiwara, Y., and Teranishi, S. (1971b). *Bull. Chem. Soc. Jap.* **44**, 578.
Moro-oka, Y., Kitamura, T., and Ozaki, A. (1969). *J. Catal.* **13**, 53.
Moss, G. P., Reese, C. B., Schofield, K., Shapiro, R., and Todd, A. R. (1963). *J. Chem. Soc., London* p. 1149.
Muller, E., and Schwabe, K. (1928). *Z. Elektrochem.* **34**, 170.
Muxfeldt, H., and Rogalski, W. (1965). *J. Amer. Chem. Soc.* **87**, 933.
Muxfeldt, H., Buhr, G., and Bangert, R. (1962). *Angew. Chem., Int., Ed. Engl.* **1**, 157.
Muxfeldt, H., Hardtmann, G., Kathawala, F., Vedejs, E., and Mooberry, J. B. (1968). *J. Amer. Chem. Soc.* **90**, 6534.
Nakamura, S., and Yasui, T. (1970). *J. Catal.* **17**, 366.
Nakamura, S., and Yasui, T. (1971). *J. Catal.* **23**, 315.
Nikiforova, A. V., Moiseev, I. I., and Syrkin, Ya. K. (1963). *Zh. Obshch. Khim.* **33**, 3239.
Okada, H., Noma, T., Katsuyama, Y., and Hashimoto, H. (1968). *Bull. Chem. Soc. Jap.* **41**, 1395.
Ouellette, R. J., and Levin, C. (1968). *J. Amer. Chem. Soc.* **90**, 6889.
Ouellette, R. J., and Levin, C. (1971). *J. Amer. Chem. Soc.* **93**, 471.
Pasternak, I. S., and Vadekar, M. (1970). *Can. J. Chem.* **48**, 212.
Phillips, F. C. (1894). *Amer. Chem. J.* **16**, 255.
Platz, R., and Fuchs, W. (1967). U.S. Patent 3,333,004.
Platz, R., and Nohe, H. (1966). German Patent 1,231,230.
Post, G. G., and Anderson, L. (1962). *J. Amer. Chem. Soc.* **84**, 471.
Reiners, R. A. (1958). U.S. Patent 2,845,439.

Rottenberg, M., and Baertschi, P. (1956). *Helv. Chim. Acta* **39**, 1973.
Rottenberg, M., and Thurkauf, M. (1959). *Helv. Chim. Acta* **42**, 226.
Rylander, P. N., and Kilroy, M. (1958). Unpublished observations, Engelhard Minerals and Chemicals Corp., Engelhard Ind. Div., Menlo Park, New Jersey.
Rylander, P. N., and Kilroy, M. (1962). Unpublished observations from Engelhard Minerals and Chemicals Corp., Engelhard Ind. Div., Menlo Park, New Jersey.
Sakakibara, T., Nishimura, S., and Odaira, Y. (1969). *Tetrahedron Lett.* p. 1019.
Schleppnik, A. A., and Oftedahl, M. L. (1969). U.S. Patent 3,455,960.
Schleppnik, A. A., and Oftedahl, M. L. (1970). U.S. Patent 3,494,959.
Schuller, W. H. (1971). *Ind. Eng. Chem., Prod. Res. Develop.* **10**, 441.
Schultz, R. G., and Gross, D E. (1968). *Advan. Chem. Ser.* **70**, 97.
Schultz, R. G., and Rony, P. R. (1970). *J. Catal.* **16**, 133.
Schwerdtel, W. (1968). *Hydrocarbon Process.* **47**, 187.
Selwitz, C. M. (1970). U.S. Patent 3,542,852.
Shue, R. S. (1971). *Chem. Commun.* p. 1510.
Sih, C. J., Kupchan, S. M., Katsui, N., and El Taijeb, O. (1963). *J. Org. Chem.* **28**, 854.
Smidt, J., and Krekeler, H. (1963). *Erdoel Kohle, Erdgas, Petrochem.* **16**, 560.
Smidt, J., Hafner, W., Jira, R., Sedlmeier, J., Sieber, R., Ruttinger, R., and Kojer, H. (1959a). *Angew. Chem.* **71**, 176.
Smidt, J., Sieber, R., Hafner, W., and Jira, R. (1959b). German Patent 1,059,453.
Smidt, J., Hafner, W., Sedlmeier, J., Jira, R., and Ruttinger, R. (1964). U.S. Patent 3,131,223.
Sneeden, R. P. A., and Turner, R. B. (1955a). *J. Amer. Chem. Soc.* **77**, 130.
Sneeden, R. P. A., and Turner, R. B. (1955b). *J. Amer. Chem. Soc.* **77**, 190.
Stangl, H., and Jira, R. (1970). *Tetrahedron Lett.* p. 3589.
Stern, E. W. (1963). *Proc. Chem. Soc., London* p. 111.
Stern, E. W. (1967). *Catal. Rev.* **1**, 73.
Stern, E. W. (1970). *Chem. Commun.* p. 736.
Stern, E. W., and Spector, M. L. (1961). *Proc. Chem. Soc., London* p. 370.
Suami, T., and Ogawa, S. (1970). U.S. Patent 3,496,196.
Suga, K., Watanabe, S., and Hijikata, K. (1971). *Chem. Ind. (London)* p. 33.
Szonyi, G. (1968). *Advan. Chem. Ser.* **70**, 53.
Takao, K., Fujiwara, Y., Imanaka, T., and Teranishi, S. (1970a). *Bull. Chem. Soc. Jap.* **43**, 1153.
Takao, K., Fujiwara, Y., Imanaka, T., Yamamoto, M., Hirota, K., and Teranishi, S. (1970b). *Bull. Chem. Soc. Jap.* **43**, 2249.
Takao, K., Wayaku, M., Fujiwara, Y., Imanaka, T., and Teranishi, S. (1970c). *Bull. Chem. Soc. Jap.* **43**, 3898.
Tamura, M., and Yasui, T. (1968a). *Kogyo Kagaku Zasshi* **71**, 1859.
Tamura, M., and Yasui, T. (1968b). *Chem. Commun.* p. 1209.
Tamura, M., and Yasui, T. (1969a). *Kogyo Kagaku Zasshi* **72**, 557.
Tamura, M., and Yasui, T. (1969b). *Kogyo Kagaku Zasshi* **72** 578 and 581.
Teramoto, K., Oga, T., Kikuchi, S., and Ito, M. (1963). *Yuki Gosei Kagaku Kyokai Shi* **21**, 298.
Tetenbaum, M. T., and Stone, H. (1970). *Chem. Commun.* p. 1699.
Theissen, R. J. (1970). U.S. Patent 3,523,125.
Theissen, R. J. (1971). *J. Org. Chem.* **36**, 752.
Trenner, N. R. (1947). U.S. Patent 2,428,438.
Tsou, K.-C., and Seligman, A. M. (1952). *J. Amer. Chem. Soc.* **74**, 5605.
Turner, R. B., and Meschino, J. A. (1958). *J. Amer. Chem. Soc.* **80**, 4862.
Uchida, A., Nakazawa, T., Oh-uchi, K., and Matsuda, S. (1971). *Ind. Eng. Chem., Prod. Res. Develop.* **10**, 153.

Unger, M. O., and Fouty, R. A. (1969). *J. Org. Chem.* **34**, 18.
Vanderwerff, W. D. (1970). *Amer. Chem. Soc., Div. Petrol. Chem., Prepr.* **15**, B34.
van Helden, R., and Balder, B. (1964). U.S. Patent 3,145,237.
van Helden, R., and Verberg, G. (1965). *Rec. Trav. Chim. Pays-Bas* **84**, 1263.
van Helden, R., Kohll, C. F., Medema, D., Verberg, G., and Jonkhoff, T. (1968). *Rec. Trav. Chim. Pays-Bas* **87**, 961.
Vargaftik, M. N., Moiseev, I. I., Syrkin, Ya. K., and Yakshin, V. V. (1962). *Izv. Akad. Nauk SSSR* p. 868.
Vaska, L., and DiLuzio, J. (1961). *J. Amer. Chem. Soc.* **83**, 2784.
Vizsolyi, J. P., and Tener, G. M. (1962). *Chem. Ind. (London)* p. 263.
von Schuching, S., and Frye, G. H. (1965). *J. Org. Chem.* **30**, 1288.
Wieland, H. (1912). *Chem. Ber.* **45**, 484 and 2606.
Wieland, H. (1913). *Chem. Ber.* **46**, 3327.
Wieland, H. (1921). *Chem. Ber.* **54**, 2253.
Williams, P. H., and Lutz, E. F. (1968). U.S. Patent 3,407,220.
Witkop, B., Patrick, J. B., and Rosenblum, M. (1951). *J. Amer. Chem. Soc.* **73**, 2641.
Wright, D. (1970). British Patent 1,208,866.
Yanagawa, H., Kato, T., and Kitahara, Y. (1970). *Synthesis* **1**, 257.
Yoshimura, J., Sato, T., and Ando, H. (1969). *Bull. Chem. Soc. Jap.* **42**, 2352.

CHAPTER 4

Osmium and Ruthenium Tetroxides as Oxidation Catalysts

The differences between osmium tetroxide and ruthenium tetroxide in their reactions with organic compounds are much more striking than their similarities. Reactions with osmium tetroxide are limited mainly to olefins and certain aromatics displaying olefinic character. Ruthenium tetroxide, on the other hand, attacks many organic functions with great vigor. The same distinctions apply when these reagents are used as catalysts and for these reasons are best considered separately. Osmium tetroxide has been used much more widely than ruthenium tetroxide despite the greater versatility of the latter. This situation might be expected to change as the relatively new ruthenium reagent becomes of age.

OSMIUM TETROXIDE

Osmium tetroxide is used widely in organic synthesis, especially for hydroxylation reactions. Addition of stoichiometric quantities of osmium tetroxide to a carbon–carbon double bond followed by cleavage affords *vic*-glycols of *cis* configuration, usually in excellent yield (Fieser and Fieser, 1967). A variety of reagents has been used to cleave the osmium complex, including acidic solutions of sodium or potassium chlorate, sodium sulfite in aqueous ethanol, alkaline solutions of mannitol or formaldehyde, and hydrogen sulfide (Gunstone, 1960).

A more economical and convenient way of achieving the same end is to carry out the hydroxylation with an appropriate oxidizing agent and only catalytic amounts of osmium tetroxide. Oxidizing agents that have been used for this purpose include metal chlorates, periodate, oxygen, hydrogen peroxide, and phenyl iodosoacetate. With certain of these oxidants, the major products may be α-ketols, or aldehydes, or ketones resulting from cleavage at the carbon–carbon double bond.

Metal Chlorates as Oxidants

Olefins may be hydroxylated by metal chlorates in aqueous solution containing catalytic amounts of osmium tetroxide. Hydroxylation presumably proceeds with formation of osmate esters (Criegee, 1936; Criegee *et al.*, 1942), which undergo oxidative cleavage to diols with regeneration of osmium tetroxide. Sidgwick (1952) has suggested there may be another explanation since the oxidation potential of a potassium chlorate solution is definitely raised by traces of osmium tetroxide. Waters (1945) suggested the reaction involves free hypochlorous acid as a source of hydroxyl radicals. The mechanism seems as yet unsettled.

Zelikoff and Taylor (1950) examined in detail the osmium tetroxide-catalyzed oxidation of fumaric and maleic acids by aqueous potassium chlorate to *racemic*- and *meso*-tartaric acids. Chlorate is reduced to chloride and osmium tetroxide is regenerated. The reaction overall was kinetically of second order, first order with respect to osmium tetroxide, first order with respect to organic substrate, and zero order in water and chlorate. Osmate ester formation was assumed to be rate-determining, with oxidative hydrolysis relatively fast.

$$\begin{array}{c}\text{HOOCCH}\\ \| \\ \text{HOOCCH}\end{array} + OsO_4 \xrightarrow{\text{Slow}} \begin{array}{c}\text{HOOCCHO}\\ | \\ \text{HOOCCHO}\end{array}\!\!>\!Os\!\begin{array}{c}=O\\ \\ =O\end{array}$$

$$3\begin{array}{c}\text{HOOCHO}\\ | \\ \text{HOOCHO}\end{array}\!\!>\!Os\!\begin{array}{c}=O\\ \\ =O\end{array} + 3\,H_2O + KClO_3 \longrightarrow 3\begin{array}{c}\text{HOOCCHOH}\\ | \\ \text{HOOCCHOH}\end{array} + 3\,OsO_4 + KCl$$

Sodium, potassium (Glattfeld and Woodruff, 1927), barium, and silver chlorates have all been used with success in these oxidations. Barium and silver chlorates are useful in that the metal may be more easily removed from solution and also better yields may be obtained sometimes (Ernest, 1964; Bláha *et al.*, 1960; Braun, 1930a, b, 1932). With silver chlorate formation of chlorohydrins arising by addition of hypochlorous acid to the double bond is minimized

(Zbiral and Rasberger, 1968). For instance, in osmium tetroxide-catalyzed hydroxylation of crotonic acid with potassium or barium chlorate, about 20% of the substrate is chlorinated whereas with silver chlorate very little chlorination occurs, especially when the chlorate is added gradually and at 0°C. Hypochlorite is removed from solution by immediate decomposition of silver hypochlorite (Braun, 1929). Alcoholic silver precipitates in the presence of chlorate should be handled very carefully. Oxidations with chlorate are apt to become violent if too much catalyst is used (Glattfeld and Rietz, 1940).

$$3\ AgOCl \rightarrow AgClO_3 + 2\ AgCl$$

The use of silver chlorate with substrates that add hypochlorous acid only slowly may not be advantageous; quantitative yields of *meso*-tartaric acid are obtained by oxidation of maleic anhydride with barium chlorate, whereas silver chlorate affords only a 70% yield (Braun, 1929).

The most appropriate metal to use is determined also by solubility properties of the substrate. Both barium and silver chlorates give much better yields than potassium chlorate in oxidation of crotonic acid to *threo*-1,2-dihydroxybutyric acid; the *threo* acid forms a compound with its own potassium salt that is but slightly soluble in cold alcohol (Glattfeld and Chittum, 1933).

The results obtained in hydroxylation of unsaturated acids may depend also on the acidity of the medium. Early attempts to oxidize fumaric acid and maleic acid with potassium chlorate catalyzed by osmium tetroxide gave *racemic*- and *meso*-tartaric acids in yields of 60% and 72%, respectively (Hofmann *et al.*, 1914). The procedure was later improved by Milas and Terry (1925) to give yields approaching quantitative. Experiments carried out with the disodium salt of fumaric acid, the acid sodium salt, and the free acid gave yields of *meso*-tartaric acid of 39, 90, and 99%, respectively, showing the advantage of using the free acid. The rate of oxidation is not altered by changes in the concentration of oxidant but the yield is to some extent. Excess oxidant gives better yields; the authors comment that apparently it is important to reoxidize the catalyst rapidly. The overall rate is dependent on pH; oxidation proceeds 3 to 5 times more rapidly with acid ions than with the un-ionized acid molecule (Zelikoff and Taylor, 1950).

The course of hydroxylation is influenced also by steric considerations. Osmium tetroxide-catalyzed sodium chlorate oxidation of 1-methylene-3,4-diphenyl-2-cyclohexene proceeds with preferential attack on the least-hindered

CH_2 ... Ph, Ph, H — 1.6 moles

$\xrightarrow[\text{18 mg } OsO_4,\ \text{25 ml dioxane},\ 75°C]{\text{4.8 moles } NaClO_3}$

CH_2OH, OH ... Ph, Ph, H — 80%

bond and affords a mixture of stereoisomers in 80% yield (Zimmerman and Samuelson, 1969).

Kaufmann and Jansen (1959) prepared 9,10,11,12-tetrahydroxystearic acid by osmium tetroxide-catalyzed sodium chlorate oxidation of the corresponding diolefin, but the 9,10,11,12,13,14-hexahydroxystearic acid could not be prepared similarly from the triolefin. The hexahydroxy compound was prepared satisfactorily by the pyridine-catalyzed interaction of the triolefin and stoichiometric quantities of osmium tetroxide followed by cleavage with sodium sulfite.

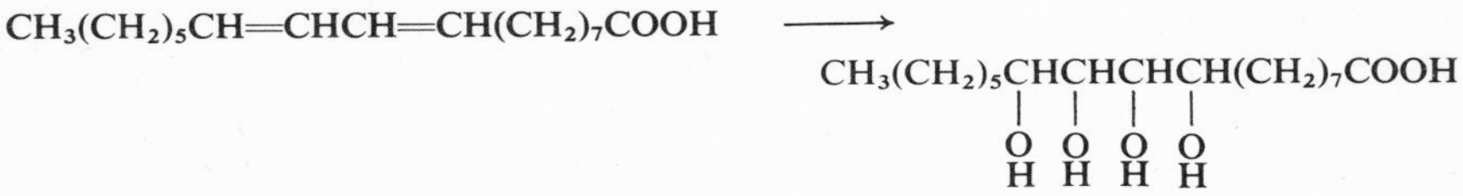

$$CH_3(CH_2)_5CH{=}CHCH{=}CH(CH_2)_7COOH \longrightarrow CH_3(CH_2)_5CH(OH)CH(OH)CH(OH)CH(OH)(CH_2)_7COOH$$

Various secondary changes may occur in osmium-catalyzed oxidations. Osmium tetroxide-catalyzed oxidation of furfural by sodium chlorate in aqueous solution affords *meso*-tartaric acid in 49% yield. The oxidation was assumed to go through 2-furoic acid and, in fact, this acid is itself converted to *meso*-tartaric acid together with a lesser amount of oxalic acid (Milas, 1927). Oxidation of hydroquinone with sodium chlorate in water catalyzed by osmium tetroxide produced a compound tentatively identified as dihydroxy dihydroquinone. This material was shown later to be actually a dimer, **I** (Anderson and Thomson, 1967). Hofmann (1912) reported the product to be quinhydrone. Other workers found quinhydrone in one experiment and in a longer experiment, quinone (Steele and Rylander, 1968). The products evidently depend to a large extent on the reaction time and on whether or not the catalyst becomes deactivated.

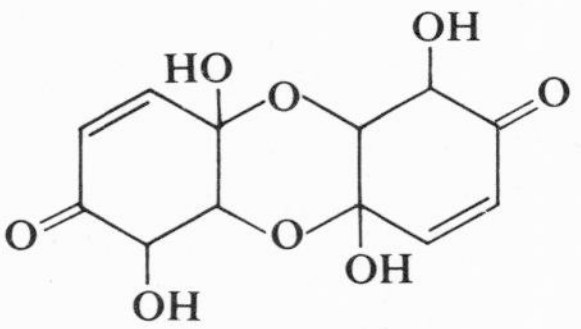

HO OH O O O OH OH

(I)

Epoxidation

In an unusual reaction short-chain olefins are oxidized to epoxides in 50% aqueous acetic acid by sodium chlorate–osmium tetroxide. Epoxidation is nonstereospecific and the epoxide oxygen is derived solely from the chlorate. Propylene at 2 atm and 0°C mixed with equal amounts of 2 *M* $NaClO_3$ and 0.1 *M* OsO_4 affords after 1 hour a solution 0.5 *M* in propylene oxide and 0.7 *M* in 1-chloropropan-2-ol, the latter compound arising by addition of hypochlorous acid to the double bond. Oxidation of a preformed osmate ester of propylene by $NaClO_3$ affords large amounts of carbon dioxide, but no epoxide (Kruse, 1968).

Peroxides as Oxidants

Hydrogen peroxide in the presence of osmium tetroxide interacts with olefins, usually giving a glycol as the primary product. Hydroxylations are carried out often with hydrogen peroxide in anhydrous *t*-butanol (Milas and Sussman, 1937), but other workers consider it unnecessary to use an anhydrous medium (Mugdan and Young, 1949). Benzene, ether, and acetone have been used also as solvents. Hydrogen peroxide–*tert*-butanol containing 5 to 8 % water is said to be favorable for hydroxylation with only minor cleavage of the double bond (Milas *et al.*, 1959).

The course of hydroxylation is complex. It has been suggested that the reactive species is peroxyosmic acid, H_2OsO_6, which breaks down to give hydroxyl groups that in turn add to the carbon–carbon double bond (Milas, 1955; Csányi, 1959). Later another intermediate was postulated based on evidence from spectra and paper chromatograms (Milas *et al.*, 1959). The same intermediate was used to account for the formation of carbonyl compounds by carbon–carbon bond cleavage in the absence of water and with excess hydrogen peroxide. Criegee (1936) postulated a slightly different intermediate in oxidative cleavage. Norton and White (1965) using spectroscopic evidence concluded that the osmium–olefin complex incorporates more than 1 mole of olefin per mole of osmium. For propylene these authors formulated the complex **II** and suggested it could decompose directly into propylene glycol or, by interaction with another hydrogen peroxide molecule, to acetol as primary product.

$$OsO_4(CH_3CH{=}CH_2)_nH_2O_2 \longrightarrow CH_3CH(OH)CH_2OH$$

$$OsO_4(CH_3CH{=}CH_2)_nH_2O_2 \xrightarrow{H_2O_2} CH_3C({=}O)CH_2OH$$

(II)

Hydroxylation of olefins with hydrogen peroxide catalyzed by osmium tetroxide may lead to glycols (Milas and Sussman, 1936), aldehydes (Criegee, 1936), or ketones (Norton and White, 1965) as the main products of the reaction. Hydrogen peroxide is fairly stable to either osmium tetroxide or to an olefin, but when all three are mixed, a vigorous reaction ensues (Norton and White, 1965). *tert*-Butyl- and *tert*-amyl alcohols are stable toward hydrogen peroxide for months at room temperature and secondary and primary alcohols are attacked very slowly. Daniels and Fischer (1963) reported that *tert*-butyl alcohol–osmium tetroxide solution was stable only if the alcohol were first shaken for 48 hours with solid potassium permanganate, filtered, and dried over magnesium sulfate and distilled directly from fresh potassium permanganate. Unless impurities were removed by this procedure, the catalyst solution was unstable; it turned black and was not suitable for hydroxylation reactions. 1,2-Glycols are oxidized slowly, but at a rate insufficient to permit glycols as precursors of aldehydes and ketones under hydroxylation conditions.

Milas and Sussman (1937) hydroxylated a number of olefins with hydrogen peroxide in *tert*-butanol and obtained *vic*-glycols in fair-to-excellent yields. An interesting exception was stilbene, which affords quantitative yields of benzaldehyde. Cyclopentadiene has been reported to be hydroxylated by 1,4-addition affording cyclopentene-2-diol-1,4, or with excess oxidant, a tetrol (Milas and Maloney, 1940). In other hands, hydroxylation afforded a diol containing 52.6% 1,2-hydroxyls (Owen and Smith, 1952). Osmium tetroxide-catalyzed oxidation of cyclohexene with hydrogen peroxide in ether affords cyclohexanediol, adipaldehyde, and adipic acid in 35, 10, and 20% yields, respectively (Mugdan and Young, 1949). Methanol in general proved a better solvent than ethanol, ether, or ethyl acetate (Cosciug, 1941). Hydroxylation of furan–maleic anhydride adducts in acetone or acetone–ether affords the *cis*-glycol, probably of *exo* configuration (Daniels and Fischer, 1963). Ring expansion may occur on osmium-catalyzed oxidation of small unsaturated rings conjugated with a carbonyl function (Boswell, 1966).

CHO $\xrightarrow[\text{OsO}_4,\ \text{aq. isopropanol}]{H_2O_2}$ HO, O, OH + O, OH, OH + O, OH

In a thorough investigation of the hydroxylation of propylene, Norton and White (1965, 1967) examined in detail the effects of reaction variables upon yields of various products under batch and continuous addition conditions. The rate of reaction with gaseous olefins was found to be very sensitive to pressure. Greater selectivity toward acetol production and higher-efficiency hydrogen peroxide consumption are obtained at higher temperatures, higher pressures, and lower hydrogen peroxide concentration. Osmium trichloride may be used as a catalyst in these reactions as well as osmium tetroxide (Norton, 1967). Presumably osmium tetroxide is formed *in situ*; almost all osmium compounds are very easily oxidized to the tetroxide (Griffith, 1967). Norton (1969) used an osmium tetroxide–*p*-dioxane complex to decrease the volatility of osmium tetroxide. The complex is prepared by pouring excess dioxane over the tetroxide crystals, cooling, and filtering under vacuum (Norton, 1969).

Osmium tetroxide–hydrogen peroxide has been used with success for hydroxylation of the carbon–carbon double bonds of α,β-unsaturated ketones, a structure hydroxylated often only with difficulty. For instance, 4-cholestenone affords two cholestan-3-one-4,5-diols in 60% yield and 1-cholestenone affords cholestan-3-one-1,2-diol (Eastham *et al.*, 1959). Treatment of testosterone acetate in ether with hydrogen peroxide and osmium tetroxide for 6 days at room temperature affords 3-oxo-4β,5β-dihydroxy-17β-acetoxyandrostane (Jeger *et al.*, 1969). Double bonds between two carbonyls are hydroxylated with ease; 4-cholestene-3,6-dione affords quantitatively a cholestane-3,6-dione-4,5-diol (Butenandt *et al.*, 1938). Steroids substituted by acetoxy or bromo in the 21-position are oxidized at the 17,20-double bond by hydrogen peroxide–osmium tetroxide directly to 17-hydroxy-20-keto compounds (Miescher and Schmidlin, 1950).

The peroxide of tertiary amines have been used similarly (Forsblad, 1969). In an example, 1.344 gm of 19-nor-9β-10α-$\Delta^{4,17(20)}$-pregnadiene-3-one in 80 ml of *tert*-butanol, 0.06 gm of osmium tetroxide, and 2 ml of pyridine was agitated for 40 minutes at room temperature, while 1.44 gm of the peroxide of triethylamine oxide was added in small portions. After recrystallization from acetone and then ethanol, 0.66 gm of 19-nor-9β-10-α-Δ^{4}-pregnene-17α-ol-3,20-dione was obtained (Bucourt *et al.*, 1968).

CH_3–CH= (ring) → CH_3–C=O, –OH (ring)

Osmium tetroxide–hydrogen peroxide hydroxylations have proved useful in carbohydrate chemistry (Moody, 1964). Attack occurs at the less-hindered face (Stevens *et al.*, 1966). Hydroxylation of **III** with *tert*-butylhydroperoxide catalyzed by osmium tetroxide gives the pentol, **IV**, in 40% yield, isolated as the pentaacetate. The double bond of **III** was extraordinarily inert to additive reagents that normally react with cyclohexenes, including bromine, hypobromous acid, performic acid, silver iododibenzoate, and hydrogen and catalyst. The authors suggested the striking inertness of the double bond may be due to deactivation through π-complex interaction with the carbonyl group of the acetoxymethyl side chain. Hydroxylation with stoichiometric osmium tetroxide in pyridine affords compound **IV** but only in 11% yield. Catalytic hydroxylation was achieved by adding dropwise 2.16 gm of **III** in 8 ml of *tert*-butyl alcohol to a stirred solution of 4.0 ml of 30% hydrogen peroxide in 6 ml of *tert*-butyl alcohol containing 0.03 gm of osmium tetroxide. Stirring was continued for 5 days at 25°C. After evaporation, the product was acetylated with pyridine and acetic anhydride to afford the pentaacetate (McCasland *et al.*, 1968).

CH_2OAc AcO— —OAc (III) $\xrightarrow[\text{2. } (CH_3CO)_2O]{\text{1. } OsO_4,\ \text{—O—O—}}$ CH_2OAc AcO— —OAc AcO OAc (IV)

Aromatic hydrocarbons are oxidized very slowly by hydrogen peroxide–osmium tetroxide. After 10 days at room temperature, benzene in *tert*-butanol was converted in 23% yield to phenol, probably as a result of dehydration of the initial glycol, 1,2-dihydroxycyclohexadiene-3,5 (Milas and Sussman, 1937). After several months, allomucic acid and *meso*-tartaric acid are formed (Cook and Schoental, 1950). Naphthalene is converted to phthalic acid, and phenanthrene affords phenanthrenequinone, *cis*-9,10-dihydro-9,10-dihydroxyphenanthrene, and diphenic acid. The products in oxidations of aromatics depend on the reaction times, which may be as long as one year (Cook and Schoental, 1950). Pyrene is reported to afford pyrene-4,5-quinone in low yield (Oberender and Dixon, 1959); other workers found the 1,6-quinone and the 3,6-quinone, but no pyrene-4,5-quinone (Cook and Schoental, 1950). Anthracene, 1,2-benzanthracene, and 1,2-5,6-dibenzanthracene are each oxidized to the 9,10-quinones. A mechanism involving free hydroxyl radicals was tentatively postulated to account for these results (Cook and Schoental, 1950).

Air as Oxidant

Several workers have used air as an oxidant in the osmium tetroxide-catalyzed oxidation of olefins. Air oxidation of Os^{VI} to Os^{VIII} in aqueous solution is possible, but highly dependent on pH. Oxidation is rapid at pH 11, but above pH 12.5 the rate decreases markedly. Below pH 8, Os^{VI} disproportionates to Os^{IV} and Os^{VIII} (Perichon *et al.*, 1963). Osmium-catalyzed air reduction in a trisodium phosphate and disodium hydrogen phosphate buffer (pH 12.3) of allyl alcohol, ethylene, ethylene glycol, glyoxal, glycerol, glucose, and cyclohexene affords oxalic acid and carbon dioxide. Benzoic acid arises from styrene oxidation. Reaction ceases, due to formation of an osmium-containing solid, if the concentration of allyl alcohol, glycerol, and glucose exceeds 2 moles of substrate per mole of osmium tetroxide. The limitation does not apply to cyclohexene, 1-octene, ethylene, and sucrose; large molar excesses of these substrates can be used (Cairns and Roberts, 1968).

Osmium-catalyzed interaction of ethylene and oxygen in alkaline solution has been examined in considerable detail (Stautzenberger *et al.*, 1967). The oxidation involves formation of osmate ester, hydrolysis, and reoxidation of Os^{VI}. The rate of Os^{VI} oxidation in alkaline solution reached a maximum around pH 9.5 [k (min^{-1} = 1.20], and was sharply lower at both pH 9.0 ($k = 0.39$) and pH 10.0 ($k = 0.51$). Ethylene was hydroxylated by circulating two volumes of ethylene and one volume of oxygen through 0.0025 M OsO_4 at 50°C and 1 atm. The maximum gas absorption rate is at pH 9.5 and falls as the reaction progresses. The decline in rate was attributed to the inhibiting effect of ethylene glycol, which was believed to complex with osmium tetroxide; addition of excess ethylene glycol to the reaction mixture causes gas absorption to cease. At elevated pressures, the rate of gas absorption is higher and greater concentrations of ethylene glycol can be tolerated.

Amino acids have been oxidatively deaminated by air in alkaline solution with osmium tetroxide as catalyst. The rate of ammonia liberation increases with increasing alkali concentration up to a 2 N solution and then gradually decreases (Nyilasi and Somogyi, 1964).

Periodate as Oxidant

Periodate has the power to cleave glycols, and oxidations carried out with this reagent are apt to proceed beyond the glycol stage. Osmium tetroxide-catalyzed periodate oxidation of pyrene results in formation of two major products, both arising from attack at the 4,5-bond, the bonds of lowest electron density. Conversion of pyrene rises as the ratio of oxidizing agent to hydrocarbon is increased.

CHOH O C=O + O O

24% 23%

The authors (Cook and Schoental, 1950) pointed out that since both the quinone and the lactol could lead to formation of 4,5-dialkylphenanthrenes, this method is probably the best available synthesis of these sterically hindered compounds. Osmium tetroxide–hydrogen peroxide oxidations were less successful; most of the starting material was recovered unchanged. Ruthenium dioxide–periodate oxidation was less specific than osmium-catalyzed oxidation and a complex mixture was formed arising from attack at positions other than the 4,5-double bond.

Aldehydes and ketones may be produced from olefins in good yields by combining the hydroxylating qualities of osmium tetroxide with glycol-cleaving properties of periodate (Pappo *et al.*, 1956; Ireland and Kierstead, 1966; Seelye and Watkins, 1969; Sandermann and Bruns, 1969; Schmalzl and Mirrington, 1970; Cross and Webster, 1970; Ando *et al.*, 1970; MacSweeney and Ramage, 1971; Kochetkov *et al.*, 1971). This reaction, for instance, provided a key intermediate in the total synthesis of Aflatoxin-M_1. The isomeric aldehydes were obtained in 60% overall yield from the ketone (Buchi and Weinreb, 1969).

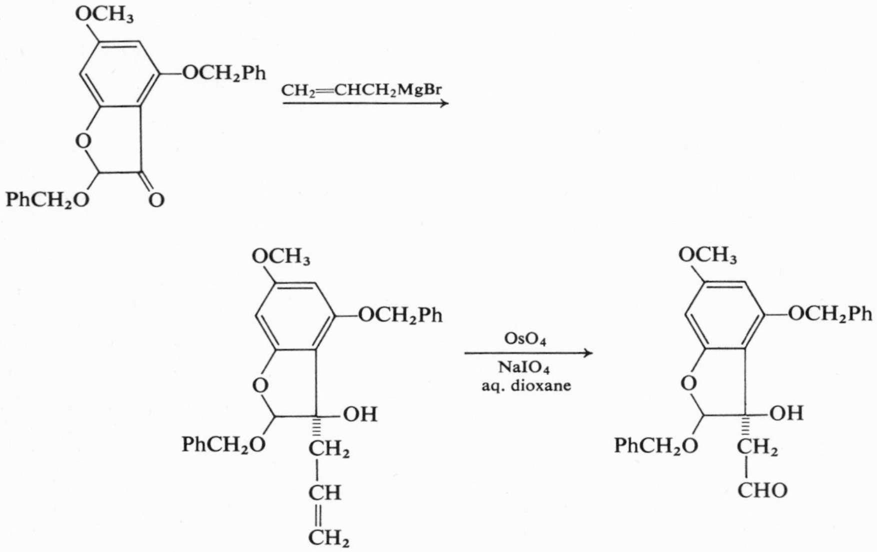

$CH_2{=}CHCH_2MgBr$

OsO_4, $NaIO_4$, aq. dioxane

The preferred experimental technique depends on the substrate. 1-Dodecene and *trans*-stilbene, oxidized at 25°C in 75% aqueous dioxane with 1 mole percent of osmium tetroxide and 210 mole percent of sodium periodate, affords undecanal and benzaldehyde in 68 and 85% yields, respectively. Olefins giving aldehydes sensitive to self-condensation are oxidized better by the use of two immiscible phases, for example, ether and water. In this solvent system cyclohexene and cyclopentene afford adipaldehyde and glutaraldehyde in 77 and 76% yields, respectively (isolated as the 2,4-dinitrophenylhydrazones).

Oxidation and cleavage of the methylene group of 2β,6β-dimethyl-2α-(2′-methylallyl)6α-phenylcyclohexanone (6.45 gm) in 650 ml of dioxane, 64 ml of acetic acid, and 129 ml of water containing 65 mg of osmium tetroxide by 22.8 gm of paraperiodic acid resulted in formation of the diketone, 2α-acetonyl-2β,6β-dimethyl-6α-phenylcyclohexanone in 85% yield (Ireland *et al.*, 1970).

The batch size may determine the preferred technique for oxidation. The unsaturated ester, **V**, in amounts less than 5 gm was oxidized in high yield to the aldehyde, **VI**, by catalytic quantities of osmium tetroxide and two molar equivalents of sodium periodate. However, with batch sizes above 5 gm, the yield of **VI** constantly decreased. Satisfactory yields were obtained with batches up to 50 gm when the original procedure of adding small amounts of solid sodium periodate was abandoned, and instead a dilute solution of the oxidant in water was added to the olefin through a capillary tip over approximately 10 hours (Shamma and Rodriguez, 1968).

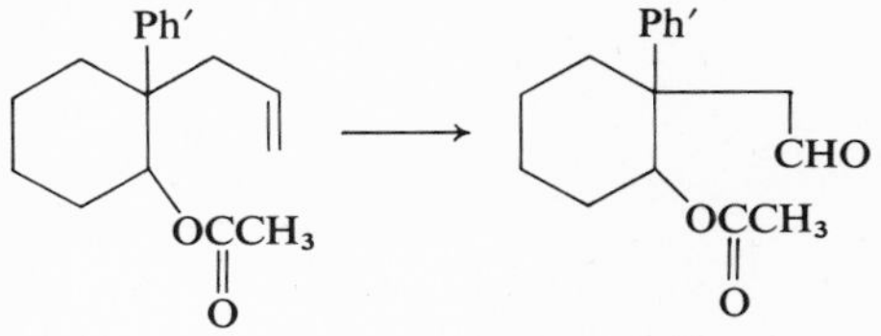

(V) → (VI)

The products of oxidation may depend on the buffer present. Osmium-catalyzed periodate oxidation of the 18:18a enol double bond in **VII** affords the glycol, **VIII**, almost quantitatively when the oxidation is carried out in

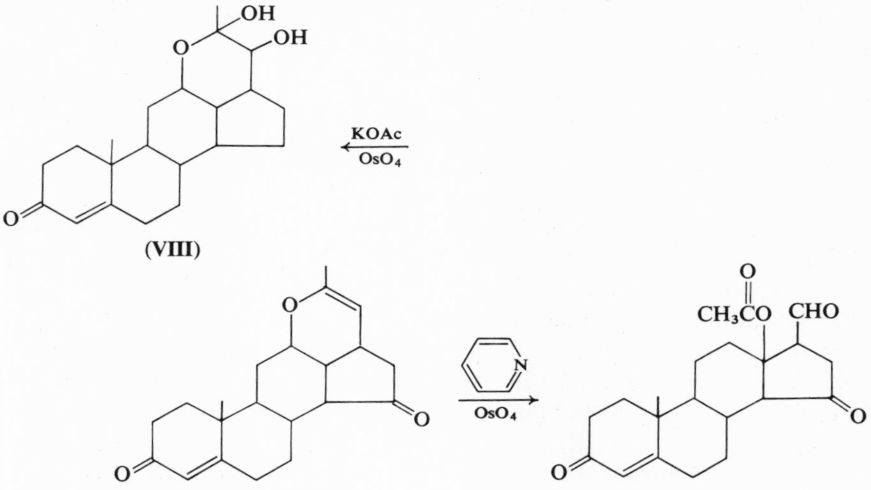

(VIII) ← KOAc, OsO_4

(VII) → pyridine, OsO_4 → (IX)

aqueous solution buffered by potassium acetate, whereas with pyridine instead of potassium acetate, the 11β-acetoxy-18-oxo compound, **IX**, is formed directly (Wettstein *et al.*, 1961).

Osmium tetroxide-catalyzed sodium periodate oxidations have been used to introduce a carbonyl group through the following sequence (Henry and Hoff, 1969):

$$O_2N\text{-(imidazo-pyrrolidine)} \xrightarrow[OH^-]{PhCHO} O_2N\text{-(imidazo-pyrrolidine)=CHPh} \xrightarrow{932 \text{ mg } NaIO_4,\ 19 \text{ mg } OsO_4} O_2N\text{-(imidazo-pyrrolidine)=O}$$

500 mg

Formyl derivatives were prepared similarly, affording, for example, 1-(2-acetoxyethyl)-2-formyl-5-nitroimidazole (Henry and Hoff, 1969).

Hypochlorite as Oxidant

Sodium hypochlorite with catalytic amounts of osmium tetroxide may be used to hydroxylate olefins in fair yield. The reaction is carried out best at pH 9 to 11 to avoid free chlorine and with slow addition of the reagent to avoid over-oxidation. Chlorination may accompany hydroxylation, but the halogen compounds can be hydrolyzed increasing the overall yield. Oxidation in the dark of allyl alcohol dissolved in *tert*-butanol with sodium hypochlorite affords glycerol in 78% yield; the yield is raised to 98% by heating the oxidation products for 2 hours at 100°C in a solution maintained at pH 12.2 (Cummins, 1970).

Hexacyanoferrate as Oxidant

Osmium-catalyzed oxidation by hexacyanoferrate ions has not found much synthetic use, but the system does make a convenient one for kinetic studies. The oxidation of acetone and ethyl methyl ketone by alkaline hexacyanoferrate(III) ion catalyzed by osmium tetroxide is believed to proceed via an activated complex between enolate and osmium tetroxide which rapidly decomposes, followed by a fast reaction between the reduced osmium moiety and ferricyanide (Singh *et al.*, 1969). Mandelic acid is oxidatively decomposed to benzoate by hexacyanoferrate(III) ion in alkaline medium in the presence of catalytic amounts of osmium tetroxide; without osmium there is no appreciable oxidation (Singh *et al.*, 1968).

Mayell (1968) utilized for the first time the ferro–ferricyanide couple to produce glycols quantitatively from olefins in aqueous alkaline solutions containing osmium tetroxide as catalyst. Potassium ferricyanide and alkali consumed in producing the glycol were regenerated electrochemically using an oxygen-depolarized cathode.

Disproportionation

Certain oxidations may be achieved through disproportionation reactions. For example, sulfones may be made by disproportionation of sulfoxides using osmium tetroxide as a catalyst. In disproportionation of dimethyl sulfoxide at 130°–150°C, the effective catalyst life of osmium tetroxide was about 10,000 gm of product per gram of catalyst (British Patent 853,623, Nov. 9, 1960).

RUTHENIUM TETROXIDE

Ruthenium tetroxide, a powerful oxidizing agent, oxidizes sulfides to sulfoxides and sulfones (Djerassi and Engle, 1953), sulfinates to sulfonates (Ree and Martin, 1970), primary alcohols to acids, secondary alcohols to ketones, ethers to esters, amides to imides, olefins to aldehydes, ketones, or acids (Berkowitz and Rylander, 1958), amines to amides (Felix *et al.*, 1970), and cyclic ethers to lactones (Weinstock and Wolff, 1960; Wettstein *et al.*, 1968; Rylander and Berkowitz, 1966; Moriarty *et al.*, 1970). Oxidation of alcohols to ketones (Corey *et al.*, 1963, 1964; House and Blankley, 1967; Ramey *et al.*, 1967) has proved especially useful with carbohydrates (Beynon *et al.*, 1963, 1966, 1968; Butterworth *et al.*, 1968, 1969; Collins *et al.*, 1969; Ezekiel *et al.*, 1969; Flaherty *et al.*, 1966; Follman and Hogenkamp, 1970; Tronchet and Tronchet, 1968). An interesting oxygen insertion reaction occurred in oxidation of osuloses, probably resulting from a Baeyer-Villiger reaction involving peracetic acid (Nutt *et al.*, 1965, 1968). Cleavage of olefins has been useful in oxidation of 3-alkylidenegrisens to grisen-3-ones (Dean and Knight, 1962) and in determining the location of isolated double bonds in steroids and triterpenes (Snatzke and Fehlhaber, 1963).

The above reactions might be carried out with catalytic quantities of ruthenium tetroxide if the reactions are conducted in the presence of an oxidizing agent of sufficient strength to reoxidize the reduced ruthenium. Ruthenium tetroxide has been prepared from compounds of ruthenium in the lower oxidation state in acid, neutral, and alkaline media. Suitable oxidizing agents in alkaline media are chlorine and metal hypochlorites, whereas in acidic media bromates, permanganates, periodates, perchloric acid, sodium bismuthate, chromic acid, lead tetraacetate, and other oxidants have been used (Avtokratova, 1963). Reoxidation of reduced ruthenium to the tetroxide stage may not always be necessary; the catalytic cycle might be carried on through intermediate ruthenates and perruthenates (MacLean, 1969).

If ruthenium dioxide is used to begin the cycle, it should be in the dihydrate form, since other modifications may not be oxidized under mild conditions.

Oxidation of Aromatics

Ruthenium tetroxide attacks aromatic systems vigorously, but it leaves paraffins untouched. Caputo and Fuchs (1967) have shown that this distinction can be used in synthesis by converting *p-tert*-butylphenol to pivalic acid, phenylcyclohexane to cyclohexanecarboxylic acid, and *cis*-3- and *cis*- and *trans*-2-phenylcyclobutanecarboxylic acid to *cis*-1,3- and *cis*- and *trans*-1,2-cyclobutanedicarboxylic acids. The latter conversions were used to establish the stereochemistry of phenylcyclobutanecarboxylic acid (Caputo and Fuchs, 1968). Ruthenium tetroxide was used in catalytic quantities with sodium periodate as the oxidizing agent, the ruthenium dioxide formed in the reaction being reoxidized in a cycle process to the tetroxide. In an example, 4.0 gm of phenylcyclohexane in 20 ml of carbon tetrachloride was oxidized with 32 gm of sodium periodate in 250 ml of water with 3 mg of ruthenium tetroxide as catalyst. The mixture was stirred for 10 days at 60°C until the yellow color of ruthenium persisted. The very slow reaction can be attributed to the use of minute amounts of catalyst.

Oberender and Dixon (1959) used periodate–ruthenium tetroxide to oxidize pyrene. The reaction was much less specific than osmium-catalyzed oxidations; the latter reagent attacked only the 4,5-bond, whereas ruthenium tetroxide attacked nonspecifically. The products were pyrene-4,5-quinone (11%), pyrene-1,6-quinone (2%), the lactol of 4-formylphenanthrene-5-carboxylic acid (1%), and an unidentified aldehyde (10%). Both α- and β-naphthols are oxidized to phthalic acid by catalytic amounts of ruthenium tetroxide using either periodate or hypochlorite as oxidant. Methyl α-naphthyl ether is oxidized more slowly and is also converted to phthalic acid. Controlled oxidation of 5-methoxy-1-naphthol affords 3-methoxyphthalic acid in 50% yield (Ayres and Hossain, 1972).

OH, OCH_3 ⟶ COOH, COOH, OCH_3

Aromatic Steroids

Interesting use of ruthenium tetroxide–sodium periodate has been made in oxidation of steroids having ring A aromatic (Piatak *et al.*, 1969*a*,*b*). Oxidation of estrone (**X**) or 1,17β-dihydroxy-4-methylestra-1,3,5(10)triene, or 1,17β-dihydroxyestra-1,3,5(10)-triene each affords the diacid, **XI** with degradation of ring A terminating at carbons 5 and 10.

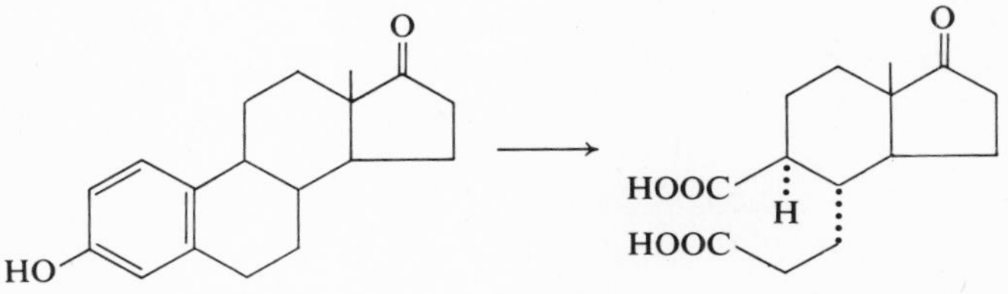

(**X**) (**XI**)

Typically, 1 gm of estrone in 100 ml of acetone is added to a stirred, yellow ruthenium tetroxide mixture obtained by combining 400 mg of ruthenium dioxide in 50 ml of acetone with 3 gm of sodium periodate in 15 ml of water. The reaction is kept yellow by adding portionwise a freshly prepared solution of 11.5 gm of sodium periodate in 115 ml of 50% aqueous acetone. After 4.5 hours, the reaction is terminated by addition of a few drops of isopropanol. Isolation of the acid fraction with sodium bicarbonate affords 670 mg of **XI**. Similar oxidations have been carried out by Wolff and Zanati (1969).

A totally different type of reaction occurs during oxidation of estradiol diacetate (**XII**). The major product, **XIII**, obtained in 40% yield is derived by introduction of two additional oxygen atoms.

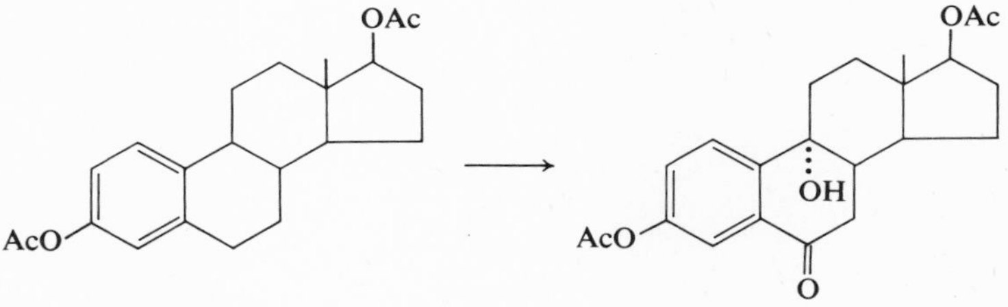

(**XII**) (**XIII**)

Oxidation of 1-methylestradiol diacetate under the same reaction conditions results, on the other hand, in formation of a diacid with destruction of the aromatic ring. The course of reaction and types of products derived is evidently quite sensitive to structure (Piatak *et al.*, 1969b).

Oxidation of Alcohols

Primary alcohols are oxidized smoothly to acids by ruthenium tetroxide–sodium periodate (House and Blankley, 1968), and secondary alcohols are converted to ketones (Ramey *et al.*, 1967). Oxidation of steroid alcohols to the corresponding ketones may give good results (Nakata, 1963). For example, 5α-androstane-3β-ol-17-one is oxidized to the corresponding 1,17-dione in 82% yield by a large excess of periodate and 1 mole percent ruthenium tetroxide. These catalytic oxidations may also be carried out with lead tetraacetate in glacial acetic acid.

Striking success was achieved with ruthenium dioxide–sodium periodate in the difficult oxidation of the hydroxylactone, **XIV**, to the corresponding keto-lactone, **XV**. The reaction failed with 15 other oxidizing procedures (Moriarty *et al.*, 1970).

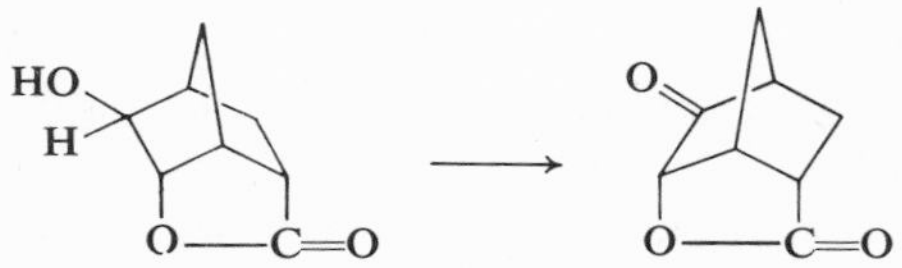

(XIV) (XV) 80%

The reaction is carried out in two phases by adding with vigorous stirring a 10% aqueous solution of 40 mmoles of sodium periodate to a mixture of 30 mmoles of the hydroxylactone in water and 250 mg of ruthenium dioxide in carbon tetrachloride. The oxidation is complete when the yellow color of ruthenium tetroxide persists. The excess oxidizing agent is destroyed by addition of isopropanol. These reactions proceed under neutral conditions with the following stoichiometry:

$$RuO_2 + 2\,IO_4^- \longrightarrow RuO_4 + 2\,IO_3^-$$

$$2 \;{>}CHOH + RuO_4 \longrightarrow 2 \;{>}C{=}O + RuO_2 + 2\,H_2O$$

Prior alkaline hydrolysis of lactones to the hydroxy acid salt permits oxidation to the keto acid (Moriarty *et al.*, 1970; Gopal *et al.*, 1972).

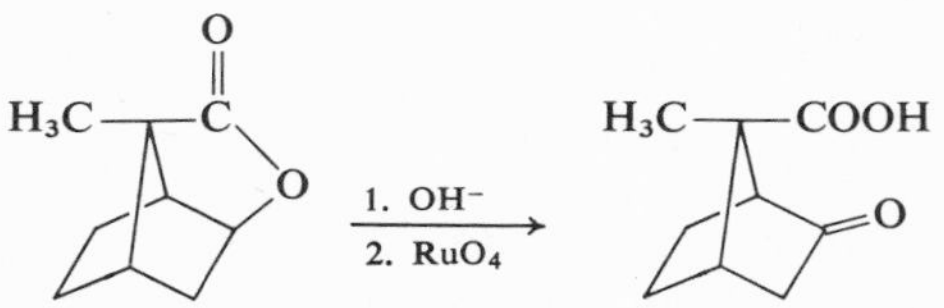

Carbohydrates

Stoichiometric use of ruthenium tetroxide in the oxidation of carbohydrates has met with considerable success (Butterworth and Hanessian, 1971). Catalytic use of ruthenium has been limited primarily to periodate as an oxidant, but presumably other oxidants might work as well. Certain oxidations of carbohydrates with this oxidizing system may be sluggish (Howarth *et al.*, 1970). Parikh and Jones (1965) have applied ruthenium dioxide–sodium periodate oxidations to the preparation of substituted keto sugars with excellent results. Typically, the carbohydrate derivative is dissolved in carbon tetrachloride or chloroform with 20 mg of ruthenium dioxide per gram of carbohydrate and a 5% aqueous solution of sodium periodate is added dropwise to the mixture with vigorous stirring at or below room temperature. The pH is controlled meticulously between 6 and 7 with a pH meter by periodic additions of small quantities of dilute sodium bicarbonate solution. Usually 1.3 molar equivalents of periodate is required. The end of the reaction is signaled by a color change from black ruthenium dioxide to yellowish ruthenium tetroxide. Excess periodate is destroyed by addition of *n*-propanol and ruthenium dioxide is removed by filtration. Oxidation of aromatic substituents is very slow relative to oxidation of the hydroxyl function. Quantitative yields of 6-*O*-benzoyl-1,2:4,5-di-*O*-isopropylidene-*threo*-glycero-3-hexulose are obtained by oxidation of 6-*O*-benzoyl-1,2:4,5-di-*O*-isopropylidene dulcitol. Similarly, but with less success, oxidation of 1,2-isopropylidene-3-benzyl-6-trityl-α-D-glucofuranose in carbon tetrachloride with ruthenium tetroxide–sodium periodate afforded 1,2-isopropylidene-3-benzyl-6-trityl-5-keto-*a*-D-glucofuranose in 25% yield (Inouye and Ito, 1970).

In a slightly different oxidation procedure, a 15% solution of the sugar in ethanol-free chloroform and an equal volume of water is stirred with 0.24 mole potassium carbonate, 1.3 moles potassium periodate, and 0.05 mole ruthenium dioxide for each mole of substrate. The yields ranged from 83 to 95% in the five examples reported (Lawton *et al.*, 1969). Dichloromethane has been used in place of chloroform in this type of reaction (zu Reckendorf and Bischof, 1970).

Amines are attacked by ruthenium tetroxide but the reagent has been applied successfully nonetheless to the oxidation of a hydroxyl function in an amino sugar. Oxidation of 9-(3′,5′-*O*-isopropylidene-β-D-xylofuranosyl)-adenine with ruthenium tetroxide in a mixture of carbon tetrachloride and aqueous sodium bicarbonate gave 9-(3′,5′-*O*-isopropylidene-2′-keto-D-xylofuranosyl)adenine in 74% yield. The stoichiometric amount of sodium periodate as a 5% solution must be added dropwise to the reaction mixture (Rosenthal *et al.*, 1970).

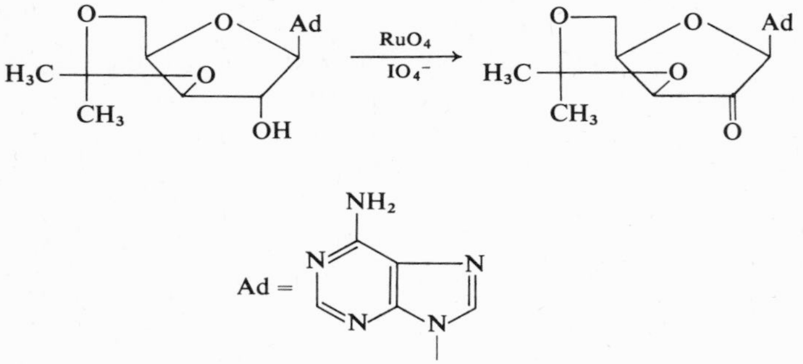

RuO_4

IO_4^-

Ad =

OXIDATION OF OLEFINS

Olefins are oxidized by ruthenium tetroxide–periodate to acids or ketones (Sobti and Dev, 1970), depending on the structure of the olefin (Pappo and Becker, 1956). Typically, oxidations are carried out with 4 moles of sodium periodate and 2 mole percent of ruthenium dioxide in aqueous acetone.

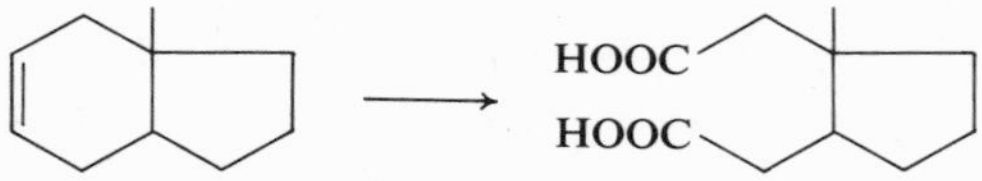

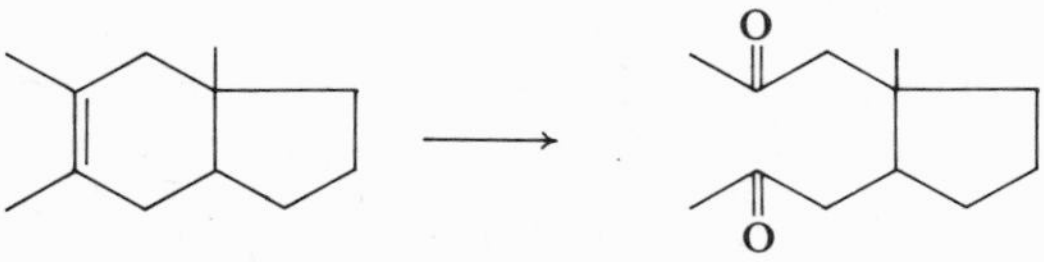

Lead tetraacetate may also be used in these oxidations, illustrated below with an unsaturated keto steroid. The oxidation is carried out with 6 moles of lead tetraacetate and 5 mole percent of ruthenium dioxide in 90% acetic acid.

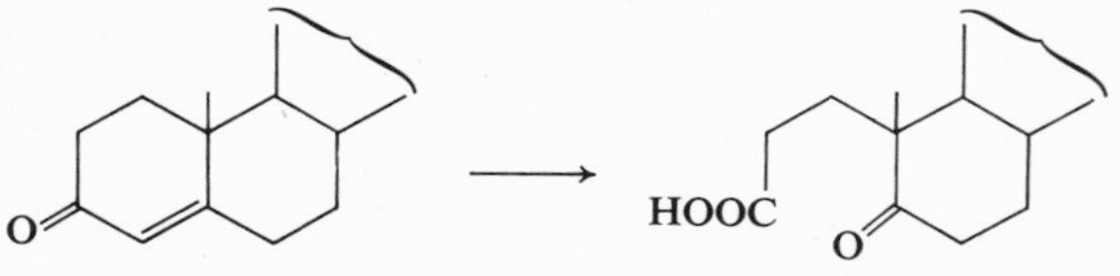

85%

Another application in the steroid field is as follows:

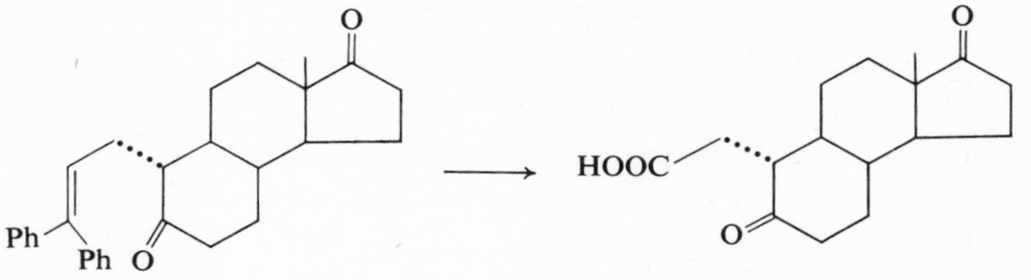

Ph Ph O → HOOC O

The tetroxide catalyst is prepared from 200 mg of ruthenium dioxide and 2 gm of sodium periodate in 100 ml of water, the solid components dissolving to give a yellow solution. Six grams of substrate is oxidized with 200 mg of ruthenium tetroxide in 50 ml of acetone. More periodate is added whenever the solution loses its yellow color and turns black (Sondheimer *et al.*, 1964). Satisfactory results were obtained with ruthenium dioxide obtained by the method of Muller and Schwabe (1929).

A similar oxidation of the diphenylethylene compound, **XVI**, with ruthenium tetroxide–sodium periodate gave excellent yields of the keto acid, **XVII**. The procedure was superior to ozonolysis (Stork *et al.*, 1963).

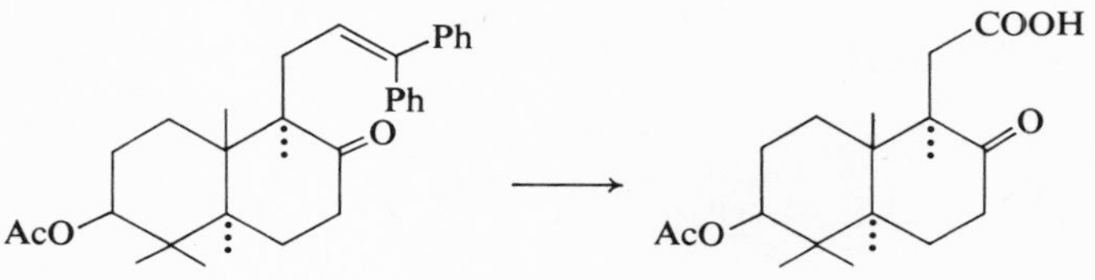

AcO Ph Ph O → AcO COOH O

(XVI) **(XVII)**

Sarel and Yanuka (1959) applied this oxidation procedure to the conversion of 3α-acetoxy-24,24-diphenylchol-23-ene to 3α-acetoxynorcholanic acid. The oxidation proceeds smoothly, affording the desired product in 78–83% yield. Oxidations with chromic acid in acetic acid are erratic and highly temperature-dependent. Osmium tetroxide is completely ineffective. Oxidations with ruthenium tetroxide have also been applied successfully to the synthesis of steroidal pyrimidines (Caspi and Piatak, 1963).

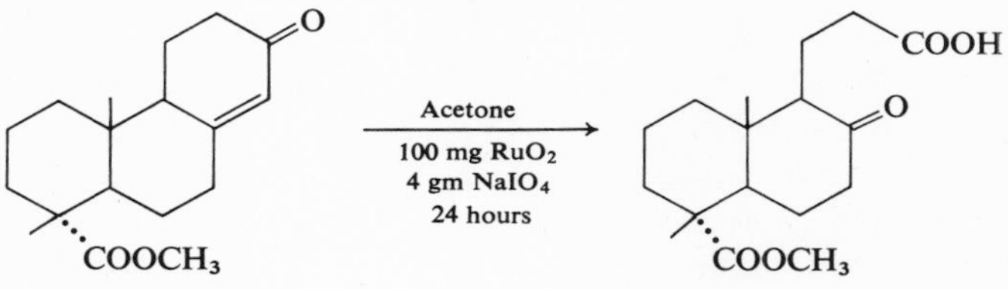

$COOCH_3$ O, 500 mg → (Acetone; 100 mg RuO_2; 4 gm $NaIO_4$; 24 hours) COOH O $COOCH_3$ 80%

Excellent yields of keto acids were obtained also by ruthenium tetroxide–sodium periodate oxidation of α,β-unsaturated ketones in the resin acid series (Pelletier *et al.*, 1970).

Oxidation of conjugated and cross-conjugated steroidal ketones with ruthenium tetroxide–sodium periodate has been investigated in some detail (Piatak *et al.*, 1969*a,b*). Oxidation of ring A or ring C α,β-unsaturated ketones generally gives the expected diacids or keto acids in very good yield. Oxidation of 3β-acetoxy-5β-pregnan-16-en-20-one does not, however, yield the expected diacid, but rather the seco-17,20-diketo-16-carboxylic acid.

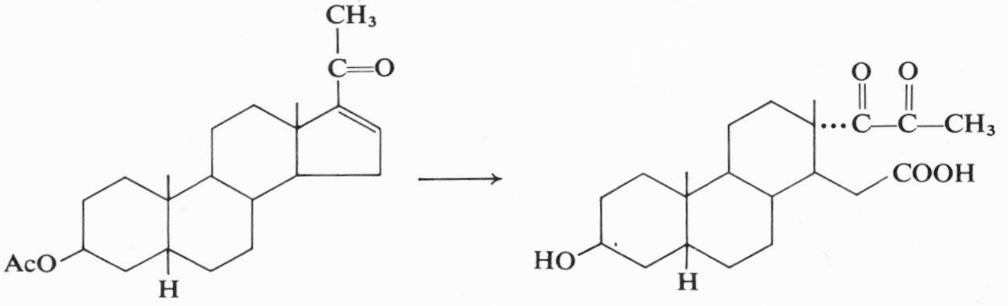

Oxidation of Acetylenes

Acetylenes are oxidized by ruthenium tetroxide–sodium periodate to the corresponding α-diketones (Pappo and Becker, 1956).

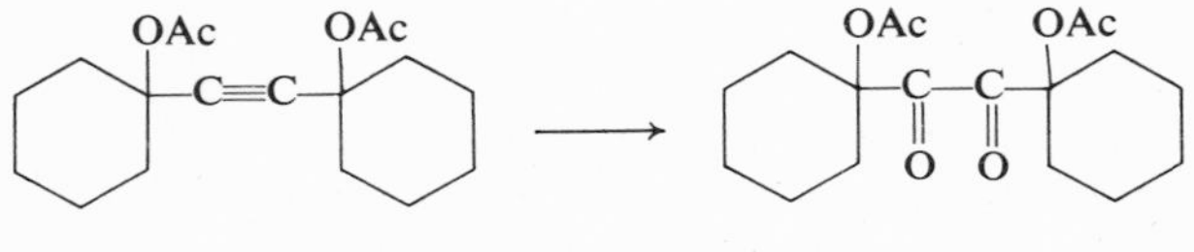

Ruthenium Tetroxide–Sodium Hypochlorite

Keblys and Dubeck (1968) oxidized olefins in aqueous alkaline solution with sodium hypochlorite and catalytic quantities of ruthenium salts. In an example, 27 parts of 2-butyl-1-octene containing 1.65 parts of ruthenium trichloride was oxidized at 17°–22°C by addition over a 7 hour period of 475 parts of 1.4 *M* sodium hypochlorite containing 13 parts sodium hydroxide. Undecanone-5 was isolated in 83% yield. A similar oxidation of 1-dodecene affords undecanoic acid in 90% yield. The reagent has been applied also to the oxidation of alcohols and aromatic rings (Wolfe *et al.*, 1970). Cyclohexanol affords cyclohexanone in 95% yield on oxidation at 0°C. Further oxidation leads to adipic acid. The reagent appears superior to ruthenium periodate for degradation of aromatic rings. Oxidation of β-phenylpropionic acid with ruthenium hypochlorite at room temperature for 3 hours affords succinic acid in 94% yield and benzoic acid in 6% yield according to the stoichiometry.

$$28[O] + C_6H_5\text{—}CH_2CH_2COOH \longrightarrow HOOCCH_2CH_2COOH + C_6H_5\text{—}COOH$$

453 224 94% 6%

REFERENCES

Anderson, H. A., and Thomson, R. H. (1967). *J. Chem. Soc. C* p. 2152.

Ando, M., Nanaumi, K., Nakagawa, T., Assao, T., and Takase, K. (1970). *Tetrahedron Lett.* p. 3891.

Avtokratova, T. D. (1963). "Analytical Chemistry of Ruthenium." Daniel Davey, New York.

Ayres, D. C., and Hossain, A. M. M. (1972). *Chem. Commun.* p. 428.

Berkowitz, L. M., and Rylander, P. N. (1958). *J. Amer. Chem. Soc.* **80**, 6682.

Beynon, P. J., Collins, P. M., and Overend, W. G. (1963). *Proc. Chem. Soc., London* p. 342.

Beynon, P. J., Collins, P. M., Doganges, P. T., and Overend, W. G. (1966). *J. Chem. Soc., C* p. 1131.

Beynon, P. J., Collins, P. M., Gardiner, D., and Overend, W. G. (1968). *Carbohyd. Res.* **6**, 431.

Bláha, L., Weichet, J., Žvaček, J., Šmolik, S., and Kakáč, B. (1960). *Collect. Czech. Chem. Commun.* **25**, 237.

Boswell, G. A., Jr. (1966). U.S. Patent 3,265,738.

Braun, G. (1929). *J. Amer. Chem. Soc.* **51**, 228.

Braun, G. (1930a). *J. Amer. Chem. Soc.* **52**, 3176.

Braun, G. (1930b). *J. Amer. Chem. Soc.* **52**, 3188.

Braun, G. (1932). *J. Amer. Chem. Soc.* **54**, 1133.

Buchi, G., and Weinreb, S. M. (1969). *J. Amer. Chem. Soc.* **91**, 5048.

Bucourt, R., Costerousse, G., Nomine, G., Pierdet, A., and Tessier, J. (1968). U.S. Patent 3,383,385.

Butenandt, A., Wolz, H., von Dresler, D., and Meinerts, U. (1938). *Ber Deut. Chem. Ges. B* **71**, 1483.

Butterworth, R. F., and Hanessian, S. (1971). *Synthesis* **1**, 70.

Butterworth, R. F., Overend, W. G., and Williams, N. R. (1968). *Tetrahedron Lett.* p. 3239.

Butterworth, R. F., Collins, P. M., and Overend, W. G. (1969). *Chem. Commun.* p. 378.

Cairns, J. F., and Roberts, H. L. (1968). *J. Chem. Soc., C* p. 640.

Caputo, J. A., and Fuchs, R. (1967). *Tetrahedron Lett.* p. 4729.

Caputo, J. A., and Fuchs, R. (1968). *J. Org. Chem.* **33**, 1959.

Caspi, E., and Piatak, D. M. (1963). *Experientia* **19**, 465.

Collins, P. M., Doganges, P. T., Kolarikal, A., and Overend, W. G. (1969). *Carbohyd. Res.* **11**, 199.

Cook, J. W., and Schoental, R. (1950). *J. Chem. Soc., London* p. 47.

Corey, E. J., Casanova, J., Jr., Vatakencherry, P. A., and Winter, R. (1963). *J. Amer. Chem. Soc.* **85**, 169.

Corey, E. J., Ohno, M., Mitra, R. B., and Vatakencherry, P. A. (1964). *J. Amer. Chem. Soc.* **86**, 478.
Cosciug, T. (1941). *Ann. Sci. Univ. Jassy* [Sect. 1.] **27**, 303.
Criegee, R. (1936). *Justus Liebigs Ann. Chem.* **522**, 75.
Criegee, R., Manchard, B., and Wannowius, H. (1942). *Justus Liebigs Ann. Chem.* **550**, 99.
Cross, B. E., and Webster, G. R. B. (1970). *J. Chem. Soc., C* p. 1839.
Csányi, L. J. (1959). *Acta Chim. (Budapest)* **21**, 35.
Cummins, R. W. (1970). U.S. Patent 3,488,394.
Daniels, R., and Fischer, J. L. (1963). *J. Org. Chem.* **28**, 320.
Dean, F. M., and Knight, J. C. (1962). *J. Chem. Soc., London* p. 4745.
Djerassi, C., and Engle, R. R. (1953). *J. Amer. Chem. Soc.* **75**, 3838.
Eastham, J. F., Miles, G. B., and Krauth, C. A. (1959). *J. Amer. Chem. Soc.* **81**, 3114.
Ernest, I. (1964). *Collect. Czech. Chem. Commun.* **29**, 266.
Ezekiel, A. D., Overend, W. G., and Williams, N. R. (1969). *Tetrahedron Lett.* p. 1635.
Felix, A. M., Fryer, R. I., and Sternbach, L. H. (1970). U.S. Patent 3,546,212.
Fieser, L. F., and Fieser, M. (1967). "Reagents for Organic Synthesis," p. 759. Wiley, New York.
Flaherty, B., Overend, W. G., and Williams, N. R. (1966). *J. Chem. Soc., C* p. 398.
Follman, H., and Hogenkamp, H. P. C. (1970). *J. Amer. Chem. Soc.* **92**, 671.
Forsblad, I. B. (1969). U.S. Patent 3,455,973.
Glattfeld, J. W. E., and Chittum, J. W. (1933). *J. Amer. Chem. Soc.* **55**, 3663.
Glattfeld, J. W. E., and Rietz, E. (1940). *J. Amer. Chem. Soc.* **56**, 974.
Glattfeld, J. W. E., and Woodruff, S. (1927). *J. Amer. Chem. Soc.* **49**, 2309.
Gopal, H., Adams, T., and Moriarty, R. M. (1972). *Tetrahedron* **28**, 4259.
Griffith, W. P. (1967). "The Chemistry of the Rarer Platinum Metals," p. 69. Wiley (Interscience), New York.
Gunstone, F. D. (1960). *Advan. Org. Chem.* **1**, 103.
Henry, D. W., and Hoff, D. R. (1969). U.S. Patent 3,472,864.
Hofmann, K. A. (1912). *Ber. Deut. Chem. Ges.* **45**, 3329.
Hofmann, K. A., Ehrhart, O., and Schneider, O. (1914). *Ber. Deut. Chem. Ges.* **46**, 1667.
House, H. O., and Blankley, C. J. (1967). *J. Org. Chem.* **32**, 1741.
House, H. O., and Blankley, C. J. (1968). *J. Org. Chem.* **33**, 53.
Howarth, G. B., Szarek, W. A., and Jones, J. K. N. (1970). *J. Chem. Soc., C* p. 2218.
Inouye, S., and Ito, T. (1970). U.S. Patent 3,519,683.
Ireland, R. E., and Kierstead, R. C. (1966). *J. Org. Chem.* **31**, 2543.
Ireland, R. E., Grand, P. S., Dickerson, R. E., Bordner, J., and Rydjeski, D. R. (1970). *J. Org. Chem.* **35**, 570.
Jeger, O., Wehrli, U., and Schaffner, K. (1969). U.S. Patent 3,484,456.
Kaufmann, H. P., and Jansen, H. (1959). *Ber. Deut. Chem. Ges.* **92**, 2789.
Keblys, K. A., and Dubeck, M. (1968). U.S. Patent 3,409,649.
Kochetkov, N. K., Usov, A. I., and Adamyants, K. S. (1971). *Tetrahedron* **27**, 549.
Kruse, W. (1968). *Chem. Commun.* p. 1610.
Lawton, B. T., Szarek, W. A., and Jones, J. K. N. (1969). *Carbohyd. Res.* **10**, 456.
McCasland, G. E., Furuta, S., and Durham, L. J. (1968). *J. Org. Chem.* **33**, 2835.
MacLean, A. F. (1969). U.S. Patent 3,479,403.
MacSweeney, D. F., and Ramage, R. (1971). *Tetrahedron* **27**, 1481.
Mayell, J. S. (1968). *Ind. Eng. Chem., Prod. Res. Develop.* **7**, 129.
Miescher, K., and Schmidlin, J. (1950). *Helv. Chim. Acta* **33**, 1840.
Milas, N. A. (1927). *J. Amer. Chem. Soc.* **49**, 2005.
Milas, N. A. (1955). *In* "The Chemistry of Petroleum Hydrocarbons" (B. T. Brooks *et al.*, eds.), Vol. II, Chapter 37, p. 399. Van Nostrand-Reinhold, Princeton, New Jersey.

Milas, N. A., and Maloney, L. S. (1940). *J. Amer. Chem. Soc.* **62**, 1841.
Milas, N. A., and Sussman, S. (1936). *J. Amer. Chem. Soc.* **58**, 1302.
Milas, N. A., and Sussman, S. (1937). *J. Amer. Chem. Soc.* **59**, 2345.
Milas, N. A., and Terry, E. M. (1925). *J. Amer. Chem. Soc.* **47**, 1412.
Milas, N. A., Trepagnier, J. H., Nolan, J. T., Jr., and Iliopulus, M. I. (1959). *J. Amer. Chem. Soc.* **81**, 4370.
Moody, G. J. (1964). *Advan. Carbohyd. Chem.* **19**, 172.
Moriarty, R. M., Gopal, H., and Adams, T. (1970). *Tetrahedron Lett.* p. 4003.
Mugdan, M., and Young, D. P. (1949). *J. Chem. Soc., London* p. 2988.
Muller, E., and Schwabe, K. (1929). *Z. Electrochem.* **35**, 165.
Nakata, H. (1963). *Tetrahedron* **19**, 1959.
Norton, C. J. (1967). U.S. Patent 3,335,174.
Norton, C. J. (1969). U.S. Patent 3,424,765.
Norton, C. J., and White, R. E. (1965). *Advan. Chem. Ser.* **51**, 10.
Norton, C. J., and White, R. E. (1967). U.S. Patent 3,337,635.
Nutt, R. F., Arison, B., Holly, F. W., and Walton, E. (1965). *J. Amer. Chem. Soc.* **87**, 3273.
Nutt, R. F., Dickinson, M. J., Holly, F. W., and Walton, E. (1968). *J. Org. Chem.* **33**, 1789.
Nyilasi, J., and Somogyi, P. (1964). *Ann. Univ. Sci. Budapest. Rolando Eotvos Nominatae, Sect. Chim.* **6**, 139.
Oberender, F. G., and Dixon, J. A. (1959). *J. Org. Chem.* **24**, 1226.
Owen, L. N., and Smith, P. N. (1952). *J. Chem. Soc., London* p. 4026.
Pappo, R., and Becker, A. (1956). *Bull. Res. Counc. Isr., Sect. A* **5**, 300.
Pappo, R., Allen, D. S., Jr., Lemieux, R. U., and Johnson, W. S. (1956). *J. Org. Chem.* **21**, 478.
Parikh, V. M., and Jones, J. K. N. (1965). *Can. J. Chem.* **43**, 3452.
Pelletier, S. W., Iyer, K. N., and Chang, C. W. J. (1970). *J. Org. Chem.* **35**, 3535.
Perichon, J., Palous, S., and Buvet, R. (1963). *Bull. Soc. Chim. Fr.* [5] p. 982.
Piatak, D. M., Bhat, H. B., and Caspi, E. (1969a). *J. Org. Chem.* **34**, 112.
Piatak, D. M., Herbst, G., Wicha, J., and Caspi, E. (1969b). *J. Org. Chem.* **34**, 116.
Ramey, K. C., Lini, D. C., Moriarty, R. M., Gopal, H., and Welsh, H. G. (1967). *J. Amer. Chem. Soc.* **89**, 2401.
Ree, B. R., and Martin, J. C. (1970). *J. Amer. Chem. Soc.* **92**, 1660.
Rosenthal, A., Sprinzl, M., and Baker, D. A. (1970). *Tetrahedron Lett.* p. 4233.
Rylander, P. N., and Berkowitz, L. M. (1966). U.S. Patent 3,278,558.
Sandermann, W., and Bruns, K. (1969). U.S. Patent 3,427,328.
Sarel, S., and Yanuka, Y. (1959). *J. Org. Chem.* **24**, 2018.
Schmalzl, K. J., and Mirrington, R. N. (1970). *Tetrahedron Lett.* p. 3219.
Seelye, R. N., and Watkins, W. B. (1969). *Tetrahedron* **25**, 447.
Shamma, M., and Rodriguez, H. R. (1968). *Tetrahedron* **24**, 6583.
Sidgwick, N. V. (1952). "The Chemical Elements and their Compounds," Vol. 2, p. 1500. Oxford Univ. Press (Clarendon), London and New York.
Singh, N. P., Singh, V. N., and Singh, M. P. (1968). *Aust. J. Chem.* **21**, 2913.
Singh, V. N., Singh, H. S., and Saxena, B. L. (1969). *J. Amer. Chem. Soc.* **91**, 2643.
Snatzke, G., and Fehlhaber, H. W. (1963). *Justus Liebigs Ann. Chem.* **663**, 123.
Sobti, R. R., and Dev, S. (1970). *Tetrahedron* **26**, 649.
Sondheimer, F., Mechoulam, R., and Sprecher, M. (1964). *Tetrahedron* **20**, 2473.
Stautzenberger, A. L., MacLean, A. F., and Hobbs, C. C. (1967). *Amer. Chem. Soc., Div. Petrol. Chem.*, 1967 Paper D-85.
Steele, D. R., and Rylander, P. N. (1968). Unpublished observations, Engelhard Minerals and Chemicals Corp., Engelhard Ind. Div., Newark, New Jersey.

Stevens, C. L., Filippi, J. B., and Taylor, K. G. (1966). *J. Org. Chem.* **31**, 1292.
Stork, G., Meisels, A., and Davies, J. E. (1963). *J. Amer. Chem. Soc.* **85**, 3419.
Tronchet, J. M. J., and Tronchet, J. (1968). *C. R. Acad. Sci., C* **267**, 627.
Waters, W. A. (1945). *Annu. Rep. Progr. Chem.* **42**, 152.
Weinstock, J., and Wolff, M. E. (1960). U.S. Patent 2,960,503.
Wettstein, A., Anner, G., Heusler, K., Ueberwasser, H., Wieland, P., Schmidlin, J., and Billeter, J. R. (1961). U.S. Patent 2,994,694.
Wettstein, A., Jeger, O., Anner, G., Heusler, K., Kalvoda, J., Meystre, C., and Wieland, P. (1968). U.S. Patent 3,385,848.
Wolfe, S., Hasan, S. K., and Campbell, J. R. (1970). *Chem. Commun.* p. 1420.
Wolff, M. E., and Zanati, G. (1969). *J. Med. Chem.* **12**, 629.
Zbiral, E., and Rasberger, M. (1968). *Tetrahedron* **24**, 2419.
Zelikoff, M., and Taylor, H. A. (1950). *J. Amer. Chem. Soc.* **72**, 5039.
Zimmerman, H. E., and Samuelson, G. E. (1969). *J. Amer. Chem. Soc.* **91**, 5307.
zu Reckendorf, W. M., and Bischof, E. (1970). *Tetrahedron Lett.* p. 2475.

CHAPTER 5

Isomerization

Condon (1958) pointed out that a 500-page book was needed in 1942 (Egloff *et al.*, 1942) to cover the subject of isomerization of hydrocarbons; the field has burgeoned since then. Much of the impetus for study of isomerization stems from its importance in major petroleum refining processes, and on the theoretical side, as a means of elucidating mechanisms of metal catalysis. Both of these areas have been reviewed frequently and are not emphasized in the present discussion, which focuses on the use of noble metal-catalyzed isomerization reactions useful to synthetic organic chemists. Isomerization here includes double-bond migration, *cis–trans* isomerization, configurational changes, allylic rearrangements, and various skeletal changes.

DOUBLE-BOND MIGRATION

Double-bond migration in olefins has been observed in the presence of all noble metals with and without hydrogen and both heterogeneously and homogeneously (Hubert and Reimlinger, 1970). Much attention has been devoted to mechanistic aspects of these processes (Orchin, 1966; Davies, 1967; Hallman *et al.*, 1967; Hartley, 1969; Lyons, 1971a), but they will be considered here only insofar as the mechanistic study clarifies the choice and handling of catalytic systems. Historically most isomerizations of double bonds by metals have been carried out with heterogeneous catalysts, but more recently homogeneous catalysts have come to the fore, a situation brought about by the newness of homogeneous metal catalysis, by interest in theoretical aspects of isomerization, and by the possibility of achieving novel results.

Homogeneous Catalysts

A wide variety of homogeneous metal catalysts have been employed for double-bond migrations. Various generalities concerning the functioning of these catalysts have proved useful. Cramer and Lindsey (1966) pointed out three frequent characteristics of isomerization systems: (1) a co-catalyst is often required, (2) regardless of the co-catalysts, isomerization is facilitated by acid, (3) isomerization in deuterated solvent yields deuterated olefin. Transition metal compounds become catalysts for isomerization only when they are converted to hydrides. Isomerization proceeds by addition of metal hydride to coordinated olefin followed by isomerizing elimination. Among the reactions available for hydride generation are oxidation of the carbon compound by co-catalysts, disproportionation of hydrogen, oxidation of the metal with hydrogen or proton, and displacement of hydrogen from coordinated olefin. Each of these reactions furnishes initiating hydride but the exchange of olefin protium and solvent deuterium that accompanies isomerization in CH_3OD or CH_3COOD indicates that hydride is regenerated from solvent proton; proton is therefore the working catalyst, a supposition confirmed by the inhibiting effect of added base. Reversible oxidation of metal by solvent proton serves as the means of exchanging olefinic hydrogen with solvent proton (Cramer and Lindsey, 1966; Cramer, 1966).

All acids are not equivalent in their promoting effects. Isomerization of 1-butene with lithium palladous chloride is strongly promoted by trifluoroacetic acid, less so by perchloric acid, and diminished by hydrochloric acid due to the chloride ion (Cramer and Lindsey, 1966). Promotion effects are not general and the chemistry of the isomerizing catalyst has to be considered in each case. The rate of olefin isomerization catalyzed by $RuCl_2(Ph_3P)_3$ is greatly accelerated in the presence of small amounts of hydroperoxides, which convert the ruthenium complex into a catalytically active compound having a CO ligand. Small amounts of compounds such as benzyl formate and phenylacetaldehyde which are decarbonylated easily also are effective promoters. Carbon monoxide itself functions as a promoter, but the results are complicated due to the formation of an insoluble dicarbonyl complex, $RuCl_2(CO)_2(Ph_3P)_2$ (Lyons, 1971a, b). Oxygen has a strong promoting effect in olefin isomerization catalyzed by $(Ph_3P)_3RhCl$ (Augustine and Van Peppen, 1970).

Isomerization leads to an equilibrium distribution of isomers, but initial products are kinetically controlled, a fact that permits wide variation in product composition. Isomerization of 1-hexene in the presence of platinum(II) complexes or rhodium trichloride is characterized by preferential formation of the *cis*-2 isomer in amounts far greater than its equilibrium concentration. The percentage of *cis*-2-hexene in rhodium chloride-catalyzed isomerization reaches a maximum of about 45% and then falls to 16% as the isomerization continues. Similarly, the *cis* isomer predominates in isomerization of 1-pentene by rhodium chloride and triphenylphosphine (Horner *et al.*, 1968). In contrast to these results, isomerization of 1-pentene by a platinum–tin chloride complex results in a high initial ratio of *trans*- to *cis*-2-pentene of about 10 (Bond and Hellier, 1965a, 1967).

Migration from the 2- to the 3-position is much slower than from the 1- to the 2-position, a phenomenon permitting formation of a single positional isomer in high yield. Starting with an equimolar mixture of *cis*- and *trans*-3-heptene, no significant isomerization occurred with rhodium chloride in the relatively very long time of 24 hours. The rate of isomerizations is increased by the presence of co-catalysts that include alcohols, ether, ketones, and carboxylic acids. Of the co-catalysts tested, only water was ineffective. Isomerization by iridium trichloride proceeds only in the presence of a co-catalyst (Harrod and Chalk, 1964).

In an interesting extension of this work using vinyl- and allyl-deuterated 1-olefins, movement of deuterium was followed as isomerization proceeded. Isomerization of deuterio olefins by rhodium trichloride was found to be accompanied by a redistribution of deuterium onto all carbon atoms of the allylic system as well as by a facile intermolecular exchange of deuterium with initially unlabeled olefin. Isomerization catalyzed by bis(benzonitrile)dichloropalladium differed significantly from rhodium-catalyzed isomerization in that with the palladium catalyst, double-bond shift was accompanied by a movement of hydrogen or deuterium from C-3 to C-1, and relatively little intermolecular exchange occurred. Results with rhodium were rationalized in terms of a reaction mechanism involving a rapid, reversible addition of hydride to coordinated olefin and those with palladium by a modified π-allyl mechanism. The authors pointed out that although different catalyst systems may use different mechanisms to effect isomerization, different substrates or different reaction media may also cause changes in mechanism for the same metal-ion catalyst (Harrod and Chalk, 1966).

Isomerizations with homogeneous catalysts may be very temperature-sensitive. Isomerization of 1-hexene with tris(triphenylphosphine)chlororhodium in benzene at room temperature is very slow, whereas at 40°C, the reaction is rapid. This sensitive dependence on temperature was attributed to the metal–carbon bond being too stable to undergo rapid reaction at room temperature (Baird *et al.*, 1967). The authors pointed out that the borderline for stability is apt to be critically dependent on the nature of the ligands, and perhaps also on the charge on the complexes.

Catalyst Inactivation

Harrod and Chalk (1965) noted in discussing mechanisms of isomerization that simple palladium π-allyl complexes are ineffective isomerization catalysts under conditions where palladium olefin complexes are effective. The life of an isomerization catalyst may therefore be inversely related to the ease with which the substrate olefin or its isomers are converted to π-allyl compounds (Cramer and Lindsey, 1966; Ketley and Braatz, 1968). In a group of four hexenes, the one with most branching deactivated the catalyst most rapidly (Sparke *et al.* 1965). Bond and Hellier (1965b), on the other hand, favor π-allylic intermediates in isomerization at 70°C and suggest the mechanism may change with temperature.

Supported Complex Catalysts

In isomerization, as in other areas, efforts have been made to combine the advantages of homogeneous and heterogeneous catalysts. For example, butenes have been isomerized over noble metal complexes supported on silica gel. Bis(acetylacetonato)palladium is about 20 times more active than the corresponding platinum complex. The acidity of the silica gel contributes to the overall activity; palladium complexes supported on sodium-exchanged silica gel are much less active than on gels not so treated (Misono *et al.*, 1968). The higher activity of palladium over platinum complexes demonstrated here is general (Conti *et al.*, 1971).

Synthetic Applications

Isomerization of olefinic double bonds by noble metals has proved a valuable synthetic tool. The synthetic utility of isomerization was increased by the advent of homogeneous catalysis, which offered the possibility of obtaining products differing from those obtained heterogeneously. The results of isomerization of 4-vinylcyclohexene by different homogeneous catalysts illustrate the variability in product that may sometimes be obtained by appropriate choice of catalyst. By refluxing 4-vinylcyclohexene in an ethanolic solution of rhodium trichloride, predominantly equal amounts of the 1- and 3-isomer are obtained together with ethylbenzene and a lesser amount of ethylcyclohexane. The products are believed to have arisen by a sequence of two 1,3-hydride shifts to give a mixture of ethylcyclohexadienes that undergo disproportionation. Rhodium trichloride was appreciably more effective in this reaction than several complex rhodium catalysts (Attridge and Wilkinson, 1971).

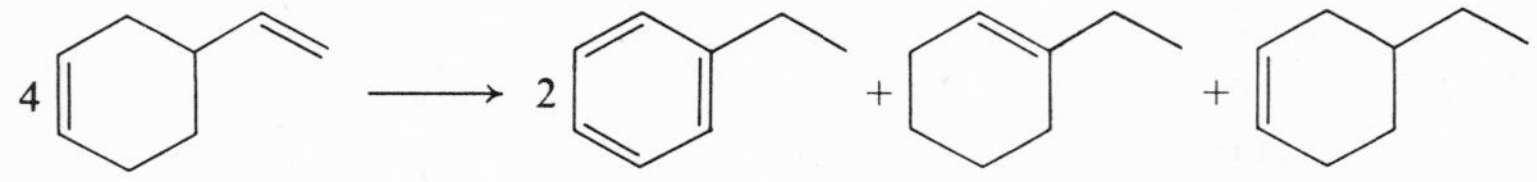

4 [4-vinylcyclohexene] ⟶ 2 [ethylbenzene] + [1-ethylcyclohexene] + [4-ethylcyclohexene]

The above results contrast to those obtained with $PtCl_2(Ph_3P)_2 \cdot SnCl_2$ which affords a mixture of nonconjugated isomers in 65% yield (Lyons, 1971a, b). Similar results are obtained with a mixture of $(Ph_3P)_3RhCl$ and $(C_2H_5)AlCl_2$ as catalyst (Hughes, 1970). The ruthenium complex, $RuCl_2(Ph_3P)_3$, is substantially inactive for isomerization of pure 4-vinylcyclohexene unless small amounts of air or hydroperoxides are added. All these catalysts retain the double bond selectively in the exocyclic position in contrast to conventional acid or base isomerization catalysts (Lyons, 1971a, b).

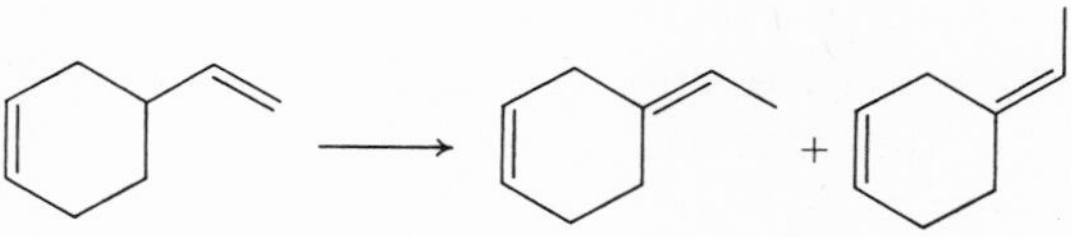

Isomerization of 1,5-cycloctadiene to the 1,3-isomer is readily achieved with a bis(benzonitrile)palladium dichloride catalyst. Ten grams of 1,5-cyclooctadiene refluxed 90 minutes with 0.3 gm of the palladium catalyst affords 1,3-cyclooctadiene in 99.5% yield (Zuech, 1968). The same reaction may be achieved with a number of other catalysts (Arnet and Pettit, 1961; Lafont and Vivant, 1963; Tayim and Bailar, 1967; Tayim and Vassilian, 1970), but rhodium chloride, on the other hand, permits quantitative isomerization in the reverse sense through formation of a dimer $(1,5\text{-}C_8H_{12})RhCl_2$. Pure 1,5-cyclooctadiene can be obtained by treating the dimer with aqueous potassium cyanide (Rinehart and Lasky, 1964).

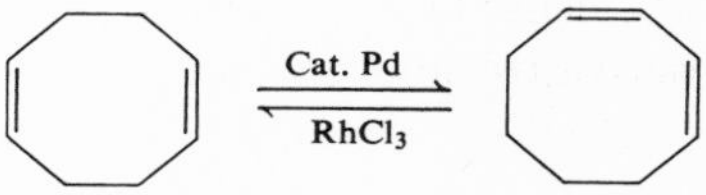

Cat. Pd
$RhCl_3$

Under other conditions, bis(benzonitrile)palladium dichloride converts 1,5-cyclooctadiene to the metal complex of 1,2-divinylcyclohexane (Trebellas *et al.*, 1966).

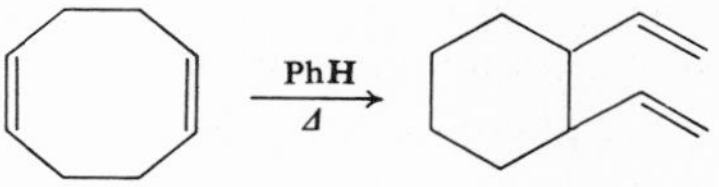

PhH
Δ

This catalyst has also been used for the isomerization of 4-phenyl-1-butene to a mixture of *cis*- and *trans*-1-phenyl-2-butene and *trans*-1-phenyl-1-butene. Large portions of 1-phenyl-1-butene are formed without passing through 1-phenyl-2-butene (Davies *et al.*, 1968).

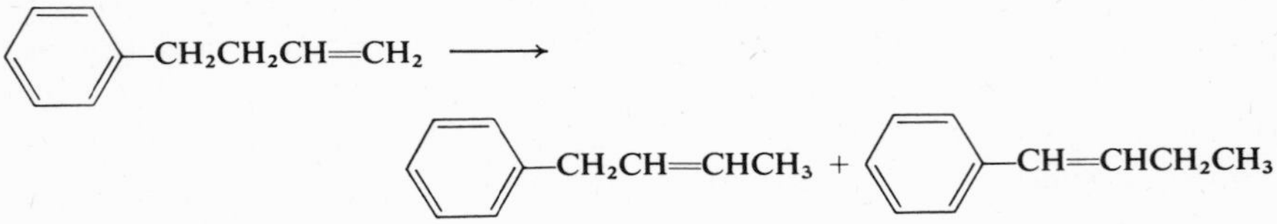

$C_6H_5-CH_2CH_2CH{=}CH_2 \longrightarrow C_6H_5-CH_2CH{=}CHCH_3 + C_6H_5-CH{=}CHCH_2CH_3$

Rhodium chloride has been used for a variety of isomerizations. Terpenes such as myrcene, myrcenol, and dihydromyrcene are isomerized in good yield and with minimum polymerization to ocimene, ocimenol, and allo-dihydromyrcene, respectively. The reaction is preferably carried out in a nitrogen atmosphere and in the presence of hydrogen chloride and an oxidation inhibitor. For example, 300 gm of myrcene, 15 gm of 2,6-di-*tert*-butyl-4-methylphenol, 1.5 gm of rhodium trichloride, and 3 gm of concentrated aqueous hydrochloric acid, heated under nitrogen 2 hours at 105°C, affords 245 gm of ocimene in 80% yield. Supported rhodium catalysts may be used as well (Lemberg, 1967).

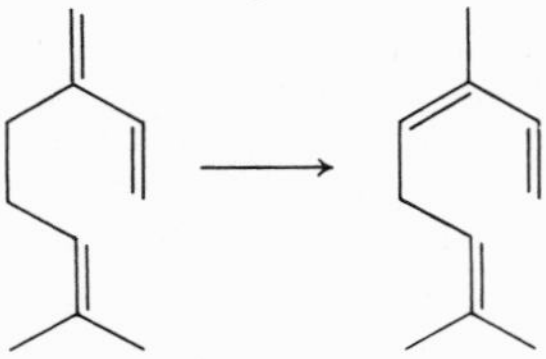

Rhodium chloride has been used for isomerization of 2,5-dihydrothiophene-1,1-dioxide to the 2,3-derivative and of furfuryl alcohol to 2,3-dihydro-2-furaldehyde (Rinehart, 1969).

$CH_2OH \longrightarrow CHO$

Diallyl ether (**I**) containing a few percent of allyl alcohol is rapidly converted to the cyclized product, **II**, when heated with catalytic amounts of rhodium trichloride. The conversion is thought to involve two insertions with a rhodium hydride intermediate. Allyl phenyl ether is isomerized by this catalyst to a mixture of *cis*- and *trans*-phenyl propenyl ethers (Bright *et al.*, 1971).

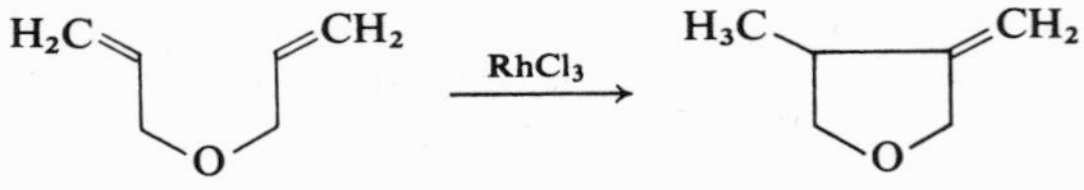

$H_2C{=}$ $=CH_2$ $\xrightarrow{RhCl_3}$ H_3C $=CH_2$

O O

(I) **(II)**

Isomerization of 1,4-dienes to 1,3-dienes may be achieved cleanly under mild conditions with tris(triphenylphosphine)rhodium chloride (MacSweeney and Ramage, 1971). 1-Methoxycyclohexa-1,4-diene is equilibrated with the more stable conjugated isomer, 1-methoxycyclohexa-1,3-diene, by refluxing in chloroform for 2 hours with 1% catalyst by weight. No isomerization was observed at room temperature. Rhodium-catalyzed isomerization of **III** affords **IV** after only refluxing for 5 minutes in chloroform, whereas base-catalyzed isomerization is arrested at **V** (Birch and Subba Rao, 1968).

OCH_3 C_2H_5 **(V)** $\xleftarrow{\text{base}}$ OCH_3 C_2H_5 **(III)** $\xrightarrow{\text{Rh}}$ OCH_3 $CHCH_3$ **(IV)**

Heterogeneous Catalysts

Isomerization of olefins occurs over all noble metals. A comparison of catalysts used in isomerization of olefins revealed the following decreasing order of activity: Pd > Ni > Rh > Ru > Os ≃ Ir ≃ Pt (Obeschchalova *et al.*, 1968; Abubaker *et al.*, 1968). The order is somewhat different than the relative order assigned for isomerization of 1-pentene under hydrogenation conditions with is, Pd ≫ Ru > Rh > Pt ≫ Ir (Bond and Rank, 1965), but in both comparisons palladium is the most active catalyst. Two metals together may have synergistic effects (Plonsker and McEuen, 1967). Platinum has been used quite effectively despite its relatively inefficiency (Maurel *et al.*, 1966; Levina and Petrov, 1937). A variety of catalyst supports may be used in isomerization. At elevated temperatures, acid sites on the support may also contribute to the isomerization. Wells and Wilson (1967) after examining butene isomerization in the absence of hydrogen over all noble metals-on-alumina concluded that the support appears to be an essential constituent of the catalyst. Hydrogen atoms were thought to migrate from the support to the metal to initiate isomerization.

Isomerization in the Presence of Hydrogen

Isomerization is a frequent concomitant of hydrogenation reactions and in many cases controls the outcome of reduction. Such varied factors as selectivity among different types of olefins, migration of olefins to inaccessible positions, changes in the functional group, catalyst inhibition, stereochemistry, and loss of optical activity can all be related to an isomerization prior to saturation (Rylander, 1971). The properties of partially hydrogenated fats depends in large measure on the geometrical and positional isomer content that arises during the hydrogenation due to isomerization. The use of noble metal catalysts in fat hydrogenation has been reviewed recently (Rylander, 1970a). Isomerization also has an important influence on the products of aromatic hydrogenation (Rylander, 1970b).

Hydrogen is a powerful co-catalyst for isomerization of olefins over noble metals, its effectiveness for isomerization being limited only by its tendency to enter into the reaction products. The ratio of isomerization activity to hydrogenation activity can be influenced by various additives and by reaction conditions. Isomerization is favored by conditions giving hydrogen-poor catalysts, i.e., active catalysts, low pressure, high temperature, and poor agitation (Zajcew, 1960a). Hydrogenation activity of palladium catalysts can be sharply curtailed by the addition of a sulfur compound to the substrate, whereas the isomerization activity is little affected. Effective isomerization of 1-olefins to internal olefins can be achieved by passing a stream containing terminal olefins over a palladium-on-alumina catalyst at 160° to 200°C in the presence of hydrogen and small amounts of a sulfur-containing compound (Garner and Benedict, 1970). Heavy metals, alkalis, and amines decrease isomerization activity relative to hydrogenation activity (Huntsman *et al.*, 1963). Certain steric requirements must be met for double-bond migration to occur. The allylic hydrogen to be removed must be sterically accessible to the catalyst and on the same side of the molecule as the entering hydrogen (Bream *et al.*, 1957; Fischer and Mabry, 1968).

Configurational Changes

Certain compounds may undergo racemization in the presence of hydrogen and noble metals without hydrogenation taking place. For example, the protoberberine, (–)-coreximine, and derivatives when hydrogenated in the presence of platinum oxide undergoes racemization at C-13. Labeling experiments with deuterium established that hydrogen exchange at C-13 accompanied racemization. No exchange and no racemization occurred in hydrogenations using palladium oxide or palladium-on-carbon (Kametani and Ihara, 1968).

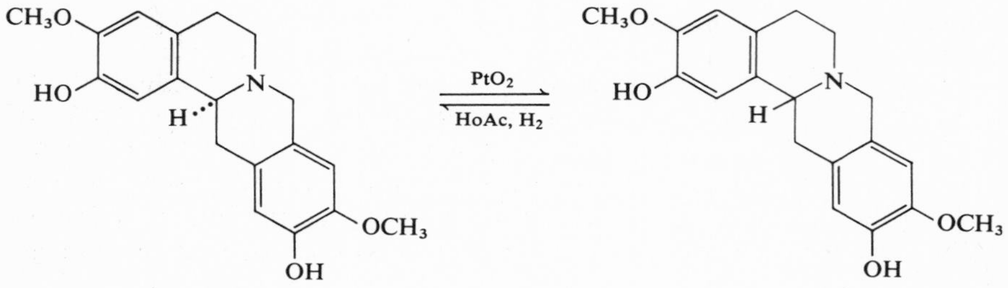

CH_3O, HO, N, H, OCH_3, OH

$$\xrightarrow[HoAc,\ H_2]{PtO_2}$$

CH_3O, HO, N, H, OCH_3, OH

Treatment of L-(−)-norlaudanosine (**VI**) for 48 hours with platinum oxide and hydrogen at room temperature and pressure affords the racemate, but (−)-1,2,3,4-tetrahydro-6,7-dihydroxy-1-(3,4,5-trimethoxybenzyl)isoquinoline (**VII**), which has vicinal phenolic hydroxy groups in the isoquinoline moiety, is not racemized. The infrared adsorption spectrum of **VII** showed a betaine-type structure, a structure unable to form a double bond between a carbon and nitrogen atom, which the authors postulated as a necessary intermediate in the racemization. No racemization of norlaudanosine occurred over platinum oxide in ethanol in the presence of sodium ethoxide nor over palladium black, palladium-on-carbon, or W_2-Raney nickel (Kametani *et al.*, 1968). Similarly, platinum oxide in acetic acid was effective for the racemization of (−)-coreximine, whereas palladium oxide in acetic acid and palladium-on-carbon or W_2-Raney nickel in ethanol were not (Kametani and Ihara, 1968).

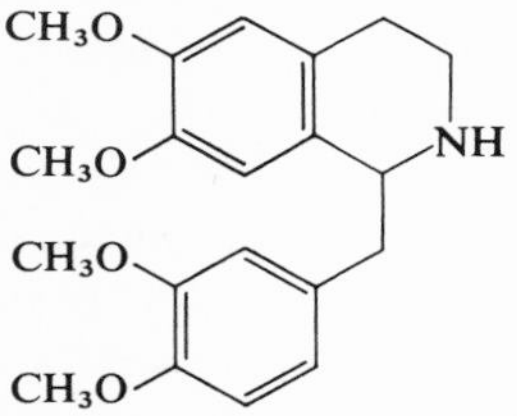

(**VI**)

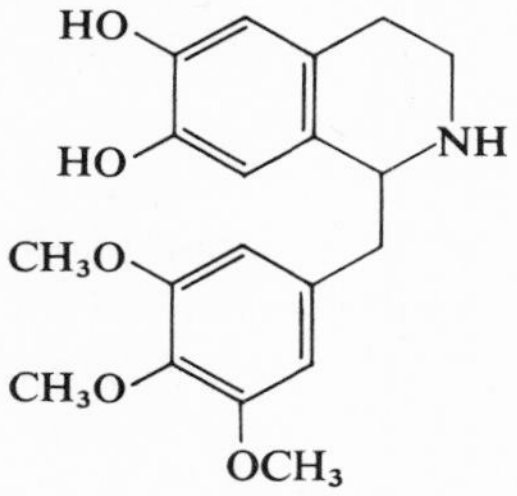

(**VII**)

Cis–Trans Isomerization

Olefins in the presence of noble metal catalysts can undergo *cis–trans* isomerization; the process is speeded tremendously if carried out in the presence of hydrogen. Isomerization may occur without a shift in the position of the double bond or it may occur as a consequence of migration (Dutton *et al.*, 1968). The geometrical isomer formed as a result of migration depends on the conformation of adjacent carbons at the time double-bond migration occurs. A parallel in catalytic activity for double-bond migration and *cis–trans* isomerization need not necessarily exist, although if most of the isomerization is a consequence of migration, it is reasonably expected. For instance, the increasing order for noble metals in both geometrical and positional isomerization of tall oil fatty acids is Pt < Ir < Ru < Rh < Pd (Zajcew, 1960b).

Geometrical isomerization without migration is best seen in those molecules such as stilbene where migration is impossible, or in molecules such as 3-methyl-2-pentene, where the double bond is already in a preferred position. The increasing order for activity in *cis–trans* isomerization of 3-methyl-2-pentene in alcohol is Pt < Ni < Pd (Gostunskaya *et al.*, 1964). In isomerization of *cis*-stilbene in ethanol in the presence of hydrogen, the order of increasing activity is $\mathrm{Ru} < \mathrm{Pt} < \mathrm{Pd} < \mathrm{Rh}$. No isomerization at room temperature occurs without hydrogen present except over ruthenium. *cis*-Cinnamic acid is isomerized in refluxing ethanol without hydrogen with the following order of increasing activity: $\mathrm{Pd} < \mathrm{Rh} < \mathrm{Ru}$; with hydrogen present Rh is the most active isomerization catalyst (Bellinzona and Bettinetti, 1960). Isomerization relative to hydrogenation may increase with increasing temperature (Pietra and Bertoglio, 1956). *cis*-2-Styrylpyridine is converted quantitatively to the *trans* isomer by heating in *p*-cymene with a trace of 15% palladium-on-carbon. Under the same condition in the absence of catalyst, no conversion occurs (Williams *et al.*, 1961).

$$\text{2-Pyridyl(H)C=C(H)Ph } (cis) \longrightarrow \text{2-Pyridyl(H)C=C(H)Ph } (trans)$$

Isomerization of the highly hindered olefin *cis*-di-*tert*-butylethylene to the *trans* isomer by acidic catalysis is unusually difficult, but the reaction can be accomplished readily by refluxing the substrate for 2 hours over 5% palladium-on-alumina (Puterbaugh and Newman, 1959).

$$\mathrm{(CH_3)_3C(H)C{=}C(H)C(CH_3)_3}\ (cis) \longrightarrow \mathrm{(CH_3)_3C(H)C{=}C(H)C(CH_3)_3}\ (trans)$$

Homogeneous noble metal catalysts are also effective for *cis–trans* isomerization. *cis-trans*-1,5-Cyclodecadiene is isomerized by rhodium chloride in proton-donor solvents to *cis-cis*-1,6-cyclodecadiene. Isomerization is thought to occur via a hydrido-alkyl intermediate (Trebellas *et al.*, 1967). Rhodium chloride also rapidly converts a 96% *cis*-polybutadiene to a 69% *trans* isomer in a water emulsion with sodium dodecylbenzene sulfonate at 50° to 80°C (Rinehart *et al.*, 1962). *cis, trans, trans*-Cyclododeca-1,5-9-triene is isomerized to the all-*trans* isomer by $\mathrm{RuCl_2(Ph_3P)_2}$ and $\mathrm{IrCl(CO)(Ph_3P)}$ but $\mathrm{Co(Ph_3P)_2Cl_2}$, dichlorobis(cycloocta-1,5-diene)dirhodium, cycloocta-1,5-dieneplatinum(II) dichloride, and π-cyclododeca-1,5,9-trienylrhodium(III) chloride are inactive (Attridge and Maddock, 1971). 2-Phenyl-*trans*-butene is isomerized almost quantitatively to the *cis* isomer in the presence of palladium acetate and refluxing benzene–acetic acid (Danno *et al.*, 1969).

$$\mathrm{H_3C(H)C{=}C(Ph)CH_3} \longrightarrow \mathrm{H_3C(H)C{=}C(CH_3)Ph}$$

Enol esters undergo *cis–trans* isomerization in the presence of Pd(II) as a result of an exchange reaction. The kinetics and stereochemical results are consistent with an oxypalladation–deoxypalladation mechanism for ester exchange. Every exchange involves a *cis–trans* isomerization. As expected from this mechanism, exchange of cyclic enol esters in which rotation of the oxypalladation product is impossible, fails to occur (Henry, 1971).

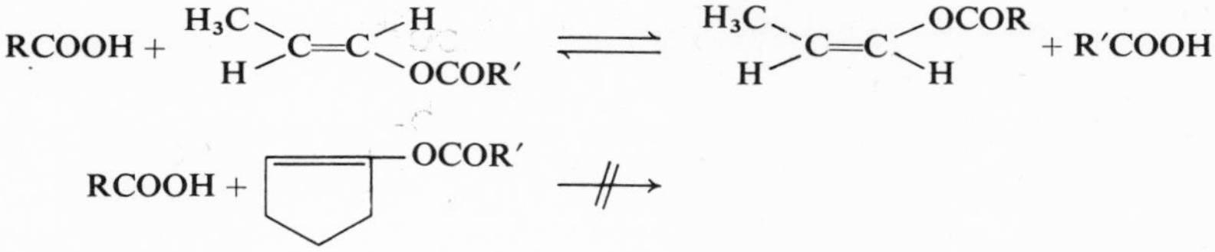

$$\mathrm{RCOOH} + (H_3C)(H)C{=}C(H)(OCOR') \rightleftharpoons (H_3C)(H)C{=}C(OCOR)(H) + \mathrm{R'COOH}$$

$$\mathrm{RCOOH} + \text{1-cyclopentenyl-}OCOR' \not\rightarrow$$

Aromatization

Hydrogenations of multiple unsaturated compounds are sometimes incomplete due to migration of one or more double bonds into a hydrogenation-resistant aromatic system. An example is the conversion of the coumarin, poncitrin, to tetrahydroponcitrin (Tomimatsu *et al.*, 1969).

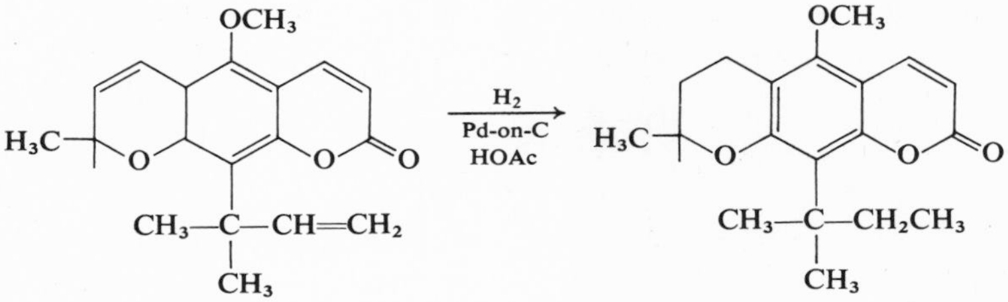

OCH_3; H_3C; O; O; O; $CH_3{-}C{-}CH{=}CH_2$; CH_3 $\xrightarrow[\text{Pd-on-C, HOAc}]{H_2}$ OCH_3; H_3C; O; O; O; $CH_3{-}C{-}CH_2CH_3$; CH_3

Another example is the conversion of 4,5-acenaphthenequinonedibenzenesulfonimide to 4,5-dibenzenesulfonamidoacenaphthene (Richter and Weberg, 1958). Such isomerizations during hydrogenation are best prevented by the use of homogeneous catalysts (see chapter on homogeneous hydrogenation) or, if heterogeneous catalysts are used, by the avoidance of palladium, the best of the noble metals for promoting isomerization.

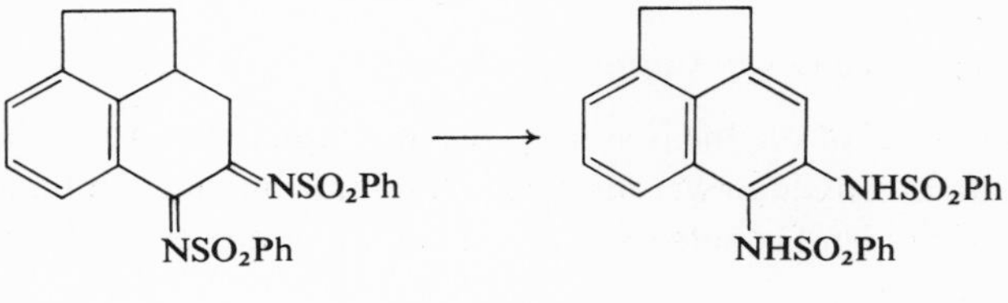

NSO_2Ph; NSO_2Ph $\longrightarrow$ $NHSO_2Ph$; $NHSO_2Ph$

Aromatization by isomerization may occur following other changes in the molecule. Hydrogenation of desmethoxy-β-erythroidine over platinum oxide in neutral medium results in formation of the aromatic compound, allo-dihydrodes-methoxy-β-erythroidine. The reaction probably proceeds by hydrogenolysis of the carbon–nitrogen bond allylic to three olefinic bonds, aromatization by isomerization, and ring closure by 1,4-addition to the conjugated lactone. The formation of this product may be prevented by conducting the hydrogenation in acidic medium (Boekelheide *et al.*, 1953).

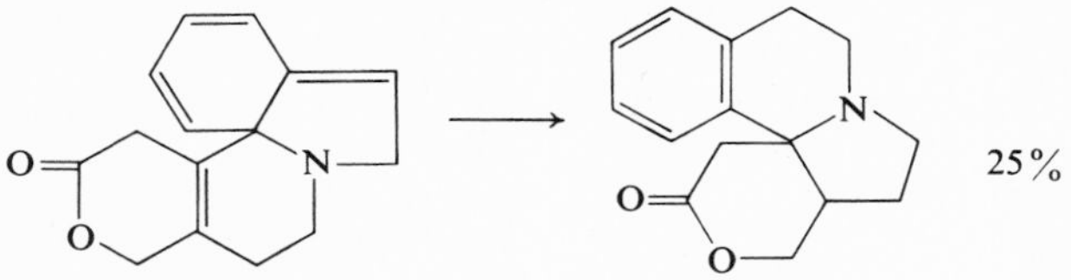

25%

Migration into Inaccessible Positions

Facile migration of a double bond into a tetrasubstituted position, a position hydrogenated only with some difficulty, permits the use of hydrogen in isomerization reactions without fear of concomitant saturation (Rylander, 1967). The triterpenoid, mexicanol (**VIII**), was converted almost quantitatively into isomexicanol (**IX**) when shaken with platinum oxide and hydrogen in acetic acid. This isomerization together with facile oxidation of the $\Delta^{7,9(11)}$-diene helped place the original bond at position 7((8) (Connolly *et al.*, 1967).

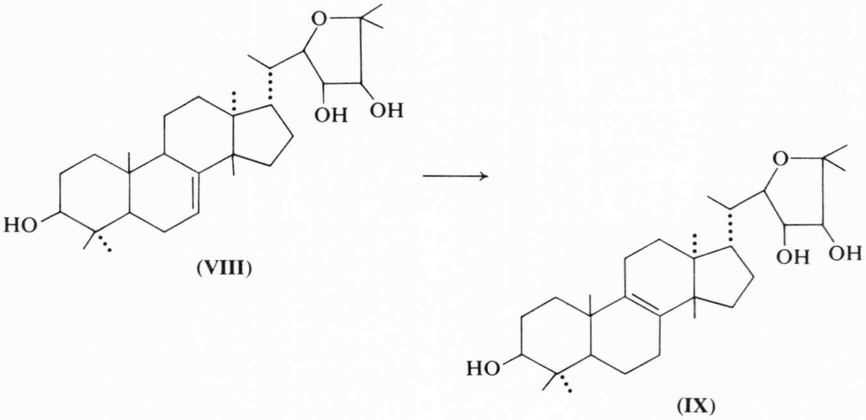

(VIII)

(IX)

Conversions of Alcohols to Carbonyls

Hydrogenation of olefins is sometimes incomplete due to conversion of an alcohol to an aldehyde or ketone through migration of a double bond. When the isomerizing metal is palladium, the reaction comes to a virtual halt at this stage inasmuch as palladium is a relatively poor catalyst for the saturation of aliphatic carbonyl compounds. With saturation sufficiently inhibited, the isomerization gains synthetic utility. Hydrogenation of 3-epiisotelekin (**X**) over 10% palladium-on-carbon in ethyl acetate affords the saturated ketone, **XII**, as the major product. The authors ruled out isomerization of **X** to the enol **XI** as source of the saturated ketone, with the observation that treatment of **X** with catalyst without hydrogen resulted mostly in unchanged substrate plus 5–6% of an oxidation product, **XIII** (Herz *et al.*, 1968). However, since generally isomerization over platinum metal catalysts occurs very much more readily in the presence of hydrogen than in its absence, this comparison may be misleading.

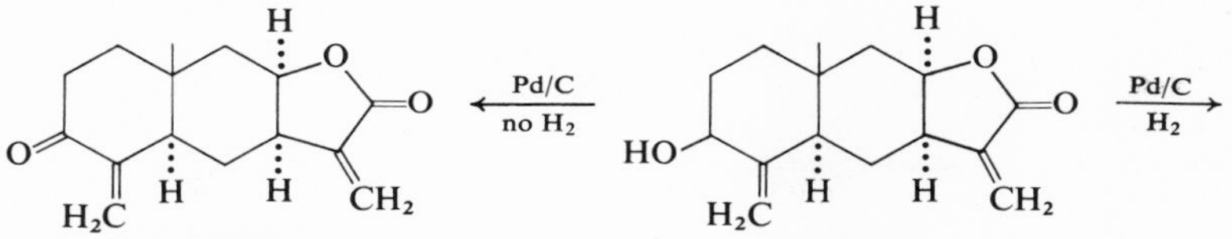

(XIII) ← Pd/C, no H_2 — (X) — Pd/C, H_2 →

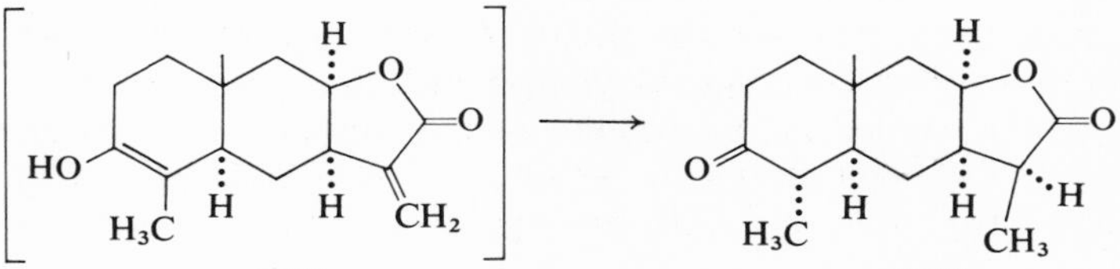

[(XI)] → (XII)

A similar type of rearrangement is found in the hydrogenation of the kaurene derivative, **XIV**, in which an allylic alcohol is converted to a saturated aldehyde, **XV** (Barnes and MacMillan, 1967). Allylic migrations of this type when the double bonds are sterically impeded is well documented (McQuillin, 1963).

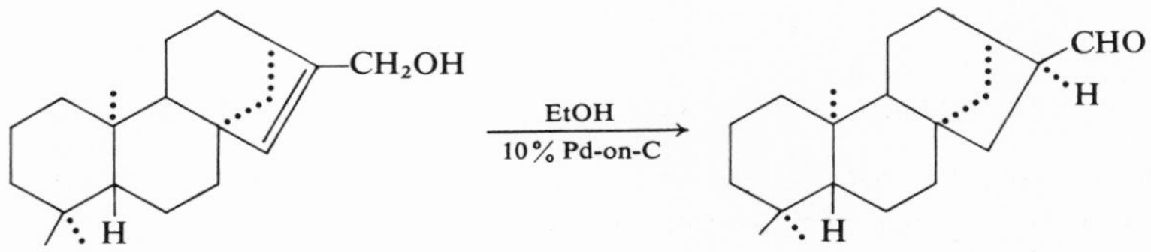

(XIV) — EtOH, 10% Pd-on-C → (XV)

An unusual reaction occurs when morphine (**XVI**) is heated in acid in the presence of a noble metal catalyst with or without hydrogen (Rodd, 1960). Major products are dihydromorphinone (**XVII**) and *O*-desmethylthebainone (**XVIII**) formed in proportions that depend on the experimental conditions. The highest yield, 60%, of **XVIII** is obtained with large amounts of 10% palladium-on-carbon. The proportion of **XVIII** decreases and **XVII** increases as the quantity of catalyst is decreased. The use of palladium black also favors **XVII** (Weiss and Weiner, 1949).

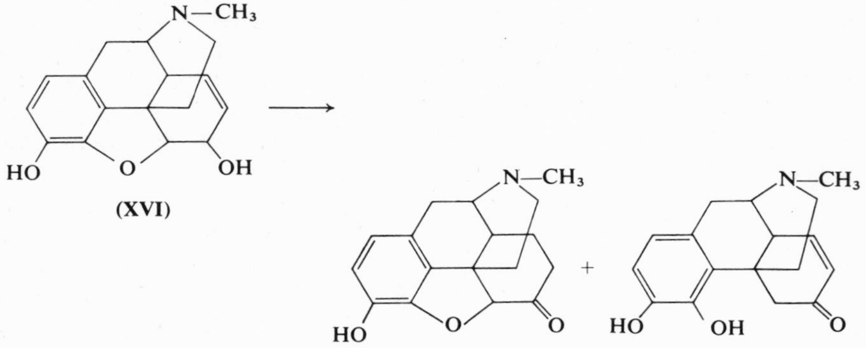

(XVI) → (XVII) + (XVIII)

Acetylenic alcohols may also be converted to saturated carbonyl compounds during hydrogenation through isomerization of the intermediate unsaturated alcohol. Under suitable conditions, the saturated carbonyl compound can be made the major product of the reaction. The propargyl alcohols **XIX** and **XX** when vaporized with water and hydrogen and passed over a palladium–iron–kieselguhr catalyst containing added sulfur are converted to propionaldehyde and methyl ethyl ketone, respectively (Reppe *et al.*, 1955).

$$\underset{\textbf{(XIX)}}{HC{\equiv}CCH_2OH} + H_2 \xrightarrow{105°C} CH_3CH_2CHO$$

$$\underset{\textbf{(XX)}}{HC{\equiv}C\underset{\substack{|\\OH}}{C}HCH_3} + H_2 \xrightarrow{105°C} CH_3CH_2\underset{\substack{\|\\O}}{C}CH_3$$

Carbonyl Migrations

Bis-acetylenic ketones undergo an extremely facile intramolecular isomerization in the presence of platinum(IV) chloride.

$$RC{\equiv}C{-}\overset{\overset{O}{\|}}{C}{-}C{\equiv}CR \xrightarrow{PtCl_4} RC{\equiv}C{-}C{\equiv}C{-}\overset{\overset{O}{\|}}{C}R$$

A solution of 1 mmole of 3-oxo-1,5-diphenylpentadiyne (R = Ph) in 10 ml of ethyl acetate containing 0.35 mmole of $PtCl_4$ affords after 3 minutes, 5-oxo-1,5-diphenylpentadiyne in 72% yield (Müller and Segnitz, 1970).

Migration in the Absence of Hydrogen

Double-bond migration occurs over all platinum metal catalysts without added hydrogen but the reaction is much slower in its absence and usually proceeds at a convenient rate only at elevated temperatures. Isomerization without hydrogen has the advantage that concomitant saturation is avoided except through disproportionation reactions. Disproportionation has been evoked to explain reactions that on the surface appear to be isomerizations. *N*-Styrylindolin (**XXI**) when treated with palladium-on-carbon in refluxing xylene is transformed into *N*-(β-phenylethyl)indole (**XXII**) (Mabry *et al.*, 1967). If the reaction proceeds by isomerization, the intermediate would be similar to the quaternary nitrogen formed when dimethylaniline is converted to cyclohexanone through reductive hydrolysis (Kuhn and Haas, 1958).

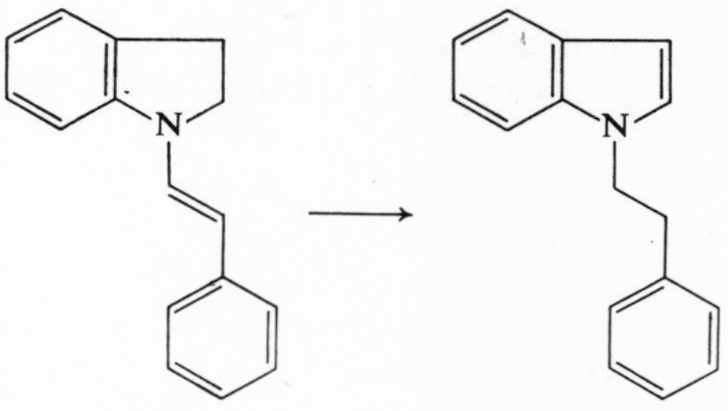

N N

(XXI) (XXII)

Isomerization of allylic alcohols to the corresponding ketones can be achieved by heating with noble metal catalysts. Special procedures may be needed to achieve satisfactory results. Conversion of the glycol, **XXIII**, to the corresponding ketones is minimal when the substrate is heated under reflux at 130°C and 15 mm Hg pressure over 5% Pd-on-$CaCO_3$, but when the system is arranged to distill the ketones as formed, the reaction proceeds smoothly. Presumably the ketones are too strongly adsorbed on the catalyst to permit any substantial conversion (Pascal and Vernier, 1969).

$$\underset{\text{(XXIII)}}{CH_3CH{=}CH\underset{\text{HO}}{CH}\underset{\text{OH}}{CH}CH{=}CHCH_3} \longrightarrow CH_3(CH_2)_2\underset{\text{O}}{\overset{}{C}}{-}\underset{\text{O}}{C}(CH_2)_2CH_3 + CH_3(CH_2)_3\underset{\text{O}}{C}{-}\underset{\text{OH}}{CH}CH{=}CHCH_3$$

Some of the advantages of isomerization in the presence of hydrogen may be obtained without hydrogen by carrying out the reaction in refluxing alcohol, or even with small amounts of alcohol present. Apparently, catalytic quantities of hydrogen are provided by dehydrogenation of the hydroxyl function. For instance, over 5% palladium–5% ruthenium-on-carbon, 1-dodecene was 74% converted to 2-dodecene by refluxing 11 hours in a nitrogen atmosphere, but the same conversion was achieved in 30 minutes with *n*-octanol present. Similar comparisons were made with a variety of catalysts (McEuen *et al.*, 1968).

Configurational Changes

Isomerization of paraffins may involve only a change in configuration or extensive changes involving rearrangement of the carbon skeleton. The former change occurs more readily and isomerizations of this type may be achieved easily without skeletal rearrangement. Equilibration of configurational isomers over noble metals has proved a useful means of establishing their equilibrium concentrations (Allinger and Cole, 1959). The equilibrations are carried out at elevated temperature with or without added hydrogen in the presence of noble metal catalysts, frequently palladium. The technique has been applied to decalins (Allinger and Cole, 1959), hydrindanes (Allinger and Cole, 1960), perhydroazulenes (Allinger and Zalkow, 1961), isopropylcyclohexanes (Allinger and Hu, 1962), *t*-butylcyclohexanone (Allinger *et al.*, 1966), dimethylcyclohexanes (Allinger *et al.*, 1968), and all C_8 and C_9 di- and trialkylcyclohexanes (Mann, 1968). The equilibrations are thought to proceed by the dehydrogenation–hydrogenation activity of the catalysts (Allinger and Cole, 1960). Both *cis*- and *trans*-bicyclo[5.2.0]nonane are left virtually unchanged under normal reaction conditions. The resistance to equilibration is interpreted as due to the reluctance of a carbon atom in the four-membered ring to assume a planar configuration, as is presumably necessary for isomerization. Equilibration is facilitated by the presence of hydrogen. The rate of equilibration of 1,2-dialkylcyclopentanes and 1,3-dimethylcyclohexanes over noble metals falls steadily with increasing helium in a helium–hydrogen mixture. Supported osmium, rhodium, iridium, and ruthenium catalyst lose considerable activity even at low helium concentration (Bragin *et al.*, 1967).

Isomerization may also be carried out conveniently in a flow reactor. The technique effectively limits contact time and avoids unnecessarily long exposure to high temperatures. Isomerization of di-*endo*-trimethylenenorbornane over 2% palladium-on-silica gel in a flow reactor at 200° to 250°C provides a convenient synthesis of the *exo* isomer. The process was entirely selective at conversion greater than 90% (Quinn *et al.*, 1970). Under the same conditions, di-*endo*-2,3-dimethylbicyclo[2.2.1]heptane was converted to equimolar mixtures of the di-*exo* and *exo-endo* dimethyl compounds. The authors suggested these transformations occur by an intermediate cyclopropyl compound.

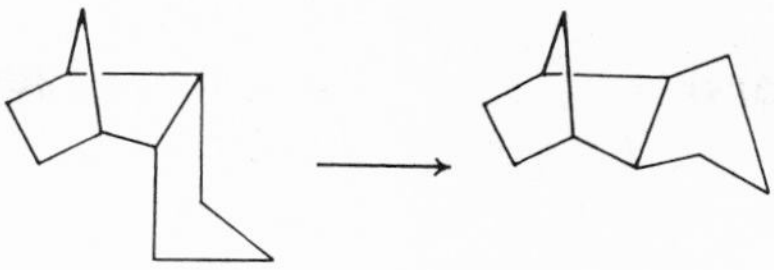

Complex Molecules

Inversion without skeletal isomerization has also been of considerable utility in synthesis of complex molecules. Epimerization at C-14 of *trans*-fused C and D rings in steroids is brought about conveniently by heating with palladium-on-carbon. The temperature can be kept low enough to avoid dehydrogenation and cleavage reactions (Bachmann and Dreiding, 1950). Dehydrogenation may occur under more vigorous conditions or at prolonged contact times. Epimerization of estrone over palladium-on-carbon in refluxing triglyme affords (+)-isoequilen in 48% yield (Pelletier *et al.*, 1971).

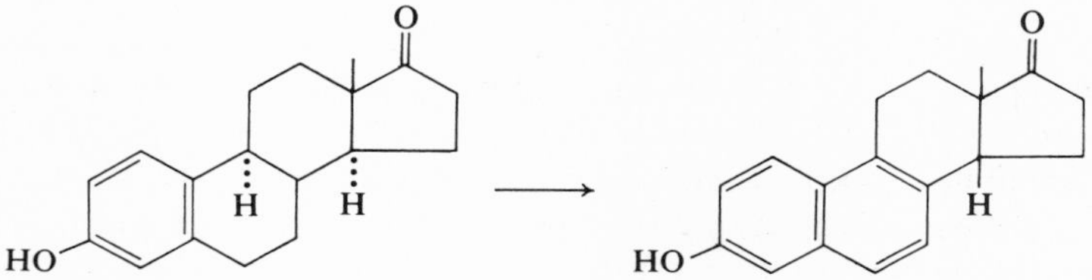

Isomerization of naturally occurring materials having the steroidal-type A/B ring juncture into the nonsteroidal (antipodal) configuration provides a convenient entry into compounds of this type. The transformation is accomplished by a two-step isomerization involving first aluminum chloride in dry benzene which gives rise to a C-10 methyl group with an α-orientation, followed by isomerization with 10% palladium-on-carbon in refluxing triglyme for 3 hours (Pelletier *et al.*, 1971).

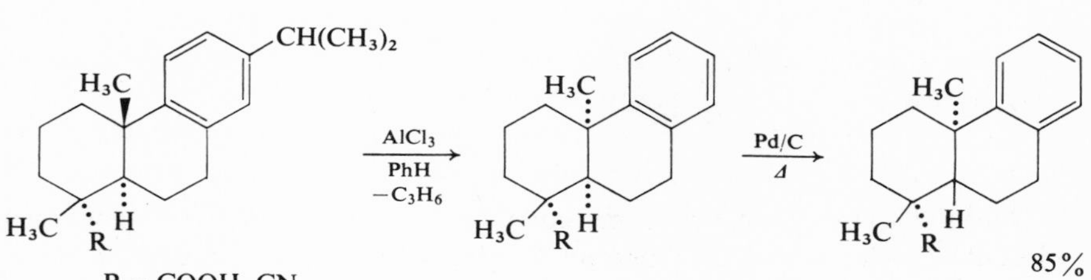

$R = COOH, CN$; $AlCl_3$, PhH, $-C_3H_6$; Pd/C, Δ ; 85%

The latter isomerization was applied to conversion of methyl 5α,10α-podocarpa-8(9)-en-15-oate (**XXIV**) to the 5β-isomer, **XXV**, in 67% yield (Pelletier *et al.*, 1971). Unexpected configurational changes of this type during dehydrogenation reactions can lead to misassignment of structure (Mathew *et al.*, 1964).

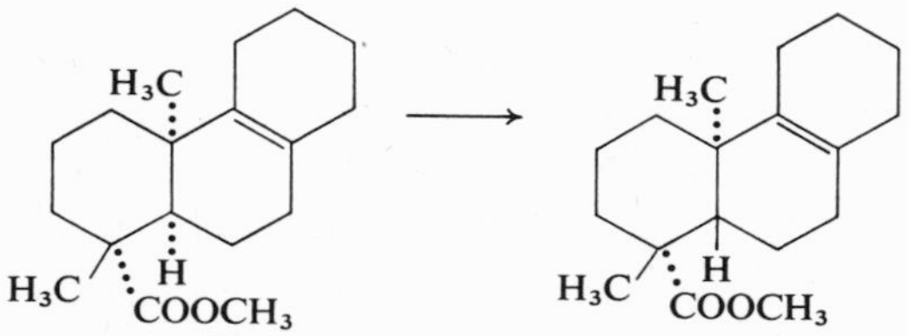

(XXIV) (XXV)

A palladium-on-zinc and -iron hydroxide catalyst that is said to be useful for effecting racemization has been described in detail. The catalyst was used in achieving partial racemization of 1-(*p*-methoxybenzyl)-2-methyl-1,2,3,4,5,6,7,8-octahydroisoquinoline with a levo rotation into its enantiomorph. The latter compound is an intermediate in the synthesis of the antitussive dextromethorphan. Racemization is conveniently carried out in methanol at 65°C, at pH 8 with 10% catalyst loading (British Patent 804,788; Nov. 26, 1958). Presumably, the purpose of this special catalyst is to prevent the disproportionation that often occurs readily when incipient aromatic systems are treated with palladium.

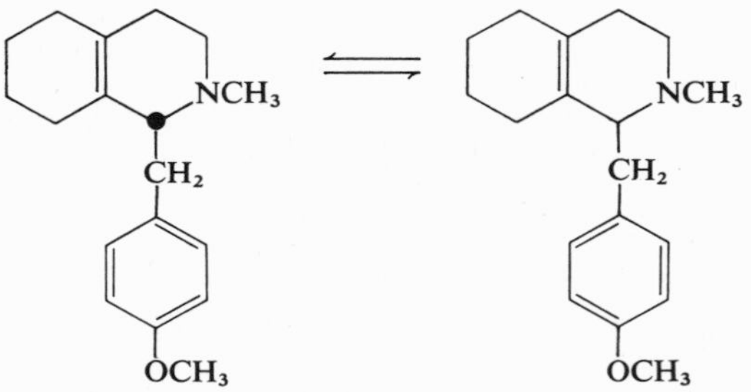

Racemization at carbon atoms carrying amino groups may sometimes present a special difficulty. Dimers are formed through dehydrogenation of the amine to an imine followed by coupling and hydrogenation to a diamine (Kindler *et al.*, 1931). This type of by-product can be prevented by carrying out the reaction in the presence of ammonia. Arthur (1964) brought an isomeric mixture of bis(4-aminocyclohexyl)methane to near equilibrium concentration by heating at 180°–250°C with ruthenium and ammonia.

Skeletal Isomerization

The importance of skeletal isomerization in modern industry cannot be overestimated. The reaction is involved in most major catalytic petroleum processes. Besides the primary purpose of increasing the octane number of gasoline, the reaction as embodied in catalytic reforming (Pollitzer *et al.*, 1970) provides a convenient route to a variety of aromatic hydrocarbons. Requirements for individual xylenes, as precursors of synthetic fibers, is met by isomerization of xylenes and ethylbenzene and separation of the resulting xylene mixture with recycle of undesired isomers (Uhlig and Pfefferle, 1970).

$$C_6H_5{-}CH_2CH_3 \xrightarrow[\substack{180\ \text{psi}\ H_2 \\ 450°C}]{\text{Pt-on-}Al_2O_3} CH_3{-}C_6H_4{-}CH_3$$

Noble metal-catalyzed isomerizations are carried out most effectively on acidic supports such as alumina or silica-alumina and additional acidic materials such as hydrogen chloride, aluminum chloride, or boria may be used as well. The presence of hydrogen often facilitates the reaction as well as helping keep the catalyst clean (Burbidge and Rolfe, 1966). Alkanes undergo skeletal isomerization with much more difficulty than alkenes, but the strong similarity between the reactions has led a number of investigators to believe that they go through the same carbonium ion intermediate (Germain, 1969).

Many supported metals have been used as isomerization catalysts. Evaluation of a metal's effectiveness depends on the reaction conditions, catalyst preparation, and support. In one study, activities for skeletal isomerization of butane and butenes increased in the order of $Co < Ni < Rh < Fe < Pd < Pt$ for the metals supported on alumina (Panchenkov *et al.*, 1970). The study supports the general consensus regarding metals, and palladium and more especially platinum are usually used for skeletal isomerization. Good results have been claimed for sulfided rhodium (Hettinger *et al.*, 1963).

The importance of skeletal isomerization in petroleum processing overshadows its much more limited utility in other areas. There are frequently only small energy differences among isomers, and complex mixtures form resulting in moderate yields and troublesome separations. An example of a synthetically useful isomerization is based on the intriguing discovery that *endo*-tetrahydrodicyclopentadiene could be isomerized to adamantane by aluminum chloride (Schleyer, 1957; Schleyer and Donaldson, 1960) which in turn led to extensive investigations of other methods of converting hydrocarbons containing ten or more carbon atoms into adamantanes. The cleanest method developed for achieving these transformations employs a chlorinated 0.5% platinum-on-alumina catalyst contained in a hot tube (165°C). The hydrocarbon and dry hydrogen chloride are passed over the catalyst and the products collected in a cold trap. This catalyst system shows a remarkable selectivity for adamantane formation with only about 1% by-product formation in the most favorable cases. *endo*- and *exo*-Tetrahydrodicyclopentadiene are each converted to adamantane in 99% yield.

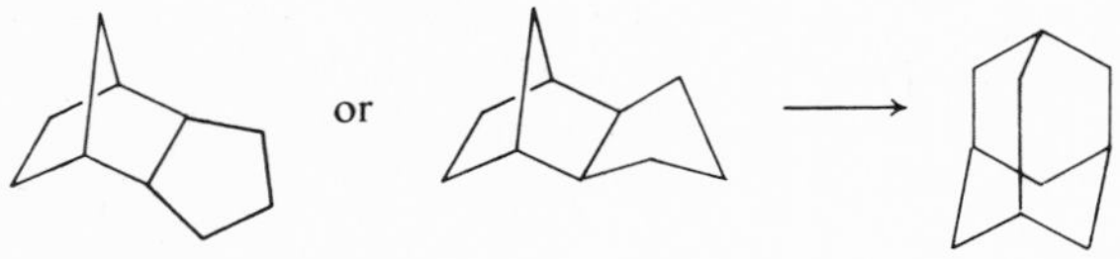

or

exo-2,3-Tetramethylenenorbornane affords 2-methyladamantane in 98% yield accompanied by 2% of the 2-isomer.

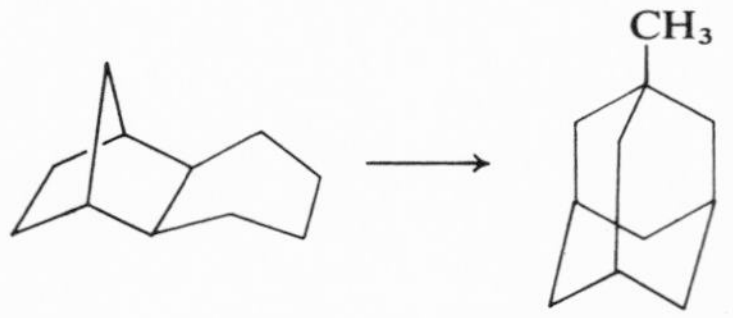

CH_3

Perhydroacenaphthene (a mixture of four isomers) affords 1,3-dimethyladamantane in 86% yield accompanied by 7% each of 1-ethyladamantane and 1,X-dimethyladamantane (Johnston *et al.*, 1971).

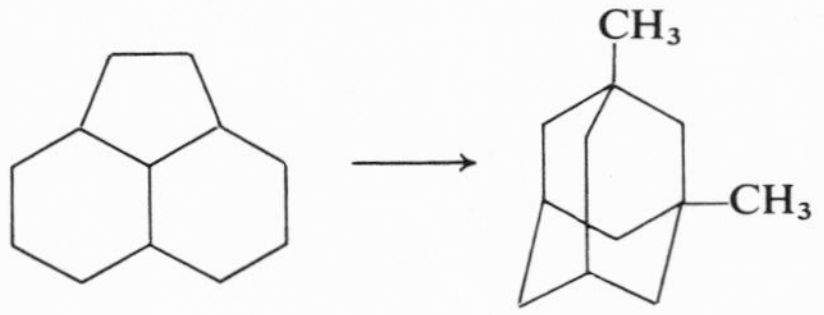

CH_3

CH_3

Heterocyclics

Isomerizations proceed more readily and often with greater specificity if heteroatoms are present in the molecule, providing a preferential point of attack. For example, a general method for the preparation of tetrahydropyran involves ring enlargement of ketofurans under hydrogenation conditions in the presence of platinum-on-carbon granules. At higher temperatures, the tetrahydropyran is isomerized to heptanone-2 in excellent yield with or without hydrogen present.

$$\text{2-methyl-5-acetylfuran} \xrightarrow[\substack{10\%\ \text{Pt-on-C} \\ 200^{\circ}\text{C} \\ \text{VHSV}=0.1}]{1\ \text{atm}\ H_2} \text{2,6-dimethyltetrahydropyran}\ (25\%) \xrightarrow[350^{\circ}\text{C}]{\text{Pt-on-C}} CH_3C(=O)(CH_2)_4CH_3\ (98\%)$$

Similarly, 2-methyl-5-propionylfuran affords 2-methyl-6-ethyltetrahydropyran in about 20% yield, which in turn can be isomerized at 320°C to a mixture of octanone-2 and octanone-3. 2,6-Diethyltetrahydropyran is obtained in about 20% yield from 2-ethyl-5-propionylfuran, and subsequently isomerized at 320°C to afford nonanone-3 in 90% yield (Belskii and Shuikin, 1961).

$$\text{2-ethyl-5-propionylfuran} \xrightarrow[200^{\circ}\text{C}]{10\%\ \text{Pt-on-C}} \text{2,6-diethyltetrahydropyran} \xrightarrow[320^{\circ}\text{C}]{10\%\ \text{Pt-on-C}} CH_3(CH_2)_5C(=O)CH_2CH_3$$

Compounds containing an allylic oxygen atom undergo isomerization under relatively mild conditions with a number of noble metal catalysts. Excellent yields of normal Claisen products are formed by isomerization of alloxypyridine and 2-crotoxypyridine on heating with small amounts of chloroplatinic acid. The authors (Steward and Seibert, 1968) suggested that formation of isomerized products in essentially quantitative yields involves complexing of the substrate with platinum as a bidentate ligand. Reactions catalyzed by boron trifluoride etherate afford, in contrast, two abnormal Claisen products.

$$\text{2-}(OCH_2CH{=}CHCH_3)\text{pyridine} \xrightarrow[125^{\circ}\text{C}]{H_2PtCl_6} \text{1-}(CH_3CH(CH{=}CHCH_3))\text{-2-pyridone}$$

$$\text{2-}(OCH_2CH{=}CHCH_3)\text{pyridine} \xrightarrow{BF_3 \cdot OEt_2} \text{1-}(CH_2CH{=}CHCH_3)\text{-2-pyridone}\ (82\%) + \text{3-}(CH_2CH{=}CHCH_3)\text{-2-pyridone (N–H)}\ (18\%)$$

Allylic esters undergo an allylic rearrangement in the presence of Hg(II) and Pd(II) salts. Henry (1971, 1972a, b) by the use of appropriate substitution and oxygen-labeling experiments established that in the case of Pd(II), the isomerization involves both an acetoxymetallation–deactoxymetallation mechanism and also one involving no exchange of ester. Other noble metal salts, such as $PtCl_2$, $IrCl_3$, $RUCl_3OsCl_3$, and $RhCl_3$ also promote the isomer ization, but much more slowly than Pd(II). Those metals are effective which activate the double bond by forming stable π-complexes and which also are capable of undergoing acetoxymetallation reactions. The failure of silver acetate to promote isomerization was attributed to its inability to undergo this reaction.

$$CH_3CH{=}CHCH_2OAc \underset{HOAc}{\overset{Pd(II)}{\rightleftharpoons}} CH_3\underset{}{\overset{\overset{OAc}{|}}{C}}HCH{=}CH_2$$

Smutny (1970) isomerized 1-phenoxy-2,7-octadiene to a mixture of *o*- and *p*-octadienylphenols by contact with palladium, platinum, or ruthenium catalysts and phenoxide anion. Various complex rhodium catalysts, such as $(allyl)_2RhCl_2$ and $(Ph_3P)_3RhCl$, are effective for isomerization of butadiene monoepoxide to 2-butanal in high yield with or without solvent at 65°C (Lini *et al.*, 1969).

$$CH_2{=}CHCH\underset{\diagdown O \diagup}{\ }CH_2 \longrightarrow CH_3CH{=}CHCHO$$

Aromatizations

Aromatization through double-bond migration provides a convenient entry into a variety of aromatic systems. The yields of products are generally good because the direction of migration is not random but strongly determined. Aromatization may or may not require or accompany a simultaneous dehydrogenation. Treatment of 1,2-bis(3-cyclohexen-1-yl)ethylene, prepared by contacting 4-vinylcyclohexene with a molybdenum catalyst, affords a mixture of 1-phenyl-2-cyclohexylethane and bibenzyl, the major product being determined largely by the temperature. On refluxing 5 gm of substrate with 0.3 gm of 10% palladium-on-carbon, a 63.5% yield of bibenzyl was obtained after 3.6 moles of hydrogen had been evolved, whereas if the refluxing is done with a xylene diluent, bibenzyl is obtained in only 14% yield and phenylcyclohexylethane in 73% yield (Norell, 1968).

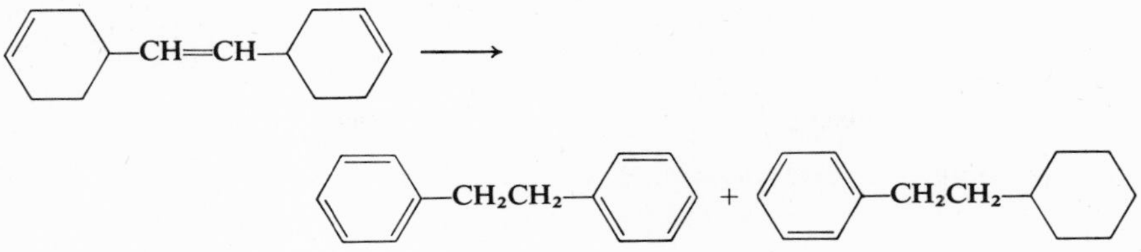

Leonard and Berry (1953) applied an isomerization reaction to the synthesis of 3,7-dibenzyltropolones. Aromatization is best carried out by heating 3,7-dibenzylidene-1,2-cycloheptanedione in triethyleneglycol at 280°C with 10% palladium-on-carbon for 2 hours. The solubility of triethylene glycol in water makes possible a facile separation of the product. Heating the substrate with palladium-on-carbon at 250°C without solvent affords an 80% yield of the tropolone, which is recovered by sublimation.

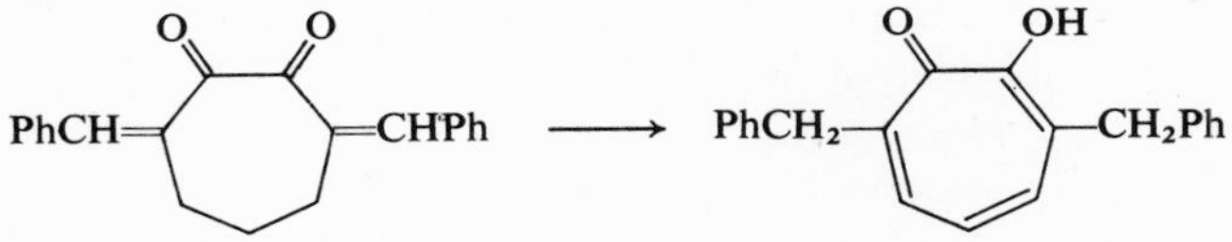

The higher homolog, 3,10-dibenzylidene 1,2-cyclodecanedione, behaves differently and affords 3,10-dibenzyl-1,2-cyclodecanedione and two transannular products, 1-benzyl-11-phenylbicyclo[6.2.1]hendecane-9,10-dione and 9-hydroxy-7,10-dibenzylbicyclo[5.3.0]10-decen-9-one. The hydrogen required for these products may arise from the catalyst or by disproportionation (Leonard and Little, 1958).

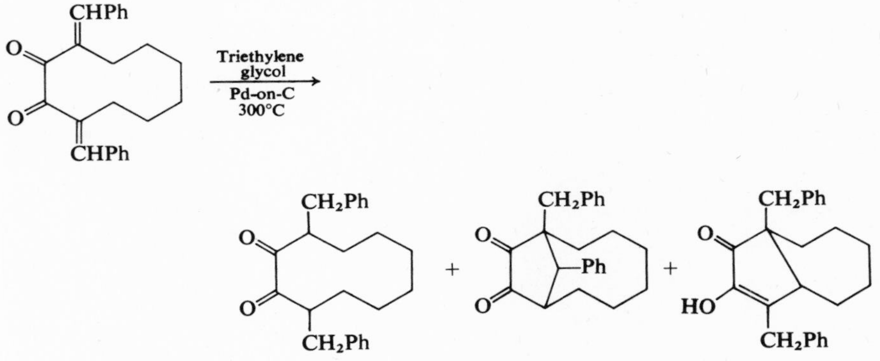

A similar isomerization was applied to the synthesis of γ-pyridones. The reaction can be carried out with or without solvent to afford high yields of product. Ethylene glycol is an ideal solvent providing a favorable temperature for the reaction. Isomerization in this system is considerably faster than the corresponding carbocyclic system (Leonard and Locke, 1955).

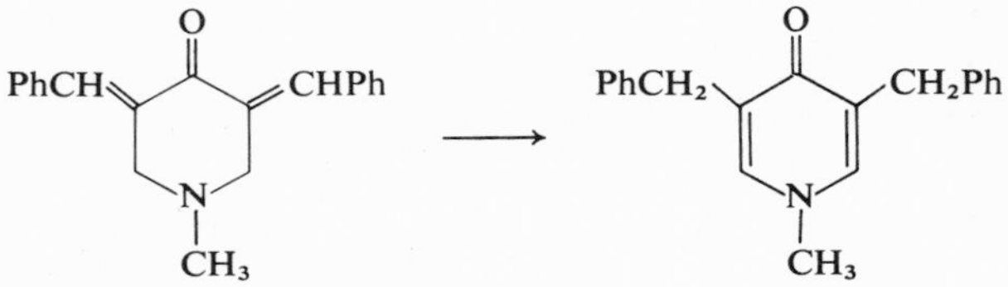

Other examples of aromatization of this type are the conversion of 2,6-dibenzylidenecyclohexanone to 2,6-dibenzylphenol (Horning, 1945; Leonard and Berry, 1953), 2,7-dibenzylidenecycloheptanone to 2,7-dibenzyltropolone (Leonard *et al.*, 1957), and 3,5-dibenzylidenetetrahydro-4*H*-pyran-4-ones to 3,5-dibenzyl-4*H*-pyran-4-ones (Leonard and Choudhury, 1957).

An interesting aromatization involving dehydrogenation and hydrogenolysis occurs when an amino-isooxazole moiety in a steroid is converted to a 2-carboxyamido-3-aminosteroid by refluxing in dioxane over 10% palladium-on-carbon (deRuggieri *et al.*, 1969).

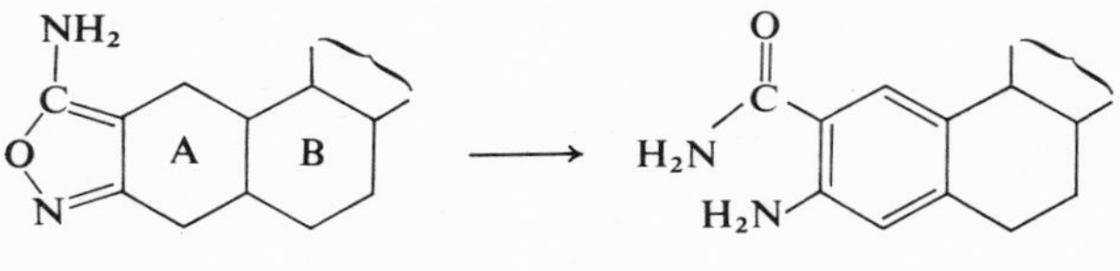

VALENCE ISOMERIZATION

Valence isomerization refers to processes in which products arise by changes in only bond configuration. The first example of metal-catalyzed valence isomerization, quadricyclane to norbornadiene, was reported in 1967 (Hogeveen and Volger, 1967) but already a number of good reviews of this burgeoning field have appeared (Mango, 1969; Halpren, 1970; Mango and Schachtschneider, 1971; Mango, 1971; Paquette, 1971).

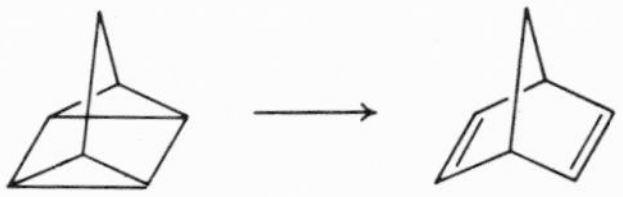

Much intense interest stems from theoretical problems connected with metal-catalyzed valence isomerizations that remove symmetry restrictions imposed by the Woodward-Hoffmann rules (1965, 1969) of orbital symmetry conservation. Synthetic utility of the reaction is limited by the tendency of the isomerization to move from the more-strained to the less-strained compound, or in general, from the less available to the more available. The reaction may be very selective and show at times a marked dependence of product on catalytic metal. The extreme sensitivity of certain strained compounds to metal catalysis has important practical implications relating to their synthesis.

Effect of Catalysts

A number of workers have noted that the products of metal-catalyzed isomerization may depend critically on the metal (Volger *et al.*, 1969; Katz and Cerefice, 1969a, b) and on the ligand (Manassen, 1970) employed in the process. Both Ag(I) and Pd(II) effect the same rearrangement of cubane (**XXVI**) into **XXVII** whereas rhodium(I) affords **XXVIII** (Cassar *et al.*, 1970).

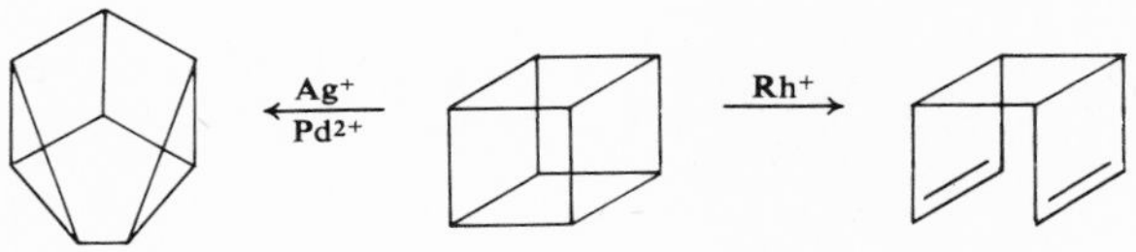

Ag^+, Pd^{2+} ← → Rh^+

(XXVII) **(XXVI)** **(XXVIII)**

Isomerization of *exo,exo*- and *exo,endo*-2,4-dimethylbicyclobutanes (**XXIX**) in the presence of catalytic amounts of bis(benzonitrile)dichloropalladium affords 2-methylpenta-1,3-diene (**XXX**) as a major product, whereas Ag(I) affords only hexa-1,3-dienes, **XXXI** and **XXXII** (Sakai *et al.*, 1971).

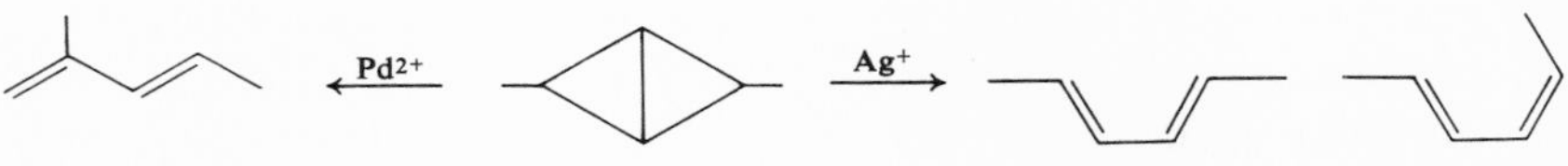

Pd^{2+} ← → Ag^+

(XXX) **(XXIX)** **(XXXI)** **(XXXII)**

Similarly, tricyclo[4.1.0.^{2,7}]heptane (**XXXIII**) is converted to cyclohepta-1,3-diene (**XXXIV**) in a silver-catalyzed reaction (Paquette *et al.*, 1970), whereas catalysis by palladium(II) affords 3-methylenecyclohexene (**XXXV**) in 90% yield (Sakai *et al.*, 1971).

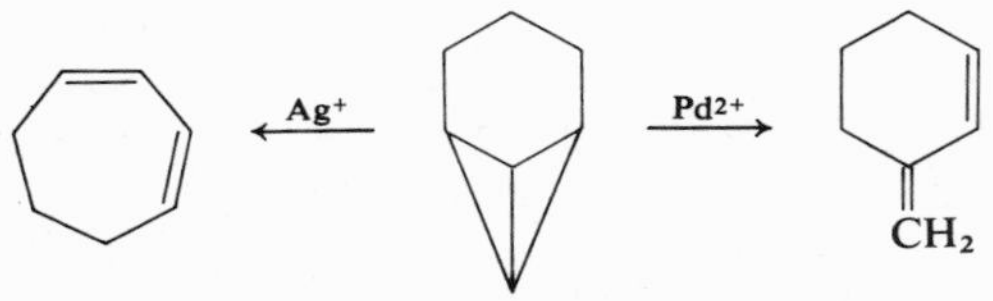

Ag^+ Pd^{2+} CH_2

(XXXIV) (XXXIII) (XXXV) 90%

Synthetic Applications

Synthetic applications of metal-catalyzed valence isomerization have lagged behind theoretical advances inasmuch as the isomerization usually moves from the exotic to the mundane. However, a few useful syntheses have been reported. A convenient new route to the azulene nucleus involves as the critical reaction a rhodium dicarbonyl chloride dimer-catalyzed rearrangement of 1-methyl-2,2-diphenylbicyclo[1.1.0]butane (Gassman and Nakai, 1971).

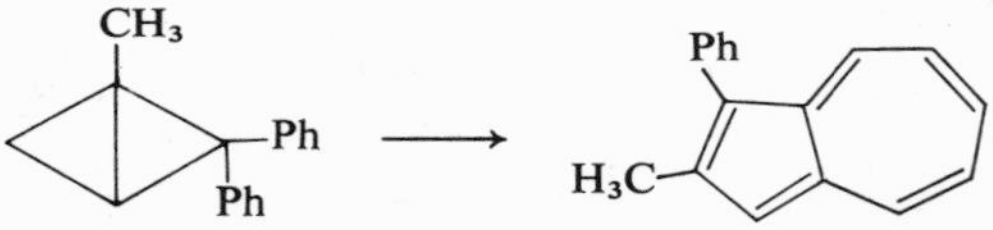

CH_3 Ph Ph Ph H_3C

A practical synthesis of bicyclo[4.2.2]deca-2,4,7,9-tetraene (**XXXVII**) is through catalytic conversion of bullvalene (**XXXVI**) in the presence of bis-(benzonitrile)palladium dichloride. At low conversion, **XXXVII** is the sole product of the isomerization. With longer reaction times, naphthalene and 9,10-dihydronaphthalene can be detected (Vedejs, 1968).

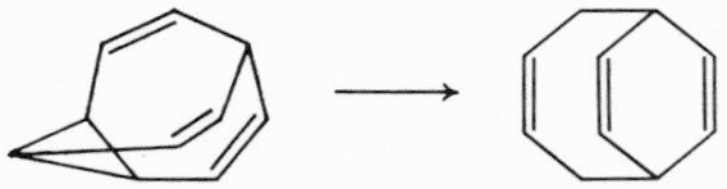

(XXXVI) (XXXVII)

Inadvertent Catalysis

The sensitivity of highly strained systems to catalysis by traces of metal can be extreme and have untoward effects in synthesis. The stability of both **XXXVIII** and **XXXIX** was dependent on the history of the sample. When **XXXIX** was purified by fractionation through a stainless steel spinning band column, the product was unstable and rapidly reverted to the diene, **XL**. A stable product was obtained by use of a Teflon spinning band (Gassman *et al.*, 1968; Gassman and Patton, 1968).

CH_3O—OCH_3 (XXXVIII)

CH_3O—OCH_3, CH_3, CH_3 (XXXIX) ⟶ CH_3O—OCH_3, CH_3, CH_3 (XL)

REFERENCES

Abubaker, M., Gostunskaya, I. V., and Kazanskii, B. A. (1968). *Vestn. Mosk. Univ., Khim*, Ser. II, **23**, pp. 105 and 148.

Allinger, N. L., and Cole, J. L. (1959). *J. Amer. Chem. Soc.* **81**, 4080.

Allinger, N. L., and Cole, J. L. (1960). *J. Amer. Chem. Soc.* **82**, 2553.

Allinger, N. L., and Hu, S.-E. (1962). *J. Org. Chem.* **27**, 3417.

Allinger, N. L., and Zalkow, V. B. (1961). *J. Amer. Chem. Soc.* **83**, 1144.

Allinger, N. L., Blatter, H. M., Freiberg, L. A., and Karkowski, F. M. (1966). *J. Amer. Chem. Soc.* **88**, 2999.

Allinger, N. L., Szkrybalo, W., and Van Catledge, F. A. (1968). *J. Org. Chem.* **33**, 784.

Arnet, J. E., and Pettit, R. (1961). *J. Amer. Chem. Soc.* **83**, 2954.

Arthur, W. J. (1964). U.S. Patent 3,155,724.

Attridge, C. J., and Maddock, S. J. (1971). *J. Chem. Soc., C* p. 2999.

Attridge, C. J., and Wilkinson, P. J. (1971). *Chem. Commun.* p. 620.

Augustine, R. L., and Van Peppen, J. F. (1970). *Chem. Commun.* p. 495.

Bachmann, W. E., and Dreiding, A. S. (1950). *J. Amer. Chem. Soc.* **72**, 1323.

Baird, M. C., Mague, J. T., Osborn, J. A., and Wilkinson, G. (1967). *J. Chem. Soc., A* p. 1347.

Barnes, M. F., and MacMillan, J. (1967). *J. Chem. Soc., C* p. 361.

Bellinzona, G., and Bettinetti, F. (1960). *Gazz. Chim. Ital.* **90**, 426.

Belskii, I. F., and Shuikin, N. I. (1961). *Dokl. Akad. Nauk. SSSR* **136**, 591.

Birch, A. J., and Subba Rao, G. S. R. (1968). *Tetrahedron Lett.* p. 3797.

Boekelheide, V., Anderson, A. E., Jr., and Sauvage, G. L. (1953). *J. Amer. Chem. Soc.* **75**, 2558.

Bond, G. C., and Hellier, M. (1965a). *Chem. Ind. (London)* p. 35.

Bond, G. C., and Hellier, M. (1965b). *J. Catal.* **4**, 1.

Bond, G. C., and Hellier, M. (1967). *J. Catal.* **7**, 217.

Bond, G. C., and Rank, J. S. (1965). *Proc. Int. Congr. Catal., 3rd, 1964* Vol. 2, p. 1225.

Bragin, O. V., Lun-Syan, T., and Liberman, A. L. (1967). *Kinet. Katal.* **8**, 931.

Bream, J. B., Eaton, D. C., and Henbest, H. B. (1957). *J. Chem. Soc., London* p. 1974.

Bright, A., Malone, J. F., Nicholson, J. K., Powell, J., and Shaw, B. L. (1971). *Chem. Commun.* p. 712.

Burbidge, B. W., and Rolfe, J. R. K. (1966). *Hydrocarbon Process.* No. 7, p. 168.

Cassar, L., Eaton, P. E., and Halpren, J. (1970). *J. Amer. Chem. Soc.* **92**, 3515 and 6366.

Condon, F. E. (1958). *Catalysis* **6**, 43.

Connolly, J. D., Handa, K. L., McCrindle, R., and Overton, K. H. (1967). *Tetrahedron Lett.* p. 3449.

Conti, F., Donati, M., and Pregaglia, G. F. (1971). *J. Organometal. Chem.* **30**, 421.

Cramer, R. (1966). *J. Amer. Chem. Soc.* **88**, 2272.

Cramer, R., and Lindsey, R. V., Jr. (1966). *J. Amer. Chem. Soc.* **88**, 3534.

Danno, S., Moritani, I., and Fujiwara, Y. (1969). *Tetrahedron* **25**, 4809.

Davies, N. R. (1967). *Rev. Pure Appl. Chem.* **17**, 83.
Davies, N. R., DiMichiel, A. D., and Pickles, V. A. (1968). *Aust. J. Chem.* **21**, 385.
deRuggieri, P., Gandolfi, C., and Guzzi, U. (1969). U.S. Patent 3,458,502.
Dutton, H. J., Scholfield, C. R., Selke, E., and Rohwedder, W. K. (1968). *J. Catal.* **10**, 316.
Egloff, G., Hulla, G., and Komarewsky, V. I. (1942). "Isomerization of Pure Hydrocarbons." Van Nostrand-Reinhold, Princeton, New Jersey.
Fischer, N. H., and Mabry, T. J. (1968). *Tetrahedron* **24**, 4091.
Garner, J. W., and Benedict, B. C. (1970). U.S. Patent 3,531,545.
Gassman, P. G., Aue, D. H., and Patton, D. S. (1968). *J. Amer. Chem. Soc.* **90**, 7271.
Gassman, P. G., and Nakai, T. (1971). *J. Amer. Chem. Soc.* **93**, 5897.
Gassman, P. G., and Patton, D. S. (1968). *J. Amer. Chem. Soc.* **90**, 7276.
Germain, J. E. (1969). "Catalytic Conversion of Hydrocarbons," p. 156. Academic Press, New York.
Gostunskaya, I. V., Leonova, A. I., and Kazanskii, B. A. (1964). *Neftekhimiya* **4**, 379.
Hallman, P. S., Evans, D., Osborn, J. A., and Wilkinson, G. (1967). *Chem. Commun.* p. 305.
Halpren, J. (1970). *Accounts Chem. Res.* **3**, 392.
Harrod, J. F., and Chalk, A. J. (1964). *J. Amer. Chem. Soc.* **86**, 1776.
Harrod, J. F., and Chalk, A. J. (1965). *Nature* (*London*) **205**, 280.
Harrod, J. F., and Chalk, A. J. (1966). *J. Amer. Chem. Soc.* **88**, 3491.
Hartley, F. R. (1969). *Chem. Rev.* **69**, 799.
Henry, P. M. (1971). *J. Amer. Chem. Soc.* **93**, 3853.
Henry, P. M. (1972a). *J. Org. Chem.* **37**, 2443.
Henry, P. M. (1972b). *J. Amer. Chem. Soc.* **94**, 5200.
Herz, W., Subramaniam, P. S., and Geissman, T. A. (1968). *J. Org. Chem.* **33**, 3743.
Hettinger, W. P., Keith, C. D., and Lorenc, W. F. (1963). U.S. Patent 3,092,676.
Hogeveen, H., and Volger, H. C. (1967). *J. Amer. Chem. Soc.* **89**, 2486.
Horner, L., Büthe, H., and Siegel, H. (1968). *Tetrahedron Lett.* p. 4023.
Horning, E. C. (1945). *J. Org. Chem.* **10**, 263.
Hubert, A. J., and Reimlinger, H. (1970). *Synthesis* **1**, 405.
Hughes, W. B. (1970). U.S. Patent 3,514,497.
Huntsman, W. D., Madison, N. L., and Schlesinger, S. I. (1963). *J. Catal.* **2**, 498.
Johnston, D. E., McKervey, M. A., and Rooney, J. J. (1971). *J. Amer. Chem. Soc.* **93**, 2798.
Kametani, T., and Ihara, M. (1968). *J. Chem. Soc., C* p. 191.
Kametani, T., Ihara, M., and Shima, K. (1968). *J. Chem. Soc., C* p. 1619.
Katz, T. J., and Cerefice, S. (1969a). *J. Amer. Chem. Soc.* **91**, 2405.
Katz, T. J., and Cerefice, S. (1969b). *Tetrahedron Lett.* p. 2561.
Ketley, A. D., and Braatz, J. A. (1968). *Chem. Commun.* p. 169.
Kindler, K., Peschke, W., and Dehn, W. (1931). *Justus Liebigs Ann. Chem.* **485**, 113.
Kuhn, R., and Haas, H. J. (1958). *Justus Liebigs Ann. Chem.* **611**, 57.
Lafont, P., and Vivant, G. (1963). French Patent 1,337,889.
Lemberg, S. (1967). U.S. Patent 3,344,171.
Leonard, N. J., and Berry, J. W. (1953). *J. Amer. Chem. Soc.* **75**, 4898.
Leonard, N. J., and Choudhury, D. (1957). *J. Amer. Chem. Soc.* **79**, 156.
Leonard, N. J., and Little, J. C. (1958). *J. Amer. Chem. Soc.* **80**, 4111.
Leonard, N. J., and Locke, D. M. (1955). *J. Amer. Chem. Soc.* **77**, 1852.
Leonard, N. J., Miller, L. A., and Berry, J. W. (1957). *J. Amer. Chem. Soc.* **79**, 1482.
Levina, R. Y., and Petrov, D. A. (1937). *Zh. Obshch. Khim.* **7**, 747.
Lini, D. C., Ramey, K. C., and Wise, W. B. (1969). U.S. Patent 3,465,043.

Lyons, J. E. (1971a). *Chem. Commun.* p. 562.
Lyons, J. E. (1971b). *J. Org. Chem.* **36**, 2497.
Mabry, T. J., Wyler, H., Parikh, I., and Dreiding, A. S. (1967). *Tetrahedron* **23**, 3111.
McEuen, J. M., Plonsker, L., and Dubeck, M. (1968). U.S. Patent 3,367,988.
McQuillin, J. (1963). *Tech. Org. Chem.* **9**, 498.
MacSweeney, D. F. and Ramage, R. (1971). *Tetrahedron*, **27**, 1481.
Manassen, J. (1970). *J. Catal.* **18**, 38.
Mango, F. D. (1969). *Advan. Catal.* **20**, 291.
Mango, F. D. (1971). *Chem. Tech.*, 758.
Mango, F. D., and Schachtschneider, J. H. (1971). *In* "Transition Metals in Homogeneous Catalysis," p. 223. Dekker, New York.
Mann, G. (1968). *Tetrahedron* **24**, 6495.
Mathew, C. T., Sen Gupta, G., and Dutta, P. C. (1964). *Proc. Chem. Soc., London* p. 337.
Maurel, R., Giusnet, M., Marco, M., and Germain, J. E. (1966). *Bull. Soc. Chim. Fr.* [5] p. 3082.
Misono, M., Saito, Y., and Yoneda, Y. (1968). *J. Catal.* **10**, 200.
Müller, E., and Segnitz, A. (1970). *Synthesis* **3**, 147.
Norell, J. R. (1968). U.S. Patent 3,387,051.
Obeschchalova, N. V., Feldblyum, V. S., Basner, M. E., and Dzyuba, V. S. (1968). *Zh. Org. Khim.* **4**, 574.
Orchin, M. (1966). *Advan. Catal.* **16**, 1.
Panchenkov, G. M., Volokhova, G. S., and Zhorov, Yu. M. (1970). *Neftekhimiya* **10**, 178.
Paquette, L. A. (1971). *Accounts Chem. Res.* **4**, 280.
Paquette, L. A., Allen, G. R., Jr., and Henzel, R. P. (1970). *J. Amer. Chem. Soc.* **92**, 7002.
Pascal, Y. L., and Vernier, F. (1969). *C. R. Acad. Sci.*, **268**, *Ser. C* p. 1177.
Pelletier, S. W., Ichinohe, Y., and Herald, D. L., Jr. (1971). *Tetrahedron Lett.* p. 4179.
Pietra, S., and Bertoglio, C. (1956). *Gazz. Chim. Ital.* **86**, 1129.
Plonsker, L., and McEuen, J. M. (1967). U.S. Patent 3,352,938.
Pollitzer, E. L., Hayes, J. C., and Haensel, V. (1970). *Advan. Chem. Ser.* **97**, 20.
Puterbaugh, W. H., and Newman, M. S. (1959). *J. Amer. Chem. Soc.* **81**, 1611.
Quinn, H. A., McKervey, M. A., Jackson, W. R., and Rooney, J. J. (1970). *J. Amer. Chem. Soc.* **92**, 2922.
Reppe, W., *et al.* (1955). *Justus Liebigs Ann. Chem.* **596**, 25.
Richter, H. J., and Weberg, B. C. (1958). *J. Amer. Chem. Soc.* **80**, 6446.
Rinehart, R. E. (1969). U.S. Patent 3,433,808.
Rinehart, R. E., and Lasky, J. S. (1964). *J. Amer. Chem. Soc.* **86**, 2516.
Rinehart, R. E., Smith, H. P., Witt, H. S., and Romeyn, H., Jr. (1962). *J. Amer. Chem. Soc.* **84**, 4145.
Rodd, E. H. (1960). "Chemistry of Carbon Compounds," Vol. IVC, p. 2069. Elsevier, Amsterdam.
Rylander, P. N. (1967). "Catalytic Hydrogenation over Platinum Metals," p. 98. Academic Press, New York.
Rylander, P. N. (1970a). *J. Amer. Oil. Chem. Soc.* **47**, 482.
Rylander, P. N. (1970b). *Amer. Chem. Soc., Div. Petrol. Chem., Prepr.* **15**, No. 1, B115.
Rylander, P. N. (1971). *Advan. Chem. Ser.* **98**, 150.
Sakai, M., Yamaguuchi, H., and Masamune, S. (1971). *Chem. Commun.* p. 486.
Schleyer, P. von R. (1957). *J. Amer. Chem. Soc.* **79**, 3292.
Schleyer, P. von R., and Donaldson, M. M. (1960). *J. Amer. Chem. Soc.* **82**, 4645.
Smutny, E. J. (1970). U.S. Patent 3,518,318.
Sparke, M. B., Turner, L., and Wenham, A. J. M. (1965). *J. Catal.* **4**, 332.

Steward, H. F., and Seibert, R. P. (1968). *J. Org. Chem.* **33**, 4560.
Tayim, H. A., and Bailar, J. C., Jr. (1967). *J. Amer. Chem. Soc.* **89**, 3420.
Tayim, H. A., and Vassilian, A. (1970). *Chem. Commun.* p. 630.
Tomimatsu, T., Hashimoto, M., Shingu, T., and Tori, K. (1969). *Chem. Commun.* p. 168.
Trebellas, J. C., Olechowski, J. R., and Jonassen, H. B. (1966). *J. Organometal. Chem.* **6**, 412.
Trebellas, J. C., Olechowski, J. R., Jonassen, H. B., and Moore, D. W. (1967). *J. Organometal. Chem.* **9**, 153.
Uhlig, H. F., and Pfefferle, W. C. (1970). *Advan. Chem. Ser.* **93**, 204.
Vedejs, E. (1968). *J. Amer. Chem. Soc.* **90**, 4751.
Volger, H. C., Hogeveen, H., and Gaasbeck, M. M. P. (1969). *J. Amer. Chem. Soc.* **91**, 218.
Weiss, U., and Weiner, N. (1949). *J. Org. Chem.* **14**, 194.
Wells, P. B., and Wilson, G. R. (1967). *J. Catal.* **9**, 70.
Williams, J. L. R., Webster, S. K., and Van Allan, J. A. (1961). *J. Org. Chem.* **26**, 4893.
Woodward, R. B., and Hoffmann, R. (1965). *J. Amer. Chem. Soc.* **87**, 395.
Woodward, R. B., and Hoffmann, R. (1969). *Angew. Chem., Int. Ed. Engl.* **8**, 781.
Zajcew, M. (1960a). *J. Amer. Oil Chem. Soc.* **37**, 11.
Zajcew, M. (1960b). *J. Amer. Oil Chem. Soc.* **37**, 473.
Zuech, E. A. (1968). U.S. Patent 3,387,045.

CHAPTER 6

Oligomerizations, Telomerizations and Condensations

Oligomerization of Olefins

Olefins in the presence of metal catalysts can be condensed to compounds ranging in molecular weight from less than a dimer to high molecular weight polymers. The majority of work in this area has been concerned with base metal catalysts, but in recent years, increasing attention has been given to noble metals as appreciation of their uniqueness grows. Most studies have been concerned with small olefins, but presumably some of the catalytic systems should be applicable to formation of carbon–carbon bonds in more complex molecules as well.

Ethylene

Ethylene is dimerized by rhodium trichloride in ethanol at 40°C to a mixture of linear butenes in greater than 99% yield; there is very little higher hydrocarbon. The author (Cramer, 1965) with considerable evidence suggests the following sequence of steps for the dimerization. A bis(ethylene) complex of rhodium(I) is rapidly converted by hydrogen chloride to an ethylrhodium(III) compound, that then rearranges in a slow, rate-determining chain-growth reaction to a *n*-butylrhodium(III) complex. This complex decomposes rapidly through loss of hydrogen chloride to give a 1-butene complex of rhodium(I). Coordinated 1-butene and solvent are rapidly displaced from this complex by ethylene regenerating the initial rhodium(I) complex. This fast exchange relative to the rate of overall synthesis limits the products to substantially only butenes (Cramer, 1968). Methanol and water are also satisfactory solvents, but in glyme, diglyme, tetrahydrofuran, or nitrobenzene a tarry second layer separates that contains most of the rhodium (Cramer, 1965).

Ethylene is converted to chloroform solvent (100 ml) by 1 gm of rhodium trichloride trihydrate in 5.5 hours at 10 atm pressure and 25°C into 48 gm of butenes, consisting almost exclusively of 2-butene with a 2.6 *trans* to *cis* ratio. The water of hydration strongly affects the reaction rate; hydrated rhodium trichloride is at least 20 times more active than the anhydrous salt. Without any solvent, the reaction is very slow (Ketley *et al.*, 1967). Ruthenium tribromide and platinic chloride are also catalysts for dimerization of ethylene, but they are not so effective as rhodium chloride. Ruthenium chloride at 50°C gives a mixture of butenes; at 150°C about 30% of hexenes and octenes are formed as well (Alderson *et al.*, 1965).

Ethylene is dimerized by palladium chloride under mild conditions in a number of solvents. Palladium fluoride, bromide, iodide, and nitrate are ineffective. Palladium cyanide in dichloromethane dimerizes ethylene at about half the rate of palladium chloride, and forms in addition to dimer, a highly crystalline polyethylene. With 1.5 gm of palladium chloride slurried in chloroform (stabilized by 0.03 M of ethanol) ethylene at 50°C and 10 atm is converted into 52% *trans*-2-butene, 47% *cis*-2-butene, and 1% 1-butene at a rate of about 0.1 mole/hour. Without ethanol present in the chloroform only di-μ-chloro-dichlorobis(ethylene)dipalladium is formed. Palladium chloride does not remain an active catalyst indefinitely. After several days at 40°C, no more butene is produced, due probably to conversion of an active π-complex into an inactive π-allyl complex (Ketley *et al.*, 1967). The solvent has a marked effect on the life of a palladium chloride catalyst. The catalyst turnover decreases with solvent in the order acetic acid, dichloroethane, tetrachloroethane, chlorobenzene, benzene, methyl acetate, triethylene glycol, butyl acetate, chloroform, cyclohexane, and ethanol with the extremes differing by a factor of 15 (Kusunoki *et al.*, 1966). Improved rates of dimerization are obtained in ethanol by carrying out the dimerization in the presence of a sulfone co-catalyst (Klein, 1967). Improved rates are also obtained in the presence of bidentate ligands with catalysis by di-μ-chloro-bis(ethylene)dipalladium, but the rate of reduction to metallic palladium is also increased (Kawamoto *et al.*, 1968, 1969). Substantial amounts of butenes are formed as by-products in oxidations of ethylene with air, palladium and copper (Schaeffer, 1966). Platinum catalysts are ineffective in the dimerization of ethylene.

In acetic acid–acetic anhydride solvent containing hydrochloric acid propylene is produced as well as butenes (Aguilo and Stautzenberger, 1969). Propylene was unexpectedly the major product in an attempted dimerization of ethylene with palladium chloride in benzonitrile containing sodium fluoride.

Potassium fluoride is about equivalent to sodium fluoride, but lithium and cesium fluoride inhibit propylene formation. The reaction is so far of limited utility, for when applied to propylene no higher molecular weight hydrocarbon was found (Crano *et al.*, 1968).

Propylene

Propylene is dimerized over various noble metal catalysts with results that depend strongly on the solvent. In chloroform or dichloromethane solvent, the products of palladium chloride-catalyzed dimerization at 40°C and 8 atm pressure are about 65% 2-hexene and 3-hexene and about 35% isomers of 2-methylpentene and 4-methylpentene. In anisole about 90% of the product is linear dimers. When the palladium chloride–propylene complex, $(C_3H_6)_2$-Pd_2Cl_4, is used instead of palladium chloride, only straight-chain hexenes are formed. The branched dimers were assumed to arise through palladium chloride acting as a weak Friedel-Crafts catalyst (Ketley *et al.*, 1967). Dimerization of propylene with rhodium chloride at 50°C or ruthenium chloride at 207°C gives low conversion to C_6 olefins (Alderson *et al.*, 1965), but with rhodium trichloride trihydrate in a number of solvents, a mixture consisting of about 43% linear hexenes and 57% branched hexenes is obtained. The authors (Ketley *et al.*, 1967) believe that the mechanism is completely different from that of palladium chloride-catalyzed dimerization, based on the observation that the rhodium reaction is acid-catalyzed, unaffected by large excesses of ethanol, and gives predominately branched isomers. An insertion mechanism proposed earlier by Cramer (1965) was used to account for the results.

Methylcyclopropene

Highly strained 1-methylcyclopropene (**I**) undergoes thermal oligomerization affording mainly **II**, but in the presence of certain palladium catalysts a mixture of **III** and **IV** is formed in high yield. Active palladium catalysts include the chloride, bromide, iodide, nitrate, π-allylpalladium chloride, and dichlorobis-(benzonitrile) palladium(II). Dichlorobis(triphenylphosphine)palladium(II) is

(II) $\xleftarrow{\Delta}$ (I) $\xrightarrow{Pd}$ (III) + (IV)

ineffective, as are chlorides of platinum(II), osmium(II), ruthenium(III), rhodium(III), and rhenium(V). The authors (Weigert *et al.*, 1970) noted this type of cyclodimerization is more typical of acetylenes than olefins.

Terpenes

Dunne and McQuillin (1970) reported several interesting dimerizations of terpenes over disodium tetrachloropalladate(II) and tetrakis(triphenylphosphine)palladium. As carried out, the reactions did not appear catalytic with respect to palladium, but presumably could be made so.

Ethylene and Styrene

Ethylene adds readily to styrene under mild conditions in the presence of di-μ-chloro-dichlorobis(styrene)dipalladium(II). By bubbling ethylene into a styrene solution at 50°C containing the palladium catalyst, a mixture of codimers including *trans*-1-phenyl-1-butene, *cis*-1-phenyl-1-butene, and *trans*-1-phenyl-2-butene is formed in 1278, 57, and 170% yields, respectively, based on palladium(II) used. No dimers of ethylene or styrene are formed, but in the absence of ethylene, styrene is converted to *trans*-1,3-diphenyl-1-butene (Kawamoto *et al.*, 1970).

$$PhCH{=}CH_2 + CH_2{=}CH_2 \longrightarrow$$

$$\underset{Ph}{\overset{H}{>}}C{=}C\underset{H}{\overset{CH_2CH_3}{<}} + \underset{Ph}{\overset{H}{>}}C{=}C\underset{CH_2CH_3}{\overset{H}{<}} + \underset{PhCH_2}{\overset{H}{>}}C{=}C\underset{H}{\overset{CH_3}{<}}$$

Oligomerization of Substituted Olefins

A number of substituted olefins have been oligomerized over noble metals with varying degrees of success. Much of the research effort in this area has been directed toward formation of products with industrial potential; the reaction as a general synthetic tool seems relatively unexplored.

Acrylic Esters

Methyl acrylate is dimerized end-to-end by rhodium or ruthenium chlorides to afford dimethyl 2-hexenedioate. A mixture of 79 gm of methanol, 129 gm of methyl acrylate, 3 gm of rhodium chloride, and 2 gm of hydroquinone, heated to 140°C for 10 hours in a 400 ml pressure vessel, affords mainly unchanged material together with 9 gm of dimethyl 2-hexenedioate. Dimerization with ruthenium chloride occurs at 210°C, but addition of a small amount of ethylene permits use of 150°C temperatures and gives a 56% yield of dimer. This catalytic effect of ethylene is of interest in regard to the mechanism of dimerization. A mixture of 40 gm of methanol, 144 gm of methyl acrylate, 2 gm of ruthenium trichloride, and 10 gm of ethylene heated to 150°C for 16 hours in a 400 ml pressure vessel, affords 83 gm of unchanged methyl acrylate, 4 gm of methyl 3-pentenoate, and 34 gm of dimethyl 2-hexenedioate (Alderson *et al.*, 1965).

$$2H_2C{=}CHCOOCH_3 \longrightarrow CH_3O\underset{\underset{O}{\|}}{C}CH{=}CHCH_2CH_2\underset{\underset{O}{\|}}{C}OCH_3$$

Acrylamides

Acrylamide and its *N*-monosubstituted derivatives are dimerized over rhodium chloride to *trans*-α-hydromuconamides.

$$2H_2C{=}CH\underset{\underset{O}{\|}}{C}NHR \xrightarrow[\text{Alcohol}]{RhCl_3} \overset{\overset{O}{\|}}{RNHC}(H)C{=}C(H)CH_2CH_2\overset{\overset{O}{\|}}{C}NHR$$

The yield of dimer is usually only 2.5 to 5.5 moles per gram atom of rhodium, but it can be improved by addition of redox reagents, such as *p*-benzoquinone. With ruthenium trichloride as catalyst, only tars are obtained and with palladium chloride, only starting material. *N,N*-Disubstituted acrylamides cleave at the amide group over rhodium chloride and produce secondary amines and polymers of the olefinic part (Kobayashi and Taira, 1968).

Acrylonitrile

Dimerization of acrylonitrile to afford dicyanobutene or adiponitrile by addition of hydrogen has received much attention because of the industrial importance of these products as precursors of hexamethylenediamine. Both homogeneous and heterogeneous catalysts have been used in this reaction, with some form of ruthenium generally the most successful catalyst. Formation of dicyanobutene from acrylonitrile does not require the addition of hydrogen, but dimerizations have been successful only when carried out under hydrogen pressure, a consequence of which is that hydrogenation of the substrate to propionitrile becomes a major competing reaction.

Acrylonitrile is dimerized end-to-end in the presence of ruthenium trichloride and hydrogen to afford a mixture of *cis*- and *trans*-1,4-dicyanobutene and propionitrile as the main products together with some adiponitrile. Ten milliliters of acrylonitrile, 20 ml of ethanol, 0.200 gm of ruthenium trichloride trihydrate heated 30 minutes under 30 atm of hydrogen at 150°C affords 51%

propionitrile, 4% adiponitrile, 18% *cis*- and 22% *trans*-1,4-dicyano-1-butene (Misono *et al.*, 1968a).

$$CH_2{=}CHCN \rightarrow CH_3CH_2CN + NC(CH_2)_4CN + NCCH{=}CHCH_2CH_2CN$$

This reaction is sensitive to the catalyst, solvent, pressure, temperature, and various additives. High pressure favors a high rate of reaction and formation of propionitrile and adiponitrile, whereas a lower pressure favors dicyanobutene. The rate of reaction is very slow below 120°C, between 140° and 180°C the rate increases with temperature but the composition is unchanged, above 180°C, the conversion begins to decrease. Dichlorotetrakis(acrylonitrile)-ruthenium (Chabardes *et al.*, 1969), dichloro(dodeca-2,6,10-triene-1,12-diyl)-ruthenium(IV), and ruthenium(III) acetylacetonate have activities in this reaction similar to ruthenium trichloride. Dichloro(dicarbonyl)bis(pyridine)-ruthenium(II) is ineffective as a catalyst, whereas $RuCl_2(Ph_3PO)_4$, $RuCl_2(Ph_3P)_4$, or $RuCl_2$ affords propionitrile as the main product (Misono *et al.*, 1967, 1968b). A vapor-phase dimerization of acrylonitrile to 1,4-dicyano-1-butene over $PdCl_2$—Ph_3Bi has been reported (Japanese Patent Appl. 25,726/71).

Various improvements in the reaction can be effected by use of appropriate solvents and additives. Reactions catalyzed by $RuCl_2(CH_2{=}CHCN)_4$ carried out in the presence of small amounts of *N*-methylpyrrolidine proceed at several times the rate achieved in the absence of the amine, but the composition of the product is little affected. The yield of dimers is about 62 to 66% at 150 psi and 50 to 53% at 600 psi hydrogen pressure. The dimers consist of about 45% *cis*- and 45% *trans*-dicyanobutenes, 9% adiponitrile, and less than 1% methyleneglutaronitrile (McClure *et al.*, 1968). Dimerization of acrylonitrile may be carried out at convenient rate at low hydrogen pressures, about 85 psi, if ruthenium chloride is used with the co-catalysts $SnCl_2$ or Et_4NSnCl_3 together with bifunctional amines such as *N*-methylmorpholine and/or bifunctional alcohols, such as Methyl Cellosolve. Dimerization at low pressure does not occur in isopropanol, aqueous or anhydrous 1,2-dimethoxyethane, or ethylene glycol, but it does occur in propane-1,3- or butane-1,4-diol. The authors suggested the solvent is functioning as a bridging ligand (Billig *et al.*, 1968).

A number of other improvements in ruthenium-catalyzed systems have been suggested, but short of an extensive investigation it is impossible to judge the merits of these modifications especially in regard to repeated reuse of the catalytic system. Improved rates of dimerization of acrylonitrile have been claimed for the use of ruthenium trichloride in conjunction with cadmium powder and several other metals (Shinohara *et al.*, 1970), $RuO(OH)_2$ (Chabardes *et al.*, 1968, 1969) or cuprous chloride and a Lewis base such as triethylamine or triphenylphosphine (Belgian Patent 696,911). Heterogeneous dimerization catalysts may be derived from ruthenium trichloride and an acid salt supported on carbon powder (Linn and Stiles, 1971).

Allyl Alcohol

Allyl alcohol undergoes an oxidative dimerization and forms a mixture of **V** and **VI** when heated with catalytic quantities of palladium chloride. During the reaction, some allyl alcohol is oxidized to acrolein. By the use of *p*-benzoquinone in the system, allyl alcohol oxidation is inhibited and the chain length is greatly increased. For example, 3 mmoles of palladium chloride, 970 mmoles of allyl alcohol, and 120 mmoles of *p*-benzoquinone afforded 224 mmoles of **V** and **VI** (Urry and Sullivan, 1969).

$H_2C{=}CHCH_2OH \longrightarrow$ **(V)** + **(VI)**

(V): ring H_2C–O–C(H)(CH_2OH)–CH_2–C(=CH_2)– ; (VI): ring H_2C–O–C(H)(CH_2OH)–CH=C(CH_3)–

(V) **(VI)**

Haloolefins

An interesting type of coupling reaction of vinyl halides to butadienes has been described by Jones (1967). Vinyl chloride can be quantitatively converted to butadiene at room temperature by Sn(II) solutions containing catalytic amounts of platinum chloride co-catalyzed by cesium fluoride. The reaction is performed best by stirring a mixture of 2 molar equivalents of $(C_2H_5)_4$-$NSnCl_3$, 1 equivalent of $PtCl_2$, and CsF in dimethylformamide containing 2% water under an atmosphere of vinyl chloride. Acetonitrile and dimethyl sulfoxide are also suitable solvents. The function of cesium fluoride, which is essentially insoluble, is obscure; neither cesium chloride nor sodium fluoride have a co-catalytic effect. Vinyl fluoride couples about as readily as vinyl chloride. Coupling of larger vinyl halide molecules is less successful. Allyl chloride is converted to 1,5-hexadiene in 50% conversion under the above conditions.

Telomerization of Olefins*

Noble metals catalyze the addition of a variety of substances to olefinic, and more especially, diolefinic linkages. The yields and catalyst life depend markedly on the reaction and conditions. Presumably, these relatively unexplored reactions with monoolefins could be much improved with further development.

* The term "telomerization" was originally applied to reactions in which a second substance entered a growing polymer chain. As used here, perhaps incorrectly, the term is meant to imply formation of adducts between one or more olefinic molecules and a second substance.

Hydrogen Cyanide

Saturated nitriles are obtained by palladium-catalyzed addition of hydrogen cyanide to olefins. Zero-valent $Pd[P(OPh)_3]_4$, in the presence of excess triphenyl phosphite, is an effective catalyst; without excess of the phosphorus ligand, the reaction rate is very low. Best results are obtained by the use of equimolar amounts of triphenyl phosphite and hydrogen cyanide. The need for large amounts of triphenyl phosphite in large-scale runs imposed by this latter condition can be circumvented by the technique of adding hydrogen cyanide portionwise at such a rate that the unreacted concentration always remains below the concentration of the phosphite.

A typical small-scale experiment is illustrated by formation of *exo*-2-cyanobicyclo[2.2.1]heptane from hydrogen cyanide addition to bicyclo[2.2.1]-heptene. A mixture of 0.675 gm (0.5 mmole) of tetrakis(triphenyl phosphite)-palladium(0), 5.85 gm (62 mmoles) of bicyclo[2.2.1]heptene, 4.96 gm (16 mmoles) of triphenyl phosphite, and 33 mmoles of hydrogen cyanide in 25 ml of benzene heated to 120° for 18 hours affords *exo*-2-cyanobicyclo[2.2.1]-heptane in 82% yield (Brown and Rick, 1969).

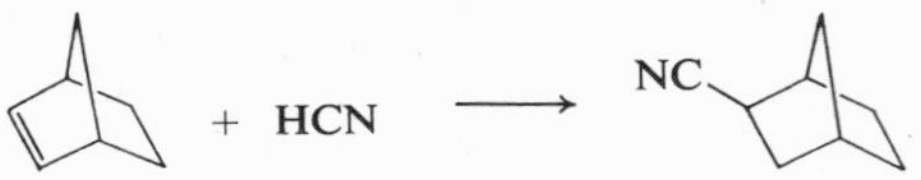

+ HCN ⟶ NC

Hydrogen cyanide addition to 2-vinylbicyclo[2.2.1]hept-5-ene produces three distinct mixtures comprised of 5- and 6-cyano-2-vinylbicyclo[2.2.1]-heptane, *cis* and *trans* isomers of 5- and 6-cyano-2-ethylidenebicyclo[2.2.1]-heptane, and *endo*-2-cyanotricyclo[4.2.1.0(3,7)]nonane (Brown and Rick, 1969).

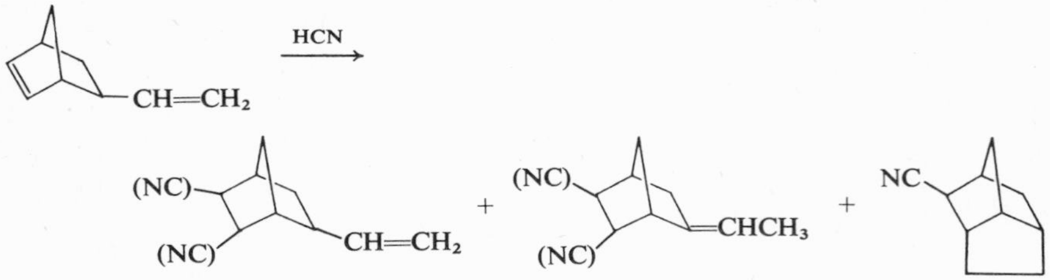

$CH{=}CH_2$ $\xrightarrow{HCN}$

(NC) (NC) $CH{=}CH_2$ + (NC) (NC) $=CHCH_3$ + NC

Earlier workers (Odaira *et al.*, 1965) added cyanide to olefins in polar solvents via stoichiometric palladium cyanide. In non-polar solvents, palladium cyanide converts ethylene to high molecular weight polymers (Blackham, 1965).

Formaldehyde

Formaldehyde interacts with branched olefins in the presence of catalytic amounts of palladium chloride and cupric chloride affording 1,3-dioxanes in good yield. Straight-chain olefins give poor yields. Palladium nitrate and acetate are also catalytically active, but palladium complexes of triphenylphosphine, pyridine, dimethylglyoxime, acetylacetone, and diethyl sulfide are inactive (Sakai *et al.*, 1967).

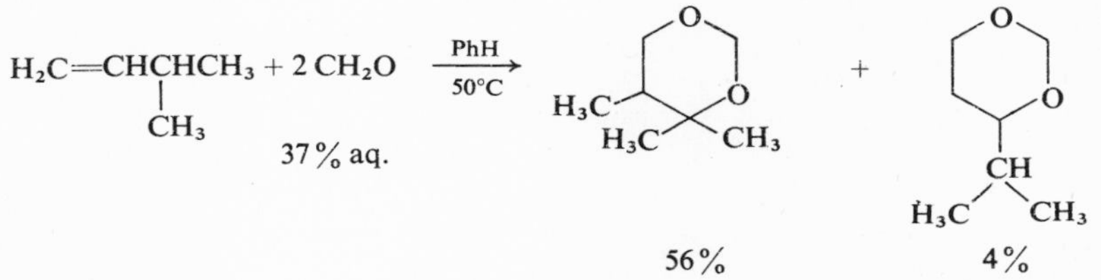

$$H_2C{=}CHCH(CH_3)CH_3 + 2\,CH_2O\ (37\%\ aq.) \xrightarrow[50^\circ C]{PhH} \text{4,4,5-trimethyl-1,3-dioxane } (56\%) + \text{4-isopropyl-1,3-dioxane } (4\%)$$

Butyrolactone

Interaction of ethylene and butyrolactone in the presence of cupric chloride and palladium metal or palladium chloride affords a mixture of three esters.

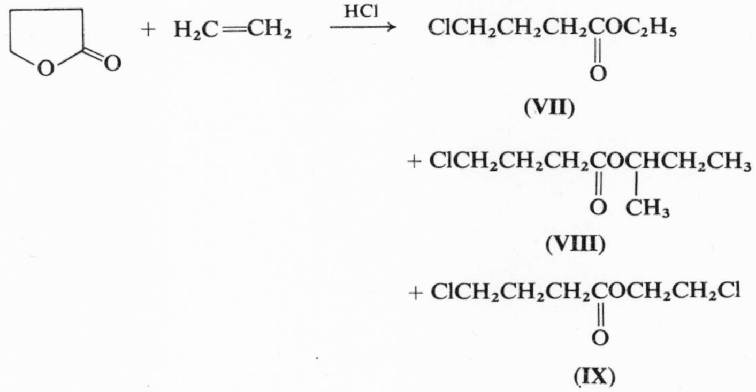

$$\text{butyrolactone} + H_2C{=}CH_2 \xrightarrow{HCl} ClCH_2CH_2CH_2COOC_2H_5\ \textbf{(VII)}$$

$$+\ ClCH_2CH_2CH_2COOCH(CH_3)CH_2CH_3\ \textbf{(VIII)}$$

$$+\ ClCH_2CH_2CH_2COOCH_2CH_2Cl\ \textbf{(IX)}$$

Of these products only 2-chloroethyl 4-chlorobutyrate (**IX**) depends on the presence of cupric chloride. The authors suggest that formation of esters **VII** and **VIII** depends on the addition of ethylene and 2-butene respectively, to 4-chlorobutyric acid formed by ring opening of the lactone with hydrochloric acid. The origin of **IX** is in more doubt. Yields of these esters in total are fair, but conversions are low. The authors commented that improved conversions would be expected if the reaction mixture had been stirred (Saegusa *et al.*, 1967).

Sulfur Dioxide

In a reaction analogous to carbonylation of ethylene, ethylene and sulfur dioxide interact in the presence of palladium chloride to afford ethyl *trans*-but-2-enyl sulfone. About 6 moles of product are formed per mole of palladium chloride and the reaction terminates with precipitation of palladium metal (Klein, 1968). With potassium chloride as a co-catalyst, ethyl vinyl sulfone is formed (Klein, 1969).

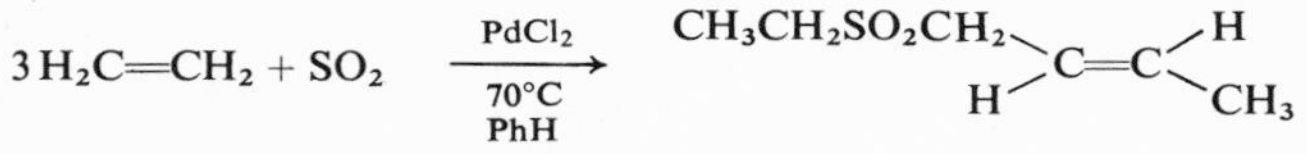

$$3\,H_2C{=}CH_2 + SO_2 \xrightarrow[\substack{70^\circ C\\ PhH}]{PdCl_2} CH_3CH_2SO_2CH_2(H)C{=}C(H)CH_3$$

Maleic Anhydride

Bicyclo[2.1.0]pentane (**X**) adds to maleic anhydride in the presence of rhodium dicarbonyl chloride dimer to afford **XI**, which is also obtained from cyclopentene and maleic anhydride. Bicyclopentane is isomerized rapidly and quantitatively by this rhodium catalyst to cyclopentene (Yates, 1971). Extensive scrambling of deuterium occurs when deuterated bicyclopentanes are subjected to this isomerization. Scrambling occurs at an intermediate stage for 1-deuteriocyclopentene is not isomerized under identical conditions (Gassman *et al.*, 1971).

$$\text{(X)} + 2\ \text{maleic anhydride} \xrightarrow{[Rh(CO)_2Cl]_2} \text{(XI)} + \text{succinic anhydride}$$

Active Methylene Compounds

Allylic alcohols, amines, and esters will interact with active methylene compounds such as acetylacetone in the presence of palladium catalysts to produce carbon-allylated products in good yield. For example, 0.05 mole of acetylacetone, 0.043 mole of allyl alcohol, 0.25 mmole of palladium acetylacetonate, and 0.75 mmole of triphenylphosphine heated together at 85°C for 3 hours affords 3-allylacetylacetone in 70% yield and 3,3-bis(allyl)acetylacetone in 26% yield.

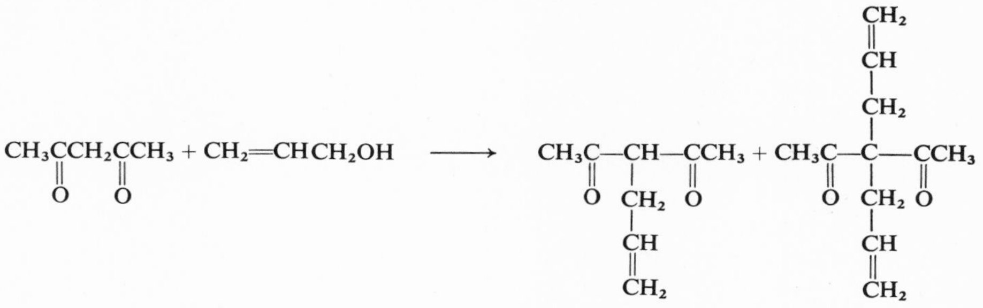

$$CH_3\underset{\substack{\|\\O}}{C}CH_2\underset{\substack{\|\\O}}{C}CH_3 + CH_2{=}CHCH_2OH \longrightarrow CH_3\underset{\substack{\|\\O}}{C}{-}\underset{\substack{|\\CH_2\\|\\CH\\\|\\CH_2}}{CH}{-}\underset{\substack{\|\\O}}{C}CH_3 + CH_3\underset{\substack{\|\\O}}{C}{-}\overset{\substack{CH_2\\\|\\CH\\|\\CH_2\\|}}{\underset{\substack{|\\CH_2\\|\\CH\\\|\\CH_2}}{C}}{-}\underset{\substack{\|\\O}}{C}CH_3$$

Allylic esters react more slowly than the corresponding allylic alcohols, but the rate with esters can be accelerated by addition of a tertiary amine to the reaction mixture. Phenylacetone and phenylacetonitrile also can be allylated by this catalyst system; interaction of these compounds with 2,7-octadien-1-ol affords 3-phenyl-5,10-undecadiene-2-one and 2-phenyl-4,9-decadienenitrile, respectively (Atkins *et al.*, 1970). All reactions were performed in a nitrogen atmosphere.

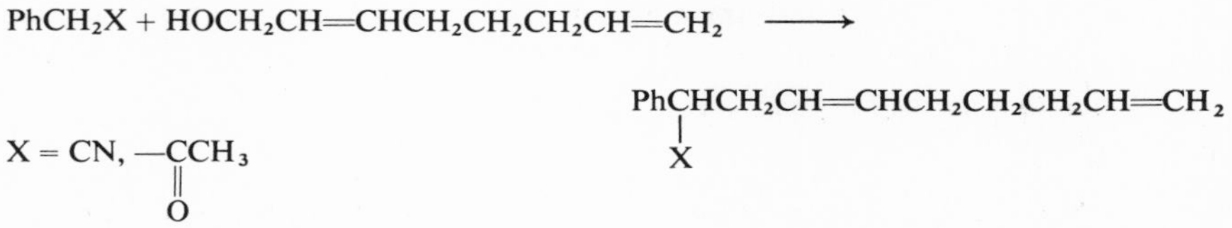

$$PhCH_2X + HOCH_2CH{=}CHCH_2CH_2CH_2CH{=}CH_2 \longrightarrow Ph\underset{\substack{|\\X}}{C}HCH_2CH{=}CHCH_2CH_2CH_2CH{=}CH_2$$

$$X = CN, \; {-}\underset{\substack{\|\\O}}{C}CH_3$$

Oligomerization of Dienes

Dienes are oligomerized readily by noble metal catalysts to afford materials ranging from dimers to high molecular weight polymers. Polymerization is beyond the scope of this work, but it is worth noting that certain noble metal complexes produce highly stereospecific materials (Rinehart *et al.*, 1961; Smith and Wilkinson, 1962; Canale *et al.*, 1962; Canale and Hewett, 1964; Dauby *et al.*, 1967; Morton and Das, 1969; Rinehart, 1969).

Allene

Allene, in the presence of certain rhodium catalysts, is converted with high specificity to a single tetramer, 1,4,7-trimethylenespiro[4.4]nonane, and a single pentamer of unknown structure, a result contrasting with the complex mixture formed in thermal oligomerization. A typical experiment uses 150 gm of allene, 1 gm of tris(triphenylphosphine)chlororhodium, and 0.1 gm of hydroquinone in 75 gm of chloroform contained at 80°C for 16 hours in a silver-lined tube. Polymer formation is suppressed by the presence of hydroquinone.

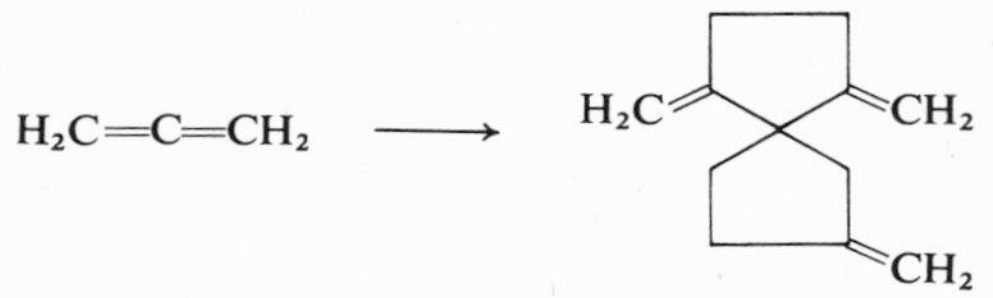

$$H_2C{=}C{=}CH_2 \longrightarrow$$

H_2C CH_2

CH_2

Similar results were obtained using the more active $(Ph_3P)_2Rh(CO)Cl$, but the chance of violent explosion using this catalyst is high (Jones and Lindsey, 1968). The same tetramer was obtained in 70–80% yield by $Rh_2Cl_2(C_2H_4)_4$ and triphenylphosphine. The role of the phosphine ligand is very specific; no trialkyl or triaryl phosphine or arsines were effective other than triphenylphosphine (Otsuka *et al.*, 1969a). In the absence of triphenylphosphine, allene is converted to a monocyclic pentamer (Otsuka *et al.*, 1969b). Addition of π-acceptor ligands such as carbon monoxide or an alkyl isocyanide cause polymerization rather than tetramerization. Several ruthenium complexes also promoted polymerization of allene, whereas the palladium complexes tested were inactive (Otsuka and Nakamura, 1967). In the presence of triaryl phosphines or arsenes, palladium salts catalyze the head-to-head polymerization of allene (Shier, 1969a).

Butadiene

Butadiene forms a number of different oligomers often with high selectivity. There is frequently a marked dependency on the solvent and reaction conditions. Dimerization of butadiene to 1,3,7-octatriene has been achieved over a number of noble metal catalysts in the presence of sodium phenoxide and triphenylphosphine (Smutny, 1966). Other workers obtained the octatriene in 85% yield by heating 13 gm of dried butadiene, 20 ml of acetone, and 219 mg of bis(triphenylphosphine)(maleic anhydride)palladium at 115°C for 7 hours. Benzene and tetrahydrofuran are also suitable solvents and bis(triphenylphosphine)(*p*-benzoquinone)palladium and tetrakis(triphenylphosphine)palladium are suitable catalysts (Takahashi *et al.*, 1967, 1968b). The only other dimer is a small amount of the thermal product, 4-vinyl-1-cyclohexene. An unusual condition of synthesis of the octatriene involves dimerization over $(Ph_3P)_3Pt$ under 400 psig of carbon dioxide pressure. Without carbon dioxide, the product is 90–97% 4-vinyl-1-cyclohexene. Oxygen must be carefully excluded from the system, otherwise the product is mainly vinylcyclohexene. Probably oxygen and carbon dioxide convert the catalyst to the carbonato complex $(Ph_3P)_2(CO_3)Pt$, which was demonstrated to be a catalyst for vinylcyclohexene formation. Similar results were obtained with $(Ph_3P)_4Pd$ as the catalyst (Kohnle *et al.*, 1969; Kohnle and Slaugh, 1969).

$$H_2C{=}CHCH{=}CHCH_2CH_2CH{=}CH_2 \xleftarrow{400\text{ psi }CO_2} H_2C{=}CHCH{=}CH_2 \longrightarrow \text{4-vinylcyclohexene }(CH{=}CH_2)$$

In contrast to most palladium catalysts, bis(π-allyl)palladium slowly converts butadiene at 40°C to a mixture containing 30% *n*-dodecatetraene in higher oligomers, with traces of vinylcyclohexene, the only other product. The corresponding nickel catalyst, bis(π-allyl)nickel, on the other hand, converts butadiene to cyclododecatriene. The authors tentatively concluded that the greater atomic volume of palladium prevents the ring closure observed with the nickel catalyst. Butadiene does not oligomerize at all over the more stable bis(π-allyl)platinum under these conditions, but at higher temperatures mainly trimers and tetramers are formed (Lazutkin *et al.*, 1973). Various donors can alter the course of the reaction. A catalyst formed from 1 mole of phosphorus trichloride per mole of bis(π-allyl)palladium converts butadiene at 90°C exclusively to vinylcyclohexene (Wilke *et al.*, 1966).* Similarly, π-allylpalladium acetate converts butadiene in benzene into the linear trimer, *n*-dodecatetraene, but the linear dimer becomes the major product when triphenylphosphine is added to the system (Medema and Van Helden, 1971).

Oligomerization of butadiene in formic acid–dimethylformamide in the presence of palladium catalysts, such as diacetatopalladium(II) follows an unusual course and a reduced product, octa-1,6-diene, is formed in 90% yield. Octa-1,3,7-triene and octa-2,7-dienyl formate were ruled out as intermediates in formation of octa-1,6-diene. With platinum salts, such as lithium tetrachloroplatinate, in formic acid–dimethylformamide, the main product is octa-1,7-diene obtained in 80% yield (Gardner and Wright, 1972).

$$H_2C{=}CH(CH_2)_4CH{=}CH_2 \xleftarrow[HCOOH]{Pt} 2H_2C{=}CHCH{=}CH_2 \xrightarrow[HCOOH]{Pd} H_2C{=}CH(CH_2)_3CH{=}CHCH_3 + CO_2$$

Norbornadiene

Mrowca and Katz (1966) reported an unusual cycloaddition reaction of norbornadiene catalyzed by rhodium-on-carbon. A variety of transition metal complexes had been used previously in this type of reaction, but catalysis by a metal itself was rare. Norbornadiene when refluxed with 5% rhodium-on-carbon is converted quantitatively to a product containing 70–80% dimers and 20–30% trimers (Katz and Acton, 1967). Reaction also occurs over the

* This reference is a good review of reactions over other transition metal catalysts.

rhodium catalyst at room temperature but much more slowly. No reaction at all could be effected over 30% palladium-on-carbon. The dimers are mainly stereoisomers of **XII** together with smaller amounts of **XIII** [see Katz *et al.* (1969) for assignment of **XII**].

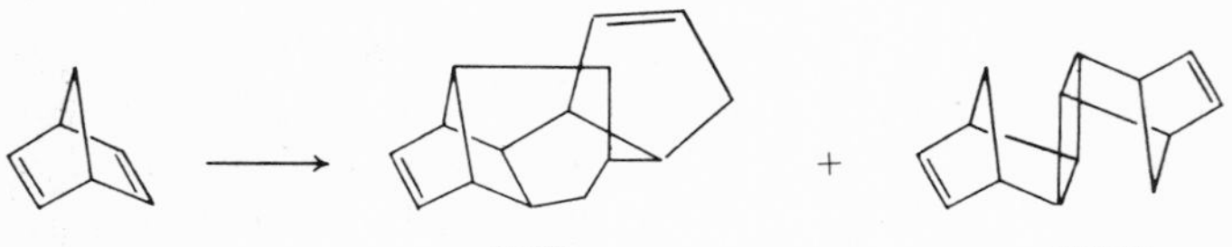

(XII) (XIII)

Similar dimers are formed together with trimers, some of which have greatly altered skeletons (Katz *et al.*, 1969), by heating norbornadiene with the rhodium complexes $(Ph_3P)_3RhCl$ or $(Ph_3P)_2(CO)RhCl$. The authors suggest that the same active catalytic species is involved regardless of the initial rhodium complex (Acton *et al.*, 1972).

Benzonorbornadiene (**XIV**) when heated with 5% rhodium-on-carbon, gives the dimer, **XV**, in 80% yield (Katz *et al.*, 1967). Norbornene is not dimerized by this catalyst.

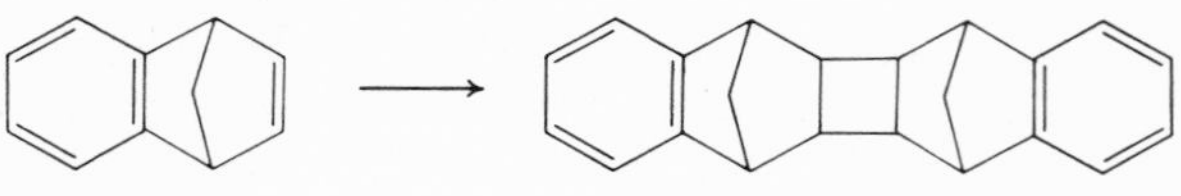

(XIV) (XV)

A different type of dimer is derived from norbornadiene over catalysts capable of forming active dimeric metal–metal-bonded fragments. Dimerization proceeds "head-to-head" in the presence of certain rhodium, cobalt, and iridium catalysts and Lewis acid co-catalysts such as boron trifluoride etherate to afford "Binor-S" (1,2,4:5,6,8-dimetheno-*s*-indancene). Suitable catalysts for the dimerization include $CoBr_2(Ph_3P)_2$, $CoI_2(Ph_3P)_2$, and $RhCl(Ph_3P)_3$, all with $BF_3O(C_2H_5)_2$ as the co-catalyst. In a typical dimerization, 20 ml of freshly distilled norbornadiene, 0.2 gm of catalyst, and 1 ml of boron trifluoride etherate are placed in a flask equipped with a reflux condenser, and dimerization is initiated by gently heating. Occasional cooling may then be required to moderate the reaction, which at times may proceed with explosive violence. The yields are nearly quantitative except over iridium catalysts, which give not more than 10% (Schrauzer *et al.*, 1970).

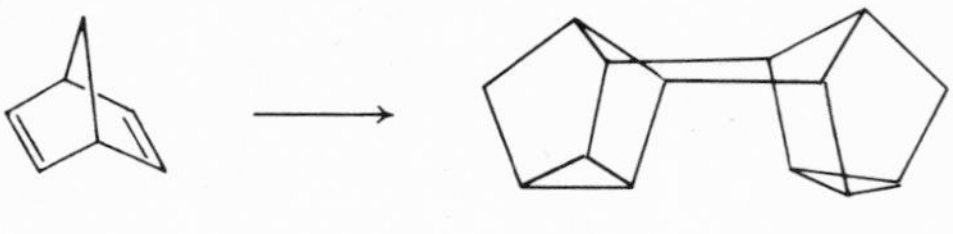

Addition of Dienes to Olefins

Olefins add readily to dienes in the presence of certain noble metal catalysts. The type of product formed depends on the substrate structure, catalyst, and conditions.

Butadiene, Piperylene, and Isoprene

Ethylene and butadiene combine under mild conditions in the presence of rhodium trichloride trihydrate to give 1,4-hexadiene in high yields. 1,3-Butadiene (3.1 moles), 2.7 moles of ethylene, 0.2 gm of rhodium trichloride trihydrate, and 1 ml of ethanol, held for 16 hours at 50°C afford 2% butene, 91% hexadienes (of which 20% are conjugated), 3% octadiene, and 4% higher olefins at 54% conversion. The *trans–cis* ratio of the product as well as the conversion increases with the amount of alcohol present in the system. With little or no alcohol present, the ratio is about 2.5–3.4, whereas with gross amounts the ratio rises to 20. Anhydrous rhodium chloride is ineffective as a catalyst unless small amounts of water are added to the system; large amounts of water strongly inhibit the reaction. Best results are obtained with ethylene–butadiene ratios near 1. Butene and octadienes are both favored by employing excesses of ethylene (Alderson *et al.*, 1965). Rhodium chloride catalysts deactivated in this reaction are easily reactivated by treatment of catalyst residue with hydrochloric acid and ethanol (British Patent 948,041, Jan. 29, 1964). Reduced rhodium on a support such as carbon or asbestos is also said to be effective in this reaction (Japanese Patent 8053/66). Rhodium(III) salts in combination with amides, phosphoramides, or phosphine oxides are said to allow control of the *cis–trans* ratio (Su, 1969).

Cramer (1967) examined this synthesis with care and elucidated its major features. The reaction shows an induction period during which time the catalyst is converted to an active form. The steps consist of reduction of rhodium(III) to rhodium(I), oxidative addition of hydrogen chloride to afford a rhodium(III) hydride, and reaction of this species with coordinated olefin to give a rhodium(III) alkyl (Cramer, 1965). Hexadienes are not formed in the absence of halide. The discrimination for addition of ethylene to butadiene is accounted for by the relatively high stability of the necessary complex. An olefin insertion reaction leads to 1,4-hexadiene released in a rate-determining step. Although initial reduction of Rh(III) is required for activation, the reaction scheme does not demand a cyclic valence change for rhodium, such as was invoked for butene synthesis (Cramer, 1968).

In reactions of higher 1,3-dienes, the vinyl group is found on the more hindered position, presumably because the structure of the oligomer is determined by the structure of the π-allylic intermediate. Ethylene and 1,3-pentadiene afford mostly 3-methyl-1,4-hexadiene (Cramer, 1967). In the presence of rhodium trichloride trihydrate, the reaction occurs with facility. A mixture of 0.2 gm of rhodium trichloride trihydrate, 1 ml of ethanol, 68 gm of 1,3-pentadiene, and 42 gm of ethylene, heated to 50°C for 16 hours affords 89 ml of 3-methyl-1,4-hexadiene.

$$H_2C{=}CHCH{=}CHCH_3 + H_2C{=}CH_2 + RhH \longrightarrow$$

$$[CH_3{-}CH{\cdots}CH{\cdots}CH{-}CH_3\ (\pi\text{-allyl}){-}Rh(CH_2{=}CH_2)] \longrightarrow H_2C{=}CHCH(CH_3)CH{=}CHCH_3$$

Interaction of ethylene and isoprene occurs much more slowly. The major product is 4-methyl-1,4-hexadiene, corresponding to insertion of ethylene in the complex **XVII** and not **XVI** (Cramer, 1967).

$$[CH_2{\cdots}CH{\cdots}C(CH_3)_2\ (\pi\text{-allyl}){-}Rh(CH_2{=}CH_2)] \quad \textbf{(XVI)}$$

$$[CH_2{\cdots}C(CH_3){\cdots}CH{-}CH_3\ (\pi\text{-allyl}){-}Rh(CH_2{=}CH_2)] \quad \textbf{(XVII)} \longrightarrow H_2C{=}CHCH_2C(CH_3){=}CHCH_3$$

Butadiene and ethylene interact in the presence of tributylphosphine, palladous chloride, and diisobutylaluminum chloride at 80°C to afford *trans*-1,4-hexadiene in 75% yield. At higher temperatures, the rate is faster but polymerization and isomerization to 2,4-hexadiene occurs (Schneider, 1972).

$$H_2C{=}CHCH{=}CH_2 + CH_2{=}CH_2 \longrightarrow H_2C{=}CHCH_2(H)C{=}C(H)CH_3\ (trans)$$

Under similar conditions, butadiene interacts with propylene to afford a mixture of 2,3-dimethyl-1,4-pentadiene, 2-methyl-1,4-hexadiene, and 2,5-heptadiene in yields of 12, 62, and 26%, respectively. Piperylene and ethylene interact to afford a mixture of 83% 3-methyl-1-*trans*-4-hexadiene and 1-*trans*-4-heptadiene (Schneider, 1972).

$$H_2C{=}CHCH{=}CHCH_3 + CH_2{=}CH_2 \longrightarrow$$

$$\underset{83\text{–}86\%}{H_2C{=}CHCH(CH_3){-}CH{=}CH{-}CH_3\ (trans)} + \underset{14\text{–}17\%}{H_2C{=}CHCH_2{-}CH{=}CH{-}C_2H_5\ (trans)}$$

Propylene adds to butadiene in a rhodium-catalyzed reaction to afford 2-methyl-1,4-hexadiene. A mixture of 2 gm of rhodium chloride, 100 ml of ethanol, and 126 gm of propylene is heated to 50°C while 87 gm of butadiene is injected over a 6-hour period. The mixture when held at 50°C an additional 10 hours and then fractionally distilled, affords 142 gm of 2-methyl-1,4-hexadiene consisting of 93% *trans* isomer and 7% *cis* isomer (Alderson *et al.*, 1965).

$$H_2C{=}CHCH{=}CH_2 + CH_3CH{=}CH_2 \longrightarrow H_2C{=}C(CH_3){-}CH_2CH{=}CHCH_3$$

Oligomerization of Acetylenes

Various acetylenes are transformed in the presence of noble metal catalysts to dimers, trimers, or higher polymers, the type of product depending markedly on the substrate, catalyst, and solvent (Brailovskii *et al.*, 1968; Whitesides and Ehmann, 1969).

Acetylene Carboxylates

Dimethyl acetylenedicarboxylate, refluxed over 10% palladium-on-carbon in benzene for 72 hours affords hexamethyl mellitate in 93% yield; no trace of the trimer was found in the absence of the catalyst (Bryce-Smith, 1964). The same reaction occurs in the presence of certain metalocyclopentadiene complexes of iridium and rhodium through a stepwise sequence represented by **XVIII–XXI**. The iridocycle **XIX** was prepared in high yield by treating **XVIII** with a mole of dimethyl acetylenedicarboxylate (Collman and Kang, 1967). In support of this overall sequence was the observation that completely deuterated **XIX** when treated with unlabeled dimethyl acetylenedicarboxylate affords only two hexamethyl mellitates; one contained two labeled acetylenes and one contained no labeled acetylenes. Evidence for incorporation of an acetylene in a latent coordination site giving **XX** is the observation that trimerization is completely inhibited by a carbon monoxide pressure of 60 psig (Collman *et al.*, 1968). The rhodium complex, $(Ph_3P)_2(CO)RhCl$, on the other hand, is an

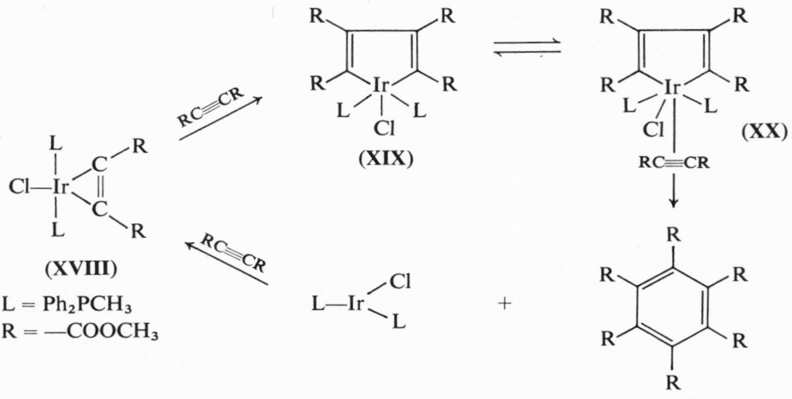

(XVIII) (XIX) (XX) (XXI)

L = Ph_2PCH_3
R = —$COOCH_3$

effective catalyst for trimerization (Collman and Kang, 1967). Methyl propiolate is trimerized over $(Ph_3P)_2(CO)RhCl$ to a mixture of 1,3,5- and 1,2,4-tri-(carbomethoxy)benzenes (Collman and Kang, 1967), whereas over 10% palladium-on-carbon, polymers are formed (Bryce-Smith, 1964). Ethyl propiolate (and other acetylenes) is converted to polymers with a number of Group **VIII** metals plus a hydridic reducing agent (Luttinger and Colthup, 1962).

Acetylenic Carbinols

Certain monosubstituted α-hydroxyacetylenes are dimerized readily by $(Ph_3P)_3RhCl$. For example, 3-methylbut-1-yn-3-ol is converted to *trans*-2,7-dimethyloct-3-en-5-yne-2,7-diol in end-to-end dimerization.

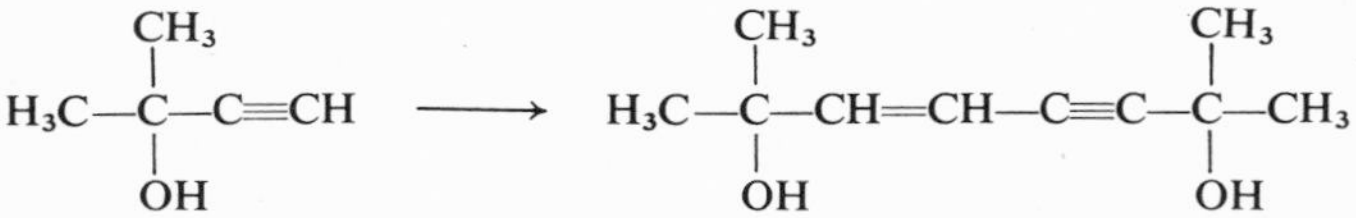

$$H_3C{-}C(CH_3)(OH){-}C{\equiv}CH \longrightarrow H_3C{-}C(CH_3)(OH){-}CH{=}CH{-}C{\equiv}C{-}C(CH_3)(OH){-}CH_3$$

The yield depends on the amount of catalyst, reaction time, and solvent. The best yield of dimer, 73%, was obtained by refluxing 2 ml of 3-methylbut-1-yn-3-ol in 30 ml of benzene with 48 mg $(Ph_3P)_3RhCl$ for 6 hours. In ethanol, the yield was less than 2%. The complexes *trans*-$RhCl(CO)(Ph_3P)_2$, $(Ph_3P)_3RuCl_2$, and $(Ph_3P)_3Pt(0)$ are not effective catalysts for the dimerization. Chlorodicarbonylrhodium(I) trimerizes this substrate to 1,2,4-tris(1-hydroxy-1-methylethyl)benzene in small yield (Chini *et al.*, 1967).

Successful dimerization of acetylenic alcohols by $(Ph_3P)_3RhCl$ depends critically on their structure. 3-Methylpent-1-yn-3-ol is easily and selectively dimerized, whereas but-1-yn-3-ol gives only linear polymers. 1-Ethynlcyclohexan-1-ol gives high yields of the dimer, *trans*-1,4-di(1-hydroxycyclohex-1-yl) but-1-en-3-yne, whereas acetylated 3-methylbut-1-yn-3-ol, butynone, but-3-yn-1-ol, or hex-1-yne, none of which carry an α-hydroxyl, could be dimerized. The authors (Singer and Wilkinson, 1968) proposed a mechanism for the dimerization which takes into account the direction of addition, the requirement of an α-hydroxy function, and the restriction to monosubstituted acetylenes.

$$\text{C}_6\text{H}_{10}(\text{OH})\text{—C}\equiv\text{CH} \longrightarrow \text{C}_6\text{H}_{10}(\text{OH})\text{—C}\equiv\text{C—CH}=\text{CH—C}_6\text{H}_{10}(\text{OH})$$

Phenylacetylenes

The products derived from oligomerization of phenylacetylene depend markedly on the catalyst. In the presence of $RhCl(Ph_3P)_3$ at 50°–70°C phenylacetylene is converted without solvent or co-catalyst almost entirely to *trans*-1,4-diphenylbutenyne. The author (Kern, 1968) viewed phenylacetylene as difunctional, the triple bond acting as one function, the carbon–hydrogen bond as the second, and both enter the coordination sphere of rhodium. In benzene solvent, *trans*-1,4-diphenylbutenyne can be isolated in 22% yield, but the product is mainly polymeric (Singer and Wilkinson, 1968).

$$2\,\text{PhC}\equiv\text{CH} \longrightarrow [\text{H, C}\equiv\text{CPh, PhC}\equiv\text{CH on Rh}] \longrightarrow \text{PhC}\equiv\text{C—CH}=\text{CHPh}$$

Phenylacetylene in benzene is trimerized by $Rh_4(CO)_{12}$ to a mixture of 1,2,4- and 1,3,5-triphenylbenzene (ratio 2.2 to 1, total yield 40%). With $Co_4(CO)_{12}$ as catalyst, the yield is much lower (11–16%) and the ratio of isomers much higher (8.9 to 1) (Iwashita and Tamura, 1970). In the presence of dichlorobis(benzonitrile)palladium, phenylacetylene is converted to labile polymeric material not easily characterized (Dietl and Maitlis, 1968). Phenylacetylene is dimerized slowly at 100°C to 1-phenylnaphthalene over palladium black (prepared by sodium borohydride reduction of palladous chloride) (Brown and Brown, 1962), but over 10% palladium-on-carbon a vigorous reaction ensues at room temperature affording a complex mixture of linear polymers (Bryce-Smith, 1964). Some activity was found also with platinum, iridium, ruthenium, rhodium, and nickel catalysts, but none were so active as palladium. Phenylacetylene is converted by palladium chloride in benzene into a mixture of 1,2,4-triphenylbenzene (0.2 gm), *trans*-1,3,5-triphenyl-1-hexyne-2,5-diene (0.3 gm), and 1.1 gm of a polyene (molecular weight 966) (Odaira *et al.*, 1965).

Methylphenylacetylene, unlike phenylacetylene, is converted smoothly at 40°C over dichlorobis(benzonitrile)palladium in methylene chloride to a mixture of three trimers.

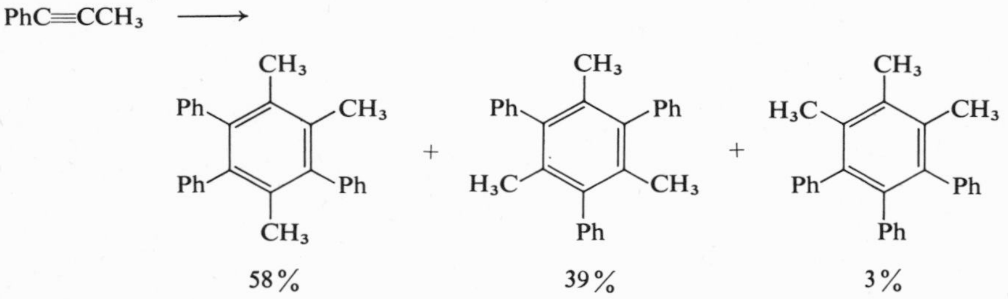

$$PhC{\equiv}CCH_3 \longrightarrow$$

58% + 39% + 3%

Formation of 1,2,3-trimethyl-4,5,6-triphenylbenzene was unexpected; its formation requires, formally at least, cleavage of a carbon–carbon triple bond (Dietl and Maitlis, 1968). Other workers obtained mainly poly(methylacetylene) (Odaira *et al.*, 1965). In benzene solvent, the major product is a yellow labile complex, $[(CH_3C_2Ph)_3PdCl_2]_2$ (Dietl *et al.*, 1970). Diphenylacetylene is trimerized smoothly by palladium complexes in nonhydroxylic solvents, such as benzene, chloroform, and acetone, to hexaphenylbenzene. The reaction provides an easy route to this hydrocarbon (Blomquist and Maitlis, 1962).

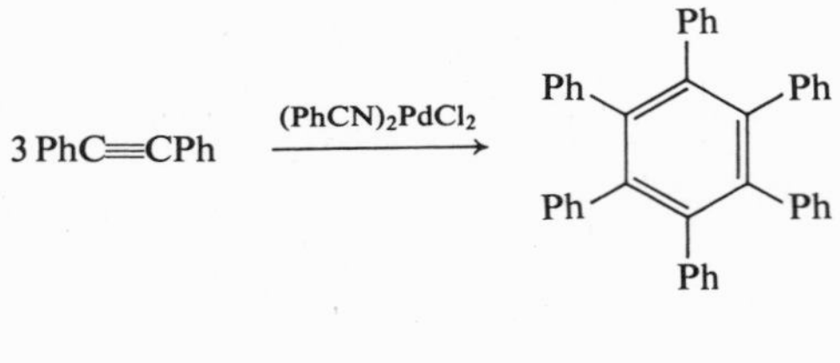

$$3\,PhC{\equiv}CPh \xrightarrow{(PhCN)_2PdCl_2} C_6Ph_6$$

Telomerization of Dienes

Dienes interact in the presence of noble metal catalysts with a variety of compounds forming adducts containing one or more molecules of the diene. In general, the reaction proceeds more readily than similar reactions with monoolefins.

Allene

Allene in glacial acetic acid at 50°C and 1 atm is converted in the presence of a palladium acetate catalyst to a mixture of allyl acetate, 2,3-dimethylbuta-1,3-diene, 3-methyl-2-hydroxymethylbuta-1,3-diene acetate, and 2,3-dihydroxymethylbuta-1,3-diene diacetate in the approximate ratio 7:19:100:14. Carbon–carbon bonds are formed between internal carbon atoms of allene and carbon–oxygen bonds are formed at the terminal allene carbon atoms (Shier, 1967).

$$H_2C{=}C{=}CH_2 + HOAc \longrightarrow \underset{7}{H_2C{=}CHCH_2OAc} + \underset{19}{H_2C{=}C(CH_3){-}C(CH_3){=}CH_2}$$

$$+ \underset{100}{H_2C{=}C(CH_2OAc){-}C(CH_3){=}CH_2} + \underset{14}{H_2C{=}C(CH_2OAc){-}C(CH_2OAc){=}CH_2}$$

Interaction of a 1,2-alkadiene with a 1-alkyne and a carboxylic acid in the presence of a palladium salt produces a conjugated en-yne carboxylic acid and other material. Again a carbon–carbon bond is established at the center atom of the allene structure. A mixture of 0.25 gm of palladium nitrate dihydrate, 0.5 gm of potassium acetate, 0.25 gm of *o*-phthalic acid, 0.1 gm of 3,5-dinitrocatechol, 55 gm of acetic acid, 12 gm of methylacetylene, and 36 gm of allene, heated to 65°C for 64 hours affords 2.7 gm of 2-methylpent-1-ene-3-yne and 4 gm of 2-acetoxymethylpent-1-ene-3-yne (Shier, 1969b).

$$H_2C{=}C{=}CH_2 + CH_3C{\equiv}CH + HOAc \longrightarrow$$

$$H_3C{-}C{\equiv}C{-}C(CH_3){=}CH_2 + CH_2{=}C(CH_2OAc){-}C{\equiv}CCH_3$$

Allene interacts with compounds such as acetylacetone, ethyl acetoacetate, diethyl malonate, ethyl cyanoacetate, and malononitrile in the presence of Pd(0) and Rh(I) catalysts to afford a mixture of mono diene and bis diene. Palladium(II) and rhodium(III) catalysts are completely ineffective in this reaction although they are effective in the corresponding telomerization of amines and allene. A solution of 30.6 ml of diethylmalonate, 8 gm of allene, and 1.46 gm of bis(triphenylphosphine)(maleic anhydride)palladium in 25 ml of tetrahydrofuran heated to 100°C for 6 hours affords **XXII** in 86% yield (Coulson, 1972).

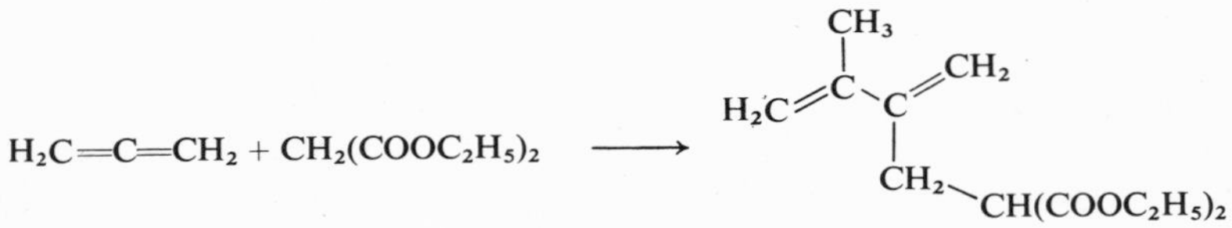

$$H_2C{=}C{=}CH_2 + CH_2(COOC_2H_5)_2 \longrightarrow H_2C{=}C(CH_3){-}C({=}CH_2){-}CH_2{-}CH(COOC_2H_5)_2$$

(XXII)

Allene interacts with amines in the presence of various palladium or rhodium compounds, such as $PdCl_2$, $RhCl_3$, $(Ph_3P)_4Pd$, $(Ph_3P)_2Pd \cdot$olefin, to afford derivatives of 3-methyl-2-methylene-3-butenylamine.

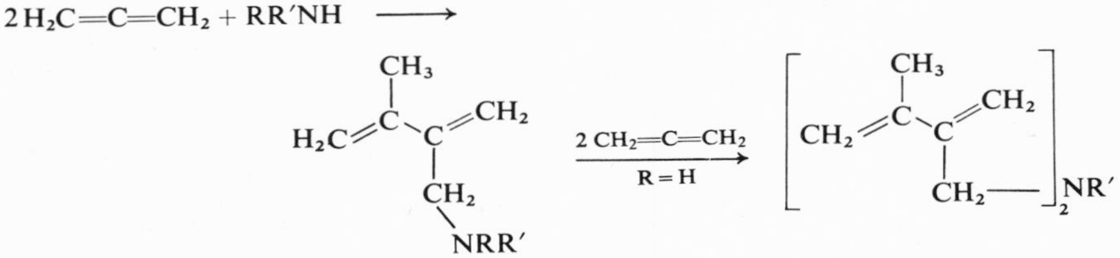

$$2\,H_2C{=}C{=}CH_2 + RR'NH \longrightarrow H_2C{=}C(CH_3){-}C({=}CH_2){-}CH_2{-}NRR' \xrightarrow[R=H]{2\,CH_2{=}C{=}CH_2} \left[CH_2{=}C(CH_3){-}C({=}CH_2){-}CH_2{-}\right]_2 NR'$$

When the amine substrate carries two hydrogens either the mono diene or bis diene can be made to predominate by varying the reaction temperature or mole ratio of reactants. The reactions can be carried out at atmospheric pressure by passing allene into a solution of the amine and catalyst in hexamethylphosphoramide at 70°–90°C or more conveniently in lower-boiling solvent in a sealed vessel at 100°–140°C. The reaction offers considerable synthetic potential since moderate to high yields of product usually are obtained. Interaction of allene with *N*-deuteriopiperidine affords the monodiene, **XXIII**, in 79% yield with $95 \pm 3\%$ of the deuterium atoms located in the methyl group (Coulson, 1972).

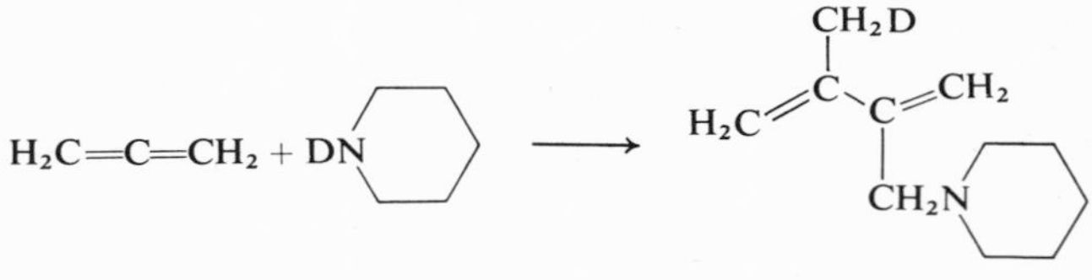

$$H_2C{=}C{=}CH_2 + DNC_5H_{10} \longrightarrow H_2C{=}C(CH_2D){-}C({=}CH_2){-}CH_2NC_5H_{10}$$

(XXIII)

Butadiene and Nucleophiles

Butadiene in the presence of various noble metal catalysts and a nucleophile, such as an alcohol, phenol, acid, amine, active methylene compound, or silane (Takahashi *et al.*, 1969) undergo condensation affording 2,7-octadienyl derivatives. The yields are at times very good, but the reaction is sensitive to the catalyst, solvent, and various modifiers.

Phenols

Condensation of phenol and butadiene occurs readily over a number of noble metal catalysts in the presence of sodium phenoxide to afford 1-phenoxy-2,7-octadiene. The reaction is remarkable in that the chain cleanly terminates after oligomerizing only two butadiene units and produces a single product in high yield (Smutny and Chung, 1969). It can be run with equal success in the presence or absence of solvents (Smutny, 1968). The base has a profound effect on the reaction and without it, poor yields are obtained. Phenoxide anion seems specific; ammonia, pyridine, or triethylamine in place of phenoxide give poor yields of product (Smutny, 1970b,c). The type of catalyst is not critical; palladium chloride, platinum chloride, 10% platinum-on-carbon, 5% palladium-on-barium sulfate, and various π-allylic palladium complexes all gave uniformly high yields; ruthenium catalysts were inferior, however. Various complex rhodium catalysts may also be used with the yield depending importantly on the ligands (Table I) (Dewhirst *et al.*, 1970).

TABLE I
FORMATION OF 1-PHENOXY-2,7-OCTADIENE

Rhodium catalyst	Conversion	Selectivity
Chlorotris(triphenylphosphine)rhodium(I)	95	85
Bis[chlorodicarbonylrhodium(I)]	65	60
Bis[chlorodiethylenerhodium(I)]	20	45

The yields of product are sensitive to solvent. A very low yield was obtained in dimethyl sulfoxide (6%), moderate yields (60–67%) in tetrahydrofuran, benzene, or dimethylformamide, and high yields (100%) in chloroform or acetonitrile (Smutny, 1970b). The reaction can also be carried out successfully without solvent, preferably in an excess of butadiene. Phenol (0.4 mole) and butadiene (1.7 moles) heated together at 100°C in the presence of 0.00056 mole palladium chloride and 0.014 mole of sodium phenoxide affords *trans*-1-phenoxyoctadiene in 91% yield at 96% conversion of the phenol, accompanied by about 4% of the *cis* isomer and about 5% of 3-phenoxy-1,7-octadiene (Smutny, 1967b). A solvent is useful when working with substituted phenols not soluble in butadiene (Smutny *et al.*, 1969).

$$C_6H_5\text{—}OH + 2CH_2\text{=}CHCH\text{=}CH_2 \longrightarrow C_6H_5\text{—}OCH_2CH\text{=}CHCH_2CH_2CH_2CH\text{=}CH_2$$

Condensations of butadiene and phenol carried out in the presence of a tertiary aromatic phosphine or excess phenol take a different course and afford octadienyl phenols. For example, 0.425 mole of phenol, 1.66 moles of butadiene, 1.0 gm of palladium chloride, 2.0 gm of sodium phenoxide, 2.3 gm of triphenylphosphine, heated 2.5 hours at 90°C affords *o*-[1-(2,7)octadienyl)]-phenol in 69% yield based on phenol charged and the corresponding *para* isomer in 31% yield (Smutny, 1970b). Yields of 2,7-octadienylphenols above 90% have been obtained using phenol, *p*-chlorophenol, *p*-methylphenol, and α-naphthol as substrates (Smutny *et al.*, 1969).

$$C_6H_5{-}OH + 2\,CH_2{=}CHCH{=}CH_2 \longrightarrow HO{-}C_6H_4{-}CH_2CH{=}CHCH_2CH_2CH_2CH{=}CH_2$$

1,3,7-Octatriene can be obtained in high yield from phenoxyoctadiene by addition of small amounts (0.004 mole) of triphenylphosphine to the reaction product before distillation of the reaction mixture under reduced pressure (Smutny, 1967a). Sodium phenoxide, already present, is a necessary component of this elimination reaction (Smutny, 1966). The role of triphenylphosphine is believed to be maintenance of catalyst integrity; without phosphine present, palladium rapidly forms a mirror in the reactor (Smutny and Chung, 1969).

$$C_6H_5{-}OCH_2CH{=}CHCH_2CH_2CH_2CH{=}CH_2 \longrightarrow H_2C{=}CHCH{=}CHCH_2CH_2CH{=}CH_2 + C_6H_5{-}OH$$

Availability of large quantities of 1,3,7-octatriene by this route permits facile syntheses of higher oligomers. 1-Phenoxy-2,7,11-dodecatriene is prepared in 75% yield with greater than 90% purity by interaction at room temperature of octatriene, butadiene, and phenol in 4 to 4 to 1 mole ratio with π-allylpalladium–sodium phenoxide as a catalyst. Excess octatriene is used to minimize phenoxyoctadiene formation. Elimination of phenol to produce

$$C_6H_5{-}OH + CH_2{=}CHCH{=}CH_2 + CH_2{=}CHCH{=}CHCH_2CH_2CH{=}CH_2 \longrightarrow C_6H_5{-}OCH_2CH{=}CHCH_2CH_2CH_2CH{=}CHCH_2CH_2CH{=}CH_2$$

1,3,7,11-dodecatetraene and repetition of the above sequence permits synthesis of 1,3,7,11,15-hexadecapentaene (Smutny, 1973).

Phenoxyoctadiene and other allyl ethers (Hata *et al.*, 1970) undergo facile exchange reactions with active hydrogen compounds such as alcohols, carboxylic acids, primary and secondary amines, and active methylene compounds in the presence of various complex palladium catalysts. Alkyl allyl ethers undergo exchange more slowly, but the reaction can be catalyzed by the presence of phenol which presumably functions through a double exchange mechanism, affording a phenoxy allyl intermediate (Takahashi *et al.*, 1972b).

$$PhOCH_2CH{=}CH(CH_2)_3CH{=}CH_2 + PhNH_2 \xrightarrow{PdCl_2(Ph_3P)_2-PhONa} PhNHCH_2CH{=}CH(CH_2)_3CH{=}CH_2$$

Thiophenols

The palladium-catalyzed interaction between butadiene and thiophenol does not afford octadienyl derivatives analogous to those derived from phenol. Instead, the major product is the 1,2-adduct, 3-butenyl phenyl sulfide. If the reactants are heated without palladium present, the predominant product is 2-butenyl phenyl sulfide, derived by 1,4-addition (Smutny, 1973).

$$C_6H_5SCH_2CH{=}CHCH_3 \xleftarrow{\Delta} C_6H_5SH + CH_2{=}CHCH{=}CH_2 \xrightarrow{Pd} C_6H_5SCH_2CH_2CH{=}CH_2$$

Alcohols

Oligomerization of butadiene in methanol at 70°C in the presence of bis(triphenylphosphine)(maleic anhydride)palladium gives a 1-methoxy-2,7-octadiene in 85% yield together with smaller amounts of 1,3,7-octatriene and 3-methoxy-1,7-octadiene. The reaction presumably gives through a π-allylic intermediate for when carried out in CH_3OD, the main product is 1-methoxy-6-deuterio-2,7-octadiene (Takahashi *et al.*, 1968a).

$$2\,CH_2{=}CHCH{=}CH_2 + CH_3OD \rightarrow CH_3OCH_2CH{=}CHCH_2CH_2CHDCH{=}CH_2$$

In ethanol, the corresponding ethoxy derivative is formed, whereas in isopropanol, the major product is 1,3,7-octatriene. The alkoxy derivatives are also obtained in methanol or ethanol using bis(triphenylphosphine)(*p*-benzoquinone)palladium or tetrakis(triphenylphosphine)palladium as catalysts. The latter is almost inactive in isopropanol (Takahashi *et al.*, 1967, 1968b). The same ethers may be obtained using palladium, platinum, or ruthenium chlorides and a metal alkoxide as catalysts (Smutny, 1970a) or palladium or platinum salts and a tertiary phosphine without alkoxides (Shryne, 1970). An analogous ether is formed using α,α-bis(trifluoromethyl)benzyl alcohol. Distillation of the ether reforms the alcohol and 1,3,7-octatriene is carried overhead. A variety of complex palladium catalysts containing phosphine ligands have been taught for these processes (Dewhirst, 1970) which have been adapted to a large-scale synthesis of octadienes (Smutny, 1973).

Alcohols add to dienes in the presence of rhodium trichloride, to afford 1:1 adducts. A solution of 1.5 gm of rhodium trichloride trihydrate in 300 ethanol and 27 gm of isoprene held at room temperature for 2 days affords the ethers **XXIV** and **XXV** in 58 and 21% yields respectively at 53% conversion. At 60°C, ether **XXV** is the predominate product formed by isomerization of initially formed **XXIV**. This reaction is not catalyzed by dilute hydrochloric or sulfuric acids.

$$C_2H_5OH + CH_2{=}C(CH_3){-}CH{=}CH_2 \longrightarrow \underset{\textbf{(XXIV)}}{C_2H_5OC(CH_3)_2CH{=}CH_2} + \underset{\textbf{(XXV)}}{C_2H_5OCH_2CH{=}C(CH_3)_2}$$

Methanol, propanol, and isopropanol also add to isoprene, but more slowly and in diminished yields. Ethanol adds to butadiene in a manner similar to isoprene to give ethers **XXVI** and **XXVII** in 33 and 6% yields, respectively, and in addition a smaller quantity of an ether of butadiene dimer, **XXVIII**.

$$C_2H_5OH + CH_2{=}CH{-}CH{=}CH_2 \longrightarrow \underset{\textbf{(XXVI)}}{C_2H_5OCH(CH_3)CH{=}CH_2} + \underset{\textbf{(XXVII)}}{C_2H_5OCH_2CH{=}CHCH_3} + \underset{\textbf{(XXVIII)}}{C_2H_5OCH(CH_3){-}C({=}CH_2)CH_2CH{=}CHCH_3}$$

The author (Dewhirst, 1967) rationalizes these results in terms of an initial diene–rhodium hydride complex formed by reduction of rhodium(III) by ethanol (Dewhirst, 1966).

Water

Water scarcely reacts with butadiene under conditions where alcohols readily add; the only products from the water reactions of butadiene in the presence of palladium complexes are octatrienes. Various solvents do little to change the rate or products. However, if the reaction is carried out in the presence of carbon dioxide, octa-2,7-dien-1-ol is found as the major product in 56–69% yield together with several other products. Palladium compounds such as Pd(acac)$_2$, $(Ph_3P)_4Pd$ and $(Ph_3P)_2PdCO_3$ are all effective catalysts. The role of carbon dixoide in this reaction is unclear, but following others (Kohnle *et al.*, 1969) the authors (Atkins *et al.*, 1971) suggested carbon dioxide probably directly affects the catalyst.

+ H_2O + CO_2 →

1.0 mole 2.0 mole 0.5 mole

85°C
2–4 hours
2.0 mole *t*-butanol
2 mmole Pd(acac)$_2$
6 mmole Ph_3P

OH + OH + + Octadienyl ethers

69% 19% 4% 4%

Acids

A convenient synthesis of 2,7-alkadienyl esters involves interaction of a conjugated diene and a carboxylic acid in the presence of platinum, palladium, or ruthenium catalysts and a phenoxide promoter (Smutny, 1968). For example, interaction of 0.5 mole of acetic acid and 1.6 moles of butadiene at 100°C for 22 hours in the presence of 1 gm of palladium chloride and 2 gm of sodium phenoxide affords 2,7-octadienyl acetate in 45% yield at 37% conversion of acetic acid. Isoprene and 2,3-dimethylbutadiene undergo similar reactions. Bis(triphenylphosphine)(maleic anhydride)palladium and tetrakis(triphenylphosphine)palladium have been used with moderate success in this type of reaction without phenoxide promoter (Takahashi *et al.*, 1968b). Palladium chloride without a promoter forms mainly butadienyl acetate (Stern and Spector, 1961). Palladium chloride, or better, palladium acetate, in the presence of alkali metal acetates afford 1,7-octadiene-3-yl acetate in 25% yield and *trans*-2,7-octadiene-1-yl acetate in 48% yield from acetic acid and butadiene at 90°C (Bryant and McKeon, 1970). Palladium acetate is also an effective catalyst in the absence of a promoter (Shryne, 1971).

Butadiene and acetic acid interact in the presence of palladium acetylacetonate and triphenylphosphine to afford a mixture of octadienyl and butenyl esters. The reaction is greatly accelerated by the presence of molar quantities of tertiary amines and the yield of octadienyl esters increases. Interaction of 4.0 moles of butadiene and 4.0 moles of acetic acid in the presence of 4.0 moles of 2-(*N*,*N*-dimethylamino)ethanol, 3.0 mmoles of palladium acetylacetonate and 3.0 mmoles of triphenylphosphine gives complete conversion in 2 hours at 90°C and affords only dimeric products. No butenyl esters were found.

$$H_2C{=}CHCH{=}CH_2 + HOAc \longrightarrow \underset{71\%}{H_2C{=}CH(CH_2)_3CH{=}CHCH_2OAc} + \underset{21\%}{H_2C{=}CH(CH_2)_3\underset{OAc}{\underset{|}{C}}HCH{=}CH_2} + \underset{8\%}{H_2C{=}CHCH{=}CH(CH_2)_2CH{=}CH_2}$$

The ratio of octadienyl esters is little affected by changes in the palladium-to-triphenylphosphine ratios. Much more substantial changes in the product composition can be achieved by employing triphenyl phosphite as ligand. The above experiment, repeated with triphenyl phosphite replacing triphenylphosphine, afforded octadienyl acetates in 93% yield with the ratio of primary to secondary acetate 92 to 8. The use of trimethylolpropane phosphite as ligand completely eliminated the formation of 1,3,7-octatriene. For example, 8.0 moles of butadiene, 4.0 moles of acetic acid, 1.0 mole of *N*,*N*,*N*′,*N*′-tetramethyl-1,3-butanediamine, 4.0 mmoles of palladium acetylacetonate, and 4.0 mmoles of the phosphite gives 2,7-octadien-1-yl acetate in 81% yield and 1,7-octadien-3-yl acetate in 9% yield after 2 hours at 50°C. Variation in the ratio of primary to secondary octadienyl acetates was shown to be due to the formation of the primary acetate in a kinetically controlled reaction followed by isomerization to the secondary acetate. The facile allylic functional group isomerization over a palladium catalyst is unusual (Walker *et al.*, 1970), but other examples have been reported.

A different course is followed when formic acid without phosphines present is used in place of acetic acid; the product is octa-1,6-diene. A mixture of formic acid and triethylamine heated at 50°C in an autoclave with butadiene and diacetatopalladium(II) as catalyst affords octa-1,6-diene in better than 90% yield. The product is free from other C_8 olefins. Weaker bases such as dimethylformamide are also effective in forming highly pure octa-1,6-diene, but a complex mixture results when undiluted formic acid is used (Gardner and Wright, 1972).

$$H_2C{=}CHCH{=}CH_2 + HCOOH \xrightarrow[Pd]{(C_2H_5)_3N} H_2C{=}CH(CH_2)_3CH{=}CHCH_3 + CO_2$$

If the above reaction is carried out with platinum salts such as lithium tetrachloroplatinate(II) instead of palladium salts, the main product (80% yield) is octa-1,7-diene. Palladium salts complexed with phosphines afford a mixture of octa-1,6-diene and octa-1,7-diene in ratios that depend on the particular phosphine (Gardner and Wright, 1972).

$$H_2C{=}CHCH{=}CH_2 + HCOOH \xrightarrow[Pt]{(C_2H_5)_3N} H_2C{=}CH(CH_2)_4CH{=}CH_2$$

Amines

Amines condense with butadiene affording a number of products in ratios that depend on the catalyst used. A variety of catalysts, such as palladium, platinum, ruthenium, or rhodium halides as well as a number of complex catalysts have been used in these reactions, especially in the presence of small amounts of sodium phenoxide (Smutny, 1967b). For example, 4 ml of butadiene, 2 ml of piperidine, 3 ml of benzene, 0.1 gm of π-crotylbis(triphenylphosphine)rhodium(I), and 13 mg of phenol heated in a sealed ampoule at 70°C for 60 hours afford *N*-(2,7-octadienyl)piperidine in 92% yield and 80% conversion (Dewhirst *et al.*, 1970). Other workers (Takahashi *et al.*, 1968b; Shryne, 1970) have used tetrakis(triphenylphosphine)palladium, or bis-(triphenylphosphine)(maleic anhydride)palladium, to condense morpholine, piperidine, or diisopropylamine with butadiene to afford octadienylamines. Primary amines in the reaction are converted to mixtures of $RNH(C_8H_{13})$ and $RN(C_8H_{13})_{12}$.

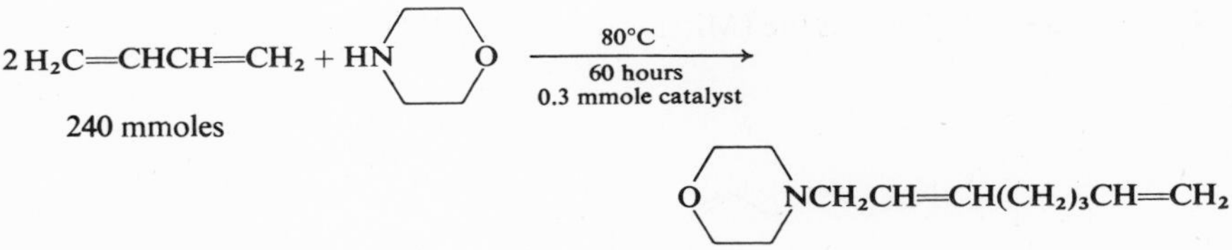

$$2\,H_2C{=}CHCH{=}CH_2 + HN\langle\text{morpholine ring}\rangle O \xrightarrow[\substack{60\text{ hours}\\ 0.3\text{ mmole catalyst}}]{80°C} O\langle\text{morpholine ring}\rangle NCH_2CH{=}CH(CH_2)_3CH{=}CH_2$$

240 mmoles

In sharp contrast to the above reaction, 1:1 adducts are formed with primary or secondary amines and butadiene in the presence of palladium catalysts carrying bidentate ligands, such as $Ph_2PCH_2CH_2PPh_2$ (Takahashi *et al.*, 1972b). Similar marked differences between unidentate and bidentate ligands had been noted previously (Iwamoto and Yuguchi, 1966; Hata and Miyake, 1968). The effect probably arises through changes in the number of coordination sites available to the reactants.

$$>N + H_2C{=}CHCH{=}CH_2 \longrightarrow >NCH(CH_3)CH{=}CH_2 + >NCH_2CH{=}CHCH_3$$

The activity of the catalyst is sharply increased by addition of small amounts of phenol; larger amounts of phenol alter the distribution of isomers. The rate of amine addition is related to their basicity. More basic amines are more reactive (Takahashi *et al.*, 1972b).

With rhodium, at least, inclusion of triphenylphosphine in the reaction system encourages the formation of octadienyl adducts (Baker and Halliday, 1972). For instance, interaction of butadiene and morpholine in the presence of rhodium chloride affords 1:1 adducts, whereas with added triphenylphosphine, the octadienyl derivatives predominate.

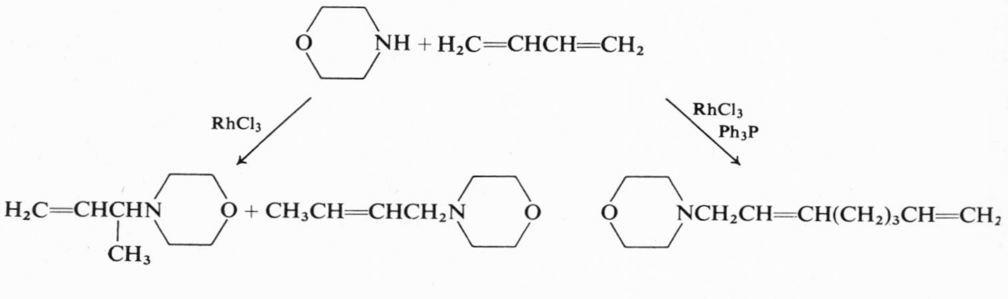

$$O(CH_2CH_2)_2NH + H_2C{=}CHCH{=}CH_2$$

$$\xrightarrow{RhCl_3} H_2C{=}CHCH(CH_3)N(CH_2CH_2)_2O + CH_3CH{=}CHCH_2N(CH_2CH_2)_2O$$

$$\xrightarrow{RhCl_3,\ Ph_3P} O(CH_2CH_2)_2NCH_2CH{=}CH(CH_2)_3CH{=}CH_2$$

Under appropriate conditions, ammonia reacts smoothly with butadiene affording dioctadienyl and trioctadienyl amines. Typically, a mixture of 5 gm of 28% aqueous ammonia, 32 gm of butadiene, 60 ml of acetonitrile, 63 mg of palladium acetate, and 261 mg of triphenylphosphine, heated to 80°C for 10 hours affords on distillation 1.2 gm of diocta-2,7-dienylamine and 29 gm of triocta-2,7-dienylamine (Mitsuyasu *et al.*, 1971).

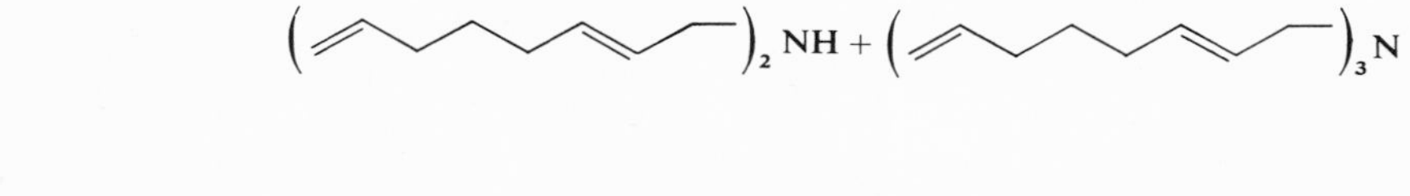

$$H_2C{=}CHCH{=}CH_2 + NH_3 \longrightarrow (H_2C{=}CH(CH_2)_3CH{=}CHCH_2)_2NH + (H_2C{=}CH(CH_2)_3CH{=}CHCH_2)_3N$$

Active Methylene Compounds

A variety of active methylene compounds including dialkyl malonates, α-formyl ketones and esters, nitroalkanes, and α-cyano and α-nitro esters (Hata *et al.*, 1971) undergo facile condensations with dienes in the presence of noble metal catalysts. The products obtained depend importantly on the catalytic metal, ligands, and ratio of reactants. Unidentate phosphine complexes of palladium produce octa-2,7-dienyl derivatives (Hata *et al.*, 1969),

$$\underset{0.1\text{ mole}}{CH_3COCH_2COOC_2H_5} + \underset{0.3\text{ mole}}{H_2C{=}CHCH{=}CH_2} \xrightarrow[\substack{25\text{ min}\\ 2\times10^{-5}\text{ mole }PdCl_2(Ph_3P)_2\\ 2\times10^{-3}\text{ mole PhONa}}]{85°C}$$

$$(H){-}\underset{\substack{|\\ C{=}O\\ |\\ OC_2H_5}}{\overset{\substack{CH_3\\ |\\ C{=}O\\ |}}{C}}{-}[CH_2CH{=}CH(CH_2)_3CH{=}CH_2]_n$$

$n = 1$ 78%
$n = 2$ 12%

whereas bidentate phosphine complexes of palladium, such as $PdBr_2(Ph_2PCH_2CH_2PPh_2)_2$ and $Pd(Ph_2PCH_2CH_2PPh_2)_2$ give 1:1 adducts. Evidently coordination of the ditertiary phosphine makes possible coordination of only one molecule of 1,3-diene. In a typical experiment, ethyl acetoacetate (0.2 mole) and butadiene (0.3 mole) heated at 142°–154°C for 2 hours in the presence of $PdBr_2(Ph_2PCH_2CH_2PPh_2)_2$ (0.00025 mole) and sodium phenoxide (0.0025 mole) afford two isomers in nearly equal amounts in 70% overall yield (Takahashi *et al.*, 1971, 1972a).

$$C_2H_5O\underset{\substack{\|\\ O}}{C}CH_2\underset{\substack{\|\\ O}}{C}CH_3 + H_2C{=}CHCH{=}CH_2 \longrightarrow$$

$$CH_3\overset{\substack{O\\ \|}}{C}\underset{\substack{|\\ COOC_2H_5}}{C}HCH_2CH{=}CHCH_3 + CH_3\overset{\substack{O\\ \|}}{C}\underset{\substack{|\\ COOC_2H_5}}{C}H\overset{\substack{CH_3\\ |}}{C}HCH{=}CH_2$$

Similarly, isoprene, penta-1,3-diene, or hexa-2,4-diene form 1:1 adducts with ethyl acetoacetate. The reaction has also been carried out with 2-formylcyclohexanone and 2-(ethoxycarbonyl)cyclopentanone as the active methylene components. The mechanism of these insertion reactions has been discussed by Hughes and Powell (1971).

Certain platinum catalysts afford branched as well as linear condensation products. Interaction of butadiene and acetylacetone in the presence of $PtCl_2(Ph_3P)_2$ and sodium phenoxide affords, in addition to linear octadienes, the branched adduct, $CH_2{=}CHCH_2CH_2CH_2CH(CH{=}CH_2)CH(COOCH_3)_2$ (Hata *et al.*, 1969). The work was later greatly elaborated to include a variety of α-methylene compounds (Hata *et al.*, 1971).

Nitroalkenes and butadiene interact in the presence of palladium catalysts to afford products in which the α-hydrogens of the nitroalkane are replaced by octa-2,7-dienyl groups. For example, a mixture of 9 gm of 1-nitro-propane, 25 gm of butadiene, 500 mg of dichlorobis(triphenylphosphine)palladium, 950 mg of sodium phenoxide, and 25 ml of *n*-butanol, allowed to stand at room temperature for 20 hours, affords 10.7 gm of 9-nitroundeca-1,6-diene and 12.8 gm of 9-ethyl-9-nitro-1,6,11,16-heptadecatetraene (Mitsuyasu *et al.*, 1971). The reaction may be applied to other nitroalkanes such as nitromethane, nitroethane, and nitrocyclohexane in a similar manner and the nitro compounds so formed subsequently hydrogenated to amines, producing a primary amino group at the middle of a carbon chain (Mitsuyasu *et al.*, 1971).

$$CH_3CH_2CH_2NO_2 + H_2C{=}CHCH{=}CH_2 \longrightarrow H_2C{=}CHCH_2CH_2CH_2CH{=}CHCH_2CH(C_2H_5)NO_2 + (H_2C{=}CHCH_2CH_2CH_2CH{=}CHCH_2)_2C(C_2H_5)NO_2$$

Allyl derivatives of active methylene compounds can be formed readily in an exchange reaction between an allyl ether and an active methylene compound. Palladium catalysts are more effective than platinum. Suitable catalysts are bis(triphenylphosphine)palladium chloride plus sodium phenoxide, palladium acetate plus triphenylphosphine, as well as various zero-valent palladium complexes (Takahashi *et al.*, 1972b).

$$PhOCH_2CH{=}CH(CH_2)_3CH{=}CH_2 + CH_3COCH_2COOCH_3 \longrightarrow \underset{84\%}{CH_3COO(CH_3CO)CHCH_2CH{=}CH(CH_2)_3CH{=}CH_2} + PhOH$$

Carbonyl Compounds

Aldehydes and butadiene undergo a telomerization in the presence of palladium catalysts to afford a linear and cyclic dimer of butadiene incorporating one molecule of aldehyde. Benzaldehyde (11 gm) and 25 ml of butadiene heated at 80°C for 10 hours in the presence of π-allylpalladium chloride (0.18 gm), sodium phenoxide (0.8 gm), and triphenylphosphine (0.26 gm) afford 1-phenyl-2-vinyl-4,6-heptadiene-1-ol (10 gm) and four stereoisomers of 2-phenyl-3,6-divinyltetrahydropyran (3.9 gm). The product ratio depends in large measure on the molar ratio of palladium to triphenylphosphine. At molar ratios of triphenylphosphine to palladium near 1, the unsaturated alcohol is predominant, whereas when the phosphine is greater than palladium, the pyran is the main product. The reaction is applicable to both aromatic (R = Ph) and aliphatic aldehydes (R = alkyl), including formaldehyde (R = H) (Ohno *et al.*, 1971). Dramatic increases in rate of reaction are obtained using methanol or ethanol as solvents. There is no reaction of the alcohol to yield octadienyl ethers even though alcohols are known to interact with butadiene over this and other palladium catalyst systems (Manyik *et al.*, 1970). The reaction of butadiene and formalin has also been carried out in the presence of $(Ph_3P)_4Pd(0)$ affording 2,5-divinyltetrahydropyran (R = H) as the major product together with small amounts of the 3,5-isomer. A number of supported palladium catalysts are ineffective in this reaction as well as catalysts based on platinum, rhodium, ruthenium, and iridium (Haynes, 1970).

$$RCHO + 2\,CH_2{=}CHCH{=}CH_2 \longrightarrow$$

$$\underset{\displaystyle RCHOH}{H_2C{=}CH\overset{|}{C}HCH_2CH{=}CHCH{=}CH_2} \quad + \quad \text{(tetrahydropyran ring: } H_2C{=}CH\text{, R, O, } CH{=}CH_2\text{)}$$

The interaction of formaldehyde and butadiene proceeds differently in the presence of palladium chloride–cupric chloride. Fifteen grams (0.5 mole) of paraformaldehyde, 42 ml (0.5 mole) of 1,3-butadiene, 50 ml of diethyl ether, and a catalyst consisting of 0.5 gm of palladium chloride and 2.5 gm of cupric chloride, heated in a 300 ml glass-lined autoclave for 3 hours affords 0.023 mole of 4-vinyl *meta*-dioxane in 98% yield based on reacted material (Inglis, 1970).

$$H_2C{=}CHCH{=}CH_2 + CH_2O \longrightarrow \text{(1,3-dioxane ring: O, O, } CH{=}CH_2\text{)}$$

Palladium-catalyzed telomerization of butadiene and ketones does not take place as readily as the corresponding reaction with aldehydes. The reaction with simple ketones produces only 1,3,7-octatriene, but certain active ketones, such as perfluoroacetone, benzil, and biacetyl afford pyran derivatives. The reactivity of these ketones was explained in terms of their increased strength of coordination to the catalyst during the initial step of the reaction (Ohno *et al.*, 1972).

Isocyanates

Isocyanates undergo a co-cyclization reaction with butadiene in the presence of palladium–triphenylphosphine complexes to afford divinylpiperidones. Isoprene reacts similarly. Typically, a benzene solution of 12 gm of phenyl isocyanate and 19 gm of isoprene heated at 100°C for 20 hours in the presence of 300 mg of bis(triphenylphosphine)(maleic anhydride)palladium affords 21 gm of a mixture of equal amounts of *cis*- and *trans*-3,6-diisopropenyl-1-phenyl-2-piperidone.

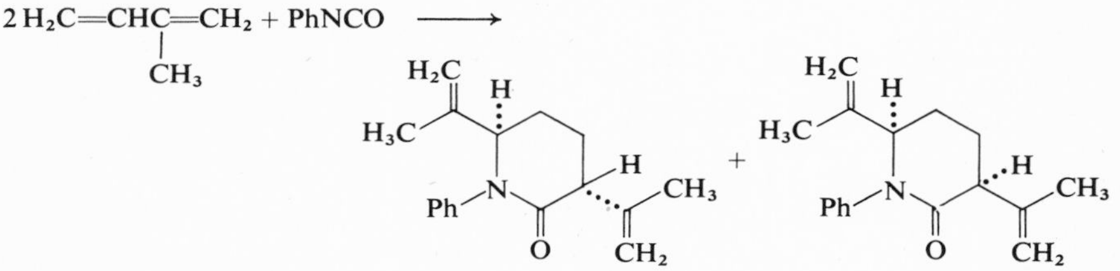

$$2\,H_2C{=}CHC(CH_3){=}CH_2 + PhNCO \longrightarrow$$

H₂C H H₃C N Ph O H CH₃ CH₂ + H₂C H H₃C N Ph O H CH₃ CH₂

Reaction of phenyl isocyanate with butadiene affords equal amounts of isomeric 3-ethylidene-1-phenyl-6-vinyl-2-piperidones. In the reaction, one double bond of butadiene migrates into conjugation with the carbonyl function (Ohno and Tsuji, 1971).

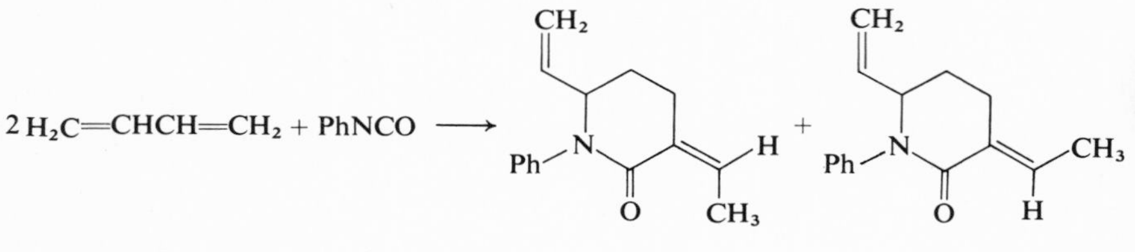

$$2\,H_2C{=}CHCH{=}CH_2 + PhNCO \longrightarrow$$

CH₂ N Ph O H CH₃ + CH₂ N Ph O CH₃ H

CONDENSATIONS

There are a number of noble metal-catalyzed reactions loosely described as "condensations," that have the general form,

$$2MX \rightarrow M-M+(X)$$

in which M – M may be the consequence of several discrete steps.

Condensation of Alcohols

Normal operating temperatures of alkali catalyzed Guerbet reactions can be sharply lowered and side reactions minimized by use of various supported palladium, platinum, rhodium, and ruthenium catalysts (Pregaglia and

Gregorio, 1970). Butanol (161 gm) containing 1.3 gm of dissolved sodium and 0.5 gm of 10% palladium-on-carbon, refluxed 5 hours in a flask containing a water trap, is converted to 2-ethylhexanol in about 95% yield and 32% conversion. Hexanol is similarly converted to 2-butyloctanol in 90% yield and 47% conversion (British Patent 1,188,233). The reaction may also be carried out with homogeneous catalysts. For example, 3 gm of rhodium trichloride, 9 gm of tributylphosphine, 3.5 gm of sodium hydroxide refluxed 14 hours with 162 gm of *n*-butanol in an apparatus equipped with a water separator affords 70.5 gm of 2-ethylhexanol (Pregaglia *et al.*, 1969).

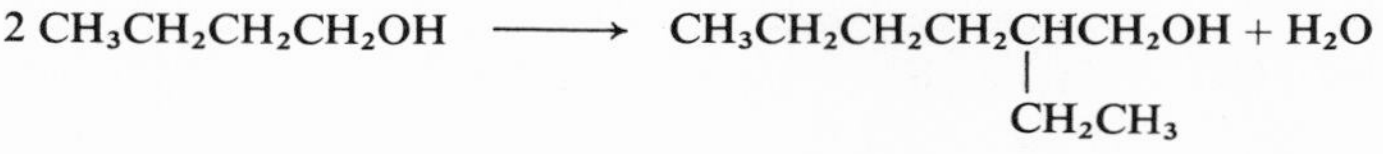

$$2\ CH_3CH_2CH_2CH_2OH \longrightarrow CH_3CH_2CH_2CH_2\underset{\substack{|\\ CH_2CH_3}}{C}HCH_2OH + H_2O$$

Similar condensations affording ketones occur at elevated temperatures over various noble metal catalysts. Isopropanol is converted to mixtures of 4-methyl-2-pentanone and 2,6-dimethyl-4-heptanone plus trimer and tetramer by passage over palladium-, platinum-, or rhodium-on-alumina at 220°C. The authors suggested the following scheme rather than aldol-type condensation to form mesityl oxide followed by hydrogenation (Davis and Venuto, 1969):

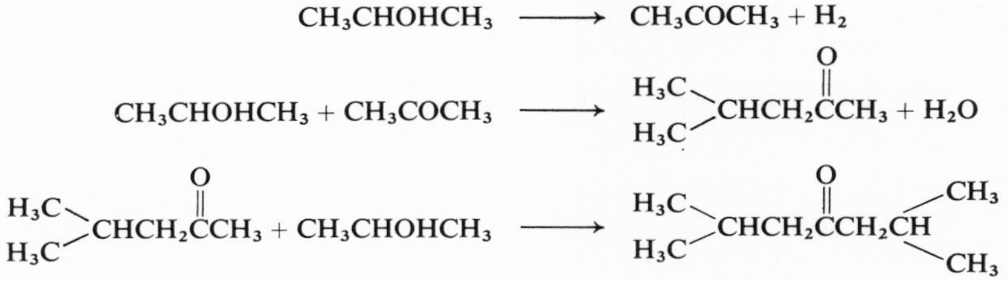

$$CH_3CHOHCH_3 \longrightarrow CH_3COCH_3 + H_2$$

$$CH_3CHOHCH_3 + CH_3COCH_3 \longrightarrow (H_3C)_2CHCH_2\overset{\substack{O\\ \|}}{C}CH_3 + H_2O$$

$$(H_3C)_2CHCH_2\overset{\substack{O\\ \|}}{C}CH_3 + CH_3CHOHCH_3 \longrightarrow (H_3C)_2CHCH_2\overset{\substack{O\\ \|}}{C}CH_2CH(CH_3)_2$$

Acetone is converted to mesitylene in 40 to 50% yields and 30 to 40% conversions at 400 psig, 260°C, and a liquid hourly space velocity of 1, by passage over a catalyst containing 0.5% palladium, 5.0% molybdena-on-alumina pellets (Hwang *et al.*, 1967). The function of palladium here is unclear.

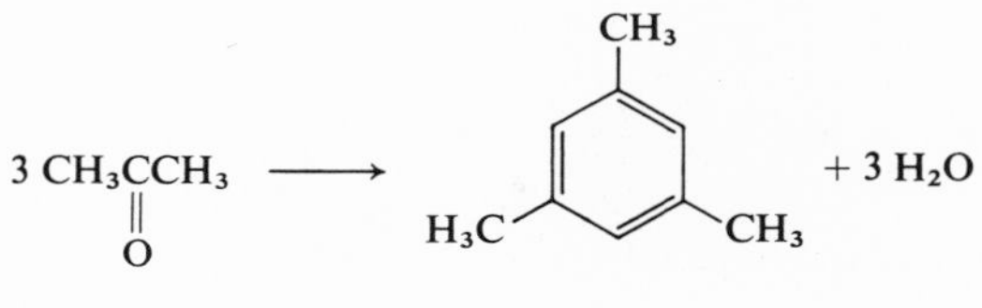

$$3\ CH_3\underset{\substack{\|\\ O}}{C}CH_3 \longrightarrow \text{1,3,5-}(CH_3)_3C_6H_3 + 3\ H_2O$$

Condensation by Dehalogenation

There are a number of examples of carbon–carbon bond formation achieved with elimination of halogen using noble metals as catalysts.

Hydrogenation of hexachlorocyclopentadiene in the presence of palladium proceeds with elimination of hydrogen chloride and formation of bis(pentachlorocyclopentadienyl). The double bonds are not saturated, due probably to steric hindrance offered by the halogens. Twenty grams of hexachlorocyclopentadiene and 1 gm of 5% palladium-on-carbon are heated to 30—40°C in a 2 liter round-bottom flask while hydrogen is bubbled through the mixture. After 6 hours, the reaction is stopped and 5.6 gm of bis(pentachlorocyclopentadienyl) is obtained after recrystallization from isopropanol (Rucker, 1959). Platinum catalysts favor dehalogenation without coupling as might be expected from earlier work (Busch and Schmidt, 1929; Busch *et al.*, 1936).

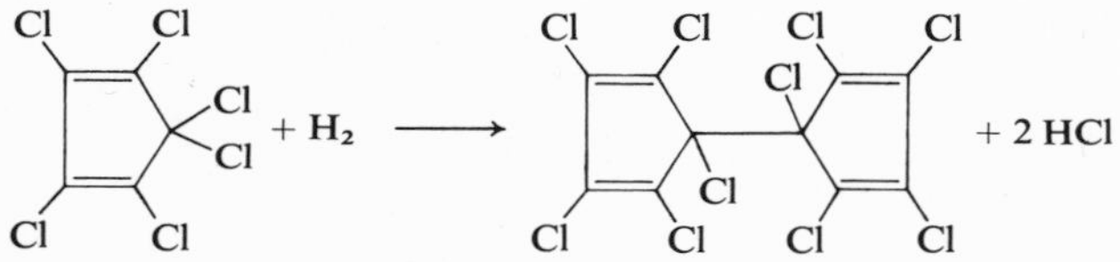

$+ H_2 \longrightarrow$ $+ 2\,HCl$

3,3′-Bipyridazines are prepared readily by interaction at room temperature of 3-halopyridazines with hydrazine hydrate in ethanolic sodium ethoxide solution in the presence of palladium-on-calcium carbonate (Igeta *et al.*, 1969). Chloro compounds give better yields than bromo compounds.

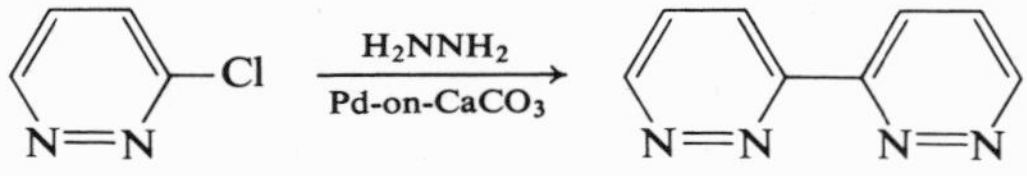

$\xrightarrow[\text{Pd-on-CaCO}_3]{H_2NNH_2}$

Biphenyls may be prepared similarly from halobenzenes using either hydrogen or hydrazine as the coupling agent. The yield of biphenyl increases with the halogen in the order Cl < Br < I (Busch and Schmidt, 1929). Methanol may also be used as a reagent for these reactions, providing both a source of hydrogen and a solvent. Methanol seems to be specific; ethanol, isopropanol, or dioxanes are ineffective (Mayo and Hurwitz, 1949).

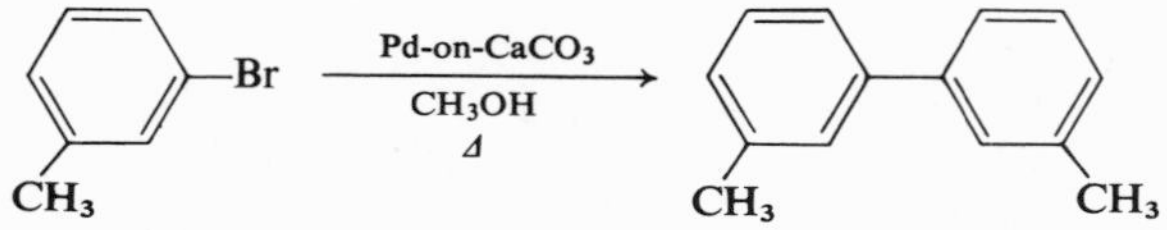

$\xrightarrow[CH_3OH,\ \Delta]{\text{Pd-on-CaCO}_3}$

Olefins undergo arylation with aryl iodides in the presence of catalytic amounts of palladium metal or palladium(II) dichloride and stoichiometric amounts of halogen acceptor.

$$PhI + CH_2{=}CHX + CH_3COOK \xrightarrow{120°C} PhCH{=}CHX + CH_3COOH + KI$$

$$X = H, Ph, CH_3, COOCH_3$$

Pyridine, triethylamine, or potassium benzoate may also be used as acceptors but are not as effective as potassium acetate. The yields are sometimes excellent; *trans*-stilbene and methyl cinnamate are derived from styrene and methyl acrylate in yields of 90 and 97%, respectively (Mizoroki *et al.*, 1971).

REFERENCES

Acton, N., Roth, R. J., Katz, T. J., Frank, J. K., Maier, C. A., and Paul, I. C. (1972). *J. Amer. Chem. Soc.* **94**, 5446.

Aguilo, A., and Stautzenberger, L. (1969). *Chem. Commun.* p. 406.

Alderson, T., Jenner, E. L., and Lindsey, R. V., Jr. (1965). *J. Amer. Chem. Soc.* **87**, 5638.

Atkins, K. E., Walker, W. E., and Manyik, R. M. (1970). *Tetrahedron Lett.* p. 3821.

Atkins, K. E., Walker, W. E., and Manyik, R. M. (1971). *Chem. Commun.* p. 330.

Baker, R., and Halliday, D. E. (1972). *Tetrahedron Lett.* p. 2773.

Billig, E., Strow, C. B., and Pruett, R. L. (1968). *Chem. Commun.* p. 1307.

Blackham, A. U. (1965). U.S. Patent 3,194,800.

Blomquist, A. T., and Maitlis, P. M. (1962). *J. Amer. Chem. Soc.* **84**, 2329.

Brailovskii, S. M., Kaliya, O. L., Temkin, O. N., and Flid, R. M. (1968). *Kinet. Katal.* **9**, 177.

Brown, E. S., and Rick, E. A. (1969). *Amer. Chem. Soc., Div. Petrol. Chem., Prepr.* **14**, No. 2, B29.

Brown, H. C., and Brown, C. A. (1962). *J. Amer. Chem. Soc.* **84**, 2827.

Bryant, D. R., and McKeon, J. E. (1970). U.S. Patent 3,534,088.

Bryce-Smith, D. (1964). *Chem. Ind. (London)* p. 239.

Busch, M., and Schmidt, W. (1929). *Chem. Ber.* **62B**, 2612.

Busch, M., Weber, W., Darboven, C., Renner, W., Hahn, H. J., Mathauser, G., Strätz, F., Zitzmann, K., and Engelhardt, H. (1936). *J. Prakt. Chem.* [N. S.] **146**, 1.

Canale, A. J., and Hewett, W. A. (1964). *J. Polym. Sci., Part B* **2**, 1041.

Canale, A. J., Hewett, W. A., Shryne, T. M., and Youngman, E. A. (1962). *Chem. Ind. (London)* p. 1054.

Chabardes, P., Gandilhon, P., Grard, C., and Thiers, M. (1968). French Patent 1,546,530; *Chem. Abstr.* **71**, 49350 (1969).

Chabardes, P., Gandilhon, P., Thiers, M., and Grard, C. (1969). U.S. Patent 3,449,387.

Chini, P., Santambrogio, A., and Palladino, N. (1967). *J. Chem. Soc., C* p. 830.

Collman, J. P., and Kang, J. W. (1967). *J. Amer. Chem. Soc.* **89**, 844.

Collman, J. P., Kang, J. W., Little, W. F., and Sullivan, M. F. (1968). *Inorg. Chem.* **7**, 1298.

Coulson, D. R. (1972). *Amer. Chem. Soc., Div. Petrol. Chem., Prepr.* **17**, No. 2, B135.

Cramer, R. (1965). *J. Amer. Chem. Soc.* **87**, 4717.

Cramer, R. (1967). *J. Amer. Chem. Soc.* **89**, 1633.

Cramer, R. (1968). *Accounts Chem. Res.* **1**, 186.
Crano, J. C., Fleming, E. K., and Trenta, G. M. (1968). *J. Amer. Chem. Soc.* **90**, 5036.
Dauby, R., Dawans, F., and Tessie, P. (1967). *J. Polym. Sci., Part C* 20, 1989.
Davis, B. H., and Venuto, P. B. (1969). *J. Catal.* **13**, 100.
Dewhirst, K. C. (1966). *Inorg. Chem.* **5**, 319.
Dewhirst, K. C. (1967). *J. Org. Chem.* **32**, 1297.
Dewhirst, K. C. (1970). U.S. Patent 3,489,813.
Dewhirst, K. C., Keim, W., and Thyret, H. E. (1970). U.S. Patent 3,502,725.
Dietl, H., and Maitlis, P. M. (1968). *Chem. Commun.* p. 481.
Dietl, H., Reinheimer, H., Moffat, J., and Maitlis, P. M. (1970). *J. Amer. Chem. Soc.* **92**, 2276.
Dunne, K., and McQuillin, F. J. (1970). *J. Chem. Soc., C* pp. 2196, 2200, and 2203.
Gardner, S., and Wright, D. (1972). *Tetrahedron Lett.* p. 163.
Gassman, P. G., Atkins, T. J., and Lumb, J. T. (1971). *Tetrahedron Lett.* p. 1643.
Hata, G., and Miyake, A., (1968). *Bull. Chem. Soc. Jap.* **41**, 2672.
Hata, G., Takahashi, K., and Miyake, A. (1969). *Chem. Ind.* (*London*) p. 1836.
Hata, G., Takahashi, K., and Miyake, A. (1970). *Chem. Commun.* p. 1392.
Hata, G., Takahashi, K., and Miyake, A. (1971). *J. Org. Chem.* **36**, 2116.
Haynes, P. (1970). *Tetrahedron Lett.* p. 3687.
Hughes, R. P., and Powell, J. (1971). *J. Chem. Soc., D* p. 275.
Hwang, Y. T., Krewer, W. A., and Sander, W. J. (1967). U.S. Patent 3,301,912
Igeta, H., Tsuchiya, T., Nakajima, M., and Yokogawa, H. (1969). *Tetrahedron Lett.* p. 2359.
Inglis, H. S. (1970). British Patent 1,205,157.
Iwamoto, M., and Yuguchi, S. (1966). *J. Org. Chem.* **31**, 4290.
Iwashita, Y., and Tamura, F. (1970). *Bull. Chem. Soc. Jap.* **43**, 1517.
Jones, F. N. (1967). *J. Org. Chem.* **32**, 1667.
Jones, F. N., and Lindsey, R. V. (1968). *J. Org. Chem.* **33**, 3838.
Katz, T. J., and Acton, N. (1967). *Tetrahedron Lett.* p. 2601.
Katz, T. J., Carnahan, J. C., Jr., and Boecke, R. (1967). *J. Org. Chem.* **32**, 1301.
Katz, T. J., Acton, N., and Paul, I. C. (1969). *J. Amer. Chem. Soc.* **91**, 206.
Kawamoto, K., Imanaka, T., and Teranishi, S. (1968). *Nippon Kagaku Zasshi* **89**, 639.
Kawamoto, K., Imanaka, T., and Teranishi, S. (1969). *Bull. Chem. Soc. Jap.* **42**, 2688.
Kawamoto, K., Imanaka, T., and Teranishi, S. (1970). *Bull. Chem. Soc. Jap.* **43**, 2512.
Kern, R. J. (1968). *Chem. Commun.* p. 706.
Ketley, A. D., Fisher, L. P., Berlin, A. J., Morgan, C. R., Gorman, E. H., and Steadman, T. R. (1967). *Inorg. Chem.* **6**, 657.
Klein, H. S. (1967). U.S. Patent 3,354,236.
Klein, H. S. (1968). *Chem. Commun.* p. 377.
Klein, H. S. (1969). U.S. Patent 3,448,159.
Kobayashi, Y., and Taira, S. (1968). *Tetrahedron* **24**, 5763.
Kohnle, J. F., and Slaugh, L. H. (1969). U.S. Patent 3,444,258.
Kohnle, J. F., Slaugh, L. H., and Nakamaye, K. L. (1969). *J. Amer. Chem. Soc.* **91**, 5904.
Kusunoki, Y., Katsuno, R., Hasegawa, N., Kurematsu, S., Nagao, Y., Ishii, K., and Tsutsumi, S. (1966). *Bull. Chem. Soc. Jap.* **39**, 2021.
Lazutkin, A. M., Lazutkina, A. I., Ovsyannikova, I. A., Yurtchenko, E. N., and Mastikhin, V. M. (1973). *Proc. Int. Congr. Catal., 5th, 1972*, in press.
Linn, W. J., and Stiles, A. B. (1971). U.S. Patent 3,562,181.
Luttinger, L. B., and Colthup, E. C. (1962). *J. Org. Chem.* **27**, 3752.
Manyik, R. M., Walker, W. E., Atkins, K. E., and Hammack, E. S. (1970). *Tetrahedron Lett.* p. 3813.
Mayo, F. R., and Hurwitz, M. D. (1949). *J. Amer. Chem. Soc.* **71**, 776.

McClure, J. D., Owyang, R., and Slaugh, L. H. (1968). *J. Organometal. Chem.* **12**, P8.
Medema, D., and Van Helden, R. (1971). *Rec. Trav. Chim. Pay-Bas* **90**, 324.
Misono, A., Uchida, Y., Hidai, M., and Kanai, H. (1967). *Chem. Commun.* p. 357.
Misono, A., Uchida, Y., Hidai, M., Shinohara, H., and Watanabe, J. (1968a). *Bull. Chem. Soc. Jap.* **41**, 396.
Misono, A., Uchida, Y., Hidai, M., and Inomata, I. (1968b). *Chem. Commun.* p. 704.
Mitsuyasu, T., Hara, M., and Tsuji, J. (1971). *Chem. Commun.* p. 345.
Mizoroki, T., Mori, K., and Ozaki, A. (1971). *Bull. Chem. Soc. Jap.* **44**, 581.
Morton, M., and Das, B. (1969). *J. Polym. Sci.*, Part C, No. 27, p. 1.
Mrowca, J. J., and Katz, T. J. (1966). *J. Amer. Chem. Soc.* **88**, 4012.
Odaira, Y., Hara, M., and Tsutsumi, S. (1965). *Technol. Rep. Osaka Univ.* **16**, 325.
Ohno, K., and Tsuji, J. (1971). *Chem. Commun.* p. 247.
Ohno, K., Mitsuyasu, T., and Tsuji, J. (1971). *Tetrahedron Lett.* p. 67.
Ohno, K., Mitsuyasu, T., and Tsuji, J. (1972). *Tetrahedron* **28**, 3705.
Otsuka, S., and Nakamura, A. (1967). *J. Polym. Sci., Part B* **5**, 973.
Otsuka, S., Nakamura, A., and Minamida, H. (1969a). *Chem. Commun.* p. 191.
Otsuka, S., Nakamura, A., Tani, K., and Ueda, S. (1969b). *Tetrahedron Lett.* p. 297.
Pregaglia, G., and Gregorio, G. (1970). U.S. Patent 3,514,493.
Pregaglia, G., Gregorio, G., and Conti, F. (1969). U.S. Patent 3,479,412.
Rinehart, R. E. (1969). *J. Polym. Sci., Part C* **27**, 7.
Rinehart, R. E., Smith, H. P., Witt, H. S., and Romeyn, H., Jr. (1961). *J. Amer. Chem. Soc.* **83**, 4864.
Rucker, J. T. (1959). U.S. Patent 2,908,723.
Saegusa, T., Tsuda, T., and Isayama, K. (1967). *Tetrahedron Lett.* p. 3599.
Sakai, S., Kawashima, Y., Takahashi, Y., and Ishii, Y. (1967). *Chem. Commun.* p. 1073.
Schaeffer, W. D. (1966). U.S. Patent 3,285,970.
Schneider, W. (1972). *Amer. Chem. Soc., Div. Petrol. Chem., Prepr.* **17**, No. 2, B105.
Schrauzer, G. N., Ho, R. K. Y., and Schlesinger, G. (1970). *Tetrahedron Lett.* p. 543.
Shier, G. D. (1967). *J. Organometal. Chem.* **10**, P15.
Shier, G. D. (1969a). U.S. Patent 3,442,883.
Shier, G. D. (1969b). U.S. Patent 3,458,562.
Shinohara, H., Watanabe, Y., and Suzuki, T. (1970). Japanese Patent 7,004,048; *Chem. Abstr.* **72**, 121037n. (1970).
Shryne, T. M. (1970). U.S. Patent 3,530,187.
Shryne, T. M. (1971). U.S. Patent 3,562,314.
Singer, H., and Wilkinson, G. (1968). *J. Chem. Soc., A* p. 849.
Smith, H. P., and Wilkinson, G. (1962). U.S. Patent 3,025,286.
Smutny, E. J. (1966). U.S. Patent 3,267,169.
Smutny, E. J. (1967a). *J. Amer. Chem. Soc.* **89**, 6793.
Smutny, E. J. (1967b). U.S. Patent 3,350,451.
Smutny, E. J. (1968). *J. Amer. Chem. Soc.* **89**, 6793.
Smutny, E. J. (1970a). U.S. Patent 3,518,315.
Smutny, E. J. (1970b). U.S. Patent 3,518,318.
Smutny, E. J. (1970c). U.S. Patent 3,499,042.
Smutny, E. J. (1973). *Ann. N.Y. Acad. Sci.* (in press).
Smutny, E. J., and Chung, H. (1969). *Amer. Chem. Soc., Div. Petrol. Chem., Prepr.* **14**, B112.
Smutny, E. J., Chung, H., Dewhirst, K. C., Keim, W., Shryne, T. M., and Thyret, H. E. (1969). *Amer. Chem. Soc., Div. Petrol. Chem., Prepr.* **14**, No. 2, B100.
Stern, E. W., and Spector, M. L. (1961). *Proc. Chem. Soc., London* p. 370.
Su, A. C. L. (1969). U.S. Patent 3,640,898.

Takahashi, K., Miyake, A., and Hata, G. (1971). *Chem. Ind.* (*London*) p. 488.
Takahashi, K., Miyake, A., and Hata, G. (1972a). *Bull. Chem. Soc. Jap.* **45**, 230.
Takahashi, K., Miyake, A., and Hata, G. (1972b). *Bull. Chem. Soc. Jap.* **45**, 1183.
Takahashi, S., Shibano, T., and Hagihara, N. (1967). *Tetrahedron Lett.* p. 2451.
Takahashi, S., Yamazaki, H., and Hagihara, N. (1968a). *Bull. Chem. Soc. Jap.* **41**, 254.
Takahashi, S., Shibano, T., and Hagihara, N. (1968b). *Bull. Chem. Soc. Jap.* **41**, 454.
Takahashi, S., Shibano, T., and Hagihara, N. (1969). *Chem. Commun.* p. 161.
Urry, W. H., and Sullivan, M. F. (1969). *Amer. Chem. Soc., Div. Petrol. Chem., Prepr.* **14**, B131.
Walker, W. E., Manyik, R. M., Atkins, K. E., and Farmer, M. L. (1970). *Tetrahedron Lett.* p. 3817.
Weigert, F. J., Baird, R. L., and Shapley, J. R. (1970). *J. Amer. Chem. Soc.* **92**, 6630.
Whitesides, G. M., and Ehmann, W. J. (1969). *J. Amer. Chem. Soc.* **91**, 3800.
Wilke, G., Bogdanovic, B., Hardt, P., Heimbach, P., Keim, W., Kröner, M., Oberkirch, W., Tanaka, K., Steinrücke, E., Walter, D., and Zimmermann, H. (1966). *Angew. Chem. Int. Ed. Engl.* **5**, 151.
Yates, K. (1971). *Accounts Chem. Res.* **14**, 136.

CHAPTER 7

Carbonylation and Hydroformylation

Addition of carbon monoxide to organic molecules is among the most studied of all reactions. The scope of this reaction is very wide, with acetylenes, alcohols, amines, ethers, olefins, and nitro compounds among the compounds susceptible to attack by carbon monoxide. Cobalt and to a lesser extent nickel and iron have received the greatest attention as catalysts, but more recently the platinum group metals, particularly rhodium and palladium, have come under increasing study (Tsuji, 1969; Bittler *et al.*, 1968; Maitlis, 1971). This review is limited to applications of noble metal catalysts; a number of more general reviews are available (Orchin and Wender, 1957; Wender *et al.*, 1957; Bird, 1967; Falbe, 1970).

CATALYSTS

A great variety of complex noble metal catalysts have been used in carbon monoxide chemistry (Evans *et al.*, 1965, 1968; Osborn *et al.*, 1965; Hallman *et al.*, 1967; Church and Mays, 1968; Kehoe and Schell, 1970b; Sibert, 1970; Booth *et al.*, 1971; Lawrenson, 1971a, b; Lawrenson and Green, 1971; Wilkinson, 1971). Carbonylation reactions are affected importantly by both the metal and complexing ligand, providing strong impetus for further catalyst development. Numerous specific examples of the effect of catalyst on the reaction are given in the following sections. The mechanism of functioning of carbonylation catalysts has been discussed in some detail by a number of investigators (Yagupsky *et al.*, 1970; Henrici-Olive and Olivé, 1971; Paulik, 1972; Orchin and Rupilius, 1972).

Catalyst Recovery

The use of noble metal catalysts on an industrial scale demands recovery of the catalyst. Because of the attractiveness of these catalysts in carbonylation chemistry, considerable attention has been given to this problem and a variety of filtration (Falbe and Weber, 1970a; Olivier, 1970a; Goldup *et al.*, 1971), extraction (Falbe and Weber, 1970b; Olivier, 1970a, b), adsorption (Olivier and Snyder, 1970), and distillation techniques for metal recovery have been developed. One pilot plant process recovered and recycled more than 99.9% of the rhodium catalyst (Olivier and Booth, 1970). Various techniques have been devised for reactivating a recovered catalyst before recycling (Olivier *et al.*, 1971; Bittler *et al.*, 1968).

In some reactions, deposition of metal may prove a problem. In one continuous hydroformylation process designed to prevent deposits of rhodium metal during the reaction, the rhodium salt is dissolved in a polar solvent of limited water content and passed together with an olefin and a mixture of carbon monoxide and hydrogen through a heated (120°C) pressure tube (200 atm) (British Patent 1,202,507, August 19, 1970).

Supported Complex Catalysts

Homogeneous catalysts entail certain difficulties in recovery and reuse, problems magnified by the cost of noble metals. Various attempts have been made to alleviate these difficulties by employing complex catalysts affixed to a solid support (Millidge, 1965a; Gladrow and Mattox, 1967; Rony, 1969; Paulik *et al.*, 1969; Čapka *et al.*, 1971; Bond, 1971a, b; Haag and Whitehurst, 1973; Allum *et al.*, 1973). One general procedure involves incorporation of the nonreacting ligands of the complex into porous solids in such a way that these ligands are part of the chemical structure of the solid; the catalytic species are then built onto these ligands. Another general method involves affixing the charged catalytic complex to porous solids that contain counterions of the complex (Haag and Whitehurst, 1971). These catalysts, besides being more easily recovered, have in principle an inherent advantage. The maximum rate obtained with a homogeneous catalyst is limited by the solubility of the catalyst, whereas the complex heterogeneous catalytic analog does not suffer this restriction and higher effective concentrations may be introduced into the system (Haag and Whitehurst, 1971).

In hydroformylation of propylene with $(Ph_3P)_2Rh(CO)Cl$-on-alumina at 148°C and 49 atm, no change in conversion or product isomer distribution occurred over 300 hours of operation. Conversion is linear with contact time, but the isomer distribution remains constant (Robinson *et al.*, 1969). Supports may be also impregnated with simple metal salts, but the metal will probably go into solution under hydroformylation conditions (Schiller, 1956; Millidge, 1965b).

ACETYLENES

Acetylenes and propargyl compounds undergo a variety of reactions with carbon monoxide in the presence of noble metal catalysts, affording some compounds obtained otherwise only with difficulty. Acetylene and carbon monoxide in tetrahydrofuran or dioxane solvents interact in the presence of $[Ru(CO)_4]_3$ to afford hydroquinone. The reaction is carried out at 200°–220°C with a carbon monoxide pressure of 75–127 atm. The yield is very sensitive to hydrogen partial pressure and the best yields, about 58%, are obtained with hydrogen pressures of 5 to 10 atm (Pino *et al.*, 1967). The catalyst is prepared easily from ruthenium trisacetylacetonate (Braca *et al.*, 1968).

$$2\,C_2H_2 + 2CO + H_2 \xrightarrow{[Ru(CO)_4]_3} \text{HO}-C_6H_4-\text{OH}$$

Somewhat higher yields, about 65%, are obtained with acetylene, carbon monoxide, and water (Pino *et al.*, 1968, 1969). Water is a better source of hydrogen in this reaction than molecular hydrogen and permits the use of lower operating pressures. The reaction seems to follow the equation

$$2\,C_2H_2 + 3\,CO + H_2O \xrightarrow{[Ru(CO)_4]_3} \text{HO}-C_6H_4-\text{OH} + CO_2$$

Various rhodium salts in a variety of solvents are also effective in the synthesis of hydroquinone from acetylene, carbon monoxide, and hydrogen. The yield depends markedly on the concentration of acetylene and the best yields are obtained when the concentration is kept below 1.5 moles of acetylene per liter (Wakamatsu *et al.*, 1969).

Acetylene is carbonylated in the presence of hydrogen chloride and a rhodium catalyst to afford a mixture of acrylyl chloride, β-chloropropionyl chloride, and succinoyl chloride. The yields of each are obscurely reported (Sauer, 1963). Other workers used bis(triphenylphosphine)palladium dichloride, bis(benzonitrile)palladium dichloride, and piperidine(triphenylphosphine)palladium dichloride as catalysts for this reaction (von Kutepow *et al.*, 1967a).

$$C_2H_2 + CO + HCl \longrightarrow CH_2{=}CHC(=O)Cl + ClCH_2CH_2C(=O)Cl + ClC(=O)CH_2CH_2C(=O)Cl$$

Carbonylation of acetylene in benzene in the presence of palladium chloride affords muconyl chloride as the major product (Tsuji *et al.*, 1964a).

$$C_2H_2 + CO \xrightarrow{PdCl_2} (ClC(=O)CH{=}CH)_2$$

Unsaturated esters are obtained by carbonylation of acetylene in alcohol solvent. Dimethyl maleate and dimethyl muconate are formed by interaction of acetylene and carbon monoxide in methanol over palladium chloride stabilized by small amounts of thiourea.

$$C_2H_2 + CO + CH_3OH \longrightarrow CH_3OC(=O)CH{=}CHC(=O)OCH_3 + (CH_3OC(=O)CH{=}CH)_2 + H_2$$

Hydrogen formed in the reaction tends to reduce acetylene and the unsaturated products, a difficulty that may be overcome by adding a small percentage of oxygen to the gaseous mixture. Acetylene, carbon monoxide, and air in volume ratios of 24.5 to 68 to 7.5 passed at a rate of 6 liters/hour for 40 hours into 130 ml of methanol containing 116 mg of thiourea and 90 mg of palladium chloride affords 3.62 gm of dimethyl maleate, 0.08 gm of dimethyl fumarate, 0.15 gm of dimethyl *cis,cis*-muconate, 0.13 gm of its *cis,trans* and *trans,trans* isomers, and 0.60 gm of heavier products. The heavier products increase as the ratio of acetylene to carbon monoxide is increased. *Trans* isomers are thought to arise by isomerization of initially formed *cis* products (Chiusoli *et al.*, 1968). This method employs reaction conditions very much milder than those described earlier (Tsuji *et al.*, 1964a).

Saturated and unsaturated mono- and dibutyl esters are obtained by carbonylation of acetylene in butanol. Sodium iodide–palladium iodide is a much more effective catalyst than the corresponding chlorides or bromides, perhaps due to the much higher solubility of the iodides in butanol (Lines and Long, 1969). Iodine or hydrogen iodide are also promoters for this reaction when catalyzed by palladium chloride, palladium sulfate, or palladium-on-carbon in alcoholic solvent (Jacobsen and Spaethe, 1962). The product distribution is changed appreciably by addition of hydrogen chloride to the reaction mixture, moving toward higher amounts of unsaturated monoester. Diesters are favored by ratios of carbon monoxide to acetylene greater than 1, but significant departures from a ratio of unity leads to a progressive decrease in conversion (Lines and Long, 1969).

Carbonylation of diphenylacetylene in alcohol followed an unexpected course and afforded a mixture of the anticipated diphenyl maleate as well as a lactone. Hydrogen chloride was necessary as a co-catalyst (Tsuji and Nogi, 1966a).

$$PhC{\equiv}CPh + CO + C_2H_5OH \xrightarrow[HCl]{PdCl_2} \text{(3,4-diphenyl lactone) } 66\% + Ph(COOC_2H_5)C{=}C(COOC_2H_5)Ph \; 30\%$$

Hydroformylation of alkynes is usually a difficult reaction, although carbonylation reactions are well known (Bird, 1963). Hydroformylation of hex-1-yne may be achieved with tris(triphenylphosphine)chlororhodium in ethanol–benzene solvent with a 1 to 4 mixture of hydrogen and carbon monoxide at 110°C and 120 atm. A mixture of *n*-heptaldehyde and 2-methylhexaldehyde is formed in about equal amounts in 15% yield (Bird *et al.*, 1967). This rhodium complex is an effective catalyst for quantitative hydrogenations of hex-1-yne to hexane and hydrogenation of the substrate competes with hydroformylation (Jardine *et al.*, 1965). Hydroformylation of alkynes proceeds smoothly, affording saturated aldehydes in high yield when rhodium carbonyl is used with a large excess of triphenylphosphine (Fell and Buetler, 1972).

Propargyl Chloride and Propargyl Alcohols

Carbonylation of propargyl compounds provides a convenient synthesis of itaconic acid and its derivatives. Carbonylation of propargyl alcohol in methanol–hydrochloric acid over palladium-on-carbon at 100°C and 1500 psig gives a mixture of methyl itaconate, methyl aconitate, and methyl 2-(methoxymethyl)acrylate in ratios that depend on the reaction conditions (Tsuji and Nogi, 1966c). Propargyl chloride in methanol is carbonylated at 100 atm pressure over palladium chloride at 20°C to afford methyl itaconate in 66% yield. Palladium-on-carbon is also a suitable catalyst but requires an elevated temperature (100°C). Extensive dicarbonylation rather than monocarbonylation is a characteristic of palladium catalysis. The authors viewed the reaction as proceeding through methyl 2,3-butadienoate, an intermediate, shown in separate experiments to afford itaconate.

$$HC{\equiv}CCH_2Cl \xrightarrow[CH_3OH]{CO} [H_2C{=}C{=}CHCOOCH_3] \longrightarrow CH_2{=}C(COOCH_3)CH_2COOCH_3$$

One, two, or three moles of carbon monoxide may be incorporated in propargyl alcohol in methanol affording, respectively, methyl 2-methoxymethylacrylate, methyl itaconate, and methyl aconitate in ratios that depend on the amount of solvent, catalyst, and concentration of hydrogen chloride.

$$\underset{5\ gm}{HC{\equiv}CCH_2OH} \xrightarrow[\substack{50\ ml\ CH_3OH \\ 18\ hours \\ 1\ gm\ PdCl_2}]{100\ atm\ CO} \underset{24\%}{H_2C{=}C(CH_2OCH_3)COOCH_3} + \underset{26\%}{H_2C{=}C(COOCH_3)CH_2COOCH_3} + \underset{5\%}{CH_3OOCCH{=}C(COOCH_3)CH_2COOCH_3}$$

Carbonylation of substituted propargyl alcohols in alcoholic solvent gives complex mixtures with the product ratio again depending on the reaction environment. Carbonylation of 2-methyl-3-butyn-2-ol in methanol gives a mixture of methyl 4-methyl-4-methoxy-2-pentenoate, methyl teraconate, dimethyl 3-methoxy-1-buten-1,2-dicarboxylate, and methyl 2,4-dihydro-2,2-dimethyl-5-oxo-3-furoate (Nogi and Tsuji, 1969). Small amounts of catalyst favor the monocarbonylation product, 4-methyl-4-methoxy-2-pentenoate, whereas with larger amounts of catalyst this material is not formed and methyl teraconate and the lactone are formed in equal amounts.

$$HC{\equiv}CC(CH_3)_2OH + CO + CH_3OH \xrightarrow{PdCl_2} CH_3C(OCH_3)(CH_3){-}CH{=}CH{-}COOCH_3 +$$

$$(H_3C)_2C{=}C(COOCH_3)CH_2COOCH_3 + CH_3C(CH_3)(CH_3O){-}C(COOCH_3){=}CHCOOCH_3 + \text{O=}\langle\text{2,2-dimethyl-3-(COOCH}_3\text{)-furan-5-one}\rangle$$

Carbonylation of substituted propargyl alcohols in benzene gives substituted itaconic anhydrides, providing a useful synthesis of this type of compound. For example, 10 gm of 1-ethynylcyclohexanol, 50 ml of dry benzene, and 1 gm of palladium chloride, pressured with carbon monoxide at 100°C for 15 hours affords cyclohexylidenesuccinic anhydride in 45% yield.

$$HC{\equiv}C\text{-}C_6H_{10}(OH) \longrightarrow C_6H_{10}{=}C(CO)(CH_2CO)O$$

Acetylenic Esters

Ethyl acetylenecarboxylate and ethyl acetylenedicarboxylate undergo carbonylation readily at room temperature (Tsuji and Nogi, 1966b). Ethyl acetylenecarboxylate affords a mixture of products in carbonylations catalyzed by palladium chloride–hydrogen chloride; dicarbonylation products increase with increasing concentrations of hydrogen chloride.

$$HC{\equiv}CCOOC_2H_5 + C_2H_5OH + CO \xrightarrow[PdCl_2\text{—}HCl]{100\ atm} (C_2H_5OCO)_2CHCH_2COOC_2H_5 +$$

$$C_2H_5OCOCH{=}CHCOOC_2H_5 + (C_2H_5OCO)_2C{=}CHCOOC_2H_5 +$$

$$(C_2H_5OCO)_2C{=}CHCH{=}C(COOC_2H_5)_2$$

These products are apparently the resultant of three competing reactions: (1) monocarbonylation, (2) dicarbonylation by concerted *cis* addition, and (3) dimerization of the substrate followed by dicarbonylation. The authors believe that dicarbonylated products arise from concerted *cis* attack of carbon monoxide on the triple bond since attempted further carbonylation of the monocarbonylated compound, ethyl fumarate, met with failure and the substrate was recovered unchanged.

Carbonylation of ethyl acetylenedicarboxylate also gives mixtures of mono- and dicarbonylated products, a higher concentration of hydrogen chloride again increasing dicarbonylation. The origin of the saturated esters in these reactions is unclear. Palladium-catalyzed carbonylation of acetylene in ethanol affords ethyl fumarate and ethyl maleate, but no ethyl succinate.

It is interesting that in these reactions no products arising by trimerization of the acetylenes were found inasmuch as trimerization of acetylenic esters occurs readily over palladium catalysts in the absence of carbon monoxide (Bryce-Smith, 1964).

AMINES

Carbonylation of amines affords isocyanates, ureas, formamides, or oxamides, the major product depending largely on reaction conditions and catalysts. Formation of isocyanates in good yield can be achieved only where further reaction between the amine substrate and isocyanate products occurs slowly. Stern and Spector (1966) achieved fair yields of isocyanates by interaction of carbon monoxide, amines, and stoichiometric quantities of palladium chloride. Palladium is reduced to the metal during the course of the reaction. Formamides, which are known to form from carbon monoxide and amines in the presence of transition metal salts (Calderazzo, 1965), were shown not be be intermediates. The mechanism of the reaction has been discussed by Stern (1967) and Tsuji and Ohno (1969). The reaction is made catalytic with respect to palladium by reoxidizing reduced palladium either *in situ* or in a subsequent step with, for instance, oxygen in the presence of copper salt (Stern and Spector, 1968), but it is not clear how well the catalytic system works.

$$RNH_2 + CO + PdCl_2 \rightarrow RNCO + Pd + 2\,HCl$$

Other products are formed under other conditions. A mixture of 7.9 gm of *n*-decylamine and 0.5 gm of palladium chloride in 30 ml of benzene, pressured with 100 atm of carbon monoxide at 180°C for 20 hours, affords 3.8 gm of *N,N'*-didecyloxamide and 2.2 gm of 1,3-didecylurea. The results depend on the nature of the catalyst and on the amine used. Palladium-on-carbon, instead of palladium chloride, gives decreased yields of the oxamide and increased yields of the formamide. With lower aliphatic amines, the yield of urea decreases and the formamide increases. Only ureas are formed from aromatic amines (Tsuji and Iwamoto, 1966).

$$RNH_2 + CO \rightarrow RHNCHO + RNHCONHR + RNHCOCONHR + H_2$$

Durand and Lassau (1969) using $Rh_2Cl_2(CO)_4$ with added phosphines were able to vary the yield of formamide obtained by carbonylation of butylamine from 35 to 100% by proper choice of phosphine and phosphine–metal ratio. Typically, 0.370 gm of $P(CH_3)_3$ (5.3 mm) in 8 ml of benzene is mixed with 1.26 mm $Rh_2Cl_2(CO)_4$ in 10 ml of benzene. This solution mixed with 56 ml of butylamine is pressured to 60 atm with carbon monoxide and heated 4 hours at 160°C. There is obtained 54.3 gm of *N*-butylformamide (98% yield) and 0.5

gm of N,N'-dibutylurea. The yield of formamide decreases to 62% when equimolar amounts of phosphine and metal are present.

Cyclic secondary amines are selectively carbonylated to the *N*-formyl derivatives as the sole products over a ruthenium-bridged acetate dicarbonyl polymer or over $Ru_3(CO)_{12}$. Carbonylation will take place under very mild conditions, 75°C and 1 atm (Byerley *et al.*, 1971).

Secondary amines can be alkylated by olefins in a process combining hydroformylation and reductive alkylation of the resulting aldehyde. The catalyst consists of a noble metal, a heterocyclic saturated amine, and an organic phosphine, arsine, or stilbine. For example, 0.3 gm of bis(triphenylphosphine)rhodium carbonyl chloride, 2 gm of triphenylphosphine, 2 gm of 1,4-diazabicyclo[2.2.2]octane, 92 gm of dibutylamine, and 100 gm of propylene in 300 gm of *t*-amyl alcohol pressured with 300 psig of carbon monoxide and 600 psig of hydrogen, and heated to 100°C for 45 minutes, affords 27 gm of tributylamine (Biale, 1970).

$$CH_3CH{=}CH_2 + CO + H_2 + (C_4H_9)_2NH \rightarrow (C_4H_9)_3N$$

Formamides have been prepared in a unique interaction between carbon dioxide, secondary amines, and hydrogen. A variety of complex rhodium, iridium, palladium, platinum, copper, and cobalt catalysts are useful in this reaction; catalyst turnovers of over 1000 moles of formamide per mole of catalyst were demonstrated for $(Ph_2PCH_2CH_2PPh_2)_2CoH$ and $(Ph_3P)_2(CO)IrCl$. Typically, the amine and catalyst are pressured to 800 psig with hydrogen and carbon dioxide and heated to 100°–125°C overnight (Haynes *et al.*, 1970).

$$R_2NH + CO_2 + H_2 \rightarrow R_2NCHO + H_2O$$

Cyclopropylamine undergoes an interesting insertion reaction on carbonylation in benzene in the presence of $Rh_6(CO)_{16}$ affording *N*-cyclopropylpyrrolidone (92%) as the major product together with *N*-allylpyrrolidone, *N*-propen-1-ylpyrrolidone, and *N*-propylpyrrolidone. When $(Ph_3P)_3RhCl$ is used as the catalyst, the major product is instead pyrrolidone (Iqbal, 1971a).

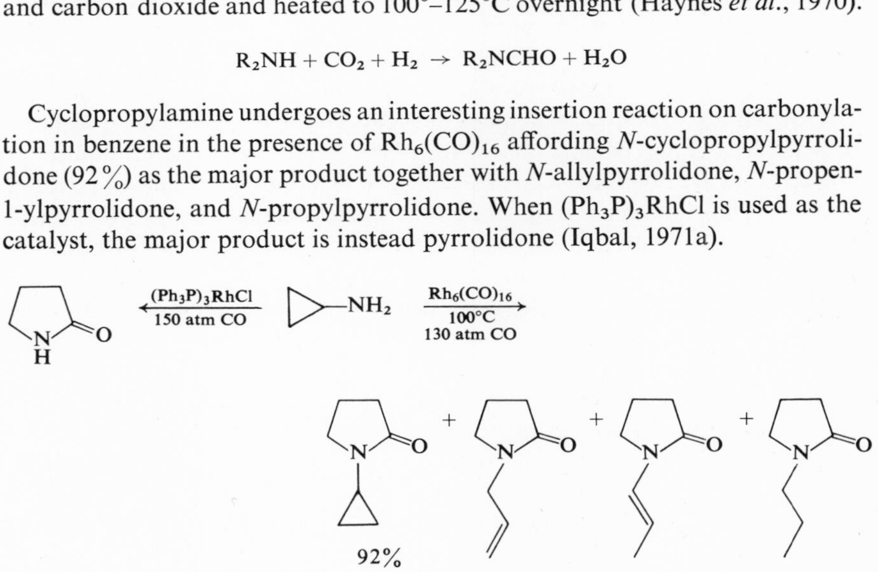

$(Ph_3P)_3RhCl$, 150 atm CO ← cyclopropyl-NH_2 → $Rh_6(CO)_{16}$, 100°C, 130 atm CO

92%

Azides

Reaction of acyl or aryl azides with an iridium(I) or rhodium(II) carbonyl complex affords organic isocyanates. Addition of carbon monoxide regenerates the original carbonyl complex, which can then react with more azide, making the whole process catalytic.

$$ArN_3 + CO \xrightarrow[\substack{M = Rh, Ir \\ L = Ph_3P}]{MCl(CO)L_2} ArNCO + N_2$$

Coordinatively saturated iridium carbonyl complexes do not react with organic azides under these mild conditions, illustrating the requirement that both an oxidizing and reducing agent be in the metal coordination sphere (Collman *et al.*, 1967). A similar heterogeneous process is the reaction of sodium azide and carbon monoxide over palladium-on-carbon to form sodium cyanate (Collman *et al.*, 1967).

Alcohols

Carbonylation of alcohols in the presence of various base metals has been investigated extensively. Reaction over these catalysts is characterized by the use of high temperature and pressures (Hohenschutz *et al.*, 1966). Noble metal catalysts, on the other hand, have been investigated relatively little, but some of the effort has met with spectacular success; the reaction provides the newest commercial synthesis of acetic acid (Roth *et al.*, 1971). The new process uses a homogeneous iodide-promoted rhodium catalyst of such extreme activity that it will convert methanol to acetic acid in 99% selectivity at pressures as low as 1 atm (Paulik and Roth, 1968). None of the various liquid by-products that are obtained in cobalt-catalyzed reactions are found when using rhodium. The rhodium component of the homogeneous catalysts may arise from a number of compounds, such as rhodium chloride, rhodium oxide, or tris(triphenylphosphine)rhodium chloride, but inasmuch as all starting components give essentially the same performance, the supposition is made that all the rhodium materials ultimately form the same active catalytic species (Roth *et al.*, 1971). The other component of the system is a halogen promoter, preferably iodide, which may be introduced into the system as hydrogen iodide, calcium iodide, methyl iodide, or as iodine. Various iridium catalysts promoted by iodine are also effective in this reaction.

The carbonylation is believed to proceed through a series of steps beginning with rapid conversion of methanol to methyl iodide. Methyl iodide then undergoes an oxidative addition reaction with a d^8 square-planar rhodium(I) complex to form a d^6 six-coordinate alkylrhodium(III) species. Rapid insertion of carbon monoxide into the rhodium–alkyl bond to afford an acylrhodium-(III) complex, followed by reaction of this complex with water to form acetic acid and the original rhodium(I) complex, completes the cycle.

Carbonylation of methanol to acetic acid may also be carried out over a rhodium-on-carbon catalyst promoted by methyl iodide. The catalyst is prepared by thermal decomposition of rhodium nitrate impregnated on activated carbon. In a flow system, 0.25 mole of methanol, 0.50 mole carbon monoxide, and 0.006 mole of methyl iodide as promoter are converted over 5.0 gm of this catalyst per hour at 215 psig and 200°C to a product of 87.6 wt% acetic acid and 9.0 wt% methyl acetate (Schultz and Montgomery, 1969).

Rhodium and iridium seem to be the best of the noble metals for carbonylation of alcohols. Two attempts to carbonylate alcohols in the presence of palladium chloride gave a mixture of products (Mador and Blackham, 1963; Graziani *et al.*, 1971).

Halogen Compounds

Carbonylations have been achieved with noble metal catalysts involving halogen compounds having no initial unsaturation. Moderate yields of β-perchloromethylpropionyl chloride are obtained by interaction of carbon tetrachloride, ethylene, and carbon monoxide in the presence of palladium, platinum, or ruthenium catalysts. The same reaction will occur without catalyst, but the yield is very much lower (Ihrman *et al.*, 1968). A patent application (Neth. application 6,408,476) describes the use of palladium chloride as catalyst for this reaction.

$$\underset{757\ \text{gm}}{CCl_4} + \underset{5000\ \text{psig}}{CO} + \underset{500\ \text{psig}}{C_2H_4} \xrightarrow[\substack{155°C \\ 17\ \text{hours}}]{1\ \text{gm}\ 5\%\ \text{Pt-on-asbestos}} CCl_3CH_2CH_2COCl$$

Noble metals also have been used as catalysts for the carbonylation of aryl halides to afford benzoyl halides (Mador and Scheben, 1969a) and palladium chloride will catalyze carbonylation of halocarbons to afford haloacetyl halides (Mador and Scheben, 1969c). The yields actually obtained are obscure. Benzoyl chloride may also be obtained by interaction of benzene and phosgene under a high carbon monoxide pressure in the presence of palladium chloride as catalyst (Long and Marples, 1965).

Vinyl halides are converted to unsaturated esters by carbonylation in the presence of a noble metal catalyst and an alkali or alkaline earth bromide or iodide. Yields are high but the conversions are low (Brewis *et al.*, 1967).

$$H_2C{=}CHCl + ROH + CO \xrightarrow{Pd} H_2C{=}CHCOOR + HCl$$

Chloramines undergo carbonylation at 50°C in the presence of palladium metal or palladium chloride to afford good yields of carbamoyl chlorides. Ureas form as the temperature is raised (Saegusa *et al.*, 1971).

$$R_2NCl + CO \xrightarrow[\substack{CO \\ 60\ atm \\ CH_3OCH_2CH_2OCH_3}]{Pd} R_2NC(=O)Cl$$

OLEFINS

Research on carbonylation reactions of olefins catalyzed by noble metals is progressing at an accelerating rate, spurred on by an appreciation of the unique characteristics of these catalysts and a realization that they can be used economically on an industrial scale. Isomer distributions and selectivities not heretofore obtainable are now possible. At their best noble metals make catalysts unprecedented for activity and selectivity.

Hydroformylation of Olefins

The principal features in hydroformylation of olefin over certain noble metal catalysts are the exceptionally mild conditions under which the reaction will proceed, the possibility of high preference for terminal addition to α-olefins, and for high selectivity to the aldehyde with little or no alcohol formation, except with special catalysts (Fell *et al.*, 1971). All of the noble metals have been examined for hydroformylation but attention has centered largely on rhodium, an exceptionally active catalyst with rates two to four orders of magnitude higher than cobalt. Palladium is inefficient because of its tendency to saturate the olefin (Tsuji *et al.*, 1965a), as do ruthenium carbonyls (Braca *et al.*, 1970).

Selectivity for terminal addition depends upon process conditions and the use of appropriate ligands. Without suitable ligands, rhodium may produce more branched aldehydes than cobalt (British Patent 801,734 Sept. 17, 1958). Preferred ligands are derived from trialkyl phosphites, triaryl phosphites, triarylphosphines, triarylstilbenes, and triarylarsines. A criterion for suitability (Pruett and Smith, 1969, 1970) is that the ligand be not too basic, which excludes trialkylphosphines and dialkylarylphosphines (Slaugh and Mullineaux, 1966). The highest selectivities are obtained with excess ligand present. Additionally, the use of excess ligand cuts down the amount of competing hydrogenation and isomerization reactions and permits the use of hydrogen-to-carbon monoxide ratios greater than 1, ratios associated with high rates and high percentages of linear aldehyde (Brown and Wilkinson, 1970). Terminal adducts are favored also by lower overall pressure, increased catalyst loadings, and increased temperatures. Reduction of aldehyde to alcohol is minimized by low catalyst loadings and the use of a carboxylate ligand (Lawrenson, 1970). The use of ammonium or alkalia metal hydroxides is said to broaden the types of solvents that may be used with these complex catalysts (Booth, 1970).

The exact nature of these complex catalysts may be a matter of conjecture (Tinker and Morris, 1971). Some rhodium complexes are altered by temperature or pressure whereas others are substantially unchanged over a wide range of conditions. For example, hydridocarbonyltris(triphenylphosphine)-rhodium(I) (Brown and Wilkinson, 1970) at 80°C and 500 psig carbon monoxide–hydrogen pressure is transformed into a mixture of rhodium carbonyl complexes, whereas chlorocarbonylbis(triphenylphosphine)rhodium remains unchanged, in contradiction to earlier suggestions that it was transformed into the former hydrido species (Evans *et al.*, 1968). The Tinker and Morris infrared technique permits identification of the predominant metal–complex species present under reaction conditions, but as they pointed out, this species may still not be the active catalyst.

The following example illustrates a condition for high selectivity to a terminal aldehyde. 1-Hexene (84 gm) in 150 gm of acetophenone containing 35 gm of triphenylphosphine and 2.0 gm of hydridocarbonyltris(triphenylphosphine)rhodium(I), $HRh(CO)(PPh_3)_3$, is pressured to and maintained at 110 psig with an equimolar mixture of hydrogen and carbon monoxide at 80°C for 35 minutes. *n*-Heptaldehyde is obtained in 96% yield at 88% conversion (Pruett and Smith, 1970).

$$CH_3(CH_2)_2CH{=}CH_2 + CO + H_2 \rightarrow CH_3(CH_2)_5CHO$$

The catalyst complex may be formed *in situ* from a supported rhodium catalyst and excess ligand. A 3-liter autoclave is charged with 200 ml of toluene, 112 gm of 1-octene, 15 gm of triphenyl phosphite, and 15 gm of 5% rhodium-on-carbon. After flushing with carbon monoxide to remove air, the reactor is pressured to 80 psig with equimolar amounts of hydrogen and carbon monoxide, heated to 90°C, and repressured as needed. At the end of 50 minutes, *n*-nonanal is obtained in 86% yield. The major by-products are formed by isomerization of 1-octene to 2-octene and subsequent hydroformylation (Pruett and Smith, 1969). Rhodium when used below 70°C is an effective catalyst for producing exclusively branched-chain aldehydes from internal olefins (Wakamatsu, 1964). Internal olefins undergo hydroformylation much more slowly than terminal olefins (Brown and Wilkinson, 1970; Heil and Marko, 1969). Hydroformylation of higher molecular weight α-olefins over rhodium followed by hydrogenation produces alcohols whose sulfates show improved detergency properties (British Patent 1,099,196, Jan. 17, 1968).

A hydroformylation process for producing aldehydes on a large scale using a rhodium catalyst has been described in some detail (Olivier and Booth, 1970). The process operates at 80°–100°C and 500 psig with hydridocarbonyltris-(triphenylphosphine)rhodium(I), $RhH(CO)(Ph_3P)_3$, as catalyst. Selectivity to the aldehyde is extremely high, reaching 98%, of which about 68% is the normal isomer. Space time yields are the order of 1500–2000 grams of product per liter per hour. Maximum rates are achieved with excess triphenylphosphine present at weight ratios of 5 to 10 parts per part of rhodium complex. Reaction rates are over 100 times greater than achievable with cobalt catalysts, permitting convenient rates at very low rhodium concentrations. More than 99.9% of the rhodium catalyst is recovered. Aldehyde is the sole reaction product and olefins are completely consumed obviating difficult separation.

The rhodium complex, chlorocarbonylbis(triphenylphosphine)rhodium(I), $(Ph_3P)_2RhCOCl$, has been examined in detail for use in hydroformylation of olefins (Craddock *et al.*, 1969). A unique feature of this catalyst is its extreme stability; the catalyst withstands repeated recycle and vacuum distillation at 180°C and 14 mm Hg. The catalyst is best used with excess triphenylphosphine in the system to increase selectivity for terminal aldehyde from α-olefins and limit variability in the product composition due to changes in reaction parameters. Without excess triphenylphosphine present, the yield of terminal aldehyde varies with the solvent and 2-methyl as well as appreciable 2-ethyl aldehydes are found in the product, whereas with excess ligand, the yields of terminal aldehyde are uniformly higher, virtually independent of the solvent, and essentially no 2-ethyl derivative is found. The use of excess ligand permits variability in both total pressure and hydrogen-to-carbon monoxide ratios without changes in either selectivity or isomer distribution. Increasing the hydrogen-to-carbon monoxide ratio from 1 to 6 increases the rate about 8 times, but even at this high ratio no paraffins or alcohols are formed. Temperature has a significant influence on rate, product distribution, and selectivity. Increased rate, branched aldehyde, and paraffin are all favored by increasing temperatures above 100°C. Reaction conditions of 100° to 125°C, 500 psig total pressure, and with excess ligand permit aldehyde selectivity in excess of 99% with a linear aldehyde content of 75 to 80%.

Internal olefins are hydroformylated much more slowly than terminal olefins; nonetheless interesting and practical use has been made of the reaction. Frankel (1971) hydroformylated methyl oleate and other oleic oils using a complex rhodium–triphenylphosphine catalyst and obtained a mixture of about equal amounts of 9 and 10 formyl compounds in very high yield. The catalyst system is remarkable for its specificity. With proper conditions, there is no isomerization of the double bond and no reduction of the formyl group such as occurs with other catalysts. Hydroformylation proceeds smoothly at 95°–110°C with a 1:1 mixture of hydrogen and carbon monoxide at 500 to

2000 psig with or without a solvent, such as toluene. The active, homogeneous catalyst is prepared *in situ* from 5% rhodium-on-carbon, alumina, or calcium carbonate in the presence of triphenylphosphine. Rhodium-on-carbon is the most active catalyst but the color of the product is not as good as that obtained with calcium carbonate or alumina. No hydrogenation of the formyl group occurs even with pure hydrogen, provided the catalyst concentration is not too high. The complex rhodium catalyst prepared by this technique stands in contrast to tris(triphenylphosphine)rhodium chloride. The latter catalyst is not only much less effective for hydroformylation of oleic oils, but it also promotes hydrogenation of the formyl group to the hydroxymethyl group. A typical experiment using this system is illustrated by the hydroformylation of safflower oil. Oleic safflower oil (998 gm) is charged to a 2 liter autoclave with 10 gm of 5% rhodium-on-support and 9 gm of triphenylphosphine. The solution is pressured under 2000 psig of 1:1 hydrogen:carbon monoxide at 110°C for 3 hours. Conversion to formylstearate is 96.2%.

Hydroformylation of Dienes

Hydroformylation of conjugated dienes usually affords only saturated monoaldehydes or alcohols. Even nonconjugated dienes afford primarily monoaldehydes since rapid isomerization into conjugation takes place under usual oxo conditions. However, Fell and Rupilius (1969a, b), using rhodium–phosphine catalyst system that had earlier been shown to effect hydroformylation of monoolefins without isomerization (Fell *et al.*, 1968), were able to obtain good yields of dialdehydes from conjugated dienes. Hydroformylation of 1,3-butadiene and 1,3-pentadiene affords 40–45% of C_6 and C_7 dialdehydes, respectively. For example, hydroformylation of butadiene carried out at 125°C and 200 atm in diethyl ether with a carbon monoxide-to-hydrogen ratio of 1 to 2 and a catalyst prepared separately from 0.02 mole percent Rh_2O_3 and 1.7 mole percent tri-*n*-butylphosphine affords 58 mole percent C_5 monoaldehydes and 42 mole percent C_6 dialdehydes in 80–90% yield. Surprisingly, the monoaldehyde was more than 96% *n*-valeraldehyde and less than 4% isovaleraldehyde, an exceptionally high percentage of straight-chain product. The dialdehyde fraction was 58% 2-methylpentandial-1,5, 29% 2-ethyl-butandial-1,4, and 9% hexandial-1,6 (Fell and Rupilius, 1969a).

Certain nonconjugated dienes can be made to form dialdehydes in good yield over rhodium catalysts even without complexing ligands present. Over rhodium oxide pimelaldehyde is obtained from 1,4-pentadiene and 2-methylheptanedial from 1,5-hexadiene as the main products. These results suggest that these diolefins form complexes with rhodium carbonyl suitable for 1,5-insertion of carbon monoxide. Over rhodium chloride catalyst, which generates hydrogen chloride during the reaction, the major product from 1,4-pentadiene is cyclohexanecarboxyaldehyde formed by intramolecular aldol condensation of pimelaldehyde. Butadiene and 1,3-pentadiene with these catalysts afford only monoaldehyde (Morikawa, 1964).

$$H_2C{=}CHCH_2CH{=}CH_2 \xrightarrow[H_2]{CO} OHC(CH_2)_5CHO \xrightarrow[H^+]{H_2} \text{(cyclohexane)}\text{-}CHO$$

In contrast to the above results, hydroformylation of *cis,cis*-1,5-cyclooctadiene can be made to afford only monosubstituted products. Cyclooctanemethanol is made in 97% yield by hydroformylation of *cis,cis*-1,5-cyclooctadiene (196 gm) in tetrahydrofuran (800 gm) over a rhodium oxide catalyst (0.2 gm). The reaction is carried out best in two stages. The first stage at 150°–170°C and 150–180 atms pressure of 1:1 carbon monoxide:hydrogen for 4 hours produces the saturated aldehyde, the second stage at 210°C for 6 hours produces the carbinol (Falbe and Huppes, 1970). At higher pressures, bis(hydroxymethyl)cyclooctane is formed in 81% yield (Falbe and Huppes, 1966). Bis(hydroxymethyl)bicycloheptane is prepared in 71% yield by hydroformylation of bicycloheptadiene over rhodium at 110 to 300 atm and 130° to 240°C (Belgian Patent 718,857, 1969). Similarly, bicyclopentadiene, 198 gm, affords 196 gm of dialcohols on hydroformylation over rhodium oxide (Falbe and Huppes, 1967a, c).

$$HOH_2C\text{-}(C_8H_{14})\text{-}CH_2OH \; (81\%) \xleftarrow[300\ atm]{CO,\ H_2} \text{(1,5-cyclooctadiene)} \xrightarrow[150\ atm]{CO,\ H_2} (C_8H_{15})\text{-}CH_2OH \; (97\%)$$

Hydroformylation of Substituted Olefins

Hydroformylation of substituted olefins is of interest in itself and because the initial products can be made to undergo various secondary changes. Other compounds of this general type are discussed in the section on allylic compounds.

Hydroformylation of acrylate ester over rhodium has been investigated by several groups and conditions for isomer control elucidated. Ethyl acrylate undergoes hydroformylation in the presence of rhodium carbonyl to afford a mixture of α- and β-carbethoxypropionaldehyde (Takegami *et al.*, 1967).

$$H_2C{=}CHCOOC_2H_5 \xrightarrow[H_2]{CO} CH_3CH(CHO)COOC_2H_5 + OHCCH_2CH_2COOC_2H_5$$

The percentage of branched product increases with decreasing temperature and with increasing partial pressures of carbon monoxide and hydrogen. The effect of partial pressure is in the same direction as has been found with cobalt, but the rhodium-catalyzed reactions are much more pressure-sensitive. By appropriate choice of conditions, the branched aldehyde percentage in the product can be varied from 15.5 to 77.6%. Conditions for these extremes in yield are shown in Table I. The rate of reaction under these conditions increases proportionally with the amount of catalyst up to about 6 mg, but is little increased by additional catalyst. The effect of the amount of catalyst on product distribution is small.

TABLE I
HYDROFORMYLATION OF ETHYL ACRYLATE BY RHODIUM CARBONYL

Temp. (°C)	Partial pressure (atm) CO	H_2	$[Rh(CO)_3]_n$ (mg)	Time (min)	Conv. (%)	β-Isomer (%)	α-Isomer (%)
90	80	80	3	240	73	22.4	77.6
120	40	80	6	120	84	84.5	15.5

The reactions were carried out with 80 ml of toluene and 20 ml of ethyl acrylate.

Hydroformylation of methacrylate esters over rhodium provides a convenient synthesis of α-methylbutyrolactone. The process is improved if carried out in the presence of a tertiary amine. In a typical experiment, 400 gm of methyl methacrylate, 400 gm of benzene, 0.2 gm of rhodium oxide, and 5 gm of pyridine is charged to a reactor which is pressured at 160°C with an equimolar mixture of hydrogen and carbon monoxide to 200 atms. After 2 hours, the catalyst is removed by filtration and the reaction product is hydrogenated over Raney nickel. The yield of lactone is 336 gm, 84% of theory. If the hydroformylation is carried out in the presence of alkyl or aryl phosphines instead of tertiary amine, attack at the α-position is strongly favored (Falbe and Huppes, 1967a, b). This reaction is also very pressure-sensitive; the percentage of terminal adduct increases markedly at lower pressures (Pruett and Smith, 1969).

$$H_2C{=}\underset{\displaystyle CH_3}{\underset{|}{C}}COOCH_3 \xrightarrow[Rh]{CO,\ H_2} OHCCH_2\underset{\displaystyle CH_3}{\underset{|}{C}}HCOOCH_3 \longrightarrow$$

CH3, O, O (lactone ring)

Terminal attack is strongly preferred in hydroformylation of 2,6-dimethyl-1-hepten-6-ol over bis(1,5-cyclooctadiene rhodium chloride) in benzene (Himmele *et al.*, 1971).

$$H_2C{=}C(CH_3)CH_2CH_2CH_2C(OH)(CH_3)CH_3 \xrightarrow[\substack{700\ atm\\100°C\\5\ hours}]{CO,\ H_2} OHCCH_2CH(CH_3)CH_2CH_2CH_2C(OH)(CH_3)CH_3 \quad 90\%$$

The low tendency for rhodium, even without stabilizing ligands, to hydrogenate aldehydes under mild hydroformylation conditions makes it suitable for converting formylcyclohexene to diformylcyclohexane (Bartlett and Hughes. 1959).

$$\text{formylcyclohexene (700 ml)} \xrightarrow[\substack{4000\ psi\ H_2,\ CO\ 1:1\\100°C}]{\substack{1200\ ml\ PhH\\0.4\ gm\ Rh_2O_3}} \text{OHC–}C_6H_{10}\text{–CHO} \quad 71\%$$

Similarly, bicycloheptencarboxaldehyde in tetrahydrofuran is converted at 130°C and 200 atm over rhodium oxide completely to a mixture of dialdehydes. The dialdehyde is reduced to the dialcohol in the same system at 240°C and 300 atm (Falbe and Huppes, 1967a; Falbe, 1971).

$$\text{bicycloheptene–CHO} \longrightarrow \text{OHC–bicycloheptane–CHO} \longrightarrow \text{HOCH}_2\text{–bicycloheptane–CH}_2\text{OH}$$

Acrylonitriles usually undergo hydroformylation at the terminal carbon, but when the reaction is carried out in the presence of an acid and an inert nonpolar diluent, 2-(hydroxymethyl)alkanenitriles are formed. For example, 15 gm of acrylonitrile, 0.1 gm of rhodium oxide, 130 ml of pentane, 0.5 ml of acetic acid, and 3000 psig of carbon monoxide and hydrogen (1:1) maintained at 127°C for 1.5 hours, affords 4.5 gm of propionitrile and 3.8 gm of 2-methyl-3-hydroxypropanenitrile. No 4-hydroxybutanenitrile is found (Kuper, 1970).

$$H_2C{=}CHCN + CO + H_2 \longrightarrow CH_3CH(CH_2OH)CN$$

Nitroalkenes can be hydroformylated in the presence of rhodium carbonyl catalysts with preservation of the nitro function. A 96% yield of isomeric 4-nitrophenylpropionaldehydes is obtained from 4-nitrostyrene and a mixture of C_4 nitroaldehydes, mainly 2-methyl-3-nitropropionaldehyde, is obtained from 3-nitropropene (Takesada and Watamatsu, 1970).

$$H_2C{=}CHCH_2NO_2 + CO + H_2 \longrightarrow OHCCH(CH_3)CH_2NO_2$$

Rhodium is the preferred catalytic metal for hydroformylation of 2-formyl-3,4-dihydro-2*H*-pyran and 2-hydroxymethy-3,4-dihydropyran.

$$\text{2-formyl-3,4-dihydro-2}H\text{-pyran (CHO)} \xrightarrow[\text{Rh}]{\text{CO, H}_2} \text{OHC–tetrahydropyran–CHO (2,6)}$$

$$\text{2-(CH}_2\text{OH)-3,4-dihydropyran} \longrightarrow \text{OHC–tetrahydropyran–CH}_2\text{OH (2,6)}$$

In a typical experiment, 337 gm of freshly distilled 2-formyl-3,4-dihydro-2*H*-pyran, 650 ml of benzene, and 150 mg of Rh_2O_3 are heated at 150°C under 10 atm of carbon monoxide to effect formation of rhodium carbonyl. After cooling to 130°C, the 2 liter reactor is pressured with 140 atm of carbon monoxide and 100 atm of hydrogen. After the pressure falls to 155 atm, the product is worked up by vacuum distillation, affording 148 gm of 2,6-diformyltetrahydropyran (70% yield, at 50% conversion). Somewhat improved yields can be obtained if the above reaction is carried out in the presence of small amounts (2 gm) of nickel boride (Fahnenstich and Weigert, 1969).

Oxidative Carbonylation of Olefins

Carbonylation of olefins carried out in the presence of an oxidant afford unsaturated acids or derivatives (Medema *et al.*, 1969). With nonstabilized catalysts, the system should preferably be anhydrous, but with catalysts stabilized by reagents such as triphenylphosphine, or arsines, or stilbenes, the need for an anhydrous system disappears. The use of these reagents also limits formation of by-products, such as carbonates, and formation of carbon dioxide through oxidation of carbon monoxide.

Oxidative carbonylation systems may be maintained anhydrous through the use of dehydrated molecular sieves, or inorganic or organic anhydrides, such as boric anhydride or acetic anhydride (Fenton and Olivier, 1967a, b). A $\frac{1}{2}$ gallon autoclave is charged with 1 gm of palladium chloride, 5 gm of lithium acetate dihydrate, 5 gm of cupric chloride, 90 gm of calcined sodium mordenite, and 500 gm of acetic acid. The autoclave is then pressured to 450 psig with ethylene and then to 900 psig with carbon monoxide and heated to 140°C while oxygen is introduced slowly in small increments over a 10 minute period. There is recovered 14 gm of acrylic acid and 43 gm of β-acetoxypropionic acid. Lithium salts are present to help catalyze the reoxidation of palladium (Fenton and Olivier, 1967b). If acetic anhydride is used in the reaction medium instead of the molecular sieve, the yield of acrylic acid is increased greatly at the expense of β-acetoxypropionic acid. This type of system can be promoted by the presence of small amounts of ferric chloride; iron was uniquely effective among many metals tried (Olivier, 1968).

$$H_2C{=}CH_2 + CO + O_2 \longrightarrow H_2C{=}CH\underset{\|}{\overset{}{C}}OH + CH_3\underset{\|}{C}OCH_2CH_2\underset{\|}{C}OH$$
$$\text{(each marked C bears =O)}$$

An anhydrous system is not needed when ligands such as triphenylphosphine are present. The following comparison illustrates the effect this ligand has on product yield and distribution. A $\frac{1}{2}$ gallon titanium-lined autoclave is charged with 0.5 gm of palladium chloride, 2.5 gm of cupric chloride, 2.5 gm of cuprous chloride, 5 gm of triphenylphosphine, 300 gm of methanol, and 111 gm of propylene and pressured to 600 psig with carbon monoxide. The autoclave is heated to and maintained at 150°C while oxygen is added in 20 psig increments. After a 15 minute reaction period, there is obtained 10.3 gm of methyl methacrylate, 14.6 gm of methyl butyrate, and 178 gm of methyl crotonate. Without triphenylphosphine present, 24.4 gm of methyl carbonate and 23.4 gm of dimethyl pyrotartarate are produced (Biale, 1970).

$$H_2C{=}CHCH_3 + O_2 + CO \xrightarrow[\substack{Pd \\ CH_3OH}]{Ph_3P} H_2C{=}\underset{CH_3}{\underset{|}{C}}COOCH_3 + CH_3CH_2CH_2\underset{O}{\underset{\|}{C}}OCH_3 + CH_3CH{=}CH\underset{O}{\underset{\|}{C}}OCH_3$$

$$H_2C{=}CHCH_3 + O_2 + CO \xrightarrow[CH_3OH]{Pd} CH_3O\underset{O}{\underset{\|}{C}}OCH_3 + CH_3O\underset{O}{\underset{\|}{C}}CH_2\underset{H_3C}{\underset{|}{C}}H\underset{O}{\underset{\|}{C}}OCH_3$$

Cupric chloride can be used as an oxidant instead of oxygen in oxidative carbonylations, but different products may be obtained. A 300 ml vessel charged with 1 gm of palladium chloride, 20 gm of anhydrous cupric chloride, 100 ml of butanol, and pressured to 500 psig with ethylene and 1000 psig with carbon monoxide, affords, after 4 hours at 120°C, 13 gm of *n*-butyl acrylate and 3 gm of *n*-butyl β-butopropionate (Fenton, 1968a, b).

$$H_2C{=}CH_2 + CO + CuCl_2 \xrightarrow[C_4H_9OH]{PdCl_2} H_2C{=}CH\overset{\overset{O}{\|}}{C}OC_4H_9 + C_4H_9OCH_2CH_2\overset{\overset{O}{\|}}{C}OC_4H_9$$

With oxygen present, succinates are formed in good yields according to the equation

$$H_2C{=}CH_2 + 2\,CO + \tfrac{1}{2}\,O_2 + 2\,ROH \xrightarrow[CuCl_2]{PdCl_2} RO\overset{\overset{O}{\|}}{C}CH_2CH_2\overset{\overset{O}{\|}}{C}OR + H_2O$$

Iron may be used as well as copper in this synthesis. Optimum results are obtained by restricting both the amount of excess proton and chloride through use of sodium acetate, or organic bases such as pyridine, to remove proton, and cuprous chloride or ferrous chloride to complex chloride ion. Water formed in the reaction has an adverse effect on yield by increasing carbon dioxide production, but this reaction can be suppressed by addition of alkyl orthoformates. Despite the stoichiometry expressed by the above equation, the best yields, 90%, are obtained when the partial pressure of ethylene is slightly higher than that of carbon monoxide (Fenton and Steinwand, 1972).

Oxalate esters are prepared by oxidative carbonylation of formate esters in the presence of palladium and iron or copper salts in alcohol solvent under 1000 psig of carbon monoxide. Oxygen is gradually introduced into the system as the reaction proceeds (Fenton and Steinwald, 1968).

$$OHC\overset{\overset{O}{\|}}{C}OCH_3 + CO + O_2 \xrightarrow{CH_3OH} CH_3O\overset{\overset{O}{\|}}{C}{-}\overset{\overset{O}{\|}}{C}OCH_3$$

Carbonylation of Olefins

One aspect of carbonylation involves interaction of an unsaturated compound, a nucleophile containing a mobile hydrogen, and carbon monoxide. Alcohols, amines, ammonia, carboxylic acids, mercaptans, and water have been used as nucleophilic components. When water is the nucleophile, the reaction is called hydrocarboxylation. Other variations of carbonylation consist of insertion of carbon monoxide into an existing bond such as formation of acids from alcohols or esters from ethers. These reactions have been investigated at length, especially with cobalt and nickel catalysts (Falbe, 1970) and more recently noble metals.

A variety of palladium catalysts including palladium metal, palladium carbonyl halides, olefin-palladium halides, and complex palladium(0) compounds have been tried for low-temperature carbonylation of olefins. The catalysts vary widely in activity. Among the best are $(Ph_3P)_2PdCl_2$; $(Ph_3P)_3PdCl_2(C_5H_{11}N)$; $(Ph_3P)PdCl_2(PhCH_2NH_2)$ and $(Ph_3P)PdClC_3H_5$. Slightly less active catalysts are $(Ph_3P)_2PbBr_2$; $(PhCN)_2PdCl_2$; $(C_5H_{11}N)_2$-$PdCl_2$; and $[(Ph_3P)PdClSPh]_2$. The anion, in addition to the ligand, contributes importantly to catalyst activity. Only chlorides and bromides confer high activity; palladium complexes that contain a nitrate, iodide, acetate, or sulfate anion are only moderately active catalysts. These catalysts may be reused by being returned to the synthesis in the form of a catalyst residue obtained after distillation of the products. It is important to add hydrogen chloride to the used catalyst to obtain an active catalyst. The authors suggest that palladium catalysts function by activation of the olefin rather than activation of carbon monoxide (Bittler *et al.*, 1968). Carbonylation of olefins in the presence of noble metal catalysts has given a variety of products including β-chloroacyl chlorides, acids, aldehydes, ketones, and saturated and unsaturated esters. The major product as well as the yields depend importantly on the catalyst, the structure of the olefin, and the reaction conditions.

Carbonylation in Aprotic Solvents

In aprotic solvent, β-chloroacyl chlorides are formed in moderate yields by interaction of carbon monoxide with an olefin and stoichiometric quantities of palladium chloride. In this reaction, terminal olefins are always attacked by carbon monoxide at the terminal carbon in contrast to attack at the secondary carbon in hydrolysis reactions (Tsuji *et al.*, 1963a, 1964b). The main factor in the formation of straight-chain acyl chlorides seems to be steric (Tsuji *et al.*, 1964b). Carbonylation of terminal olefins up to 1-pentene affords β-chloroacyl chlorides, but higher olefins give a complicated mixture of chloroesters resulting from migration of the terminal double bond. Tsuji (1969) has suggested that this reaction can be made catalytic with respect to palladium, by employing an oxidizing agent to reoxidize the reduced palladium.

$$RCH{=}CH_2 + CO + PdCl_2 \rightarrow RCH_2ClCH_2COCl + Pd$$

A modification of this carbonylation reaction is carried out in the presence of an acid acceptor to remove hydrogen chloride from the system, and with stoichiometric cupric chloride, to make the reaction catalytic with respect to palladium. The products are unsaturated acyl chlorides. For example, a reactor charged with 10 mmoles of palladium chloride, 182 mmoles of cupric chloride, 120 ml of acetonitrile, and pressured with ethylene and carbon monoxide to 72 atm total pressure at 50°C, affords 142 mmoles of acrylyl chloride, 78% yield based on cupric chloride charged (Scheben and Mador, 1969).

$$H_2C{=}CH_2 + CO \xrightarrow[PdCl_2]{CuCl_2} H_2C{=}CHC({=}O)Cl$$

Without an oxidant present, carbonylation of ethylene in acetonitrile over palladium salts of strong acids affords a mixture of homoangelica lactones. Yields are usually only several times the palladium charged (Hayden, 1969).

Carbonylation in Protic Solvents

Carbonylation of olefins in protic solvents such as alcohol or water may afford esters or acids or the solvent may function as a hydrogen donor, affording ketones as major reaction products. An example of ketone formation is the alternating oligomerization of carbon monoxide and ethylene in the presence of rhodium catalysts. In a typical experiment, 0.33 mole of ethylene and 2 moles of carbon monoxide is heated in a 100 ml of stainless steel autoclave with 50 mg of rhodium oxide in a 1:1 mixture of acetic acid and methanol. After distillation and recrystallization, octane-3,6-dione is obtained in 26% yield, undeca-3,6,9-trione and tetradeca-3,6,9,12-tetraone in 6 and 1% yield, respectively, and methyl homolevulinate in 15% yield. The active species is assumed to be $HRh(CO)_3$ arising through the reaction

$$Rh_4(CO)_{12} + 2\,CO + 2\,H_2O \rightarrow 4\,HRh(CO)_3 + 2\,CO_2$$

Water is supplied to the reaction from the solvent through either ester or ether formation. The reaction changes drastically if an organic base is present. In methanol solvent containing 0.1 mole pyridine, methyl propionate is obtained in 50% yield as the main product under the conditions described above. On the other hand, if pyridine is replaced by triethylamine, diethyl ketone is obtained in 71% yield (Iwashita and Sakuraba, 1971a).

$$CH_3CH_2C({=}O)OCH_3 \xleftarrow[CH_3OH]{C_6H_5N} CO + H_2C{=}CH_2 \xrightarrow[CH_3OH]{(C_2H_5)_3N} C_2H_5C({=}O)C_2H_5$$

Diethyl ketone is also formed in good yield over rhodium with isopropanol as hydrogen donor. A 3 liter high-pressure reactor charged with 1000 ml of isopropanol, 6.0 moles of ethylene, and 0.175 weight percent based on olefin of rhodium sesquioxide is heated to 175°C while the pressure is maintained at 3000 psig by addition of carbon monoxide. After 6 hours, the product consists of 126 gm of acetone, 170 gm of diethyl ketone, 13 gm of propionaldehyde, 396 gm of unchanged isopropanol, and 43 gm of bottoms (Hughes and Brodkey, 1958). Carbonylation of ethylene in water at 190°C in the presence of rhodium acetylacetonate affords mainly diethyl ketone together with 25% of 3,6-octanedione, whereas over ruthenium acetylacetonate at 225°C, the major product is propionic acid together with a similar amount of diethyl ketone and a lesser amount of 3,6-octanedione (Alderson and Thomas, 1962).

$$C_2H_5C(=O)C_2H_5 + CH_3CH_2COOH \xleftarrow{Ru(acac)_3} H_2C{=}CH_2 + CO \xrightarrow{Rh(acac)_3} C_2H_5C(=O)C_2H_5 + C_2H_5C(=O)CH_2CH_2C(=O)C_2H_5$$

Carbonylation of propylene in water containing hydrogen chloride and bis(triphenylphosphine)palladium dichloride at 700 atm and 120°C affords butyric acid, and similarly cyclohexene is converted to cyclohexanecarboxylic acid (von Kutepow *et al.*, 1966a). The use of hydrogen chloride in the reaction mixture permits lower operating temperatures and increases the reusability of the catalysts (von Kutepow *et al.*, 1967b).

Saturated esters are formed by carbonylation of an olefin in the presence of hydrogen chloride and catalytic quantities of palladium chloride or palladium metal in alcoholic solvent (Matsuda, 1969; Bittler *et al.*, 1968). The following sequence of reactions have been suggested to account for these results (Tsuji *et al.*, 1965b).

$$Pd + HCl \rightleftharpoons HPdCl \xrightarrow{RCH=CH_2} RCH_2CH_2PdCl \xrightarrow{CO} RCH_2CH_2COPdCl$$

$$RCH_2CH_2COPdCl \longrightarrow RCH_2CH_2COCl + Pd \xrightarrow{ROH} RCH_2CH_2COOR + HCl$$

Oxidative addition of hydrogen chloride to a palladium metal is followed by an insertion reaction leading to an alkylpalladium. Carbonylation of this complex followed by reaction with an alcohol completes the cycle. It is not clear whether an acid chloride needs to be an intermediate in the cycle; esterification might occur directly on the acylpalladium complex. A typical carbonylation is carried out as follows: A mixture of 1 gm of palladium chloride and 30 ml of 15% hydrogen chloride in alcohol pressured in a 300 ml glass-lined autoclave with 50 atm each of ethylene and carbon monoxide, affords after several hours at 80°C, 22 gm of ethyl propionate, 0.3 gm of ethyl β-ethoxypropionate, and 0.8 gm of higher-boiling ester. Chlorinated by-product forms in these reactions, becoming greater with increasing concentrations of hydrogen chloride and increasing molecular weight of the olefin. Olefins larger than ethylene give mixtures of esters. Under similar conditions, propylene gives ethyl isobutyrate and ethyl *n*-butyrate in a 2 to 1 ratio. Metallic rhenium and rhodium and their salts (Tsuji *et al.*, 1963b, c) may also be used in this reaction as well as tetra-(triphenylphosphine)palladium(0) (Neth. Appl. 6,409,121, May 15, 1964).

The system H_2PtCl_6–$SnCl_2$ gives high yields of linear esters on carbonylation of α-olefins in alcohol and suitable solvent. α-Olefins tested, except propylene, gave linearity in excess of 80% (Kehoe and Schell, 1970a). Neither component of the catalyst system was effective alone, paralleling earlier findings on hydrogenation of olefins (Cramer *et al.*, 1963). The maximum rate for carbonylation occurs at Sn:Pt ratios of 5 or greater. Acetone, methyl isobutyl ketone, or 1,2-dimethoxyethane are suitable solvents and permit operation at 200 atm in contrast to the 800–1000 atm of earlier work with this catalyst system (Jenner and Lindsey, 1959).

Saturated acids are also formed in good yield by carbonylation of olefins with complex palladium or platinum catalysts in aqueous media. The reaction is best carried out with a platinum or palladium complex of an aromatic phosphine in liquid phase. Both the rate and the ratio of normal to branched isomers are increased if the aqueous medium contains a surface-active agent and also a carboxylic acid solvent. For example, 1 gm of palladium chloride, 10 gm of triphenylphosphine, 400 gm of acetic acid, 15 gm of water, 70 gm of a mixed C_6–C_7 α-olefin, and 10 gm of stearic acid stirred at 125°C with carbon monoxide at an initial pressure of 800 psig in a $\frac{1}{2}$ gallon autoclave, affords 21.6 gm of α-methylhexanoic acid, 45 gm of heptanoic and α-methylheptanoic acid, and 10 gm of octanoic acid. The overall yield of acids is good (Fenton, 1970c). The use of a surface-active agent in this type of reaction may increase the yield of acid or the ratio of normal to isoacids or both (Fenton, 1970b). Esters are formed if alcohols are substituted for acids as solvents (Fenton and Olivier, 1971).

Some of the factors controlling the ratio of normal to branched acid have been elucidated. The α-methyl isomer predominates with an excess of hydrogen chloride present (von Kutepow *et al.*, 1969), whereas a buffer solution of carboxylate anion in carboxylic acid causes the normal acid to be the major product. Ratios of normal acid to α-methyl acid ranging from 0.25 to 10 have been obtained. A mechanistic interpretation of these effects has been advanced by Fenton and Olivier (1971). The authors noted that the pathway leading to predominately straight-chain acid is reversible. This means that carboxylic acids (and esters and anhydrides) can be converted back to olefins and carbon monoxide. For acids and esters, temperatures around 200°–250°C are needed, whereas anhydrides will react around 150°C (Fenton, 1970c).

$$RCH_2CH_2COOH \rightarrow RCH{=}CH_2 + H_2O + CO$$

Fenton and Olivier (1971) point out that this reversibility permits the conversion of unwanted to wanted isomers if the acids are heated in a closed system at 100–200 psig pressure of carbon monoxide, most favorably with a small amount of anhydride. By this means, complete conversion of an olefin to the desired isomer can be achieved if the unwanted isomer is used as a solvent.

$$CH_3\underset{\displaystyle CH_3}{\underset{|}{C}}HCOOH \underset{Ph_3P}{\overset{PdCl_2}{\rightleftharpoons}} CH_3CH_2CH_2COOH$$

Hydrocarbonylation takes a different course when rhodium is used instead of palladium; the product of the rhodium-catalyzed reaction is an aldehyde (Fenton and Olivier, 1971).

$$H_2C{=}CH_2 + 2\,CO + H_2O \xrightarrow[Ph_3P]{RhCl_3} CH_3CH_2CHO + CO_2$$

Certain complex palladium catalysts permit carbonylation of olefins under mild temperature conditions and achievement of reactions otherwise difficult with heat-sensitive compounds. Diels-Alder adducts such as the cyclohexene, I, and norbornene derivatives, **II**, that readily dissociate at high temperatures, were carbonylated for the first time with palladium catalysts (Bittler *et al.*, 1968).

$$\text{(I) cyclohexene-}(COOC_2H_5)_2 \xrightarrow[90°C,\ 700\ atm]{CO,\ C_2H_5OH} C_2H_5OOC\text{-cyclohexane-}(COOC_2H_5)_2$$

$$\text{(II) norbornene-}(COOC_2H_5)_2 \xrightarrow[90°C,\ 700\ atm]{CO,\ C_2H_5OH} C_2H_5OOC\text{-norbornane-}(COOC_2H_5)_2$$

Iwashita and Sakuraba (1971b) achieved a novel synthesis of imidazoles by interacting ammonia, carbon monoxide, and α-olefins in the presence of rhodium catalysts. Rhodium is unique for this reaction, functioning first as a carbonylation catalyst and then as a reduction catalyst; cobalt is without activity for imidazole formation. In a typical experiment, 50 mg of rhodium oxide suspended in methanol in a 300 ml autoclave containing 32 gm of propylene and 17 gm of ammonia under 250 atm pressure of carbon monoxide affords after 5 hours at 150°C, 2,4,5-tripropylimidazole in 59% yield. The triethyl and tributyl derivatives are obtained similarly from ethylene and 1-butene, respectively. Cyclohexene, an internal olefin, reacts differently and affords a mixture of cyclohexanecarbonamide, *N*,*N*-di(cyclohexylmethyl)-formamide, *N*,*N*-di(cyclohexylmethyl)methylamine, and tri(cyclohexyl-methyl)amine.

$$3\,CH_3CH{=}CH_2 + 3\,CO + 2\,NH_3 \longrightarrow$$

4,5-di($CH_3CH_2CH_2$)-2-($CH_2CH_2CH_3$)imidazole (N, NH ring)

Carbonylation of Dienes

Among noble metals palladium catalysts have figured most prominently in carbonylation of dienes. Palladium complexes permitted for the first time the smooth carbonylation of 1,5,9-cyclododecatriene into 4,8-cyclododecadiene-1-carboxylic ester. This ester is of commercial interest as it can be converted to 13-tridecanolactam. Monocarbonylation is best obtained at low temperatures (50°C) in ethanol containing hydrogen chloride with bis(triphenylphosphine)-palladium dichloride as a catalyst at 300 to 700 atm pressure of carbon monoxide. Both triphenylphosphine and hydrogen chloride are essential for good results; without either the yield is very small (von Kutepow *et al.*, 1970). Triphenylphosphine also increases catalyst life (Neth. Appl. 6,516,439, June 20, 1966). Under other conditions, two or all three of the double bonds undergo carbonylation. Extensive data relating process variables to product compositions are available together with a description of a pilot plant (Bittler *et al.*, 1968). Similar results are obtained with 1,5-cyclooctadiene, which may be carbonylated in alcohol to afford first the unsaturated ester and then the saturated diesters. These reactions are also possible with metallic palladium in the presence of hydrogen chloride (Tsuiji *et al.*, 1966; Tsuiji and Nogi. 1966a).

Carbonylation of 1,5-cyclooctadiene using $(Bu_3P)_2PdI_2$ as catalyst in tetrahydrofuran affords an unsaturated bicyclic ketone (Brewis and Hughes, 1966, 1967)

$$CO + \text{(1,5-cyclooctadiene)} \xrightarrow[1000\ atm]{150°C} \text{(bicyclic ketone, C=O)} + \text{(cyclooctene)}$$

45% 45%

whereas, 1,5-hexadiene gives a mixture of an unsaturated cyclopentanone and unsaturated bicyclic lactones, by addition of 1 and 2 moles of carbon monoxide.

$$H_2C{=}CHCH_2CH_2CH{=}CH_2 \xrightarrow{1000\text{ atm CO}}$$

H₃C CH₃ O + H₃C O O + H₃C O O

In methanol under similar conditions, 2 moles of carbon monoxide are added to afford a keto ester (Brewis and Hughes, 1965, 1969).

$$H_2C{=}CHCH_2CH_2CH{=}CH_2 \xrightarrow[CH_3OH]{CO}$$ (CH_3, =O, CH_2COOCH_3 substituted cyclopentanone)

Carbonylation of vinylcyclohexene over palladium catalysts in methanol affords either cyclohexenylpropionate or the diester depending on the reaction conditions (Bittler *et al.*, 1968).

CH_3OOC–(cyclohexyl)–$CH(CH_3)COOCH_3$ $\xleftarrow[CH_3OH,\ HCl]{CO}$ (cyclohexenyl)–$CH{=}CH_2$ $\xrightarrow[CH_3OH,\ 60°C]{CO}$ (cyclohexenyl)–$CH(CH_3)COOCH_3$

Carbonylation of butadiene in methanol at 70°C and 1000 atm pressure affords only 0.6 mole of methyl pent-3-enoate per mole of palladium with sodium chloropalladite or sodium bromopalladite as catalysts, whereas over sodium iodopalladite, 9–11 moles of ester per gram of palladium are obtained. Above 70°C, the catalyst is partly converted to palladium metal, but this can be prevented by adding ligands, such as tributylphosphine, to the reaction mixture. With ligand-stabilized catalysts, the optimum temperature is 150°C, and the yields are much higher but still dependent on the halide present. Over $[(C_4H_9)_3P]_2PdI_2$ the yield of methyl pent-3-enoate is 68% and catalyst efficiency reaches 61 moles of ester per mole of catalyst.

$$H_2C{=}CHCH{=}CH_2 \xrightarrow[CH_3OH]{CO} CH_3CH{=}CHCH_2\overset{\displaystyle O}{\overset{\|}{C}}OCH_3$$

Other conjugated dienes afford carbonylation products in fair yield over sodium iodopalladite. Penta-1,3-diene affords methyl 2-methylpent-3-enoate in 34% yield, based on diene charged; cyclopentadiene gives methyl cyclopent-2-ene-1-carboxylate in 73% yield, and 2,3-dimethylbuta-1,3-diene gives methyl

3,4-dimethylpent-3-enoate (Brewis and Hughes, 1968). Penta-1,3-diene has also been carbonylated in ethanol over palladium chloride to afford ethyl 2-methyl-3-pentenoate (Bordenca and Marsico, 1967; Hosaka and Tsuji, 1971).

$$H_2C{=}CHCH{=}CHCH_3 \xrightarrow[\substack{PdCl_2 \\ 100\text{ atm CO} \\ 25^\circ C,\ 26\text{ hours}}]{C_2H_5OH} \underset{84\%}{CH_3CH{=}CH\underset{\displaystyle CH_3}{\underset{|}{C}}COOC_2H_5}$$

Carbonylation of isoprene with palladium chloride as catalyst in ethanol affords mainly ethyl 4-methyl-3-pentenoate (82%) accompanied by ethyl 4-methyl-4-pentenoate (10%) and ethyl 4-ethoxy-4-methylvalerate (Hosaka and Tsuji, 1971).

$$\underset{20\text{ ml}}{H_2C{=}\underset{\displaystyle CH_3}{\underset{|}{C}}{-}CH{=}CH_2} + \underset{30\text{ ml}}{C_2H_5OH} + CO \xrightarrow[\substack{24\text{ hours} \\ 100\text{ atm}}]{\substack{1\text{ g }PdCl_2 \\ 25^\circ C}} \begin{array}{c} H_3C{-}\overset{\displaystyle CH_3}{\overset{|}{C}}{=}CHCH_2COOC_2H_5 \\ + \\ H_2C{=}\overset{\displaystyle CH_3}{\overset{|}{C}}{-}CH_2CH_2COOC_2H_5 \\ + \\ C_2H_5O\overset{\displaystyle CH_3}{\overset{|}{\underset{\displaystyle CH_3}{\underset{|}{C}}}}{-}CH_2CH_2COOC_2H_5 \end{array}$$

Good yields of unsaturated esters are obtained by the carbonylation of dienes over rhodium catalysts in alcohols. The yield is greatly increased by the presence of small amounts of water. In a typical example, a mixture of 1.5 gm of rhodium oxide, 0.3 gm of water, 400 ml of methanol, and 67.5 gm of butadiene was pressured with 900 psig of carbon monoxide and heated to 150°C for 2 hours. The product consists of 8.8 gm of vinylcyclohexene, 39.4 gm of methyl 3-methylbutenoate, 26.8 gm of methyl 2-methylbutenoate, 2.8 gm of methyl 1-methylbutenoate, and 3.5 gm of a mixture of methyl cinnamate and methyl hydrocinnamate (Zachry and Aldridge, 1964, 1966). The conversion of butadiene is 66% and the yield of unsaturated esters based on butadiene reacted is 73%.

Norbornadiene undergoes a 1:1 copolymerization with carbon monoxide in the presence of palladium chloride to give a high-melting polymer. The polymer has *cis* configuration with respect to the carbonyls but it was not determined whether the polymer had an *endo* or *exo* structure (Tsuji and Hosaka, 1965b).

$$\text{norbornadiene} + CO \xrightarrow[\substack{100\ \text{atm} \\ 50°C,\ 18\ \text{hours}}]{\substack{10\ \text{ml PhH} \\ 0.5\ \text{gm}\ PdCl_2}} \left[-C(=O)- \right]_n$$

Hydrocarbonylation of butadiene in the presence of a palladium catalyst affords carboxylic acids and lactones. Adipic acid is obtained ultimately from butadiene; 4-valerolactone and allyl acetic acid are intermediates and are obtained from the first stage of the reaction, conducted preferably below 100°C. Further reaction at 200°C affords adipic acid. All yields reported are quite moderate (Fenton, 1970a). Other workers using bis(triphenylphosphine)-palladium dichloride as catalyst in 16% hydrochloric acid obtained 2-butene-1-carboxylic acid in 65% yield on carbonylation of butadiene at 120°C and 700 atm (von Kutepow *et al.*, 1966a).

$$\underset{65\%}{CH_3CH{=}CHCH_2COH(=O)} \xleftarrow[\substack{HCl \\ CO,\ H_2O}]{(Ph_3P)_2PdCl_2} H_2C{=}CHCH{=}CH_2 \xrightarrow[\substack{FeCl_2 \\ HOAc \\ HCl \\ CO}]{PdCl_2} HOC(=O)(CH_2)_4COH(=O)$$

Carbonylation of butadiene in ethanol in the presence of a catalyst prepared *in situ* from palladium acetylactonate and triphenylphosphine affords ethyl nona-3,8-dienoate together with a mixture of ethoxyoctadienes (Billups *et al.*, 1971; Tsuji *et al.*, 1972).

$$\text{butadiene} + CO + C_2H_5OH \longrightarrow$$

$$CH_2{=}CH(CH_2)_3CH{=}CHCH_2C(=O)OC_2H_5 + CH_2{=}CH(CH_2)_3CH{=}CHCH_2OC_2H_5 + CH_2{=}CH(CH_2)_3CH(OC_2H_5)CH{=}CH_2$$

A mixture of 3 moles of butadiene, 6 moles of ethanol, 4 mmoles of palladium acetylacetonate, and 8 mmoles of triphenylphosphine heated at 80°C for 4 hours under 80 psig carbon monoxide in a glass reactor, affords 23 gm of ethyl nona-3,8-dienoate together with 55 gm of ethoxyoctadienes. This catalyst system is inactive for the carbonylation of 1-ethoxy-octa-2,7-diene, suggesting that ethyl nona-3,8-dienoate arises directly from butadiene and not by carbonylation of the octadienes. However, a catalyst system composed of palladium chloride and triphenylphosphine was active for carbonylation of ethoxyoctadienes. Formation of the ethoxyoctadienes is discussed further in the chapter on oligomerizations. Carbonylation of butadiene and isoprene palladium chloride complexes in ethanol and in benzene has been examined by Tsuji and Hosaka (1965a).

Butadiene is carbonylated in the presence of formaldehyde and hydrogen over rhodium or ruthenium catalysts to afford a mixture of alkanediols. Carbonylation of 30 gm of butadiene in the presence of 30 gm of paraformaldehyde, 1 gm of rhodium trichloride, 90 gm of $CH_3OCH_2CH_2OCH_2CH_2OCH_3$, and carbon monoxide and hydrogen in 2:1 mole ratio at 700 atm and 140°C afforded 15 gm of methanol, 2 gm of methyl propionate, 22 gm of *n*-pentanol, 25 gm of ethylene glycol, and 20 gm of hexamethylene glycol. The same reaction catalyzed by ruthenium trichloride affords 3 gm of methyl formate, 11 gm of methanol, and 21 gm of hexamethylene glycol. 1,5-Hexadiene affords with rhodium trichloride, octamethylene glycol as the major product (Anderson and Lindsey, 1963).

$$H_2C{=}CHCH{=}CH_2 + CO + H_2 + CH_2O \xrightarrow{RhCl_3} HOCH_2CH_2OH + CH_3(CH_2)_4OH + HO(CH_2)_6OH$$

Angelica lactone can be prepared by interaction of chloroprene, carbon monoxide, and water in the presence of palladium chloride. The yields are not given in this reference (Fenton, 1971). In alcohol, the major product of carbonylation of chloroprene in the presence of palladium chloride is *trans* ethyl 4-chloro-3-pentenoate (71%) accompanied by 7% of the *cis* isomer. The main reactions are additive carbonylations, but some products arising by substitution of the halogen are formed also (Hosaka and Tsuji, 1971).

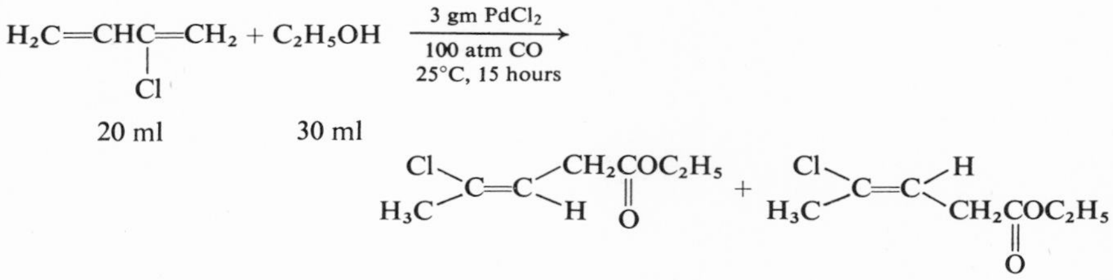
H₂C=CHC=CH₂ + C₂H₅OH
Cl
20 ml
30 ml
3 gm PdCl₂
100 atm CO
25°C, 15 hours
Cl
H₃C
C=C
CH₂COC₂H₅
H
O
+
Cl
H₃C
C=C
H
CH₂COC₂H₅
O

Allene undergoes carbonylation in alcoholic solvent to afford methacrylate esters over platinum chloride–stannous chloride catalysts (Jenner and Lindsey, 1959) or diruthenium nonacarbonyl (Benson, 1959). A 400 ml silver-lined pressure tube charged with 2 gm of diruthenium nonacarbonyl, 150 ml of methanol, and 40 gm of allene and pressured with 300 atm of carbon monoxide at 135°C for 12 hours affords 45.2 gm of methyl methacrylate (45% yield). When the reaction is carried out first at 140°C and then completed at 190°C, α,α-dimethyl-α′-methyleneglutarate is obtained in 23% yield together with methyl methacrylate in 18% yield (Kealy and Benson, 1961). Although the glutarate derivative has the composition of a dimer of methyl methacrylate, attempted dimerization of methyl methacrylate under reaction conditions failed. The glutarate is believed to arise from carbonylation of methyl α,γ-dimethylenevalerate which could arise from interaction of allene and methyl methacrylate. Noncatalytic carbonylation of allene in ethanol containing palladium chloride affords ethyl itaconate (Susuki and Tsuji, 1968).

$$H_2C{=}C{=}CH_2 \xrightarrow[CH_3OH]{CO} H_2C{=}C(CH_3)COCH_3(=O) + CH_3OC(=O){-}C(CH_3)_2CH_2C(=CH_2){-}COCH_3(=O)$$

$$H_2C{=}C(CH_3)COOCH_3 \xrightarrow{H_2C=C=CH_2} \left[CH_3OC(=O){-}C(=CH_2){-}CH_2{-}C(=CH_2){-}CH_3\right] \xrightarrow{CO,\ CH_3OH} \text{glutarate}$$

Carbonylation of allene in methanol with a ruthenium chloride–pyridine catalyst affords methyl methacrylate in 10% yield together with an ester in 12% yield which corresponds to combination of 3 moles of allene and 1 mole each of carbon monoxide, methanol, and hydrogen. In water solvent, a lactone is produced with this catalyst in about 20% yield.

$$\text{lactone} \xleftarrow[\substack{RuCl_3\cdot C_6H_5N \\ 175°C \\ 700\ atm}]{CO,\ H_2O} H_2C{=}C{=}CH_2 \xrightarrow[\substack{RuCl_3\cdot C_6H_5N \\ 175°C \\ 700\ atm}]{CO,\ CH_3OH} \text{cyclohexene ester}$$

A novel carbonylation of allene in the presence of both water and acetylene with ruthenium–pyridine catalyst affords 3-methyl-2-cyclopentenone in low yield. Carbonylation of allene at 135°C with this catalyst in the presence of cyclohexylamine affords *N*-cyclohexylmethacrylamide in 16% yield. With excess cyclohexylamine, *N*-cyclohexylformamide is formed as well (Kealy and Benson, 1961).

$$H_2C{=}C{=}CH_2 + CO + C_6H_{11}NH_2 \longrightarrow C_6H_{11}NHCHO + H_2C{=}C(CH_3)C(=O)NH{-}C_6H_{11}$$

Carbonylation of Olefinic Compounds

Carbonylation of olefins containing additional functions may lead to a variety of products inasmuch as either one or both functions may serve as a point of attack by carbon monoxide. A large amount of work has been published concerning the use of base metal catalysts (Falbe, 1970); by comparison work with noble metals is scant, with much of it recent.

Allyl Alcohols

Rhodium catalysts in the presence of large excesses of triphenylphosphine are uniquely effective hydroformylation catalysts for allyl alcohol. Both allyl alcohol and 2-butene-1,4-diol, which previously had not been successfully hydroformylated, react smoothly (Fell and Rupilius, 1969a). Carbonylation of allyl alcohol in alcohol solvent in the presence of palladium chloride or bis(triphenylphosphine)palladium dichloride affords unsaturated esters (von Kutepow *et al.*, 1966b). With ethanol, the main product is ethyl 3-butenoate accompanied by a small amount of ethyl crotonate derived presumably by isomerization of ethyl 3-butenoate.

$$H_2C{=}CHCH_2OH + CO + C_2H_5OH \rightarrow H_2C{=}CHCH_2COOC_2H_5$$

Allyl alcohol itself can serve as the source of hydroxyl in this reaction; carbonylation without additional alcohol in the presence of tris[tri(*p*-fluorophenyl)phosphine]platinum affords allyl vinylacetate as the main product. At higher temperature, the isomerized product allyl crotonate is obtained (Parshall, 1963).

$$2\,H_2C{=}CHCH_2OH + CO \xrightarrow{200°C} H_2C{=}CHCH_2\underset{\displaystyle \overset{\|}{O}}{C}OCH_2CH{=}CH_2$$

Allylic Amines

Carbonylation of allylic amines affords a variety of compounds that depend largely on the catalyst. Carbonylation of allylamine at 230°–280°C and 300 atm pressure with rhodium carbonyl affords pyrrolidone, 2-ethyl-3,5-dimethylpyridine, and 4-ethyl-3,5-dimethylpyridine in 14, 27, and 2% yields, respectively. With nickel carbonyl as catalyst, the major product was $H_2C{=}CHCH_2NHCHO$ and with iron carbonyl, 2-ethyl-3,5-dimethylpyridine (Falbe *et al.*, 1965).

$$H_2C{=}CHCH_2NH_2 \longrightarrow \text{2-pyrrolidone} + \text{3,5-dimethyl-2-ethylpyridine} + \text{4-ethyl-3,5-dimethylpyridine}$$

N-(β-Hydroxyethyl)pyrrolidone is obtained by carbonylation of *N*-(β-hydroxyethyl)allylamine. Rhodium carbonyl, rhodium oxide, or dicobalt octacarbonyl are suitable catalysts (Belgium Patent 681,405, 1966).

$$H_2C{=}CHCH_2NHCH_2CH_2OH \xrightarrow[\substack{250°C \\ 300 \text{ atm CO}}]{2000 \text{ ml THF}} \text{pyrrolidone-}NCH_2CH_2OH \quad 33\%$$

Allyl Chloride

The major product derived by carbonylation of allyl chloride is the allyl acid chloride, which may undergo further reaction with the solvent. Carbonylation of allyl chloride without solvent catalyzed by π-allylic palladium chloride (3×10^{-4} moles per mole of allyl chloride) at 110°C and 230 atm produces solely but-3-enoyl chloride (Dent *et al.*, 1964). If the reaction is prolonged, this product is isomerized in part to crotonoyl chloride. Good yields of but-3-enoyl chloride are also obtained at milder conditions (90 atm, 60°C) with higher catalyst loadings (3.55 gm of $PdCl_2$, 86.4 gm of allyl chloride) (Mador and Scheben, 1967), or with bis(π-allylpalladium chloride) in carbon tetrachloride (Neth. Appl. 6,408,476, Jan, 27, 1965). Palladium chloride, as well as palladium metal (Tsuji *et al.*, 1963b), may be used as catalysts, whereas nickel chloride, cuprous chloride, dicobalt octacarbonyl, and iron pentacarbonyl are completely inactive (Dent *et al.*, 1964). Platinum chloride and rhodium chloride are less effective than palladium chloride (Mador and Scheben, 1965). The reaction has also been carried in benzene solvent over palladium chloride; the products depend on the temperature. At elevated temperatures allyl chloride reacts with palladium chloride as a π-allyl complex, the product is but-3-enoyl chloride, and only catalytic amounts of palladium are needed, whereas at room temperature allyl chloride reacts as a simple olefin, the product is 3,4-dichlorobutroyl chloride and stoichiometric amounts of palladium chloride are needed (Tsuji *et al.*, 1964b).

$$H_2C{=}CHCH_2COCl \xleftarrow[\substack{PdCl_2 \\ CO}]{\text{elev. temp.}} H_2C{=}CHCH_2Cl \xrightarrow[\substack{PdCl_2 \\ CO}]{25°C} ClCH_2CHClCH_2COCl$$

Acetylene is a promoter for carbonylation of allylic halides over palladium catalysts. In an example, no pressure drop was noted when a mixture of allyl chloride, benzene, and 5% palladium-on-carbon were pressured with carbon monoxide at 75°C, but when the mixture was repressured with 200 psig of acetylene and 2000 psig of carbon monoxide, the reaction proceeded smoothly and afforded after hydrolysis, vinylacetic acid in 65% yield and crotonic acid in 2% yield (Closson and Ihrman, 1967).

Continuous carbonylation of allyl chloride has been practiced by passage of allyl chloride and carbon monoxide over 3% palladium-on-alumina pellets at 73°–115°C and 64 to 102 atm pressure. Conversion increases with temperature, pressure, and contact time (Mador and Scheben, 1969b).

In alcoholic solvents, carbonylation of allyl chloride affords unsaturated esters and various by-products (Closson and Ihrman, 1969). Carbonylation of 5 gm of allyl chloride in 50 ml of ethanol containing 1.0 gm of palladium chloride at 100 atm pressure and 120°C affords ethyl 3-butenoate in 47% yield and ethyl 2-butenoate in 5% yield. Under slightly different conditions, 13% of ethyl isobutyrate is formed as well. Ethyl isobutyrate presumably arises through carbonylation of propylene derived by hydrogenolysis of allyl chloride. The amounts of these products vary with the reaction conditions; temperature and the concentration of hydrogen chloride formed in the reaction seem to be important variables (Tsuji *et al.*, 1964b). Other workers found the carbonylation of allyl chloride in methanol catalyzed by di-μ-chloro-di-π-allylpalladium to be very complex. At least nine components including water were present (Dent *et al.*, 1964). Induction periods are found when palladium or palladium chloride is used as a catalyst; the true catalyst is probably a π-allylic complex. A rate enhancement occurs by addition of 1 mole of triphenylphosphine (Medema *et al.*, 1968).

$$H_2C{=}CHCH_2Cl \xrightarrow[PdCl_2]{CO} \begin{array}{c} H_2C{=}CHCH_2COOC_2H_5 \\ + \\ CH_3CH{=}CHCOOC_2H_5 \\ + \\ (CH_3)_2CHCOOC_2H_5 \end{array}$$

Carbonylation at 100 atm of allyl chloride (15 gm) in 40 ml of tetrahydrofuran over 1.8 gm of palladium chloride at 80°C proceeds by interaction of the intermediate acid chloride with the solvent and 4-chloro-*n*-butyl-3-butenoate is formed in 88% yield (Tsuji *et al.*, 1964b, 1969).

$$H_2C{=}CHCH_2Cl + CO + \text{(tetrahydrofuran)} \xrightarrow{PdCl} H_2C{=}CHCH_2COO(CH_2)_4Cl$$

Allyl chloride undergoes an interesting carbonylation in the presence of butadiene and Pd(II) to afford 3,7-octadienoyl chloride (Medema *et al.*, 1969).

$$H_2C{=}CHCH_2Cl + H_2C{=}CHCH{=}CH_2 \xrightarrow[Pd]{CO} CH_2{=}CHCH_2CH_2CH{=}CHCH_2\underset{\underset{O}{\|}}{C}Cl$$

Allyl Ethers and Allyl Esters

Allyl ethers and allyl esters are carbonylated in the presence of catalytic amounts of palladium chloride in aprotic solvent, affording β,γ-unsaturated esters and unsaturated anhydrides, respectively. The presence of chloride ion is essential; π-allylpalladium chloride or metallic palladium and hydrogen chloride are suitable catalysts. Esters do not arise from allyl ethers by simple insertion of carbon monoxide, a point proved by competitive carbonylation of equimolar mixtures of allyl ethyl ether and crotyl methyl ether. Simple carbon monoxide insertion should afford only ethyl 3-butenoate and methyl 3-pentenoate, but instead all four possible esters are formed in nearly equal amounts. This result is accounted for by π-allyl complex formation through interaction of the allyl ether and palladium chloride with cleavage of the allylic carbon–oxygen bond (Imamura and Tsuji, 1969).

$$\begin{array}{c} H_2C{=}CHCH_2OC_2H_5 \\ 3\text{ gm} \\ + \\ CH_3CH{=}CHCH_2OCH_3 \\ 3\text{ gm} \end{array} + CO \xrightarrow[\substack{150\text{ atm} \\ 130°C \\ 17\text{ gm PhH} \\ 20\text{ hours}}]{0.5\text{ gm PdCl}_2} \begin{array}{c} H_2C{=}CHCH_2COOC_2H_5 \\ + \\ H_2C{=}CHCH_2COOCH_3 \\ + \\ CH_3CH{=}CHCH_2COOC_2H_5 \\ + \\ CH_3CH{=}CHCH_2COOCH_3 \end{array}$$

Diallyl ethers react stepwise, forming first an allyl ester, and then, by insertion of a second molecule of carbon monoxide, an anhydride. The method provides a simple means of synthesizing anhydrides of β,γ-unsaturated acids (Tsuji *et al.*, 1964c).

$$H_2C{=}CHCH_2OCH_2CH{=}CH_2 \xrightarrow[PdCl_2]{CO} H_2C{=}CHCH_2\underset{\underset{O}{\|}}{C}{-}OCH_2CH{=}CH_2$$

$$\downarrow CO$$

$$H_2C{=}CHCH_2\underset{\underset{O}{\|}}{C}O\underset{\underset{O}{\|}}{C}CH_2CH{=}CH_2$$

Diallylic ethers, such as 1,4-diethoxy-2-butene, undergo both mono- and dicarbonylations and afford a complex mixture of products through secondary reactions. The authors (Imamura and Tsuji, 1969) suggested the routes indicated to account for the variety of products.

$$C_2H_5OCH_2CH{=}CHCH_2OC_2H_5 + CO + C_2H_5OH \xrightarrow{PdCl_2}$$

$$C_2H_5O\underset{\underset{O}{\|}}{C}CH_2CH{=}CHCH_2\underset{\underset{O}{\|}}{C}OC_2H_5 + C_2H_5OCH_2CH{=}CHCH_2\underset{\underset{O}{\|}}{C}OC_2H_5$$

Dicarbonylation — Monocarbonylation

$$H_2C{=}CHCH{=}CH\underset{\underset{O}{\|}}{C}OC_2H_5 + C_2H_5OCH_2\underset{\underset{OC_2H_5}{|}}{C}HCH{=}CH_2$$

Monocarbonylation followed by 1,4-elimination — Allylic rearrangement of starting material

$$H_2C{=}CH\underset{\underset{\underset{O}{\|}}{COC_2H_5}}{\underset{|}{C}}HCH_2OC_2H_5 + C_2H_5OCH_2CH{=}CHCH_3$$

Monocarbonylation — Hydrogenolysis of starting material

$$CH_3CH{=}CHCH_2\underset{\underset{O}{\|}}{C}OC_2H_5 + CH_3CH{=}CHCH_3$$

Monocarbonylation followed by hydrogenolysis — Hydrogenolysis of starting material

Carbonylations were shown to proceed through a π-allylic complex which was independently synthesized from butadiene in ethanol. The exact source of hydrogen in the hydrogenolysis reactions is not known.

$$C_2H_5OCH_2CH{=}CHCH_2OC_2H_5 + PdCl_2 \longrightarrow \text{(}\pi\text{-allyl complex: } CH(CH_2OC_2H_5)\text{—}HC\text{—}CH_2 \cdots PdCl\text{)}$$

Allylic rearrangement of 1,4-diethoxy-2-butene occurs easily in the presence of palladium; carbon monoxide or ethanol need not necessarily be present. A similar palladium-catalyzed rearrangement of 1,4-dichloro-2-butene to 1,2-dichloro-3-butene has been reported (Oshima, 1967).

Unsaturated Esters

Carbonylation of α,β-unsaturated esters in alcohol solvent in the presence of rhodium catalysts follows an unusual course, and the product, apparently arising from substitution rather than addition, is an unsaturated diester. In 400 ml of methanol with 0.3 gm of rhodium oxide, methyl acrylate (50.5 gm) was converted in 3 hours at 150°C under 1250 psig of carbon monoxide to dimethyl fumarate in 49% yield and 18% conversion (Zachry and Aldridge, 1965).

$$H_2C{=}CHCOOCH_3 + CO + CH_3OH \xrightarrow{Rh_2O_3} CH_3OOCCH{=}CHCOOCH_3$$

Carbonylation of methyl methacrylate in methanol in the presence of ruthenium carbonyl affords dimethyl 2,2,4-trimethylglutarate by hydrodimerization (Schreyer, 1967).

$$H_2C{=}C(CH_3){-}C(=O)OCH_3 + CH_3OH + CO \longrightarrow CH_3OC(=O)CH(CH_3)CH_2C(CH_3)_2{-}C(=O)OCH_3$$

NITRO COMPOUNDS

Interaction of nitro compounds and carbon monoxide affords a variety of products including urethanes, isocyanates, ureas, and pyridine derivatives in proportions that depend on the type of substrate, catalyst, and reaction conditions. High pressures of carbon monoxide have been used in all reactions reported.

Aromatic Nitro Compounds

Phenyl isocyanates have been prepared by carbonylation of nitrobenzene over 5% palladium-on-alumina at 170°C and 910 atm pressure of carbon monoxide (Belgium Patent 651,876, 1964) and over 5% rhodium-on-carbon at 500 atm. A mixture of 24.6 gm of nitrobenzene, 5.0 gm of 5% rhodium-on-carbon, 0.4 gm of anhydrous ferric chloride, and 100 ml of benzene under 500 atm pressure of carbon monoxide at 190°C for 5.5 hours affords 8.3 gm of phenyl isocyanate (35% yield) together with 6.2 gm of diphenylurea and 1,3,5-triphenylbiuret. Chlorobenzene and cyclohexane may also be used as solvents. Palladium- or rhodium-on-alumina give the best yield when used with ferric chloride, the most effective of a number of Lewis acids (Hardy and Bennett, 1967a). Partial reduction products of nitrobenzene such as nitrosobenzene, azo- and azoxybenzene may be used as well as nitrobenzene in this type of reaction (Bennett *et al.*, 1967; Bennett and Davis, 1969). All the platinum metals modified by at least two heavy metal salts have been claimed to be effective in production of aromatic isocyanates (Dodman *et al.*, 1970). Other effective modifiers are said to be nitrogen heterocyclics, such as pyridine or isoquinoline (Smith and Schnabel, 1967), organic cyano compounds (Pritchard, 1968), copper compounds (Kober *et al.*, 1967a), acyl halides (Kober *et al.*, 1967b), organophosphorus compounds (Ottmann *et al.*, 1967), oxygen heterocyclics (Smith, 1968a), and various sulfur compounds (Smith, 1968b). A modification of the carbonylation reaction producing isocyanates from nitroaromatics consists of carrying out the reaction in the presence of a halogenating agent, such as thionyl chloride, and obtaining halogenated isocyanates. Yields are only moderate (Ottmann *et al.*, 1969).

$$R\text{-}C_6H_4\text{-}NO_2 + 3\,CO \longrightarrow R\text{-}C_6H_4\text{-}NCO + 2\,CO_2$$

R = *o*-CH_3, *p*-CH_3, *m*-Cl, *m*-CF_3, *m*-CN

If the above reaction is conducted in alcohol solvent, urethanes are formed. Ten grams of nitrobenzene, 5 gm of methanol, 35 ml of toluene, 0.1 gm of rhodium trichloride, and 0.5 gm of ferric chloride, held under 80 atm pressure of carbon monoxide at 150°C for 13 hours affords methyl-*N*-phenylurethane in 81% yield (British Patent 1,080,094).

$$C_6H_5\text{-}NO_2 + 3\,CO + CH_3OH \longrightarrow C_6H_5\text{-}NHC(=O)OCH_3 + 2\,CO_2$$

Carbanilides are obtained by interaction of carbon monoxide, water, and an aromatic nitro compound in the presence of a noble metal and a Lewis acid. A preferred catalyst combination is 5% palladium-on-alumina and ferric chloride. Yields are not given in the various examples (Hardy and Bennett, 1967b). With rhodium catalysts in the presence of water and an organic tertiary amine, aromatic nitro compounds are reduced in high yield to the corresponding aniline. The basicity and nucleophilicity of the tertiary amine is important. A strongly basic amine such as *N*-methylpyrrolidine is more effective than a weakly basic amine, such as pyridine.

$$3\,CO_2 + PhNH_2 \xleftarrow[\substack{2{,}000\text{ psig CO}\\150^\circ C\\Rh}]{H_2O} PhNO_2 \xrightarrow[\substack{10{,}000\text{ psig CO}\\180^\circ C\\Pd\text{—}Fe}]{H_2O} PhNHC(=O)NHPh$$

Any rhodium salt may be used, but preformed rhodium carbonyls such as $Rh_6(CO)_{16}$ can be used under milder conditions than rhodium oxides or partially carbonylated salts of rhodium (Iqbal, 1971b).

Diphenylurea is obtained in moderate yield (54%) from nitrobenzene (1 mole) by interaction with 5 moles of carbon monoxide and 0.4 mole of hydrogen at 140°C and 150 atm in the presence of $[Ru(CO)_4]_3$ (L'eplattenier *et al.*, 1970).

Various partly reduced derivatives of nitro compounds such as nitroso (Hardy and Bennett, 1969a, b) or azo or azoxy compounds (Bennett and Davis, 1969; Hardy and Bennett, 1969c) may be used in these carbonylation reactions as well as nitro compounds.

Aliphatic Nitro Compounds

Aliphatic nitro compounds are converted to pyridines by carbon monoxide in the presence of supported palladium or rhodium and ferric chloride, the same catalyst systems that converts aromatic nitro compounds to urethanes. All components of the system must be present; if one is missing, no pyridine is found. Palladium and rhodium are equally effective supported on either carbon or alumina. The products are insensitive to temperature or carbon monoxide pressure over a wide range. Ethanol is a preferred solvent, although benzene can be used successfully at a much slower rate. In a typical experiment, 0.162 mole of 1-nitrobutane, 4.2 gm of dried 5% palladium-on-carbon, 0.013 mole of anhydrous ferric chloride, and 100 ml of anhydrous ethanol in a glass-lined autoclave were pressured to 5000 psi with carbon monoxide and heated 2 hours at 90°C. There was obtained after distillation a 39% yield of ethyl carbamate and 24% yield of 2-propyl-3,5-diethylpyridine. Similarly, 2-ethyl-3,5-dimethylpyridine was obtained in 52% yield from 1-nitropropane (Mohan, 1970).

$$CH_3CH_2CH_2CH_2NO_2 + CO \longrightarrow$$

CH_3H_2C CH_2CH_3
N $CH_2CH_2CH_3$

REFERENCES

Alderson, T., and Thomas, J. C. (1962). U.S. Patent 3,040,090.

Allum, K. G., Hancock, R. D., McKenzie, S., and Pitkethly, R. C. (1973). *Proc. Int. Congr. Catal., 5th 1972*, in press.

Anderson, T., and Lindsey, R. V. (1963). U.S. Patent 3,081,357.

Bartlett, J. H., and Hughes, V. L. (1959). U.S. Patent 2,894,038.

Bennett, R. P., and Davis, S. M. (1969). U.S. Patent 3,467,688.

Bennett, R. P., Hardy, W. B., Madison, R. K., and Davis, S. M. (1967). *Abstr. Pap., 153rd Meet., Amer. Chem. Soc. Miami Beach* p. 89.

Benson, R. E. (1959). U.S. Patent 2,871,262.
Biale, G. (1970). U.S. Patent 3,530,168.
Billups, W. E., Walker, W. E., and Shields, T. C. (1971). *Chem. Commun.* p. 1067.
Bird, C. W. (1963). *Chem. Rev.* **62**, 283.
Bird, C. W. (1967). "Transition Metal Intermediates in Organic Synthesis." Academic Press, New York.
Bird, C. W., Briggs, E. M., and Hudec, J. (1967). *J. Chem. Soc., C* p. 1862.
Bittler, K., von Kutepow, N., Neubauer, D., and Reis, H. (1968). *Angew. Chem., Int. Ed. Engl.* **7**, 329.
Bond, G. C. (1971a). Ger. Offen. 2,047,748.
Bond, G. C. (1971b). Ger. Offen. 2,055,539.
Booth, B. L., Else, M. J., Fields, R., and Haszeldine. R. N. (1971). *J. Organometal. Chem.* **27**, 119.
Booth, F. B. (1970). U.S. Patent 3,511,880.
Bordenca, C., and Marsico, W. E. (1967). *Tetrahedron Lett.* p. 1541.
Braca, G., Sbrana, G., and Pino, P. (1968). *Chem. Ind. (Milan)* **50**, 121.
Braca, G., Glauco, S., Franco, P., and Pino, P. (1970). *Chem. Ind. (Milan)* **52**, 1091.
Brewis, S., and Hughes, P. R. (1965). *Chem. Commun*, p. 489.
Brewis, S., and Hughes, P. R. (1966). *Chem. Commun.* p. 6.
Brewis, S., and Hughes, P. R. (1967). *Chem. Commun.* p. 71.
Brewis, S., and Hughes, P. R. (1968). *Chem. Commun.* p. 157.
Brewis, S., and Hughes, P. R. (1969). *Amer. Chem. Soc., Div. Petrol. Chem., Prepr.* **14**, No. 2, B170.
Brewis, S., Dent, W. T., and Williams, R. O. (1967). British Patent 1,091,042.
Brown, C. K., and Wilkinson, G., (1970). *J. Chem. Soc., A* p. 2753.
Bryce-Smith, D. (1964). *Chem. Ind. (London)* p. 239.
Byerley, J. J., Rempel, G. L., and Takebe, N. (1971). *Chem. Commun.* p. 1482.
Calderazzo, F. (1965). *Inorg. Chem.* **4**, 293.
Čapka, M., Svobada, P., Černý, M., and Hetflejš, J. (1971). *Tetrahedron Lett.* 4790.
Chiusoli, G. P., Venturello, C., and Merzoni, S. (1968). *Chem. Ind. (London)* p. 977.
Church, M. J., and Mays, M. J. (1968). *Chem. Commun.* p. 435.
Closson, R. D., and Ihrman, K. G. (1967). U.S. Patent 3,338,916.
Closson, R. D., and Ihrman, K. G. (1969). U.S. Patent 3,457,299.
Collman, J. P., Kubota, M., and Hosking, J. W. (1967). *J. Amer. Chem. Soc.* **89**, 4809.
Craddock, J. H., Hershman, A., Paulik, F. E., and Roth, J. F. (1969). *Ind. Eng. Chem., Prod. Res. Develop.* **8**, 291.
Cramer, R. D., Jenner, E. L., Lindsey, R. V., and Stolberg, U. G. (1963). *J. Amer. Chem. Soc.* 85, 1691.
Dent, W. T., Long, R., and Whitfield, G. H. (1964). *J. Chem. Soc., London* p. 1588.
Dodman, D., Pearson, K. W., and Woolley, J. M. (1970). British Patent 1,205,521.
Durand, D., and Lassau, C. (1969). *Tetrahedron Lett.* p. 2329.
Evans, D., Osborn, J. A., Jardine, F. H., and Wilkinson, G. (1965). *Nature (London)* **208**, 1204.
Evans, D., Osborn, J. A., and Wilkinson, G. (1968). *J. Chem. Soc., A* p. 3133.
Fahnenstich, R., and Weigert, W. (1969). U.S. Patent 3,441,573.
Falbe, J. F. (1970). "Carbon Monoxide in Organic Synthesis." Springer-Verlag, Berlin and New York.
Falbe, J. F. (1971). German Patent 1,618,396.
Falbe, J. F., and Huppes, N. (1966). *Brennst.-Chem.* **47**, 314.
Falbe, J. F., and Huppes, N. (1967a). U.S. Patent 3,318,913.
Falbe, J. F., and Huppes, N. (1967b). *Brennst.-Chem.* **48**, 46.

Falbe, J. F., and Huppes, N. (1967c). *Brennst.-Chem.* **48**, 182.
Falbe, J. F., and Huppes, N. (1970). U.S. Patent 3,509,221.
Falbe, J. F., and Weber, J. (1970a). French Patent 1,590,393.
Falbe, J. F., and Weber, J. (1970b). French Patent 1,598,768.
Falbe, J. F., Weitkamp, H., and Korte, F. (1965). *Tetrahedron Lett.* p. 2677.
Fell, B., and Beutler, M. (1972). *Tetrahedron Lett.* p. 3455.
Fell, B., and Rupilius, W. (1969a). *Angew. Chem., Int. Ed. Engl.* **8**, 897.
Fell, B., and Rupilius, W. (1969b). *Tetrahedron Lett.* p. 2721.
Fell, B., Rupilius, W., and Asinger, F. (1968). *Tetrahedron Lett.* p. 3261.
Fell, B., Geurts, A., and Muller, E. (1971). *Angew. Chem., Int. Ed. Engl.* **10**, 828.
Fenton, D. M. (1968a). U.S. Patent 3,397,225.
Fenton, D. M. (1968b). U.S. Patent 3,397,226.
Fenton, D. M. (1970a). U.S. Patent 3,509,209.
Fenton, D. M. (1970b). U.S. Patent 3,530,155.
Fenton, D. M. (1970c). U.S. Patent 3,530,198.
Fenton, D. M. (1971). U.S. Patent 3,564,020.
Fenton, D. M., and Olivier, K. L. (1967a). U.S. Patent 3,346,625.
Fenton, D. M., and Olivier, K. L. (1967b). U.S. Patent 3,349,119.
Fenton, D. M., and Olivier, K. L. (1971). *161st Nat. Meet., Amer. Chem. Soc., Los Angeles,* Petr. Sect., paper 5.
Fenton, D. M., and Steinwand, P. J. (1968). U.S. Patent 3,393,136.
Fenton, D. M., and Steinwand, P. J. (1972). *J. Org. Chem.* **37**, 2034.
Frankel, E. N. (1971). *J. Amer. Oil Chem. Soc.* **48**, 248.
Gladrow, E. M., and Mattox, W. J. (1967). U.S. Patent 3,352,924.
Goldup, A., Westaway, M., and Walker, G. (1971). Ger. Offen. 1,953,64.
Graziani, M., Uguagliati, P., and Carturan, G. (1971). *J. Organometal. Chem.* **27**, 275.
Haag, W. O., and Whitehurst, D. D. (1971). *N. Amer. Meet. Catal. Soc., 2nd, 1971* p. 16.
Haag, W. O., and Whitehurst, D. D. (1973). *Proc. Int. Congr. Catal., 5th, 1973* in press.
Hallman, P. S., Evans, D., Osborn, J. A., and Wilkinson, G. (1967). *Chem. Commun.* p. 305.
Hardy, W. B., and Bennett, R. P. (1967a). *Tetrahedron Lett.* p. 961.
Hardy, W. B., and Bennett, R. P. (1967b). U.S. Patent 3,335,142.
Hardy, W. B., and Bennett, R. P. (1969a). U.S. Patent 3,461,149.
Hardy, W. B., and Bennett, R. P. (1969b). U.S. Patent 3,467,687.
Hardy, W. B., and Bennett, R. P. (1969c). U.S. Patent 3,467,694.
Hayden, P. (1969). U.S. Patent 3,458,532.
Haynes, P., Slaugh, L. H., and Kohnle, J. F. (1970). *Tetrahedron Lett.* p. 365.
Heil, B., and Marko, L. (1969). *Chem. Ber.* **102**, 2238.
Henrici-Olivé, G., and Olivé, S. (1971). *Angew. Chem., Int. Ed. Engl.* **10**, 105.
Himmele, W., Hoffmann, W., Pasedach, H., and Werner, A. (1971). Ger. Offen. 1,964,962.
Hohenschutz, H., von Kutepow, N., and Himmele, W. (1966). *Hydrocarbon Process.* **45**, (No. 11) 141.
Hosaka, S., and Tsuji, J. (1971). *Tetrahedron* **27**, 3821.
Hughes, V. L., and Brodkey, R. S. (1958). U.S. Patent 2,839,580.
Ihrman, K. G., Filbey, A. H., and Zaweski, E. F. (1968). U.S. Patent 3,361,811.
Imamura, S., and Tsuji, J. (1969). *Tetrahedron* **25**, 4187.
Iqbal, A. F. M. (1971a). *Tetrahedron Lett.* p. 3381.
Iqbal, A. F. M. (1971b). *Tetrahedron Lett.* p. 3385.
Iwashita, Y., and Sakuraba, M. (1971a). *Tetrahedron Lett.* p. 2409.
Iwashita, Y., and Sakuraba, M. (1971b). *J. Org. Chem.* **36**, 3927.
Jacobson, G., and Spaethe, H. (1962). German Patent 1,138,760.

Jardine, F. H., Osborn, J. A., Wilkinson, G., and Young, J. F. (1965). *Chem. Ind.* (*London*) p. 560.
Jenner, E., and Lindsey, R. V., Jr. (1959). U.S. Patent 2,876,254.
Kealy, T. J., and Benson, R. E. (1961). *J. Org. Chem.* **26**, 3126.
Kehoe, L. J., and Schell, R. A. (1970a). *J. Org. Chem.* **35**, 2846.
Kehoe, L. J., and Schell, R. A. (1970b). U.S. Patent 3,544,635.
Kober, E. H., Schnabel, W. J., and Kraus, T. C. (1967a). U.S. Patent 3,532,963.
Kober, E. H., Schnabel, W. J., and Kraus, T. C. (1967b). U.S. Patent 3,523,965.
Kuper, D. G. (1970). U.S. Patent 3,520,914.
Lawrenson, M. J. (1970). British Patent 1,205,027.
Lawrenson, M. J. (1971a). Ger. Offen. 2,031,380.
Lawrenson, M. J. (1971b). Ger. Offen. 2,058,814.
Lawrenson, M. J., and Green, M. (1971). Ger. Offen. 2,026,926.
L'eplattenier, F., Matthys, P., and Calderazzo, F. (1970). *Inorg. Chem.* **9**, 342.
Lines, C. B., and Long, R. (1969). *Amer. Chem. Soc., Div. Petrol. Chem., Prepr.* **14**, No. 2, B159.
Long, R., and Marples, B. A. (1965). British Patent 987,516.
Mador, I. L., and Blackham, A. U. (1963). U.S. Patent 3,114,762.
Mador, I. L., and Scheben, J. A. (1965). French Patent 1,419,758.
Mador, I. L., and Scheben, J. A. (1967). U.S. Patent 3,309,403.
Mador, I. L., and Scheben, J. A. (1969a). U.S. Patent 3,423,456.
Mador, I. L., and Scheben, J. A. (1969b). U.S. Patent 3,452,090.
Mador, I. L., and Scheben, J. A. (1969c). U.S. Patent 3,454,632.
Maitlis, P. M. (1971). "The Organic Chemistry of Palladium," Vol. 2. Academic Press, New York.
Matsuda, A. (1969). *Bull. Chem. Soc. Jap.* **42**, 2596.
Medema, D., van Helden, R., and Kohll, C. F. (1968). *Symp. New Aspects Chem. Metal Carbonyls Derivatives*, Venice, Sept, 1968.
Medema, D., van Helden, R., and Kohll, C. F. (1969). *Inorg. Chim. Acta* **3**, 255.
Millidge, A. F. (1965a). French Patent 1,411,602.
Millidge, A. F. (1965b). British Patent 1,012,011.
Mohan, A. G. (1970). *J. Org. Chem.* **35**, 3982.
Morikawa, M. (1964). *Bull. Chem. Soc. Jap.* **37**, 379.
Nogi, T., and Tsuji, J. (1969). *Tetrahedron* **25**, 4099.
Olivier, K. L. (1968). U.S. Patent 3,415,871.
Olivier, K. L. (1970a). U.S. Patent 3,505,394.
Olivier, K. L. (1970b). U.S. Patent 3,547,964.
Olivier, K. L., and Booth, F. B. (1970). *Hydrocarbon Process.* **49**, (No. 4) 112.
Olivier, K. L., and Snyder, L. R. (1970). U.S. Patent 3,539,634.
Olivier, K. L., Booth, F. B., and Mears, D. E. (1971). U.S. Patent 3,555,098.
Orchin, M., and Rupilius, W. (1972). *Catal. Rev.* **6**, 85.
Orchin, M., and Wender, I. (1957). *Catalysis* **5**, 1.
Osborn, J. A., Wilkinson, G., and Young, J. F. (1965). *Chem. Commun.* p. 17.
Oshima, A. (1967). *Nippon Kagaku Zasshi* **89**, 92.
Ottman, G. F., Kober, E. H., and Gavin, D. F. (1967). U.S. Patent 3,523,962.
Ottman, G. F., Kober, E. H., and Gavin, D. F. (1969). U.S. Patent 3,481,968.
Parshall, G. W. (1963). *Z. Naturforsch. B* **18**, 772.
Paulik, F. E. (1972). *Catal. Rev.* **6**, 49.
Paulik, F. E., and Roth, J. F. (1968). *Chem. Commun.* p. 1578.
Paulik, F. E., Robinson, K. K., and Roth, J. F. (1969). U.S. Patent 3,487,112.

Pino, P., Braca, G., Settimo, S. F. a., and Sbrana, G. (1967). U.S. Patent 3,355,503.
Pino, P., Braca, G., Sbrana, G., and Cuccuru, A. (1968). *Chem. Ind.* (*London*) p. 1732.
Pino, P., Braca, G., Settimo, S. F. a., and Sbrana, G. (1969). U.S. Patent 3,459,812.
Pritchard, W. W. (1968). U.S. Patent 3,576,836.
Pruett, R. L., and Smith, J. A. (1969). *J. Org. Chem.* **34**, 327.
Pruett, R. L., and Smith, J. A. (1970). U.S. Patent 3,527,809.
Robinson, K. K., Paulik, F. E., Hershman, A., and Roth, J. F. (1969). *J. Catal.* **15**, 245.
Rony, P. R. (1969). *J. Catal.* **14**, 142.
Roth, J. F., Craddock, J. H., Hershman, A., and Paulik, F. E. (1971). *Chem. Tech.* 600
Saegusa, T., Tsuda, T., and Isegawa, Y. (1971). *J. Org. Chem.* **36**, 858.
Sauer, J. C. (1963). U.S. Patent 3,097,237.
Scheben, J. A., and Mador, I. L. (1969). U.S. Patent 3,468,947.
Schiller, G. (1956). German Patent 953,605.
Schreyer, R. C. (1967). U.S. Patent 3,322,819.
Schultz, R. G., and Montgomery, P. D. (1969). *J. Catal.* **13**, 105.
Sibert, J. W. (1970). U.S. Patent 3,515,757.
Slaugh, L. H., and Mullineaux, R. D. (1966). U.S. Patent 3,239,566.
Smith, E. (1968a). U.S. Patent 3,636,028.
Smith, E. (1968b). U.S. Patent 3,636,029.
Smith, E. and Schnabel, W. J. (1967). U.S. Patent 3,567,835.
Stern, E. W. (1967). *Catal. Rev.* **1**, 73.
Stern, E. W., and Spector, M. L. (1966). *J. Org. Chem.* **31**, 596.
Stern, E. W., and Spector, M. L. (1968). U.S. Patent 3,405,156.
Susuki, T., and Tsuji, J. (1968). *Bull. Chem. Soc. Jap.* **41**, 1954.
Takegami, Y., Watanabe, Y., and Masada, H. (1967). *Bull. Chem. Soc. Jap.* **40**, 1459.
Takesada, M., and Watamatsu, H. (1970). *Bull. Chem. Soc. Jap.* **43**, 2192.
Tinker, H. B., and Morris, D. E. (1971). *Abstr., 167nd Nat. Meet., Amer. Chem. Soc., Washington, D.C.* Inorg. 070.
Tsuji, J. (1969). *Accounts Chem. Res.* **2**, 144.
Tsuji, J., and Hosaka, S. (1965a). *J. Am. Chem. Soc.* **87**, 4075.
Tsuji, J., and Hosaka, S. (1965b). *J. Polym. Sci., Part B* **3**, 703.
Tsuji, J., and Iwamoto, N. (1966). *Chem. Commun.* p. 380.
Tsuji, J., and Nogi, T. (1966a). *Bull. Chem. Soc. Jap.* **39**, 146.
Tsuji, J., and Nogi, T. (1966b). *J. Org. Chem.* **31**, 2641.
Tsuji, J., and Nogi, T. (1966c). *Tetrahedron Lett.* p. 1801.
Tsuji, J., and Ohno, K. (1969). *Advan. Org. Chem.* **6**, 184.
Tsuji, J., Morikawa, M., and Kiji, J. (1963a). *Tetrahedron Lett.* p. 1061.
Tsuji, J., Morikawa, M., and Kiji, J. (1963b). *Tetrahedron Lett.* p. 1437.
Tsuji, J., Kiji, J., and Morikawa, M. (1963c). *Tetrahedron Lett.* p. 1811.
Tsuji, J., Morikawa, M., and Iwamoto, N. (1964a). *J. Amer. Chem. Soc.* **86**, 2095.
Tsuji, J., Imamura, S., Morikawa, M. (1964b). *J. Amer. Chem. Soc.* **86**, 4350.
Tsuji, J., Morikawa, M., and Kiji, J. (1964c). *J. Amer. Chem. Soc.* **86**, 4851.
Tsuji, J., Iwamoto, N., and Morikawa, M. (1965a). *Bull. Chem. Soc. Jap.* **38**, 2213.
Tsuji, J., Ohno, K., and Kajimoto, T. (1965b). *Tetrahedron Lett.* p. 4565.
Tsuji, J., Hosaka, S., Kiji, J., and Susuki, T. (1966). *Bull. Chem. Soc. Jap.* **39**, 141.
Tsuji, J., Kiji, J., Morikawa, M., and Imamura, S. (1969). U.S. Patent 3,427,344.
Tsuji, J., Mori, Y., and Hara, M. (1972). *Tetrahedron* **28**, 3721.
von Kutepow, N., Neubauer, D., and Bittler, K. (1966a). German Patent 1,229,089.
von Kutepow, N., Bittler, K., and Neubauer, D. (1966b). German Patent 1,221,224.
von Kutepow, N., Bittler, K., Neubauer, D., and Reis, H. (1967a). German Patent 1,237,116.

von Kutepow, N., Bittler, K., and Neubauer, D. (1967b). British Patent 1,066,772.
von Kutepow, N., Bittler, K., and Neubauer, D. (1969). U.S. Patent 3,437,676.
von Kutepow, N., Bittler, K., and Neubauer, D. (1907). U.S. Patent 3,501,518.
Wakamatsu, H. (1964). *Nippon Kagaku Zasshi* **85**, 227.
Wakamatsu, H., Takesada, M., and Sato, J. (1969). U.S. Patent 3,420,895.
Wender, I., Sternberg, H. W., and Orchin, M. (1957). *Catalysis* **5**, 73.
Wilkinson, G. (1971). British Patent 1,219,763.
Yagupsky, G., Brown, C. K. and Wilkinson, G. (1970), *J. Chem. Soc.*, (*London*) p. 1392.
Zachry, J. B., and Aldridge, C. L. (1964). U.S. Patent 3,161,672.
Zachry, J. B., and Aldridge, C. L. (1965). U.S. Patent 3,176,038.
Zachry, J. B., and Aldridge, C. L. (1966). U.S. Patent 3,253,018.

CHAPTER 8

Decarbonylation and Desulfonylation

Decarbonylations have been carried out over both supported and unsupported heterogeneous noble metal catalysts as well as a number of complex homogeneous catalysts. Most heterogeneous work has been done with palladium. A brief comparison of supported palladium, platinum, rhodium, ruthenium, and iridium for decarbonylation of benzoyl chloride, furfural, cinnamaldehyde, and benzyl alcohol established that this preference for palladium was generally justified (Rylander and Hasbrouck, 1967). Homogeneous decarbonylation catalysts have elicited considerable interest, especially in mechanistic studies. Many of these homogeneous catalysts extract carbon monoxide stoichiometrically from carbonyl compounds under very mild conditions and have been employed in synthesis in this way (Shimizu *et al.*, 1966; Dawson and Ireland, 1968; Muller *et al.*, 1969; Emery *et al.*, 1970; Sakai *et al.*, 1972); higher temperatures are required if these complexes are to function as catalysts.

Catalytic decarbonylation occurs with a number of compounds including aliphatic, aromatic, and unsaturated aldehydes, acid chlorides, acids, esters, and ketones, as well as with compounds such as alcohols and ethers that are converted to carbonyl compounds preceding decarbonylation. Certain sulfur compounds undergo desulfonylation in a reaction similar to decarbonylation, over both heterogeneous and homogeneous catalysts.

DECARBONYLATION

Aliphatic Aldehydes

Aliphatic aldehydes undergo catalytic decarbonylation at elevated temperatures. The reaction is often not clean-cut. Decarbonylation of heptaldehyde, for instance, over palladium-on-carbon slowly produces carbon monoxide, hydrogen, water, and isomeric hexenes. The reaction appears to involve dehydrogenation, isomerization, and polymerization as well as decarbonylation (Hawthorne and Wilt, 1960). Decarbonylation of butanal over a palladium film affords a mixture of propylene and propane in ratios that change as the reaction progresses. The authors (Hemidy and Gault, 1965) suggested that the olefin and paraffin are formed in parallel paths. Decarbonylation of small ring compounds may be accompanied by ring cleavage. On heating pinonic aldehyde to 220°–300°C in the presence of palladium, a mixture of pinonone, pinonenone, and several cleavage products are formed (Conia and Faget, 1964).

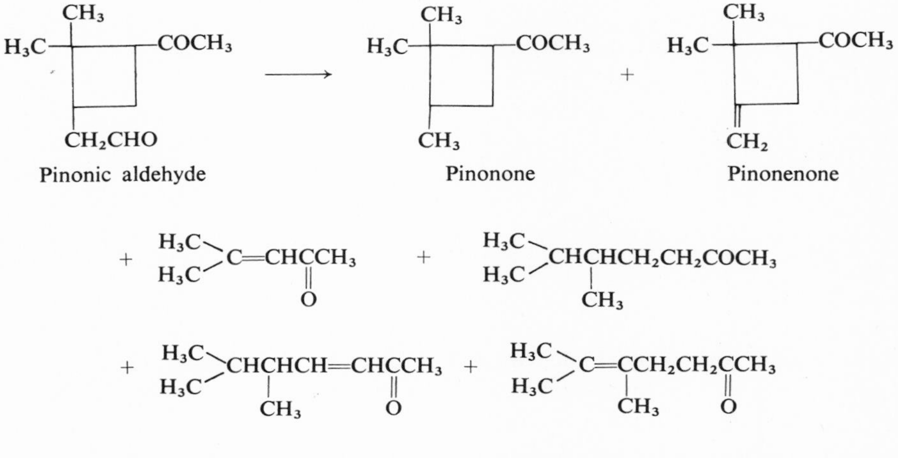

Pinonic aldehyde (cyclobutane bearing CH_3, H_3C, $COCH_3$, CH_2CHO) $\longrightarrow$ Pinonone (cyclobutane bearing CH_3, H_3C, $COCH_3$, CH_3) + Pinonenone (cyclobutane bearing CH_3, H_3C, $COCH_3$, $=CH_2$)

$$+ (H_3C)_2C{=}CHC(=O)CH_3 + (H_3C)_2CHCH(CH_3)CH_2CH_2COCH_3$$

$$+ (H_3C)_2CHCH(CH_3)CH{=}CHC(=O)CH_3 + (H_3C)_2C{=}C(CH_3)CH_2CH_2C(=O)CH_3$$

At lower temperatures, 200°–220°C, pinonic aldehyde affords a 60:40 mixture of pinonone and pinonenone (Faget *et al.*, 1964; Eschinazi, 1962).

Decarbonylation of aldehydes has proved useful in the construction of neopentyl systems. Newman and Gill (1966) decarbonylated 3,3-dimethyl-4-phenylbutanal over 10% palladium-on-carbon at 220°C and obtained neopentylbenzene in 90% yield.

$$C_6H_5CH_2C(CH_3)_2CH_2CHO \longrightarrow C_6H_5CH_2C(CH_3)_2CH_3$$

The same catalyst was applied by Wilt and Abegg (1968) to the decarbonylation of β-phenylisovaleraldehyde, a compound known to rearrange when subjected to radical-induced decarbonylation. The reaction proceeds smoothly over 10% palladium-on-carbon at 190°C to afford *t*-butylbenzene in over 80% yield accompanied by smaller amounts of rearranged and unsaturated products.

Aldehydes are decarbonylated stoichiometrically over certain complex noble metal catalysts even at room temperature; at elevated temperatures, the reaction can be made catalytic. Decarbonylation of saturated aldehydes with chlorotris(triphenylphosphine)rhodium affords the corresponding paraffin (Tsuji and Ohno, 1965), whereas over the ruthenium-diethylphenylphosphine complex, $[(Ru_2Cl_3(Et_2PhP)_6]^+Cl^-$, the product is a mixture of paraffin and olefin (Prince and Raspin, 1969). Aldehydes are decarbonylated over $RhCl(CO)(Ph_3)_2$ with a high degree of stereoselectivity and with overall retention of configuration. The authors (Walborsky and Allen, 1970) suggest that these results eliminate from consideration intervention of radical or carbonium ion intermediates and support the mechanism proposed by Tsjui and Ohno (1968a). Aldehydes deuterated at C-1 are decarbonylated with 100% of the deuterium found in the hydrocarbon formed. Since C-1-deuterated aldehydes are readily prepared (Walborsky *et al.*, 1970), this provides a good means of obtaining deuterated hydrocarbons (Walborsky and Allen, 1970). Under conditions of catalytic decarbonylation, the actual catalyst is $RhCl(Ph_3P)_2$ formed by rapid loss of carbon monoxide from $RhCl(CO)(Ph_3P)_2$ (Baird *et al.*, 1968).

Unsaturated Aldehydes

α,β-Unsaturated aldehydes undergo decarbonylation readily. The reaction has been used, for instance, as one step in a two-step synthesis of apopinene from α-pinene (Eschinazi and Pines, 1959).

$$\text{CH}_3 \xrightarrow{SeO_2} \text{CHO} \xrightarrow{Pd} \quad + \text{ CO}$$

α-Pinene Myrtenal Apopinene

Decarbonylation is carried out with 135 gm of myrtenal and 2.5 gm of 5% palladium hydroxide-on-barium sulfate in a 1 liter flask provided with a Dean-Stark distillation trap and reflux condenser. As the reaction proceeds, the reflux temperature falls from 195° to 155°C. Seventy grams of apopinene are obtained almost pure from the reaction (Eschinazi and Bergmann, 1959). The same type of decarbonylation procedure had been applied earlier to the formation of geraniolene from citral and of citronellene from citronellal (Eschinazi, 1952).

trans-Cinnamaldehydes undergo decarbonylation when heated with palladium catalysts to afford a mixture of *cis*- and *trans*-substituted styrenes. Presumably, the *cis* isomer is the sole initial product, but it is isomerized at reaction conditions to the *trans* isomer. The *cis* isomer is formed in highest yield if it is removed from the reaction mixture as it is formed. When the initial product is allowed to remain in the reaction until decarbonylation is complete, the *trans* isomer becomes the major product. On prolonged contact with the catalyst, saturated side-chain products also appear in yields up to 77%. These products are accompanied by materials of unknown structure of approximately twice the molecular weight, arising in the disproportionation (Hoffman *et al.*, 1962).

In competitive reaction, the relative rates of decarbonylation over palladium catalysts of cinnamaldehyde and *trans*-α-substituted cinnamaldehydes was observed to be cinnamaldehyde > α-phenyl- > α-methyl- > α-ethyl- > α-*n*-propyl- > α-isopropylcinnamaldehyde. The order of decreasing reactivity with α-aliphatic substituents fell as might be expected with increasing steric hindrance. Relative rates of decarbonylation were shown to be independent of the activity of the catalyst and independent of whether or not the metal was supported (Hoffman and Puthenpurackel, 1965).

Similar decarbonylations of *trans*-cinnamaldehydes have been carried out over chlorotris(triphenylphosphine)rhodium stoichiometrically and catalytically. The reaction proceeds stereoselectively and provides a useful method of making olefins of definite stereochemistry (Tsuji and Ohno, 1967b).

$$\mathrm{Ph(H)C{=}C(R)CHO} \xrightarrow{\text{Solvent}} \mathrm{Ph(H)C{=}C(R)H} + \mathrm{Ph(H)C{=}C(H)R}$$

	Solvent	%	%
$R = CH_3$	Xylene	91	9
$R = CH_3$	Benzonitrile	96	4
$R = C_2H_5$	Benzonitrile	94	6

Usually, decarbonylations proceed in benzene or methylene chloride, but certain hindered aldehydes do not react in these solvents. Higher-boiling solvents such as toluene or xylene may also be unsuitable due to the tendency of a dimeric rhodium complex to precipitate. This difficulty is overcome by using acetonitrile or benzonitrile, whose strong solvation power prevents dimer formation.

$$\text{(cyclohexenyl)}(CH_3)(CHO) \xrightarrow[\text{160°C, 1 min; benzonitrile}]{ClRh(Ph_3P)_3} \text{(cyclohexenyl)}CH_3 + ClRh(CO)(Ph_3P)_2 + Ph_3P$$

Catalytic amounts of rhodium complex can be used in these reactions at temperatures above 200°C. Rhodium trichloride also can be used as a decarbonylation catalyst (Tsuji and Ohno, 1967).

Aromatic Aldehydes

Decarbonylation of most aromatic aldehydes over palladium occurs readily. Hawthorne and Wilt (1960) examined the decarbonylation of various aromatic aldehydes and reported that usually quantitative yields of carbon monoxide were liberated in 30 to 60 minutes with a 1% loading of 5% palladium-on-carbon catalyst. There were some exceptions. 1-Naphthaldehyde readily loses carbon monoxide at 210°C to form naphthalene, whereas the 2-isomer does not decarbonylate even at 250°C. 2-Formyl-2′-biphenylcarboxylic acid is not decarbonylated at 245°C, perhaps because of conversion to 3-hydroxydiphenide. Phthalaldehydic acid, existing predominately in the cyclic form, 3-hydroxyphthalide, is dehydrated to 3,3′-oxphthalide (Hawthorne and Wilt, 1960). Decarbonylation of 1,4,5,8-tetraphenyl-2-naphthaldehyde is achieved by heating 1.1 gm of this material with 0.1 gm of 10% palladium-on-carbon for 5 hours at 270°–300°C. The tetraphenylnaphthalene is obtained in 58% yield (Bergmann *et al.*, 1964).

Homogeneous catalysts have also been employed with success in decarbonylation of aromatic aldehydes. On heating *p*-chlorobenzaldehyde or salicylaldehyde to 220°C with $RhCl(CO)(Ph_3)P_2$ at about 1% catalyst loading, Ohno and Tsuji (1968) obtained chlorobenzene in 71% yield and phenol in 80% yield.

Furfural undergoes decarbonylation readily and the reaction has been studied by a number of investigators because of the commercial importance of furan. One problem involved in the reaction is avoidance of catalyst fouling by relatively unstable furfural. Catalyst life in decarbonylation of furfural over 10% palladium-on-alumina is appreciably extended by the presence of alkali carbonates. Without added salts, 1 part of catalyst produces 5000 parts of furan whereas in the presence of sodium or potassium carbonate, the catalyst produces 10,000–30,000 parts of furan. The decarbonylation is carried out under vigorous reflux at 162°–230°C and is adjusted so that the partial pressure of furfural to furan in the vapors leaving the liquid phase is at least 5 to 1 (Copelin and Garnett, 1961). The reaction has also been carried out at 200° to 350°C under pressure in the presence of at least 0.3 mole of hydrogen per mole of furfural over a supported palladium catalyst showing an alkaline reaction (Manly and O'Halloran, 1965). Another procedure is to carry out the decarbonylation at 190°–225°C over palladium in the presence of 0.01–0.1 mole of calcium acetate per mole furfural (Dunlop and Huffman, 1966).

Other alkaline materials have also been used. A mixture of 1.8 gm of potassium, 400 ml of dibutyl phthalate, and 2gm of 5% palladium-on-carbon heated for 150°C for 20 minutes and at 200°C for 77 hours with continuous addition of 2950 gm of furfural affords 1960 gm of furan of >99.5% purity (Japanese Patent 1531/66). The alkaline ingredient may be alternatively an alkali hydroxide or alkali amide (Japanese Patent 1532/66) or potassium phthalimide or sodium butyramide (Japanese Patent 1533/66).

An unusual decarbonylation and alkylation occurs when an aromatic aldehyde and an alcohol are passed over palladium-on-titanium dioxide at 150° to 275°C, affording an alkylbenzene. Yields of the alkylated product are moderate (Dubeck and Jolly, 1970).

$$C_6H_5CHO + CH_3CH_2CH_2OH \longrightarrow C_6H_5CH_2CH_2CH_3 + CO + H_2O$$

Acid Halides

Decarbonylation of acid halides usually proceeds smoothly over either heterogeneous or homogeneous catalysts, affording olefins where it is possible and halo compounds where it is not. Benzoyl chloride, passed in a stream of dry nitrogen over 5% palladium-on-carbon at 365°C and a weight hourly space velocity of 0.38, affords chlorobenzene in 70% yield. Similarly, isophthaloyl chloride affords *m*-dichlorbenzene in 68% yield (Bain and McCall, 1965).

$$m\text{-}C_6H_4(COCl)_2 \longrightarrow m\text{-}C_6H_4Cl_2$$

Palladium chloride is also an efficient catalyst for the decarbonylation of acid halides; it is reduced rapidly to black metallic palladium soon after heating of the mixture begins. Five grams of $CH_3(CH_2)_8COCl$, heated 90 minutes with 100 mg of palladium chloride at 200°C, affords a mixture of nonenes in 90% yield. The diacid chloride of adipic acid similarly treated affords cyclopentanone in 30% yield. The acid bromide, $CH_3(CH_2)_6COBr$, affords a mixture of heptenes in 80% yield. The authors (Tsuji *et al.*, 1965) suggested that the following reactions occur between palladium, carbon monoxide, olefins, and acid chlorides. The scheme is interesting in that it relates carbonylation, decarbonylation, and the well-known Rosenmund reduction of acid chlorides.

$$\begin{matrix} Pd + HCl \\ RCH{=}CH_2 \end{matrix} \rightleftharpoons RCH_2CH_2PdCl \underset{-CO}{\overset{CO}{\rightleftharpoons}} \underset{(\mathbf{I})}{RCH_2CH_2COPdCl} \rightleftharpoons RCH_2CH_2COCl + Pd$$

$$RCH_2CH_2COPdCl \xrightarrow{ROH} RCH_2CH_2COOR$$

$$RCH_2CH_2COPdCl \xrightarrow{H_2} RCH_2CH_2CHO + Pd$$

$$\begin{matrix} RCH{=}CH_2 + H_2 + Pd \\ (RCH_2CH_3) \end{matrix} \underset{-CO}{\overset{CO}{\rightleftharpoons}} \underset{(\mathbf{II})}{RCH_2CH_2COPdH} \rightleftharpoons RCH_2CH_2CHO + Pd$$

Essential points in the scheme are the formation of the acylpalladium complexes **I** and **II**. The authors suggest that both aldehydes and acyl chlorides may form acyl–palladium bonds when contacted with palladium, and cite the observation that metallic palladium partially dissolves in heated acyl chlorides.

Decarbonylation and dehalogenation of aliphatic acid chlorides is limited in its synthetic utility because the initially formed terminal olefin is isomerized rapidly, affording a mixture of olefins. This difficulty has been circumvented by employing $[Rh(CO)_2Cl]_2$, $RhCl(Ph_3P)_3$, or $RhCl_3(Ph_3As)_3$ as catalysts in the presence of triphenylphosphine. With these systems, only terminal olefins result as isomerization is effectively inhibited. The phosphine inhibits the carbonylation reaction only slightly and even acts as a promoter in decarbonylation by $RhCl_3$. Triphenylarsine, -stibine, or -bismuthine are ineffective as isomerization inhibitors. A mixture of 61 gm of *n*-undecanoyl chloride, 40 gm of triphenylphosphine, and 0.6 gm of $[Rh(CO)_2Cl]_2$ heated under nitrogen for 30 minutes affords 1-decene of 97% purity in 76% yield. The flask is arranged so as to permit distillation of the olefin as it is formed (Blum *et al.*, 1971).

$$CH_3(CH_2)_9COCl \rightarrow CH_3(CH_2)_7CH{=}CH_2 + CO + HCl$$

Decarbonylation of certain aldehydes may occur under the conditions of the Rosenmund reduction. An attempted Rosenmund reaction of diphenylacetyl chloride gave a mixture of products, but none of the expected diphenylacetaldehyde.

$$Ph_2CHCOCl \xrightarrow[\substack{PhCH_3 \\ H_2}]{Pd/BaSO_4} \underset{47\%}{Ph_2CHCHPh_2} + \underset{13.5\%}{Ph_2CH_2} + \underset{13.5\%}{Ph_2CHCOOH}$$

Tetraphenylethane probably arises through diphenylchloromethane, which is known to decompose on heating to a mixture of tetraphenylethylene and tetraphenylethane (Burr, 1951).

Decarbonylation of aroyl chlorides has been catalyzed by various complex rhodium and ruthenium catalysts. Chlorotris(triphenylphosphine)rhodium, a compound which decarbonylates aroyl chlorides stoichiometrically at low temperatures, becomes an efficient catalyst at 200°C. The catalytic action arises through regeneration of the reagent by loss of carbon monoxide at these temperatures. Under carefully controlled conditions, the intermediate acylrhodium complex may be obtained (Tsuji and Ohno, 1966, 1968b). Other workers were unable to obtain any acylrhodium derivatives from simple aroyl halides and suggested instead the complexes are metal carbonyls (Blum *et al.*, 1967). The metal carbonyl may have arisen from the acylrhodium complex by rearrangement during attempts to isolate it (Tsuji and Ohno, 1967; Baird *et al.*, 1967).

In a typical synthesis, 5 gm of 1-naphthoyl chloride is heated with 50–100 mg of chlorotris(triphenylphosphine)rhodium for 5 minutes near the boiling point. Distillation affords 1-chloronaphthalene in 96% yield (Blum, 1966). Similar experiments were reported by Ohno and Tsuji (1968).

Aryl fluorides have been prepared by decarbonylation of aroyl fluorides with chlorotris(triphenylphosphine)rhodium. These aroyl fluorides did not undergo decarbonylation over palladium chloride or palladium-on-carbon. Chlorotris(triphenylphosphine)rhodium is catalytic only to the extent that 280–580% mole equivalent of fluoroaromatic can be formed before the activity of the catalyst is lost. The authors attributed fast deactivation of the catalyst to a halogen exchange reaction producing fluorotris(triphenylphosphine)-rhodium, which itself is inactive in decarbonylation reactions (Olah and Kreienbuhl, 1967).

$$ClRh(Ph_3P)_3 + ArCOF \rightarrow FRh(Ph_3P)_3 + ArCOCl$$

Aroyl iodides are decarbonylated readily over chlorotris(triphenylphosphine)rhodium to afford the corresponding aryliodides (Blum *et al.*, 1968). Decarbonylation of aroyl iodides over this catalyst begins at temperatures as low as 35°C.

Aroyl Cyanides

Aroyl cyanides undergo decarbonylation with chlorotris(triphenylphosphine)rhodium, affording the nitriles of aromatic acids.

The reaction is expected in view of the "pseudo-halogen" character of the cyano group (Blum *et al.*, 1967).

$$4\text{-}ClC_6H_4COCN \xrightarrow{\text{3.5 hours}} 4\text{-}ClC_6H_4CN + CO \quad 95\%$$

$$\alpha\text{-}C_{10}H_7COCN \xrightarrow{\text{1 hour}} \alpha\text{-}C_{10}H_7CN + CO \quad 87\%$$

KETONES

Ketonic carbonyls attached to quaternary carbon atoms are cleaved with loss of carbon monoxide when heated with platinum metal catalysts. For example, treatment of **III** with palladium at 180°C eliminated both carbon monoxide and hydrogen, affording **IV** (Grummitt and Becker, 1948).

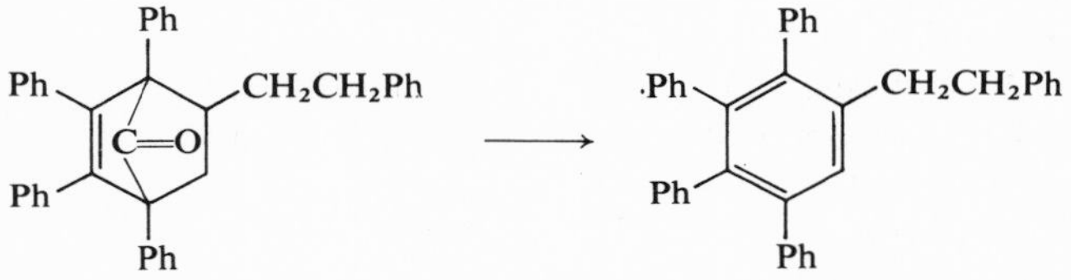

Ph Ph C=O CH_2CH_2Ph Ph Ph → Ph Ph CH_2CH_2Ph Ph Ph

(III) **(IV)**

Decarbonylation of ketones in the steroid series often accompanies dehydrogenation. (Dreiding and Pummer, 1953; Dreiding and Voltman, 1954; Gentles *et al.*, 1958; Moss *et al.*, 1958; Dreiding and Tomascewski, 1958). Decarbonylation occurs at temperatures above 300°C; at temperatures around 250°C, aromatization without cleavage occurs (Bachmann and Dreiding, 1950).

ANHYDRIDES

Decarbonylation of anhydrides has proved a useful degradative procedure (Bailey and Economy, 1955). Treatment of the cyclic anhydride, **V**, with a mixture of 10% palladium-on-carbon and copper chromite at 275°C for 3.5

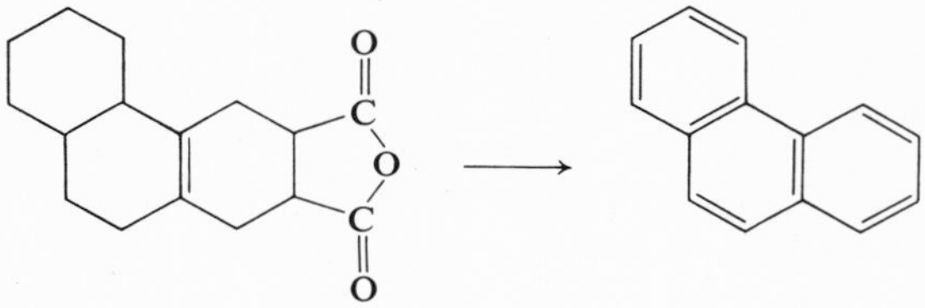

O C O C O →

(V) **(VI)**

hours affords phenanthrene **VI** in 67% yield through dehydrogenation, decarboxylation, and decarbonylation (Bailey and Quigley, 1959).

An unusual reaction of anhydrides is the chlorotris(triphenylphosphine)-rhodium-catalyzed conversion of benzoic anhydrides to fluorenones. The reaction takes place in such a way that one carbon of the new bridge originally held a carbonyl group. Thus, the product from *p*-toluic anhydride is 2,6-dimethylfluorenone. Yields, in accordance with the following equation, range from 14 to 91% (Blum and Lipshes, 1969). This reaction seems to be effectively catalyzed only by certain rhodium complexes. Attempts to use iridium, palladium, and ruthenium, catalysts gave only very low yields of fluorenones. The mechanism of this reaction has been discussed in some detail (Blum *et al.*, 1970).

H_3C CH_3 O O 3.5 gm

0.18 gm $ClRh(Ph_3P)_3$ 240°C 3.5 hours

CH_3 CH_3 O 72% + H_3C—COOH + CO

ALCOHOLS

Primary alcohols may undergo dehydrogenation and decarbonylation on heating with a catalyst, the reaction probably proceeding stepwise through the aldehyde (Newman and Mangham, 1949). Phenylethyl alcohol (122 gm) refluxed 3 hours over 100 mg of palladium-on-barium sulfate affords 40 ml of toluene and 65 gm of unchanged substrate. Decarbonylation of carbinols is a frequent concomitant of dehydrogenation of incipient aromatics. The reaction competes with hydrogenolysis of the hydroxyl function so that a mixture of products usually results. The ratio of decarbonylation to hydrogenolysis depends in large part on the position of the hydroxyl function. Newman and Zahm (1943) noted that very efficient use was made of the hydrogen in the hydrogenolysis reaction and they viewed the process as a sort of internal oxidation–reduction. The reaction is further discussed in the chapter on dehydrogenation.

Ethers

Cyclic ethers may undergo decarbonylation by being converted to aldehydes through carbon–oxygen bond cleavage. Isomerization of 2-alkyltetrahydrofurans over platinum-on-carbon catalysts at 230°–250°C affords ketones formed by cleavage of the ring at the carbon–oxygen bond farthest removed from the alkyl side chain. Small amounts of alkanes are also produced, presumably through cleavage of the carbon–oxygen bond adjacent to the alkyl substituents, followed by decarbonylation of the resulting aldehydes. In tetrahydrofuran itself, only an aldehyde is formed on ring opening, and decarbonylation to propane is a major reaction. Small amounts of *n*-butanol are formed as well (Shuikin and Belskii, 1958a, b).

Esters and Acids

Esters and acids may be decarbonylated in a reaction that is the reverse of a carbon monoxide insertion.

$$RCH_2CH_2COOH \rightarrow RCH{=}CH_2 + H_2O + CO$$

$$RCH_2CH_2COOR \rightarrow RCH{=}CH_2 + ROH + CO$$

Noble metals in general are claimed as effective catalysts, but all examples cited involve some form of palladium. For example, 75 ml of isobutyldecanoate, 1 gm of palladium chloride bis(triphenylphosphine), 3 gm of triphenylphosphine, and 2 gm of lithium chloride in a 250 ml flask equipped with a Dean-Stark trap is heated to 240°C for 2 hours. The product is 43% 1-nonene, 24% *trans*-2-nonene, 5% *cis*-2-nonene, and 14% isobutanol. The same reaction takes place without lithium chloride, but much more slowly (Fenton, 1970).

β-Ketoamides

β-Ketoamides undergo both loss of carbon monoxide and ammonia when heated with a palladium catalyst. 2-Carbamoyl-4,5-dihydro-1-acenaphthenone refluxed with 30% palladium-on-carbon in xylene affords acenaphthenone in 65% yield.

O CONH$_2$ $\xrightarrow{30\%\ Pd/C}$ O + CO + NH$_3$

Benzoylacetamide is converted under similar conditions to acetophenone (30%) and an amide having the empirical formula $C_{19}H_{16}O_2N_2$. The authors concluded that this reaction was not general to amides, but was limited to β-ketoamides (Campaigne and Bulbenko, 1961).

Formate Esters

Formate esters are smoothly transformed at 200°C over 10% palladium-on-carbon to products that depend on the starting material. Alkyl formate esters afford the corresponding alcohol in high yield through decarbonylation.

$$C_7H_{15}CH_2OCHO \rightarrow C_7H_{15}CH_2OH + CO$$

whereas benzyl formate affords mainly toluene by decarboxylation (Matthews *et al.*, 1970).

$$C_6H_5\text{—}CH_2OCHO \longrightarrow C_6H_5\text{—}CH_3 + CO_2$$

DESULFONYLATION

Arylsulfonyl halides undergo desulfonylation in the presence of either homogeneous or heterogeneous noble metal catalysts to afford aromatic halides in a reaction similar to decarbonylation of aroyl halides. Ten grams of freshly distilled p-chlorobenzenesulfonyl chloride and 0.1 gm of $ClRh(Ph_3P)_3$ heated in a Claisen flask equipped with a Vigreux column affords p-dichlorobenzene in 85% yield after 25 minutes. Desulfonylations may be conducted so that the product if sufficiently volatile, distills as formed from the reaction flask (Blum and Scharf, 1970). Palladium-on-carbon has also been used successfully at 330°C in this type reaction. Chlorobenzene is obtained in 70% yield and fluorobenzene in 31% from their respective sulfonyl halides (McCall and Cummings, 1966).

$$C_6H_5\text{—}SO_2Cl \longrightarrow C_6H_5\text{—}Cl + SO_2$$

Aromatic sulfinic acids are converted to biaryls when heated with palladium salts. The reaction, which produces sulfur dioxide and palladium, can be made catalytic with respect to palladium by the use of a reoxidizing agent such as cupric chloride.

$$2\ ArSO_2Na + Pd^{2+} \rightarrow ArAr + 2\ SO_2 + Pd + 2\ Na^+$$

Arylpalladium complexes are presumably intermediates in the formation of biaryls, a conjecture supported by isolation of the expected insertion products from arylpalladium complexes and carbon monoxide or olefins (Garves, 1970).

$$CH_3-C_6H_4-SO_2Na + CO \xrightarrow[CH_3OH,\ 60°C]{Na_2PdCl_4} CH_3-C_6H_4-C(=O)-OCH_3 \quad 29\%$$

$$CH_3-C_6H_4-SO_2Na + CH_2{=}CPh_2 \xrightarrow[dioxane-H_2O]{Li_2PdCl_4} CH_3-C_6H_4-CH{=}CPh_2 \quad 33\%$$

Nuclear fluorinated aromatic compounds may be obtained by catalytic conversion of arylfluoroformates and arylfluorothioformates. The best results are obtained by passage of the formates over platinum gauze at about 700°C with a contact time of a few seconds. Nitrogen may be used as a carrier gas. Under these conditions, fluorobenzene is obtained in 70 and 90% yields from phenylfluorothioformate and phenylfluoroformate, respectively (Christie and Pavlath, 1966).

$$PhSCOF \rightarrow PhF + CO_2 + CS_2$$

REFERENCES

Bachmann, W. E., and Dreiding, A. S. (1950). *J. Amer. Chem. Soc.* **72**, 1323.
Bailey, W. J., and Economy, J. (1955). *J. Amer. Chem. Soc.* **77**, 1133.
Bailey, W. J., and Quigley, S. T. (1959). *J. Amer. Chem. Soc.* **81**, 5598.
Bain, P. J., and McCall, E. B. (1965). U.S. Patent 3,221,069.
Baird, M. C., Mague, J. T., Osborn, J. A., and Wilkinson, G. (1967). *J. Chem. Soc., A* p. 1347.
Baird, M. C., Nyman, C. J., and Wilkinson, G. (1968). *J. Chem. Soc., A* p. 348.
Bergmann, E. D., Blumberg, S., Bracha, P., and Epstein, S. (1964). *Tetrahedron* **20**, 195.
Blum, J. (1966). *Tetrahedon Let.* p. 1605.
Blum, J., and Lipshes, Z. (1969). *J. Org. Chem.* **34**, 3076.
Blum, J., and Scharf, G. (1970). *J. Org. Chem.* **35**, 1895.
Blum, J., Oppenheimer, E., and Bergmann, E. D. (1967). *J. Amer. Chem. Soc.* **89**, 2338.
Blum, J., Rosenman, H., and Bergmann, E. D. (1968). *J. Org. Chem.* **33**, 1928.
Blum, J., Milstein, D., and Sasson, Y. (1970). *J. Org. Chem.* **35**, 3233.
Blum, J., Kraus, S., and Pickholtz, Y. (1971). *J. Organometal. Chem.* **33**, 227.
Burr, J. G., Jr. (1951). *J. Amer. Chem. Soc.* **73**, 3502.
Campaigne, E., and Bulbenko, G. F. (1961). *J. Org. Chem.* **26**, 4702.
Christe, K. O., and Pavlath, A. E. (1966). U.S. Patent 3,283,018.

Conia, J. M., and Faget, C. (1964). *Bull. Soc. Chim. Fr.* [5] p. 1963.
Copelin, H. B., and Garnett, D. I. (1961). U.S. Patent 3,007,941.
Dawson, D. J., and Ireland, R. E. (1968). *Tetrahedron Lett.* p. 1899.
Dreiding, A. S., and Pummer, W. J. (1953). *J. Amer. Chem. Soc.* **75**, 3162.
Dreiding, A. S., and Tomascewski, A. J. (1958). *J. Amer. Chem. Soc.* **80**, 3702.
Dreiding, A. S., and Voltman, A. (1954). *J. Amer. Chem. Soc.* **76**, 537.
Dubeck, M., and Jolly, J. G. (1970). U.S. Patent 3,501,541.
Dunlop, A. P., and Huffman, G. W. (1966). U.S. Patent 3,257,417.
Emery, A., Oehlschlager, A. C., and Unrau, A. M. (1970). *Tetrahedron Lett.* p. 4401.
Eschinazi, H. E. (1952). *Bull. Soc. Chim. Fr.* [5] p. 967.
Eschinazi, H. E. (1962). U.S. Patent 3,019,263.
Eschinazi, H. E., and Bergmann, E. D. (1959). *J. Amer. Chem. Soc.* **24**, 1369.
Eschinazi, H. E., and Pines, H. (1959). *J. Org. Chem.* **24**, 1369.
Faget, C., Conia, J. M., and Eschinazi, H. E. (1964). *C. R. Acad. Sci.* **258**, 600.
Fenton, D. M. (1970). U.S. Patent 3,530,198.
Garves, K. (1970). *J. Org. Chem.* **35**, 3273.
Gentles, M. J., Moss, J. B., Herzog, H. L., and Hershberg, E. B. (1958). *J. Amer. Chem. Soc.* **80**, 3702.
Grummitt, P., and Becker, E. I. (1948). *J. Amer. Chem. Soc.* **70**, 149.
Hawthorne, J. O., and Wilt, M. H. (1960). *J. Org. Chem.* **25**, 2215.
Hemidy, J. F., and Gault, F. G. (1965). *Bull. Soc. Chim. Fr.* [5] p. 1710.
Hoffman, N. E., and Puthenpurackel, T. (1965). *J. Org. Chem.* **30**, 420.
Hoffman, N. E., Kanakkanatt, A. T., and Schneider, R. F. (1962). *J. Org. Chem.* **27**, 2687.
McCall, E. B., and Cummings, W. (1966). U.S. Patent 3,256,343.
Manly, D. G., and O'Halloran, J. P. (1965). U.S. Patent 3,223,714.
Matthews, J. S., Ketter, D. C., and Hall, R. F. (1970). *J. Org. Chem.* **35**, 1694.
Moss, J. B., Herzog, H. L., and Hershberg, E. B. (1958). *J. Amer. Chem. Soc.* **80**, 3702.
Muller, E., Segnitz, A., and Langer, E. (1969). *Tetrahedron Lett.* p. 1129.
Newman, M. S., and Mangham, J. R. (1949). *J. Amer. Chem. Soc.* **71**, 3342.
Newman, M. S., and Gill, N. (1966). *J. Org. Chem.* **31**, 3860.
Newman, M. S., and Zahm, H. V. (1943). *J. Amer. Chem. Soc.* **65**, 1097.
Ohno, K., and Tsuji, J. (1968). *J. Amer. Chem. Soc.* **90**, 99.
Olah, G. A., and Kreienbuhl, P. (1967). *J. Org. Chem.* **32**, 1614.
Prince, R. H., and Raspin, K. A. (1969). *J. Chem. Soc., A* p. 612.
Rylander, P. N., and Hasbrouck, L. (1967). Unpublished work from Engelhard Industries Research Laboratory, Menlo Park, N. J.
Sakai, K., Ide, J., Oda, O., and Nakamura, N. (1972). *Tetrahedron Lett.* p. 1287.
Shimizu, Y., Mitsuhashi, H., and Caspi, E. (1966). *Tetrahedron Lett.* p. 4113.
Shuikin, N. I., and Belskii, I. F. (1958a). *Bull. Soc. Chim. Fr.* [5] p. 786.
Shuikin, N. I., and Belskii, I. F. (1958b). *Dokl. Akad. Nauk SSSR* **120**, 548.
Tsuji, J., and Ohno, K. (1965). *Tetrahedron Lett.* p. 3969.
Tsuji, J., and Ohno, K. (1966). *J. Amer. Chem. Soc.* **88**, 3452.
Tsuji, J., and Ohno, K. (1967). *Tetrahedron Lett.* p. 2173.
Tsuji, J., and Ohno, K. (1968a). *J. Amer. Chem. Soc.* **90**, 99.
Tsuji, J., and Ohno, K. (1968b). *Advan. Chem. Ser.* **70**, 155.
Tsuji, J., Ohno, K., and Kajimoto, T. (1965). *Tetrahedron Lett.* p. 4565.
Walborsky, H. M., and Allen, L. E. (1970). *Tetrahedron Lett.* p. 823.
Walborsky, H. M., Morrison, W. H., and Niznik, G. E. (1970). *J. Amer. Chem. Soc.* **92**, 6675.
Wilt, J. W., and Abegg, V. P. (1968). *J. Org. Chem.* **33**, 923.

CHAPTER 9

Silicon Chemistry

Noble metal catalysts, especially platinum and palladium, have figured prominently in the development of silicon chemistry. Reactions catalyzed by platinum group metals include addition of silanes to olefins and acetylenes, various dehydrogenations, and cleavage of carbon-silicon bonds. The mechanism of these reactions has received considerable study and many details have been elucidated (Benkeser and Hickner, 1958; Benkeser *et al.*, 1961, 1968; Chalk and Harrod, 1965; Chalk, 1970a; Citron *et al.*, 1969; Gornowicz *et al.*, 1968; Ryan and Speier, 1964; Saam and Speier, 1961; Selin and West, 1962a, b; Sommer *et al.*, 1967a, b; Sommer and Lyons, 1968; Spialter and O'Brien, 1967). The focus here is mainly on synthetic aspects of silicon chemistry involving noble metal catalysts.

Catalysts for Hydrosilylation of Olefins

Some form of platinum is the most commonly used catalyst for addition of silanes to olefins. Catalysts have included chloroplatinic acid, platinum black, and platinum-on-asbestos, -silica (Wagner, 1953a) or -carbon (Wagner, 1953b; Weyenberg and Nelson, 1965). In a comparison of the effectiveness of various catalysts for addition of methyldichlorosilane to 1-pentene, 2-pentene, and cyclohexene, chloroplatinic acid proved outstandingly effective, whereas platinum-on-carbon was much less active. Potassium chloroplatinate and platinum black were relatively ineffective, but much more active than osmium tetroxide, palladium, iridium, or ruthenium chlorides, or palladium-on-carbon (Speier *et al.*, 1957).

The nature of the active catalytic species in chloroplatinic acid hydrosilylations has been a matter of debate. Chloroplatinic acid has been assumed to be reduced to platinum because both it and platinum-on-carbon give stereoselectively *cis* addition of silanes to acetylenes resulting in *trans* products, because black particles form in the solution, and because catalysis by chloroplatinic acid shows an induction period (Benkeser *et al.*, 1961).

In one attempt to elucidate the nature of the active catalyst in silylation, a solution of ethylcyclohexane, trichlorosilane, and chloroplatinic acid in 2-propanol was refluxed 96 hours. Within the first hour an orange, platinum-containing precipitate formed on the flask and the supernatant liquid became clear. The precipitate after washing with acetone and ether was shown to catalyze addition of trichlorosilane to allylcyclohexane, but it did not catalyze addition of the silane to 1-ethylcyclohexene. Evidently, the platinum precipitate was devoid of isomerizing activity. The clear filtrate did catalyze addition of trichlorosilane to 1-ethylcyclohexene and the authors suggested that the active catalyst is a soluble platinum complex, although they could not rule out the possibility that some chloroplatinic acid remained dissolved in the filtrate. The platinum complexes, dichlorobis(ethylcyclohexene)-μ,μ'-dichlorodiplatinum(II) and dichlorobis(ethylene)-μ,μ'-dichlorodiplatinum(II) were catalysts for the addition of trichlorosilane to both allylcyclohexane and 1-ethylcyclohexene, whereas platinum-on-carbon, like the platinum-containing orange precipitate, would catalyze only addition to allylcyclohexane (Benkeser *et al.*, 1968).

In certain cases, platinum-on-carbon and chloroplatinic acid give markedly different products. Interaction of bicycloheptadiene and methyldichlorosilane affords **I**, **II**, and **III** in yields of 30, 6, and 64%, respectively, when catalyzed by chloroplatinic acid and in 94, 0, and 6% yields when catalyzed by platinum-on-carbon (Kuivila and Warner, 1964). Chloroplatinic acid-catalyzed addition

+ $HSiCl_2CH_3$ ⟶

$SiCl_2CH_3$ H **(I)** + $SiCl_2CH_3$ H **(II)** + H $SiCl_2CH_3$ **(III)** of trichlorosilane, methyldichlorosilane, and trimethylsilane to benzonorbornadiene affords predominantly *exo* isomers (Martin and Koster, 1968).

Wagner and Whitehead (1958) examined the effect of support on the activity and stability of platinum catalysts and concluded that γ-alumina was appreciably superior to α-alumina, carbon, silica, and several other supports. Some disproportionation of silicon–hydrogen and silicon–chlorine bonds occurred in the presence of platinum-on-carbon whereas no disproportionation was detected with a γ-alumina support.

Complex Catalysts

Olefin complexes of both platinum(II) and rhodium(I) are effective catalysts for addition of silanes to olefinic compounds, giving about the same results as chloroplatinic acid in terms of rate, yield, and product. Complexes of platinum(II) and rhodium(I) with chelating diolefins, such as 1,5-cyclooctadiene, are less active than complexes of monoolefins (Chalk and Harrod, 1965). The platinum chloride complex $(PtCl_2 \cdot C_3H_6)_2$ is said to be less subject to poison than other platinum catalysts (Ashby, 1964). Another complex catalyst said to be superior to chloroplatinic acid is obtained by heating an alcohol (Brown, 1968; Murphy, 1969), aldehyde, or ether with chloroplatinic acid (Lamoreaux, 1965); another is prepared from a silane, an olefin, and trimethylplatinum iodide, or hexamethyldiplatinum (Lamoreaux, 1967); and another from a platinum salt and a nitrile (Joy, 1968). A catalyst said to be more resistant to poisoning is prepared by interaction of a platinum chloride–olefin complex with a cyclic alkylvinylpolysiloxane (Modic, 1970b). Catalysts prepared by interaction of chloroplatinic acid with certain unsaturated siloxanes are said to minimize loss of platinum (Willing, 1968). Interaction of a silane or siloxane with chloroplatinic acid provides a catalyst of high and constant activity (Fish, 1971). Polymer-supported rhodium complexes have been used in an attempt to combine the advantages of homogeneous and heterogeneous catalysis (Čapka *et al.*, 1971).

Palladium–olefin complexes are not effective catalysts; palladium may be reduced readily to the metal by the silane. However, palladium stabilized by phosphine ligands does make suitable catalysts. Zero-valent palladium–phosphine complexes catalyze hydrosilylation of butadiene (Takahashi *et al.*, 1969) and monoolefins (Hara *et al.*, 1971). In the presence of triphenylphosphine bivalent palladium and even metallic palladium, prepared by reduction of palladium chloride with formic acid, are effective catalysts for hydrosilylation of olefins (Hara *et al.*, 1971).

Excellent yields of hydrosilylation products are obtained by a variety of rhodium catalysts carrying triphenylphosphine and/or carbon monoxide (Haszeldine *et al.*, 1967; Chalk, 1970b). The mechanism of hydrosilylation by complex catalysts has been discussed in detail (deCharentenay *et al.*, 1968).

Addition of Silanes to Olefins

Silicon hydrides add smoothly to carbon–carbon double bonds in the presence of appropriate noble metal catalysts with the liberation of 38 kcal/mole (Speier *et al.*, 1957). The reaction may be accompanied by double-bond migration that occurs readily in the presence of both catalyst and silane, but not with one alone. This isomerization allows an apparently exclusive formation of terminal alkylsilanes from nonterminal olefins such as 2-pentene, 3-heptene (Saam and Speier, 1958), and methylcyclohexenes. Some interesting studies have been made concerning the mechanism of isomerization. Chalk and Harrod (1965) showed that the relative rates of hydrosilylation and isomerization depend on the silane. Using 1-hexene as a typical α-olefin, they distinguished three types of reaction. In the first, hydrosilylation proceeds rapidly to completion without observable isomerization; this course occurs with trimethoxy- and triethyoxysilane. Very high yields of terminal hydrosilylation products are obtained. In the second type, hydrosilylation proceeds less rapidly to completion with concurrent extensive isomerization of excess olefin. This behavior most commonly occurs with trichloro-, ethyldichloro-, and phenyldichlorosilane. In the third type, hydrosilylation and isomerization proceed rapidly at first, but the rate of both processes soon falls to zero; this type of behavior occurs with triethyl-, tribenzyl-, and triphenylsilane and is due to thermal deactivation of the catalyst. High yields are obtained with these substrates if the reaction is allowed to proceed slowly at room temperature.

Addition of silanes to olefins is catalyzed by the rhodium complex, $(Ph_3P)_3RhCl$, with the rate of reaction being inversely related to the stability of the adduct $(Ph_3P)_2RhH(SiR_3)Cl$. For example, at 60°C, triphenylsilane and 1-hexene afford a 100% yield of *n*-hexyltriphenylsilane, whereas under comparable conditions, triethylsilane and trichlorosilane afford only 60% and 8%, respectively, of the corresponding *n*-hexylsilanes. Extensive isomerization of the double bond inward occurs during the reaction (Haszeldine *et al.*, 1969).

Phenylalkenes

Addition of silanes to phenylalkenes gives, as expected, terminal adducts; but the products also contain compounds with silicon attached to the α-carbon. These results contrast with those of Russian workers, who reported attack at the terminal carbon and at the penultimate carbon (Petrov *et al.*, 1960). Terminal adducts always predominate regardless of the silane, and the proportion of terminal adducts increases in the series $Cl_3SiH < Cl_2MeSiH < ClMe_2SiH$. The ratio of adducts is also affected by substituents on the aromatic ring and by the solvent (Musolf and Speier, 1964).

The effect of solvent may be illustrated by the addition of methyldichlorosilane to styrene catalyzed by 2% platinum-on-alumina. In tetrahydrofuran solvent, exclusively terminal addition occurs, whereas in butyl ether, the product contains 23% α-isomer and 77% β-isomer; in ethylene glycol, the distribution is 12% α-, 88% β-isomer (Pike and Borchert, 1960).

Asymmetric Hydrosilylation

Asymmetric hydrosilylation of olefins has been achieved by the use of platinum(II) complexes of chiral phosphines. Addition of 30 mmole of methydichlorosilane to 30 mmole of α-methylstyrene in the presence of 2×10^{-2} mmole *cis*-dichloro(ethylene)[(R)-benzylmethylphenyl-phosphine]-platinum(II) affords 2-phenylpropylmethyldichlorosilane in 43% yield with 5% enantiomeric excess of the R isomer. The optical yield is decreased sharply when dichlorobis[(R)-methylphenyl-*n*-propylphosphine]di-μ-chlorodiplatinum(II) is used as a catalyst. It appears advantageous to have one asymmetric center as close to the metal as possible to achieve maximum asymmetry in the product (Yamamoto *et al.*, 1971).

ADDITION OF SILANES TO DIOLEFINS

Addition of trimethylsilane to 1,5- or 1,3-cyclooctadiene catalyzed by chloroplatinic acid, *trans*-dichloro(ethylene)(pyridine)platinum(II), or 5% platinum-on-carbon gives almost exclusively 3-(trimethylsilyl)cyclooctene (Yamamoto and Kumada, 1968), in contrast to the 5-(triethylsilyl)cyclooctene reported by earlier workers in platinum-on-alumina-catalyzed addition of 1,5-cyclooctadiene and triethylsilane (Pike and McDonagh, 1963). Diolefins isolated from incomplete addition contained a mixture of 1,3- and 1,5-cyclooctadiene in 95-to-5 ratio. Yamamoto and Kumada (1968) suggested the 3-isomer is derived from 1,5-cyclooctadiene by rapid isomerization to 1,3-cyclooctadiene followed by 1,4-addition of silane.

$HSi(CH_3)_3$ → $Si(CH_3)_3$ ← $HSi(CH_3)_3$

Monosilyl and bissilyl compounds have been obtained by chloroplatinic acid-catalyzed addition of chlorosilanes to *cis,trans,trans*-1,5,9-cyclododecatriene. In monosubstitution, there is a strong preference for addition to the *trans* double bond, but the second molecule of silane adds nonselectively (Takahashi *et al.*, 1963).

Hydrosilylation of butadiene with trimethyl- or triethylsilane catalyzed by palladium complexes follows a novel course and produces silicon derivatives of butadiene dimer. Bis(triphenylphosphine)(maleic anhydride)palladium and bis(triphenylphosphine)(*p*-benzoquinone)palladium are effective catalysts, whereas tetrakis(triphenylphosphine)palladium and platinum are much less effective. Trichlorosilane and dimethylphenylsilane afford only 1:1 butadiene to silane adducts. Dimerization of butadiene is believed to proceed through a palladium hydride intermediate (Takahashi *et al.*, 1969).

$$2\,CH_2{=}CH{-}CH{=}CH_2 + R_3SiH \rightarrow R_3SiCH_2CH{=}CHCH_2CH_2CH{=}CHCH_3$$

where $R = CH_3, C_2H_5$

Addition of Silanes to Substituted Olefins

Silanes have been added to a variety of substituted olefins with the substituent having at times an influence on the course of reaction.

α,β-Unsaturated Esters

Addition of a silane to a terminally unsaturated organic compound generally proceeds with attachment of the silicon atom to the terminal carbon, but platinum-on-carbon-catalyzed addition of methyldichlorosilane to methyl acrylate occurs in the reverse sense, affording methyl α-(methyldichlorosilyl)-propionate (Goodman *et al.*, 1957). An α-methyl substituent in the α,β-unsaturated ester impedes this reverse addition and favors normal terminal addition (Curry and Harrison, 1958; Sommer *et al.*, 1957).

Considerable ester polymerization accompanies interaction of methyl methacrylate and methyldichlorosilane over 2% platinum-on-carbon at reflux. Yields of 85% may be obtained by use of hexane as a solvent and hydroquinone as a polymerization inhibitor. No adduct is obtained with rhodium-on-carbon, palladium-on-carbon, or Raney nickel as catalysts (Sommer *et al.*, 1957).

Unsaturated Nitriles

Silanes add to unsaturated nitriles at the terminal position.

$$CH_3Cl_2SiH + H_2C{=}CHCN \longrightarrow CH_3\overset{\overset{\displaystyle Cl}{|}}{\underset{\underset{\displaystyle Cl}{|}}{Si}}CH_2CH_2CN$$

In platinum-catalyzed addition of silanes to acrylonitrile, considerably improved yields are obtained if the reaction is carried out in the presence of a heterocyclic amine co-catalyst, such as pyridine, or an aminobenzonitrile, or a dialkylcyanamide (Nitzsche and Buchheit, 1965).

Allyl Ethers

Allyl ethers add silanes at the terminal carbon (Bailey, 1959a). Linear allyl ether–silicon copolymers have been made by the terminal addition of bis(trimethylsiloxy)bis(ethylhydrogensiloxane), $[Me_3SiO(EtSiHO)_2SiMe_3]$, to diallyl ether in the presence of 1% platinum-on-alumina at 150°–160°C (Bailey, 1959b).

Allyl Chlorides

Silicon hydrides interact with allylic chlorides in the presence of platinum catalysts to afford complex mixtures, arising by allylic rearrangements, elimination of chloride from allylic positions, and double-bond migrations (Smith *et al.*, 1962). Trichlorosilane, methyldichlorosilane, dimethylchlorosilane, and phenyldichlorosilane each adds to allyl chloride in the presence of chloroplatinic acid to afford propylene, and 3-chloropropyl and *n*-propylsilane derivatives. Addition to methallyl chloride, on the other hand, affords little or no isobutylene or isobutylsilane derivatives, but excellent yields of 3-chloro-2-methylpropylsilanes are obtained (Ryan *et al.*, 1960).

$$-\overset{|}{\underset{|}{Si}}H + H_2C{=}\underset{\underset{CH_3}{|}}{C}CH_2Cl \longrightarrow -\overset{|}{\underset{|}{Si}}CH_2\underset{\underset{CH_3}{|}}{C}HCH_2Cl$$

Homoallylic chloro compounds apparently react like unsubstituted olefins (Wilt and Dockus, 1970).

Unsaturated Alcohols

Silanes undergo hydrolysis and alcoholysis, but these reactions may be sufficiently inhibited in a buffered solution, pH 5 to 7, to permit platinum-catalyzed addition of silanes to hydroxylated olefins. For instance, 0.1 mole of phenyldimethylsilane, 0.1 mole of allyl alcohol, 140 ml of *t*-butanol, 7 ml of a buffer of sodium hydroxide and potassium acid phthalate, and 2.8×10^{-5} mole of chloroplatinic acid were allowed to stand at room temperature for 46 hours to afford phenyldimethyl(γ-hydroxypropyl)silane in excellent yield. Without the buffer present, a mixture of phenyldimethylallyloxysilane, phenyldimethylpropoxysilane, and phenyldimethylbutoxysilane is formed in 87% total yield (Barnes, 1968).

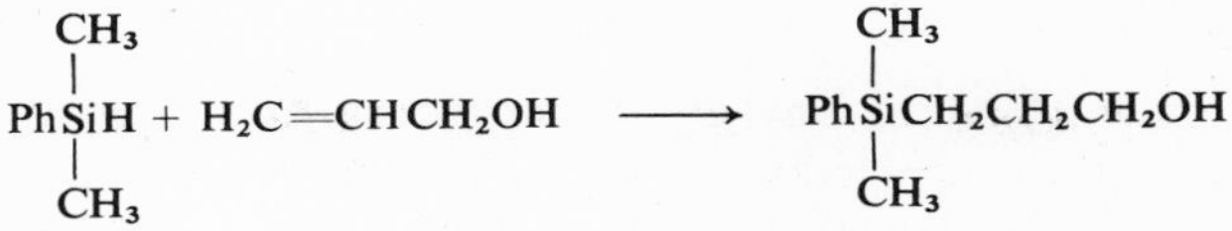

$$PhSi(CH_3)_2H + H_2C{=}CHCH_2OH \longrightarrow PhSi(CH_3)_2CH_2CH_2CH_2OH$$

Unsaturated Acetals

Silanes add smoothly to unsaturated acetals in the presence of platinum catalysts. Addition of 537 gm of acrolein dimethylacetal over a 3 hour period to 818 gm of *sym*-tetramethyldisiloxane at 90°C containing 10^{-4} mole of chloroplatinic acid, followed by 3 additional hours at 130°C affords (3,3-dimethoxypropyl)pentamethyldisiloxane in 75% yield and methoxypentamethyldisiloxane in 10% yield. Aldehydes formed by hydrolysis of the resulting products can be reductively alkylated with amines over palladium-on-carbon catalysts (Dennis and Ryan, 1970).

$$(CH_3)_3SiOSiH(CH_3)_2 + H_2C{=}CHCH(OCH_3)_2 \xrightarrow{Pt} (CH_3)_3SiOSi(CH_3)_2CH_2CH_2CH(OCH_3)_2 + (CH_3)_3SiOSi(CH_3)_2OCH_3$$

Unsaturated Isocyanurates

Silicon-substituted isocyanurates can be made by platinum-catalyzed addition of a silicon hydride to an unsaturated isocyanurate. For example, 122 gm of trimethoxysilane is added portionwise to 124 gm of triallylisocyanurate containing 0.05 gm of chloroplatinic acid at such a rate so as to maintain the temperature at 100°C. Vacuum fractionation affords 1-trimethoxysilylpropyl-3,5-diallylisocyanurate in 42% yield (Berger, 1970).

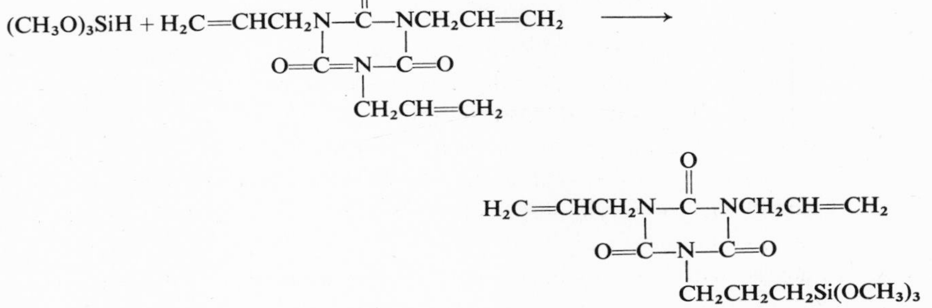

$$(CH_3O)_3SiH + \underset{\substack{|\\ O{=}C{=}N{-}C{=}O \\ | \\ CH_2CH{=}CH_2}}{H_2C{=}CHCH_2N{-}\overset{\overset{O}{\|}}{C}{-}NCH_2CH{=}CH_2} \longrightarrow$$

$$\underset{\substack{|\\ O{=}C{-}N{-}C{=}O \\ | \\ CH_2CH_2CH_2Si(OCH_3)_3}}{H_2C{=}CHCH_2N{-}\overset{\overset{O}{\|}}{C}{-}NCH_2CH{=}CH_2}$$

Unsaturated Silanes

Silanes containing both a vinyl group and hydrogen attached to silicon have been polymerized to polysilethylenes over platinum-on-carbon. For example, 29.8 gm (0.346 mole) of dimethylvinylsilane and 0.35 gm of 0.06% platinum-on-carbon heated under reflux for 4.5 hours affords polydimethylsilethylene, molecular weight about 1726, in 68% yield together with 1,1,4,4-tetramethyl-1,4-disilacyclohexane in 18% yield (Curry, 1956). The molecular weight and properties of silicon polymers may be controlled by adding perchloroethylene as a catalyst inhibitor (Nielsen, 1968).

$$H_2C{=}CH{-}SiH(CH_3)_2 \longrightarrow {-}CH_2{-}\left[{-}Si(CH_3)_2{-}CH_2{-}CH_2{-}\right]{-}Si(CH_3)_2{-} + \text{1,1,4,4-tetramethyl-1,4-disilacyclohexane } [(H_3C)_2Si(CH_2CH_2)_2Si(CH_3)_2]$$

Addition of trimethoxysilane to vinyltrimethoxysilane affords 1,2-bis-(trimethoxysilyl)ethane and, similarly, tetramethyldisiloxane adds to vinyltrimethoxysilane. These adducts are used to prepare 1,2,5-oxadisilacyclopentane heterocyclics (Frye and Collins, 1970).

$$(CH_3O)_3SiH + CH_2{=}CHSi(OCH_3)_3 \rightarrow (CH_3O)_3SiCH_2CH_2Si(OCH_3)_3$$

Hydrosilylation of Acetylenes

Acetylene adds halosilanes in the presence of a platinum catalyst to afford vinylhalosilanes. Vinyltrichlorosilane is obtained in 87% yield by interaction of acetylene and trichlorosilane in *o*-dichlorobenzene solvent containing chloroplatinic acid. The process may be carried out continuously (Gaignon and Lefort, 1968).

$$CH{\equiv}CH + Cl_3SiH \rightarrow CH_2{=}CHSiCl_3$$

Trichlorosilane adds to acetylenes in the presence of platinum-on-carbon by stereospecific *cis* addition to afford *trans*-olefins in excellent yields. The authors depicted the *cis* addition as resembling the commonly accepted picture for catalytic hydrogenation of acetylenes (Benkeser and Hickner, 1958). Similarly, 2-butyne adds methyldichlorosilane in the presence of chloroplatinic acid to afford *cis*-2-methyldichlorosilyl-2-butene (Ryan and Speier, 1966). *Cis* adducts are obtained in high yield when triphenylsilane or diphenylmethylsilane is heated with diphenylacetylene for a short time at 110°C or in cyclohexene at 80°C in the presence of either platinum-on-carbon or chloroplatinic acid. Optically active (+)-α-1-naphthylphenylmethylsilane adds stereospecifically to diphenylacetylene with complete retention of configuration to afford the *cis*-stilbene (Brook *et al.*, 1968).

$$1\text{-}C_{10}H_7\overset{*}{Si}(Ph)HCH_3 + PhC{\equiv}CPh \longrightarrow 1\text{-}C_{10}H_7\overset{*}{Si}(Ph)(CH_3)C(Ph){=}C(Ph)H$$

With appropriate reaction conditions, monoadducts of silanes and acetylenes can be virtually eliminated in favor of diadducts; diadducts are favored by increased reaction times and by increased catalyst concentration. In interaction of trichlorosilane and 1-hexyne catalyzed by chloroplatinic acid, diadduct formation was also strongly influenced by some unassessed property of the catalyst. Freshly prepared solutions of chloroplatinic acid in isopropanol were ineffective in promoting diadduct formation, but after standing 2 days, the same catalyst solution became quite effective (Benkeser *et al.*, 1967). A further example of diadduct formation is the preparation of compounds in the 1,2,5-oxadisilacyclopentane system (Polyakova *et al.*, 1965).

$$(HMe_2Si)_2O + PhC{\equiv}CPh \xrightarrow{Pt} \underbrace{Me_2Si{-}CH(Ph){-}CH(Ph){-}SiMe_2{-}O}_{\text{ring}}$$

Kraihanzel and Losee (1967) prepared a series of mixed dimetalloid acetylenes and olefins using ethynylsilanes as starting materials and chloroplatinic acid as a catalyst. Polymers are obtained by platinum-catalyzed interaction of phenylmethyldiethynylsilane and diphenylsilane.

Addition to the Nitroso Function

In a manner formally analogous to the addition to olefins, silanes add to the nitroso function. Interaction of an equimolar mixture of trimethylsilane and trifluoronitrosomethane in the dark in the presence of chloroplatinic acid (previously reduced with trimethylsilane) affords nitrogen (5%), hexafluoroazoxymethane (1%), trimethylfluorosilane (10%), *N,N*-bis(trifluoromethyl)-*O*-trimethylsilylhydroxylamine (16%), and *N*-trifluoromethyl-*O*-trimethylsilylhydroxylamine (74%) (Delany *et al.*, 1968).

$$\underset{}{(CH_3)_3SiH} + CF_3NO \rightarrow \underset{10\%}{(CH_3)_3SiF} + \underset{16\%}{(CF_3)_2NOSi(CH_3)_3} + \underset{74\%}{CF_3NHOSi(CH_3)_3}$$

Addition of Aminosilicon Hydrides to Olefins

Dennis and Speier (1970) in a pioneering study compared the effect of structure on the addition to olefins of hydrides of aminosilanes, silazanes, and trisilylamines catalyzed by platinum. The reactions were generally complex and the course unexpected.

N-Butylaminodimethylsilane interacts smoothly with 1-hexene at 100°C in the presence of chloroplatinic acid to give products whose composition depends largely on the reaction time. After 1 hour, the chief products were 2-*n*-butyl-1-hexyl-1,1,3,3-tetramethyldisilazane and butylamine, whereas after 3 hours, the product was largely *n*-butylaminohexyldimethylsilane.

$$n\text{-}C_4H_9NHSiH(CH_3)_2 \xrightarrow{C_4H_9CH{=}CH_2}$$

$$HSi(CH_3)_2\text{—}N(C_4H_9)\text{—}Si(CH_3)_2C_6H_{11} \xrightarrow{C_4H_9CH{=}CH_2} C_4H_9NHSi(CH_3)_2C_6H_{11}$$

A mixture of anilinodimethylsilane and 1-hexene affords anilinodimethylhexylsilane in 96% yield, whereas the disubstituted aminosilane, dimethylaminodimethylsilane failed to react. However, these nonreactive silanes could be made to add smoothly to the olefin if *sym*-tetramethyldisilazane were present in the reaction mixture. The authors suggested that the reaction proceeds in this case through reversible exchange of silyl groups between disilazanes and dimethylaminosilanes, an exchange that is itself platinum-catalyzed.

Tris(dimethylsilyl)amine interacts with 1-hexene to give a complex mixture of products, some of which are derived through an unprecedented exchange of hydrogen and methyl groups. Both olefins and platinum are necessary for exchange to occur.

Dehydrogenation

Silanes in the presence of platinum metal catalysts undergo a general reaction of a type

$$R_3SiH + ZOH \rightarrow R_3SiOZ + H_2$$

where Z = hydrogen, alkyl, aryl, acyl.

The reaction is useful for the preparation of silanols, alkoxy or aryloxy silanes, and silyl esters.

Sommer and Lyons (1967) investigated the stereochemistry of this general reaction by interacting optically active α-naphthylphenylmethylsilane with water, methanol, cyclohexanol, *t*-butanol, phenol, acetic acid, and benzoic acid in the presence of 10% palladium-on-carbon or Raney nickel. In every case, the reaction is stereospecific and proceeds with inversion of configuration at the silicon center. Stereospecificity is generally a little higher with nickel than with palladium-on-carbon. The reaction is also catalyzed by chloroplatinic acid and platinum ethylene chloride, but over these catalysts, the products are racemic. Platinum-on-carbon is not an effective catalyst requiring high temperatures and giving low yields.

Hydrolysis

Palladium-on-carbon, palladium-on-alumina, and ruthenium-on-carbon are excellent catalysts for hydrolysis of organosilicon hydrides to the corresponding organosilanols.

$$-\overset{|}{\underset{|}{\mathrm{Si}}}\mathrm{H} + H_2O \longrightarrow -\overset{|}{\underset{|}{\mathrm{Si}}}\mathrm{OH} + H_2$$

The method is applicable to silanols sensitive to condensation and to silanols containing siloxane linkages without cleavage or rearrangement of these linkages. The yields are very sensitive to acidity and hydrolysis is best carried out in a buffer solution to prevent the reaction mixture from becoming either acidic or basic. Palladium-on-carbon catalysts give better yields than ruthenium-on-alumina, or platinum-on-carbon or platinum-on-α-alumina (Barnes and Daughenbaugh, 1966).

Preferred catalysts for the hydrolysis of 1,3,5,7-tetramethyl-1,3,5,7-tetrahydrocyclotetrasiloxane to the corresponding tetrahydroxy compound are platinum complexes, for example, platinous ethylene chloride. The reaction is conveniently controlled by slow addition of 0.01 gm of catalyst in dioxane to the substrate (120 gm) in dioxane (150 gm)–water (40 gm) (Modic, 1970a).

$$\begin{array}{ccc} CH_3 & & CH_3 \\ | & & | \\ HSi & -O- & SiH \\ | & & | \\ O & & O \\ | & & | \\ HSi & -O- & SiH \\ | & & | \\ CH_3 & & CH_3 \end{array} + 4\,H_2O \longrightarrow \begin{array}{ccc} CH_3 & & CH_3 \\ | & & | \\ HO-Si & -O- & Si-OH \\ | & & | \\ O & & O \\ | & & | \\ HO-Si & -O- & Si-OH \\ | & & | \\ CH_3 & & CH_3 \end{array} + 4\,H_2$$

Alcoholysis

Silanes undergo alcoholysis readily in the presence of catalytic quantities of chloroplatinic acid. In a comparison of the reactivities of various alcohols, quantitative yields of $Ph(CH_3)_2SiOR$ were obtained at room temperature by reaction of a large excess of ROH in the presence of 1×10^{-4} mole of platinum per mole of $Ph(CH_3)_2SiH$. The alcohols used in descending order of reactivity were benzyl alcohol > butanol > hexanol > $(CH_3)_3SiCH_2OH$ > ethanol > methanol > *p*-chlorophenol > isopropanol > *t*-butanol (Barnes and Schweitzer, 1961).

In interaction of phenyldimethylsilane with phenol, the order of decreasing activity for metal halide catalysts was $PdCl_2 > H_2PtCl_6 > PtCl_2 > RhCl_3 > NiCl_2$ and for metal blacks Pd > Pt ≈ Raney nickel (W_2) > Au (Iwakura *et al.*, 1971).

$$Ph(CH_3)_2SiH + PhOH \rightarrow Ph(CH_3)_2SiOPh + H_2$$

Tetrasilylmethanes undergo alcoholysis in the presence of chloroplatinic acid (Merker and Scott, 1963).

$$[(CH_3)_2HSi]_4C + 4\ ROH \xrightarrow[\text{reflux}]{H_2PtCl_6} [(CH_3)_2ROSi]_4C + 4\ H_2$$

Acetolysis

A rapid reaction between *o*-tolylsilane and acetic acid takes place in the presence of platinum oxide to afford hydrogen and triacetoxytolylsilane. Similar reactions have been described using chloroplatinic acid as a catalyst (Selin and West, 1962a).

$$3\ CH_3COOH + o\text{-}CH_3C_6H_4SiH_3 \longrightarrow o\text{-}CH_3C_6H_4Si(OCOCH_3)_3 + 3\ H_2$$

Ammonolysis

Interaction of silanes and amines affords siloxamines and hydrogen. For instance, in the preparation of siloxamine polymers, 100 gm of a hydrosiloxane polymer having the average formula, $Me_3SiO(MeHSiO)_{4.5}(Me_2SiO)_{12}SiMe_3$ heated at 200°C for 3 hours with 32.6 gm of aniline and 0.013 gm of chloroplatinic acid affords 108 gm of a polymer with the average composition $MeSi[(Me)(PhNH)SiO]_{4.5}(Me_2SiO)_{12}SiMe_3$. Platinum-on-alumina may be used also for this type of reaction (Creamer, 1970). In another example, 0.2 mole of heptamethyltrisiloxane and 0.3 mole of diethylamine charged to a

300 ml pressure vessel of stainless steel with 1.3 gm of 2% platinum-on-γ-alumina and heated to 250°C for 3 hours affords the siloxamine in 86% yield after distillation at reduced pressure (Borchert, 1970).

$$[(CH_3)_3SiO]_2SiHCH_3 + (C_2H_5)_2NH \longrightarrow [(CH_3)_3SiO]_2\overset{\displaystyle CH_3}{\overset{|}{Si}}N(C_2H_5)_2 + H_2$$

Active Hydrogen Compounds

Sommer and Citron (1967) compared the activity of various catalysts in interaction of hydrogen chloride and triethylsilane. As in hydrolysis, palladium-on-carbon was the most effective. The activity sequence palladium > rhodium $\simeq$ platinum was established for this reaction. Later, exchange of a variety of halocarbons with silanes was found to be general and to proceed according to the following equation:

$$-\overset{|}{\underset{|}{C}}X + -\overset{|}{\underset{|}{Si}}H \xrightarrow{Pd/C} -\overset{|}{\underset{|}{C}}H + -\overset{|}{\underset{|}{Si}}X$$

The reaction is a convenient one for preparing silicon halides (except fluorides) in high yield, and is superior to alternative procedures (Citron *et al.*, 1969).

Organosilylthiols may be made in excellent yield by interaction of hydrogen sulfide and compounds having a silicon–hydrogen bond with palladium-on-alumina as catalyst. The reaction is very sensitive to the structure of both the silicon and sulfur reactant. Methanethiol failed to react with triethylsilane, although hydrogen sulfide reacts quantitatively. The authors interpreted these differences in terns of competition of reactants for catalyst sites (Sommer and Citron, 1967).

Stereochemistry

Interaction of optically active α-naphthylphenylmethylsilane and non-hydroxylic compounds containing an active hydrogen, such as pyrrolidine, isobutylamine, hydrogen fluoride, and hydrogen chloride, proceeds with inversion of configuration at the silicon atom. With amines stereospecificity was limited to palladium-on-alumina catalysts; palladium-on-carbon produced racemic products. However, both catalysts gave inverted products with hydrogen fluoride and hydrogen chloride. The reaction of hydrogen fluoride with compounds having a silicon–hydrogen bond provides a convenient way of forming fluorosilanes. Chloro- and bromosilanes may be made using the elemental halogen, but fluorosilanes cannot (Sommer and Citron, 1967).

ACID HALIDES

In a reaction that resembles the well-known Rosenmund reaction, acid halides are reduced by silanes in the presence of palladium catalysts to aldehydes. The reaction is easy and convenient to perform and should prove especially useful when only small amounts of aldehydes are required. Triethylsilane is the most effective reducing agent of a number of silanes tested. At least 1 mole of silane is required per mole of acid halide, but excess silane seems to have little affect on the reaction. Improved yields of aldehydes are obtained if the reaction product mixture is first water washed before distillation is attempted (Citron, 1969).

$$Et_3SiH + n\text{-}C_7H_{15}COCl \xrightarrow[\sim 100°C]{100 \text{ mg } 10\% \text{ Pd-on-}} n\text{-}C_7H_{15}CHO + Et_3SiCl$$
$$51\%$$

Various complex platinum and rhodium catalysts have been used also in this type of reaction, but over rhodium and with aroyl chlorides the reaction is apt to take a different course and produce ketones as well as aldehydes, the ketone being in many cases the principal product. Electron-releasing substituents in the aryl group favor ketone formation, whereas electron-withdrawing substituents favor aldehydes. The relative ratios of ketone and aldehyde are affected markedly by small changes in the catalyst structure (Dent *et al.*, 1970).

$$CH_3O-C_6H_4-\underset{\underset{O}{\|}}{C}Cl \longrightarrow CH_3O-C_6H_4-\underset{\underset{O}{\|}}{C}-C_6H_4-OCH_3 + CH_3O-C_6H_4-CHO$$

Catalyst		
trans-$[RhCl(CO)(PEtPh_2)_2]$	63%	2%
trans-$[RhCl(CO)(PEt_2Ph)_2]$	35%	22%
trans-$[RhCl(CO)(PEt_3)_2]$	27%	4%

CARBON–SILICON BOND CLEAVAGE

The silicon–carbon bond may be cleaved at times in the presence of platinum metal catalysts. The reaction has been used in formation of polymers, disproportionation, and removal of blocking groups.

Polymers

Silacyclobutanes and 1,1,3,3-tetramethyl-1,3-disilacyclobutane undergo a facile ring-opening polymerization in the presence of a platinum catalyst.

$$(H_3C)_2Si\langle CH_2CH_2CH_2\rangle Si(CH_3)_2 \xrightarrow[100°C]{Pt} \left[-\overset{CH_3}{\underset{CH_3}{Si}}-CH_2- \right]_n$$

These silacyclobutanes also react with a variety of silicon hydrides in the presence of platinum to afford a series of telomeric adducts. With equimolar mixtures of **IV** and a silicon hydride, yields of 80–90% of **V** and **VI** are obtained with the remainder converted to higher molecular weight telomers.

$$\underset{\textbf{(IV)}}{(H_3C)_2Si\langle CH_2CH_2CH_2\rangle} + R\overset{CH_3}{\underset{CH_3}{Si}}H \longrightarrow R\overset{CH_3}{\underset{CH_3}{Si}}-\left[-CH_2CH_2CH_2\overset{CH_3}{\underset{CH_3}{Si}}- \right]_n H$$

(V) $n = 1$
(VI) $n = 2$

These reactions are limited to silacyclobutanes which are unusually reactive because of strain arising from compression of normal bond angles; no polymerization of 1,1-dimethyl-1-silacyclopentane occurred over platinum at temperatures up to 200°C. Interaction of silacyclobutanes and silicon hydrides could involve a ligand exchange between the two silicon atoms or could involve ring-opening to an allylsilane followed by platinum-catalyzed addition of silicon hydrides to the double bond. The latter path was eliminated with the observation that interaction of 1,1-dimethyl-1-silacyclobutane and phenyldimethylsilane-*d* resulted in a product with the deuterium still attached to silicon. The ring-opening appeared to be the first example of a platinum-catalyzed exchange of an alkyl group and hydrogen on silicon (Weyenberg and Nelson, 1965).

Silanes react readily with germacyclobutanes in the presence of a trace of chloroplatinic acid to give 60–80% of **VII** plus polymer (Mazerolles *et al.*, 1967).*

$$RR'R''SiH + Bu_2Ge\langle CH_2CH_2CH_2\rangle \longrightarrow \underset{\textbf{(VII)}}{RR'R''SiCH_2CH_2CH_2GeHBu_2}$$

Disproportionation

Silanes will undergo disproportionation in the presence of a platinum catalyst (Yamamoto *et al.*, 1971). Diphenylsilane when heated with chloro-

* Hydrogermylation, with retention of configuration at the germanium atom, may be achieved with the same type of catalysts used for hydrosilylation (Corriu and Moreau, 1971).

platinic acid affords a mixture of phenylsilane, di-, tri-, and tetraphenylsilanes. At higher temperatures, 230°–300°C, disproportionation will occur in the absence of added catalyst (Gilman and Miles, 1958).

Blocking Groups

The trimethylsilyl group has been used to protect terminal ethynyl groups in oxidative coupling and hydrogenation procedures. Overall success includes removal of the protecting group by an easy cleavage of the alkynyl carbon–silicon bond. Various reagents, including alcoholic alkali, alcoholic silver nitrate, alcoholic potassium fluoride, and mercuric sulfate, have been used to effect this cleavage. More recently, Poist and Kraihanzel (1968) have shown that cleavage may be effected also by $Pt_2Cl_4(C_2H_4)_2$ in dry benzene, probably according to the equation

$$2\,RC{\equiv}CSi(CH_3)_3 + Pt_2Cl_4(C_2H_4)_2 \rightarrow 2\,(Me)_3SiCl + 2\,C_2H_4 + 2\,RC{\equiv}CPtCl$$

In refluxing alcohol, catalytic quantities of $Pt_2Cl_4(C_2H_4)_2$ or $K(PtCl_3C_2H_4)$-H_2O may be used and cleavage occurs according to the equation

$$RC{\equiv}CSi(CH_3)_3 + C_2H_5OH \xrightarrow{KPtCl_3C_2H_4} (CH_3)_3SiOC_2H_5 + RCH_2CH(OC_2H_5)_2$$

Ethynyl compounds expected in the products are rapidly converted to the ketal under the conditions of the reaction (Poist and Kraihanzel, 1968).

Vinylsilanes

Vinylsilanes undergo a series of reactions in the presence of palladium chloride that are similar to the aryl mercurial metal exchange reactions of Heck (1968). Two synthetic advantages accrue to the silane system; vinylsilanes are less toxic than organomercurial reagents and the silicon moiety is more easily removed from the product than is colloidal mercury. The reactions probably take place through a vinylpalladium intermediate which can react with itself or with other unsaturation. An example of the former is the formation of *trans,trans*-1,4-diphenyl-1,3-butadiene from interaction of β-trimethylsilylstyrene with catalytic amounts of palladium chloride and copper chloride as an oxidant.

$$\underset{4.5\text{ gm}}{Ph(H)C{=}C(H)Si(CH_3)_3} \xrightarrow[\substack{1.0\text{ gm LiCl} \\ 10\text{ gm } CuCl_2\cdot 2\,H_2O \\ 50\text{ ml } CH_3OH \\ 22\text{ hours}}]{200\text{ mg } PdCl_2} \underset{52\%\text{ yield}}{Ph(H)C{=}C(H){-}C(H){=}C(H)Ph}$$

β-Trimethylsilylstyrene adds to ethylene or to methyl acrylate in the presence of catalytic amounts of palladium chloride to afford 1-chloro-4-phenyl-3-butene and 1-phenyl-4-carbomethoxy-1,3-butadiene, respectively (Weber *et al.*, 1971).

$$\underset{H}{\overset{Ph}{>}}C{=}C\underset{Si(CH_3)}{\overset{H}{<}} + H_2C{=}CH_2 \xrightarrow[CuCl_2]{PdCl_2} \underset{H}{\overset{Ph}{>}}C{=}C\underset{CH_2CH_2Cl}{\overset{H}{<}}$$

95% *trans*

$$\underset{H}{\overset{Ph}{>}}C{=}C\underset{Si(CH_3)_3}{\overset{H}{<}} + H_2C{=}CHCOOCH_3 \xrightarrow[CuCl_2]{PdCl_2} \underset{H}{\overset{Ph}{>}}C{=}C\overset{H}{<}\underset{H}{C}{=}C\underset{COOCH_3}{\overset{H}{<}}$$

REFERENCES

Ashby, B. A. (1964). U.S. Patent 3,159,662.
Bailey, D. L. (1959a). U.S. Patent 2,888,479.
Bailey, D. L. (1959b). U.S. Patent 2,897,222.
Barnes, G. H., Jr. (1968). U.S. Patent 3,398,174.
Barnes, G. H., Jr., and Daughenbaugh, N. E. (1966). *J. Org. Chem.* **31**, 885.
Barnes, G. H., Jr., and Schweitzer, G. W. (1961). U.S. Patent 2,967,171.
Benkeser, R. A., and Hickner, R. A. (1958). *J. Amer. Chem. Soc.* **80**, 5298.
Benkeser, R. A., Burrous, M. L., Nelson, L. E., and Swisher, J. V. (1961). *J. Amer. Chem. Soc.* **83**, 4385.
Benkeser, R. A., Cunico, R. F., Dunny, S., Jones, P. R., and Nerlekar, P. G. (1967). *J. Org. Chem.* **32**, 2634.
Benkeser, R. A., Dunny, S., Li, G. S., Nerlekar, P. G., and Work, S. D. (1968). *J. Amer. Chem. Soc.* **90**, 1871.
Berger, A. (1970). U.S. Patent 3,517,001.
Borchert, R. C. (1970). U.S. Patent 3,530,092.
Brook, A. G., Pannell, K. H., and Anderson, D. G. (1968). *J. Amer. Chem. Soc.* **90**, 4374.
Brown, E. D. (1968). U.S. Patent 3,418,353.
Čapka, M., Svobada, P., Černý, M., and Hetflejš, J. (1971). *Tetrahedron Lett.* p. 4787.
Chalk, A. J. (1970a). *Trans. N.Y. Acad. Sci.* **32**, 481.
Chalk, A. J. (1970b). *J. Organometal. Chem.* **21**, 207.
Chalk, A. J., and Harrod, J. F. (1965). *J. Amer. Chem. Soc.* **87**, 16.
Citron, J. D. (1969). *J. Org. Chem.* **34**, 1977.
Citron, J. D., Lyons, J. E., and Sommer, L. H. (1969). *J. Org. Chem.* **34**, 638.
Corriu, R. J. P., and Moreau, J. J. E. (1971). *Chem. Commun.* p. 812.
Creamer, C. E. (1970). U.S. Patent 3,519,601.
Curry, J. W. (1956). *J. Amer. Chem. Soc.* **78**, 1686.
Curry, J. W., and Harrison, G. W., Jr. (1958). *J. Org. Chem.* **23**, 627.
deCharentenay, F., Osborn, J. A., and Wilkinson, G. (1968). *J. Chem. Soc., A* p. 787.
Delany, A. C., Haszeldine, R. N., and Tipping, A. E. (1968). *J. Chem. Soc., C* p. 2537.
Dennis, W. E., and Ryan, J. W. (1970). *J. Org. Chem.* **35**, 4180.
Dennis, W. E., and Speier, J. L. (1970). *J. Org. Chem.* **35**, 3879.

Dent, S. P., Eaborn, C., and Pidcock, A., (1970). *Chem. Commun.* p. 1703.
Fish, J. G. (1971). U.S. Patent 3,576,027
Frye, C. L., and Collins, W. T. (1970). *J. Org. Chem.* **35**, 2964.
Gaignon, M. H. R. J., and Lefort, M. J. C. (1968). U.S. Patent 3,404,169.
Gilman, H. and Miles, D. H. (1958). *J. Org. Chem.* **23**, 326.
Goodman, L., Silverstein, R. M., and Benitez, A. (1957). *J. Amer. Chem. Soc.* **79**, 3073.
Gornowicz, G. A., Ryan, J. W., and Speier, J. L. (1968). *J. Org. Chem.* **33**, 2918.
Hara, M., Ohno, K., and Tsuji, J. (1971). *Chem. Commun.* p. 247.
Haszeldine, R. N., Parish, R. V., and Parry, D. J. (1967). *J. Organometal. Chem.* **9**, P13.
Haszeldine, R. N., Parish, R. V., and Parry, D. J. (1969). *J. Chem. Soc., A* p. 683.
Heck, R. F. (1968). *J. Amer. Chem. Soc.* **90**, 5518, 5526, 5531, 5535, 5539, 5542, and 5546.
Iwakura, Y., Uno, K., Toda, F., Hattori, K., and Abe, M. (1971). *Bull. Chem. Soc. Jap.* **44**, 1400.
Joy, J. R. (1968). U.S. Patent 3,410,886.
Kraihanzel, C. S., and Losee, M. L. (1967). *J. Organometal. Chem.* **10**, 427.
Kuivila, H. G., and Warner, C. R. (1964). *J. Org. Chem.* **29**, 2845.
Lamoreaux, H. F. (1965). U.S. Patent 3,220,972.
Lamoreaux, H. F. (1967). U.S. Patent 3,313,773.
Martin, M. M., and Koster, R. A. (1968). *J. Org. Chem.* **33**, 3428.
Mazerolles, P., Dubac, J., and Lesbre, M. (1967). *Tetrahedron Lett.* p. 255.
Merker, R. L., and Scott, M. J. (1963). *J. Org. Chem.* **28**, 2717.
Modic, F. J. (1970a). U.S. Patent 3,540,006.
Modic, F. J. (1970b). U.S. Patent 3,516,946.
Murphy, R. A. (1969). U.S. Patent 3,458,469.
Musolf, M. C., and Speier, J. L. (1964). *J. Org. Chem.* **29**, 2519.
Nielsen, J. M. (1968). U.S. Patent 3,383,356.
Nitzsche, S., and Buchheit, P. (1965). U.S. Patent 3,167,573.
Petrov, A. D., Chernyskev, E. A., Dolgaya, N. E., Egorov, Yu. P., and Leites, L. A. (1960). *J. Gen. Chem. USSR* **30**, 376.
Pike, R. A., and Borchert, R. C. (1960). U.S. Patent 2,954,390.
Pike, R. A., and McDonagh, P. M. (1963). *J. Chem. Soc., London* p. 2831.
Poist, J. E., and Kraihanzel, C. S. (1968). *Chem. Commun.* p. 607.
Polyakova, A. M., Suchkova, M. D., Korshak, V. V., and Vdovin, V. M. (1965). *Izv. Akad. Nauk SSSR, Ser. Khim.* **7**, 1267.
Ryan, J. W., and Speier, J. L. (1964). *J. Amer. Chem. Soc.* **86**, 895.
Ryan, J. W., and Speier, J. L. (1966). *J. Org. Chem.* **31**, 2698.
Ryan, J. W., Menzie, G. K., and Speier, J. L. (1960). *J. Amer. Chem. Soc.* **82**, 3601.
Saam, J. C., and Speier, J. L. (1958). *J. Amer. Chem. Soc.* **80**, 4104.
Saam, J. C., and Speier, J. L. (1961). *J. Amer. Chem. Soc.* **83**, 1351.
Selin, T. G., and West, R. (1962a). *J. Amer. Chem. Soc.* **84**, 1856.
Selin, T. G., and West, R. (1962b). *J. Amer. Chem. Soc.* **84**, 1863.
Smith, A. G., Ryan, J. W., and Speier, J. L. (1962). *J. Org. Chem.* **27**, 2183.
Sommer, L. H., and Citron, J. D. (1967). *J. Org. Chem.* **32**, 2470
Sommer, L. H., and Lyons, J. E. (1967). *J. Amer. Chem. Soc.* **89**, 1521.
Sommer, L. H., and Lyons, J. E. (1968). *J. Amer. Chem. Soc.* **90**, 4197.
Sommer, L. H., Mackay, F. P., Steward, O. W., and Campbell, P. G. (1957). *J. Amer. Chem. Soc.* **79**, 2764.
Sommer, L. H., Michael, K. W., and Fujimoto, H. (1967a). *J. Amer. Chem. Soc.* **89**, 1519.
Sommer, L. H., Lyons, J. E., Fujimoto, H., and Michael, K. W. (1967b). *J. Amer. Chem. Soc.* **89**, 5483.

Speier, J. L., Webster, J. A., and Barnes, G. H., Jr. (1957). *J. Amer. Chem. Soc.* **79**, 974.
Spialter, L., and O'Brien, D. H. (1967). *J. Org. Chem.* **32**, 222.
Takahasi, H., Okita, H., Yamaguchi, M., and Shiihara, I. (1963). *J. Org. Chem.* **28**, 3353.
Takahashi, S., Shibano, T., and Hagihara, N. (1969). *Chem. Commun.* p. 161.
Wagner, G. H. (1953a). U.S. Patent 2,632,013.
Wagner, G. H. (1953b). U.S. Patent 2,637,738.
Wagner, G. H., and Whitehead, W. E. (1958). U.S. Patent 2,851,473.
Weber, W. P., Felix, R. A., Willard, A. K., and Koenig, K. E. (1971). *Tetrahedron Lett.* p. 4701.
Weyenberg, D. R., and Nelson, L. E. (1965). *J. Org. Chem.* **30**, 2618.
Willing, D. N. (1968). U.S. Patent 3,419,593.
Wilt, J. W., and Dockus, C. F. (1970). *J. Amer. Chem. Soc.* **92**, 5813.
Yamamoto, K., and Kumada, M. (1968). *J. Organometal Chem.* **13**, 131.
Yamamoto, K., Okinoshima, H., and Kumada, M. (1971). *J. Organometal. Chem.* **27**, C3.

Author Index

A

Abe, M., 286, *292*
Abegg, V. P., 262, *273*
Abley, P., 61, 71, 72, *74*
Abubaker, M., 151, *171*
Achard, R., 111, *115*
Acheson, R. M., 19, *53*
Acres, G. J. K., 61, *74*
Acton, N., 187, 188, *211, 212*
Adams, R., 100, *115*
Adams, R. W., 66, *74*
Adams, T., 133, 136, *142, 143*
Adamyants, K. S., 130, *142*
Adderley, C. J. R., 87, *115*
Adkins, H., 29, 30, 37, *53, 54*
Adkins, J., 38, *54*
Agar, J. H., 8, *58*
Aguiar, A. M., 73, *75*
Aguilo, A., 78. *115,* 176, *211*
Ainsworth, C. , 7, 47, *54*
Alderson, T., 176, 177, 179. 189, 191, *211,* 238, *254*
Aldridge, C. L., 243, 252, *259*
Alexander, B. H., 102, 104, *118*
Allen, D. S., Jr., 130, *143*
Allen, G. R., Jr., 106, *115,* 169, *173*
Allen, L. E., 262, *273*
Allinger, N. L., 160, *171*
Allred, E. L., 12, *54*
Allum, K. G., *254*
Andal, R. K., 61, *75*
Anderson, A. E., Jr., 156, *171*
Anderson, A. G., Jr., 2, 3, 5, 14, 26, 44, *54*
Anderson, D. G., 283, *291*
Anderson, H. A., 124, *141*
Anderson, L., 106, *115, 118*
Anderson, R. G., 2, 3, 5, 14, *54*
Anderson, T., 245, *254*
Ando, H., 103, *120*
Ando, M., 130, *141*
Angyal, S. J., 106, *115*
Anner, G., 68, *76,* 132, 133, *144*
Appell, H. R., 29, *56*
Arison, B., 133, *143*
Arnet, J. E., 149, *171*
Arpe, von H. - J., 95, *115*
Arth, G. E., 49, *58*
Arthur, W. J., 162, *171*
Arzoumanidis, G. G., 95, *115*
Asano, R., 91, 92, *115, 116, 117*
Ashby, B. A., 276, *291*
Asinger, F., 229, *256*
Assao, T., 130, *141*
Atkins, K. E., 185, 201, 202, 207, *211, 212, 214*
Atkins, T. J., 184, *212*
Attridge, C. J., 148, 154, *171*
Aue, D. H., 170, *172*
Augenstine, R. L., 146, *171*
Avtokratova, T. D., 134, *141*
Ayres, D. C., 134, *141*
Azimov, V. A., 7, *59*

B

Bach, F. L., 41, *54*
Bachmann, W. E., 3, 43, 51, *54,* 161, *171,* 268, *272*

Baddley, W. H., 61, *74*
Baertschi, P., 99, *119*
Bailar, J. C., Jr., 60, 66, *74. 75, 76,* 149, *174*
Bailey, D. L., 280, *291*
Bailey, W. J., 5, 49, *54,* 268, 269, *272*
Bain, P. J., 265, *272*
Baird, M. C., 147, *171,* 262, 267, *272*
Baird, R. L., 178, *214*
Baird, W. C., Jr., 86, *115*
Bajer, F. J., 28, *54*
Baker, D. A., 138, *143*
Baker, R., 86, *115,* 204, *211*
Baker, R. H., 39, *56*
Balaceanu, J. C., 112, *117*
Balder, B., 89, *120*
Baldwin, J. E., 84, *115*
Bangert, R., 85, *118*
Barclay, J. C., 41, *54*
Barker, S. A., 104, *115*
Barnes, G. H., Jr., 274, 277, 280, 285, 286, *291, 293,*
Barnes, M. F., 157, *171*
Barnes, R. A., 43, 51, *54*
Bartlett, J. H., 232, *254*
Bashe, R. W., 35, *56*
Basner, M. E., 151, *173*
Batley, G. E., 66, *74*
Battiste, M. A., 87, *115*
Bauer, R. S., 69, *76*
Bechter, M., 78, *117*
Becker, A., 138, 140, *143*
Becker, E. I., 268, *273*
Bell, M. R., 10, *54*
Bellinzona, G., 154, *171*
Belluco, U., 61, *74*
Belov, A. P., 83, *115*
Belskii, I. F., 10, 15, *54, 58,* 165, *171,* 270, *273*
Beltrame, P., 38, *55*
Benedict, B. C., 152, *172*
Benitez, A., 279, *292*
Benkeser, R. A., 274, 275, 282, 283, *291*
Bennett, R. P., 252, 253, 254, *254, 256*
Benson, R. E., 245, 246, *255, 257*
Berger, A., 281, *291*
Bergmann, E. D., 5, *54,* 98, *116,* 263, 264, 267, *272, 273*
Berkowitz, L. M., 62, *74,* 133, *141, 143*
Berlin, A. J., 176, 177, *212*
Bernardi, L., 25, *56*
Bernstein, S., 8, *54*
Berry, J. W., 167, 168, *172*
Bertoglio, C., 154, *173*
Bettinetti, F., 154, *171*
Beutler, M., 219, *256*
Beyler, R. E., 49, *58*
Beynon, P. J., 133, *141*
Bhandari, R. G., 27, *54*
Bhat, H. B., *143*
Bhide, G. V., 27, *54*
Biale, G., 223, 234, *255*
Bible, R. H., Jr., 3, *58*
Bieder, A., 106, *115*
Biellmann, J. F., 63, 64, 70, *74*
Biger, S., 3, *54*
Billeter, J. R., 132, *144*
Billig, E., 180, *211*
Billups, W. E., 244, *255*
Bingham, A. J., 92, *116*
Birch, A. J., 62, 67, 68, 69, 70, 71, *74,* 98, *116,* 151, *171*
Bird, C. W., 215, 219, *255*
Birkenmeyer, R. D., 114, *116*
Bischof, E., 138, *144*
Bittler, K., 215, 216, 218, 236, 238, 239, 240, 241, 242, 244, 247, *255, 258, 259*
Blackham, A. U., 182, *211,* 225, *257*
Blaha, L., 122, *141*
Bláha, L., 122, *141*
Blank, G., 95, *115*
Blankley, C. J., 133, 136, *142*
Blatter, H. M., 160, *171*
Blazejewicz, L., 108, 109, 111, *117*
Blomquist, A. T., 48, 54, 194, *211*
Blum, J., 3, 26, *54, 57,* 98, *116,* 266, 267, 269, 271, *272*
Blumberg, S., 264, *272*
Boecke, R., 188, *212*
Boekelheide, V., 18, 19, 22, *54, 55,* 113, *117,* 156, *171*
Bogdanovic, B., 187, *214*
Bond, G. C., 61, *74,* 147, 148, 151, *171,* 216, *255*
Booth, B. L., 215, *255*
Booth, F. B., 216, 227, 228, *255, 257*
Borchert, R. C., 278, 287, *291, 292*
Bordenca, C., 243, *255*
Bordner, J., 14, *58,* 131, *142*
Boswell, G. A., Jr., 28, *54,* 126, *141*
Bourne, E. J., 104, *115*

Bowyer, W. J., 52, *54*
Braatz, J. A., 87, *117*, 148, *172*
Braca, G., 217, 226, *255*, *258*
Bracha, P., 264, *272*
Bragin, O. V., 160, *171*
Brailovskii, S. M., 191, *211*
Brammer, K. W., 51, *59*
Braude, E. A., 35, 38, *54*, *57*
Braun, G., 122, 123, *141*
Bream, J. B., 152, *171*
Brewis, S., 225, 241, 242, 243, *255*
Brianza, C., 21, *55*
Briggs, E. M., 219, *255*
Bright, A., 150, *171*
Brimacombe, J. S., 101, 106, *116*
Britton, R. W., 63, *75*
Broadbent, H. S., 12, *54*
Broadbent, R. W., 51, *59*
Brodie, H. J., 69, *74*
Brodkey, R. S., 238, *256*
Brook, A. G., 283, *291*
Brown, C. A., 193, *211*
Brown, C. K., 216, 226, 227, *255*, *259*
Brown, E. D., 31, *54*, 276, *291*
Brown, E. S., 182, *211*
Brown, H. C., 193, *211*
Brown, M., 62, 67, *74*
Brown, R. G., 80, 101, *116*
Browne, P. A., 42, *54*
Bruce, M. I., 73, *74*
Bruns, K., 130, *143*
Bryant, D. R., 78, 89, 90, 96, *116*, 202, *211*
Bryce-Smith, D., 191, 192, 193, *211*, 222, *255*
Buchheit, P., 280, *292*
Buchi, G., 130, *141*
Bucourt, R., 127, *141*
Büthe, H., 71, *75*, 147, *172*
Buhr, G., 85, *118*
Bulbenko, G. F., 271, *272*
Burbidge, B. W., 163, *171*
Burmeister, J. L., 60, *75*
Burnett, J. P., Jr., 7, *54*
Burnett, R. E., 73, *75*
Burnham, J. W., 43, *55*
Burr, J. G., Jr., 267, *272*
Burrous, M. L., 274, 275, *291*
Busch, M., 210, *211*
Bushweller, C. H., 96, *116*
Butenandt, A., 127, *141*
Butler, F. P., 47, *58*
Butler, J. D., 16, *54*
Butterworth, R. F., 133, 137, *141*
Butz, E. W. J., 6, *54*
Buvet, R., 128, *143*
Buzby, G. C., 36, *54*
Bye, T. S., 43, *57*
Byerley, J. J., 223, *255*
Byrd, J. E., 79, *116*

C

Cairns, J. F., 129, *141*
Calderazzo, F., 222, 254, *255*, *257*
Campaigne, E., 8, *54*, 271, *272*
Campbell, J. R., 140, *144*
Campbell, P. G., 279, *292*
Canale, A. J., 185, *211*
Canonne, P., 20, *54*
Cantrall, E. W., 8, *54*
Čapka, M., 216, *255*, 276, *291*
Capps, D. B., 12, *54*
Caputo, J. A., 134, *141*
Carnahan, J. C., Jr., 188, *212*
Carr, R. L. K., 28, *54*
Carrà, S., 38, *55*
Carrick, W. L., 39, *56*
Carturan, G., 225, *256*
Casanova, J., Jr., 133, *141*
Caspi, E., 135, 139, 140, *141*, *143*, 260, *273*
Cassar, L., 79, *116*, 169, *171*
Castagnoli, N., Jr., 14, *58*
Cavanaugh, R., 10, 11, *55*
Cerefice, S., 169, *172*
Černý, M., 216, *255*, 276, *291*
Chabardes, P., 180, *211*
Chalk, A. J., 70, *74*, 147, 148, *172*, 274, 276, 277, *291*
Chang, C. W. J., 140, *143*
Charlton, J., 87, *116*
Charman, H. B., 3, *55*, 101, *116*
Chatterje, R. M., 15, *55*
Chatterjea, J. N., 52, *54*
Chatterjee, D. K., 15, *55*
Chatterjee, D. N., 32, 34, *55*, *58*
Chatterjee, R. M., 15, *55*
Cheeseman, G. W. H., 18, *55*
Chemerda, J. M., 11, *55*
Cheng, K. F., 61, 63, *75*
Chernyskev, E. A., 277, *292*

Chevallier, Y., 61, *74*
Chini, P., 192, *211*
Chittum, J. W., 123, *142*
Chiusoli, G. P., 218, *255*
Chou, T. S., 43, *55*
Choudhury, D. , 168, *172*
Christensen, B. E., 15, *55*
Christie, K. O., 272, *272*
Chumakov, Yu. I., 17, *55*
Chung, H., 197, 198, *213*
Church, M. J., 215, *255*
Ciccone, S., 61, *75*
Citron, J. D., 274, 287, 288, *291, 292*
Clark, D., 81, 83, 86, 87, *116*
Clark, P. W., 62, *75*
Clement, W. H., 79, 80, 81, *116*
Cline, R. E., 110, *116*
Closson, R. D., 249, *255*
Cobb, R. L., 22, *58*
Cochrane, C. C., 31, *57*
Cocker, W., 51, *55*
Coffey, R. S., 73, *74*
Cohen, E., 41, *54*
Cole, J. L., 160, *171*
Collins, P. M., 133, *141*
Collins, W. T., 282, *292*
Collman, J. P., 70, *74*, 82, 84, 85, *116*, 191, 192, *211*, 224, *255*
Colthup, E. C., 192, *212*
Condon, F. E., 145, *171*
Conia, J. M., 261, *273*
Connolly, J. D., 156, *171*
Conrow, R. B., 8, *54*
Conti, F., 148, *171*, 209, *213*
Controulis, J., 3, *54*
Cook, J. W., 128, 129, *141*
Cook, M. C., 101, 106, *116*
Cooke, D. W., 62, *75*
Coombs, M. M., 108, *116*
Cooper, B. J., 61, *74*
Copelin, H. B., 265, *273*
Corey, E. J., 133, *141, 142*
Corriu, R. J. P., 289, *291*
Cosciug, T., 126, *142*
Costerousse, G., 127, *141*
Coulson, D. R., 195, 196, *211*
Craddock, J. H., 224, 228, *255, 258*
Cramer, R. D., 60, *74*, 146, 148, *171*, 175, 176, 177, 189, 190, *211, 212*, 239, *255*
Crano, J. C., 177, *212*
Crawford, M., 20, *55*
Creamer, C. E., 286, *291*
Criegee, R., 122, 125, *142*
Cross, B. E., 51, *55*, 130, *142*
Csányi, L. J., 125, *142*
Cuccuru, A., 217, *258*
Cummings, W., 271, *273*
Cummins, R. W., 132, *142*
Cunico, R. F., 283, *291*
Cuppers, H. G. A. M., 61, *76*
Curry, J. W., 279, 282, *291*
Czajkowski, G. J., 31, *56*

D

Dang, T. P., 72, *74*
Daniels, R., 126, *142*
Danishefsky, S., 10, 11, *55*
Danno, S., 90, 91, *116, 118*, 154, *171*
Darboven, C., 210, *211*
Das, B., 185, *213*
Das, K. R., 15, 16, *55*
Das Gupta, A. K., 15, 16, *55*
Dauby, R., 185, *212*
Daughenbaugh, N. E., 285, *291*
Davidson, J. M., 80, 88, 89, 90, 101, *106*
Davies, J. E., 139, *144*
Davies, K. M., 13, *55*
Davies, N. R., 145, 149, *172*
Davis, A. G., 60, *74*
Davis, B. H., 209, *212*
Davis, G. T., 114, 115, *116*
Davis, J. W., 29, 37, 38, *53, 54*
Davis, S. M., 253, 254, *254*
Dawans, F., 185, *212*
Dawson, D. J., 260, *273*
Dawson, J. A., 61, *74*
Dean, F. M., 133, *142*
deCharentenay, F., 276, *291*
Dehn, W., 162, *172*
Delany, A. C., 283, *291*
DeLuca, E. S., 106, *115*
Dennis, W. E., 281, 284, *291*
Deno, N. C., 47, *55*
Dent, S. P., 288, *292*
Dent, W. T., 225, 248, 249, *255*
DePuy, C. H., 3, *55*
deRuggieri, P., 168, *172*
Detert, F. L., 3, 52, *55*

Dev, S., 138, *143*
de Waal, W., 63, *75*
Dewar, M. J. S., 13, *55*
Dewhirst, K. C., 61, 69, *74, 76,* 197, 198, 200, 201, 203, *212, 213*
Dhoubhadel, S. P., 52, *54*
Dickerson, R. E., 131, *142*
Dickinson, M. J., 133, *143*
Dietl, H., 193, 194, *212*
DiLuzio, J., 101, *120*
DiMichiel, A. D., 149, *172*
Dixon, J. A., 128, 134, *143*
Djerassi, C., 24, *57,* 62, 68, 69, *74, 76,* 133, *142*
Dmuchovsky, B., 64, *76*
Dockus, C. F., 280, *293*
Dodman, D., 253, *255*
Doering, W. von E., 3, *55*
Doganges, P. T., 133, *141*
Dolak, L. A., 114, *116*
Dolcetti, G., 61, *74*
Dolgaya, N. E., 277, *292*
Donaldson, M. M., 163, *173*
Donati, M., 148, *171*
Dorfman, L., 53, *58*
Dreiding, A. S., 51, *54, 55,* 159, 161, *171, 173,* 268, *272, 273*
Drenchko, P., 13, *57*
Dubac, J., 289, *292*
Dubeck, M., 140, *142,* 160, *173,* 265, *273*
Dubin, H., 49, *57*
Dunathan, H. C., 9, *59*
Duncan, W. P., 4, *56*
Dunlop, A. P., 265, *273*
Dunne, K., 178, *212*
Dunny, S., 274, 275, 283, *291*
Durand, D., 222, *255*
Durand, D. A., 12, *57*
Durham, L. J., 128, *142*
Dutta, P. C., 162, *173*
Dutton, H. J., 153, *172*
Duvall, H. M., 2, 26, *57*
Dyall, L. K., 92, *116*
Dzyuba, V. S., 151, *173*

E

Eaborn, C., 288, *292*
Eastham, J. F., 127, *142*
Eastman, R. H., 3, 52, *55*
Eaton, D. C., 152, *171*
Eaton, P. E., 79, *116,* 169, *171*
Eberhardt, G. G., 61, *74*
Eberson, L., 97, *116*
Economy, J., 49, *54,* 268, *272*
Egloff, G., 145, *172*
Egorov, Yu. P., 277, *292*
Ehmann, W. J., 191, *214*
Ehrenstein, M., 4, *55*
Ehrhart, O., 123, *142*
Eisenbraun, E. J., 4, 43, 50, *55, 56, 58*
Elderfield, R. C., 37, *55*
Else, M. J., 215, *255*
El Taijeb, O., 107, *119*
Emery, A., 260, *273*
Emken, E. A., 66, *74*
Engelhardt, H., 210, *211*
England, D. C., 29, 30, *53*
Engle, R. R., 133, *142*
Ennis, B. C., 14, *55*
Epstein, S., 264, *272*
Ercoli, A., 21, *55*
Erivanskaya, L. A., 21, *58*
Erner, W. E., 23, 24, *59*
Ernest, I., 122, *142*
Eschinazi, H. E., 261, 262, 263, *273*
Esse, R. C., 15, *55*
Evans, D., 62, *74,* 145, *172,* 215, 227 *255, 256*
Ezekiel, A. D., 133, *142*

F

Faget, C., 261, *273*
Fahnenstich, R., 233, *255*
Falbe, J. F., 215, 216, 230, 231, 232, 236, 247, *255, 256*
Fales, H. M., 26, *55*
Farkas, A., 112, *116*
Farmer, M. L., 202, *214*
Fateen, A. K., 51, *55*
Feder, J. B., 4, 26, *55*
Fedrick, J. L., 113, *117*
Fehlhaber, H. W., 133, *143*
Feldblyum, V. S., 151, *173*
Felix, A. M., 133, *142*
Felix, R. A., 291, *293*
Fell, B., 219, 226, 229, 247, *256*

Fenton, D. M., 98, *116,* 234, 235, 239, 240, 244, 245, *256,* 270, *273*
Ferebee, R., 26, *59*
Fetter, E. J., 49, *54*
Field, G. F., 15, *59*
Fields, R., 215, *255*
Fieser, L. F., 34, *55,* 121, *142*
Fieser, M., 34, *55,* 121, *142*
Filbey, A. H., 225, *256*
Filippi, J. B., 127, *144*
Fink, K., 110, *116*
Fink, R. M., 110, *116*
Fischer, J. L., 126, *142*
Fischer, N. H., 152, *172*
Fish, J. G., 276, *292*
Fisher, G. S., 47, *57*
Fisher, L. P., 87, *117,* 176, 177, *212*
Fitton, P., 88, 96, *116, 118*
Flaherty, B., 133, *142*
Flanagan, P. W., 43, 50, *55, 58*
Fleck, E. E., 39, *55*
Fleetwood, J. G., 104, *115*
Fleming, E. K., 177, *212*
Flid, R. M., 191, *211*
Follman, H., 133, *142*
Fornefeld, E. J., 29, *56*
Forsblad, I. B., 127, *142*
Foster, R. E., 113, *117*
Fotis, P., Jr., 61, *74*
Fouty, R. A., 89, *120*
Franco, P., 226, *255*
François, P., 87, *116*
Frank, J. K., 188, *211*
Frankel, E. N., 66, *74,* 228, *256*
Fraser, E., 113, *116*
Fraser, M. S., 61, *74*
Freiberg, L. A., 160, *171*
Freudewald, J. E., 32, *55*
Frevel, L. K., 89, *116*
Frew, D. W., Jr., 73, *74*
Friedel, R. A., 43, *58*
Fritze, D., 13, *56*
Frye, C. L., 282, *292*
Frye, G. H., 103, 104, *120*
Fryer, R. I., 133, *142*
Fuchs, R., 134, *141*
Fuchs, W., 82, *118*
Fuhlhage, D. W., 13, *55*
Fuji, K., 11, *55*
Fujimoto, H., 274, *292*
Fujita, E., 11, *55*
Fujiwara, Y., 82, 84, 88, 89, 90, 91, 92, *115, 116, 117, 118, 119,* 154, *171*
Furuta, S., 128, *142*
Fuson, R. C., 33, *55,* 113, *117*

G

Gaasbeck, M. M. P., 169, *174*
Gaignon, M. H. R. J., 282, *292*
Galat, A., 7, 51, *55*
Galbraith, A., 18, *55*
Gandilhon, P., 180, *211*
Gandolfi, C., 168, *172*
Gardi, R., 21, *55*
Gardiner, D., 133, *141*
Gardner, P. D., 44, *56*
Gardner, S., 187, 203, *212*
Garner, J. W., 152, *172*
Garnett, D. I., 265, *273*
Garrett, J. M., 6, *56*
Garves, K., 272, *273*
Gassman, P. G., 170, *172,* 184, *212*
Gault, F. G., 261, *273*
Gavin, D. F., 253, *257*
Geissman, T. A., 47, *56,* 157, *172*
Geller, H. H., 23, *58*
Gensler, W. J., 44, 49, *56*
Gentles, M. J., 51, *56,* 268, *273*
Georges, M. V., 33, *57*
Germain, J. E., 151, 163, *172, 173*
Gershon, R. B., 100, *117*
Gerson, F., 19, *54*
Gertner, D., 10, *54*
Geurts, A., 226, *256*
Ghiringhelli, D., 25, *56*
Gill, D. S., 61, *76*
Gill, N., 261, *273*
Gilman, H., 290, *292*
Giusnet, M., 151, *173*
Gladrow, E. M., 216, *256*
Glattfeld, J. W. E., 100, *117,* 122, 123, *142*
Glauco, S., 226, *255*
Glover, E. E., 99, *117*
Godfrey, J. C., 19, *56*
Goldblatt, L. A., 47, *57*
Goldup, A., 216, *256*
Gomez-Gonzales, L., 97, *116*
Goodall, B. L., 73, *74*
Goodman, L., 279, *292*

Gopal, H., 133, 136, *142, 143*
Gorman, E. H., 176, 177, *212*
Gornowicz, G. A., 274, *292*
Gosser, L. W., 66, *74*
Gostunskaya, I. V., 151, 153, *171, 172*
Goto, Y., 92, *117*
Gourlay, G., 82, *117*
Grand, P. S., 131, *142*
Grard, C., 180, *211*
Graziani, M., 225, *256*
Green, H. A., 22, *56*
Green, M., 215, *257*
Gregorio, G., 208, 209, *213*
Griffith, W. P., 127, *142*
Griswold, A. A., 88, *118*
Gronowitz, J. S., 69, *75*
Gronowitz, S., 69, *75*
Gross, D.E., 83, 84, *119*
Grubbs, R. H., 61, *74*
Grummitt, P., 268, *273*
Guest, H. R., 23, *56*
Guistiniani, M., 61, *74*
Gunstone, F. D., 122, *142*
Gupta, S. K., 62, *75*
Gutzwiller, J ., 62, 68, 69, *74*
Guyer, P., 13, *56*
Guzzi, U., 168, *172*

H

Haag, W. O., 216, *256*
Haas, H. J., 159, *172*
Haddad, Y. M. Y., 42, *56,* 61, *74*
Haensel, V., 163, *173*
Hafner, W., 77, 78, 80, *117, 119*
Hager, G. F., 30, 38, *54*
Hagihara, N., 186, 196, 199, 200, 201, 203, *214,* 276, 279, *293*
Hahn, H. J., 210, *211*
Hall, M. J., 4, *56*
Hall, R. F., 271, *273*
Halliday, D. E., 86, *115,* 204, *211*
Hallman, P. S., 62, *74,* 145, *172,* 215, *256*
Halpren, J., 61, 70, *74, 75,* 79, *116,* 168, 169, *171, 172,*
Hamamoto, K., 22, *59*
Hamilton, C. S., 12, *54*
Hammack, E. S., 207, *212*
Hamming, M. C.,43, 50, *55, 58*
Hancock, R. D., *254*
Handa, K. L., 156, *171*
Handford, B. O., 52, *54*
Hanessian, S., 137, *141*
Hara, M., 182, 194, 204, 206, *213,* 244, *258,* 276, *292*
Hardegger, E., 104, *117*
Hardt, P., 187, *214*
Hardtmann, G., 85, *118*
Hardy, W. B., 252, 253, 254, *254, 256*
Hargis, C. W., 78, *117*
Harkness, A. C., 70, *74*
Harmon, R. E., 62, *75*
Harris, L. E., 4, *56*
Harrison, G. W., Jr., 279, *291*
Harrison, W. F., 2, 3, 14, *54*
Harrod, J. F., 61, *75,* 147, 148, *172,* 274 276, 277, *291*
Hartley, F. R., 145, *172*
Hartwell, G. E., 62, *75*
Hasan, S. K., 140, *144*
Hasbrouck, L., 260, *273*
Hasegawa, N., 176, *212*
Hashimoto, H., 78, *118*
Hashimoto, M., 155, *174*
Haszeldine, R. N., 215, *255,* 276, 277, 283, *291,* 292
Hata, G., 199, 204, 205, 206, *212, 214*
Hatchard, W. R., 113, *117*
Hattori, K., 286, *292*
Hauser, C. R., 47, *57*
Hawthorne, J. O., 261, 264, *273*
Hay, A. S., 45, *56*
Hayashi, S., 51, *56*
Hayden, P., 81, 83, 86, 87, *116,* 237, *256*
Hayes, J. C., 163, *173*
Haynes, P., 207, *212,* 223, *256*
Heathcock, C. H., 65, *75*
Heck, R. F., 93, 94, *117,* 290, *292*
Heil, B., 73, *75,* 227, *256*
Heilbronner, E., 19, *54*
Hellier, M., 147, 148, *171*
Helmbach, P., 187, *214*
Hemidy, J. F., 261, *273*
Henbest, H. B., 42, *56,* 61, 73, *74, 76,* 152, *171*
Henrici-Olivé, G., 216, *256*
Henry, D. W., 132, *142*
Henry, P. M., 78, 81, 83, 89, *117,* 155,166, *172*
Henzel, R. P., 169, *173*
Herald, D. L., Jr., 161, *173*

Herbst, G., 135, 140, *143*
Hernandez, L., 2, *56*
Hershberg, E. B., 34, 51, *55, 56,* 268, *273*
Hershman, A., 217, 224, 228, *255, 258*
Herz, W., 47, *56,* 157, *172*
Herzog, H. L., 51, *56,* 268, *273*
Hess, B. A., Jr., 113, *117*
Hester, J. B., Jr., 8, *56*
Hetflejš, J., 216, *255,* 276, *291*
Hettinger, W. P., 163, *172*
Heusler, K., 132, 133, *144*
Hewett, W. A., 185, *211*
Heyns, K., 101, 104, 105, 107, 108, 109, 110, 111, *117*
Hickner, R. A., 274, 282, *291*
Hidai, M., 180, *213*
Hijikata, K., 86, *119*
Himelstein, N., 68, *76*
Himmele, W., 224, 232, *256*
Hinman, C. W., 43, 50, *55, 58*
Hirota, K., 84, *119*
Hise, K., 49, *57*
Hiser, R. D., 2, *56*
Ho, R. K. Y., 188, *213*
Hoa, H. A., 36, *57*
Hobbs, C. C., 129, *143*
Hodgkins, J. E., 29, *56*
Hoff, D. R., 132, *142*
Hoffman, N. E., 263, *273*
Hoffman, N. W., 70, *74*
Hoffmann, R., 169, *174*
Hoffmann, W., 232, *256*
Hofmann, K. A., 123, 124, *142*
Hogenkamp, H. P. C., 133, *142*
Hogeveen, H., 168, 169, *172, 174*
Hohenschutz, H., 224, *256*
Hokin, L. E., 107, *118*
Holden, K. G., 14, *58*
Holler, H. V., 69, *76*
Holly, F. W., 133, *143*
Honwad, V. K., 70, *76*
Horibe, I., 22, *59*
Horino, H., 94, *117*
Horner, L., 71, *75,* 147, *172*
Hornfeldt, A.-B., 69, *75*
Hornig, L., 95, *115*
Horning, E. C., 2, 8, 26, 47, 53, *56,* 168, *172*
Horning, M. G., 2, 8, 26, 47, 53, *56*
Horton, W. J., 44, *56*
Hosaka, S., 243, 244, 245, *256, 258*
Hoshino, T., 8, *56*
Hosking, J. W., 82, 84, 85, *116,* 224, *255*
Hossain, A. M. M., 134, *141*
House, H. O., 35, 47, *56,* 133, 136, *142*
Howarth, G. B., 137, *142*
Howsam, R. W., 3, *56,* 61, 62, *75*
Hoyle, K., 38, *54*
Hu, S.-E., 160, *171*
Hubert, A. J., 145, *172*
Hudec, J., 219, *255*
Hüttel, R., 78, *117*
Huffman, G. W., 265, *273*
Hughes, M. P., 29, *56*
Hughes, P. R., 241, 242, 243, *255*
Hughes, R. P., 206, *212*
Hughes, V. L., 232, 238, *254, 256*
Hughes, W. B., 149, *172*
Hui, B., 61, *75*
Hulla, G., 145, *172*
Huntsman, W. D., 152, *172*
Huppes, N., 230, 231, 232, *255, 256*
Hurwitz, M. D., 210, *212*
Husbands, J., 42, *56,* 61, *74*
Hussey, A. S., 39, *56,* 69, *75*
Hwang, Y. T., 209, *212*

I

Ichinohe, Y., 161, *173*
Ide, J., 260, *273*
Igbal, M. Z., 73, *74*
Igeta, H., 210, *212*
Ihara, M., 152, 153, *172*
Ihrman, K. G., 225, 249, *255, 256*
Ikeda, Y., 61, *75*
Ikegami, K., 88, *117*
Ikuta, M., 22, *59*
Iliopulus, M. I., 125, *143*
Illingworth, G. E., 24, *56*
Imafuku, K., 24, *57*
Imamura, S., 236, 248, 249, 250, 251, *256, 258*
Imanaka, T., 82, 84, 91, *117, 119,* 176, 178, *212*
Inglis, H. S., 207, *212*
Inomata, I., 180, *213*
Inouye, S., 137, *142*
Inque, N., 94, *117*

Ioffe, I. I., 109, *117*
Ipatieff, V. N., 29, 30, 31, *56, 58*
Iqbal, A. F. M., 223, 254, *256*
Ireland, R. E., 130, 131, *142,* 260, *273*
Irwin, W. J., 40, *56*
Isayama, K., 183, *213*
Isegawa, Y., 226, *258*
Ishii, K., 176, *212*
Ishii, Y., 183, *213*
Ismail, S. M., 60, *75*
Itatani, H., 66, *74, 75,* 89, *117*
Ito, M., 78, *119*
Ito, T., 137, *142*
Iwakura, Y., 286, *292*
Iwamoto, M., 204, *212*
Iwamoto, N., 218, 222, 226, *258*
Iwamoto, R., 17, *58*
Iwashita, Y., 193, *212,* 237, 240, *256*
Iwaski, T., 64, *76*
Iwata, R., 61, *75*
Iyer, K. N., 140, *143*
Izumi, T., 92, *117*

J

Jackman, L. M., 38, 40, *56, 57*
Jackson, G. D. F., 17, 18, *56*
Jackson, W. R., 160, *173*
Jacobson, G., 218, *256*
James, B. R., 60, 61, 68, *74, 75,* 82, *117*
Jansen, H., 124, *142*
Jardine, F. H., 60, 61, 62, *74, 75, 76,* 215, 219, *255, 257*
Jardine, I., 61, 62, 64, 72, 73, *74, 75*
Jeger, O., 127, 133, *142, 144*
Jenkins, E. F., 29, 30, *58*
Jenner, E. L., 60, *74,* 176, 177, 179, 189, 191, *211,* 239, 245, *255, 257*
Jensen, H. B., 69, *75*
Jira, R., 77, 78, 80, 81, *117, 119*
Johns, W. F., 42, *56*
Johnson, A. L., 28, *54*
Johnson, F., 49, *56*
Johnson, W. S., 130, *143*
Johnston, D. E., 164, *172*
Jolly, J. G., 265, *273*
Jonassen, H. B., 149, 154, *174*
Jones, E. W., 86, *116*
Jones, F. N., 181, 186, *212*
Jones, J. K. N., 137, *142, 143*
Jones, P. R., 283, *291*
Jones, R. G., 29, *56*
Jonkhoff, T., 82, *120*
Jouy, M., 112, *117*
Joy, J. R., 276, *292*
Jung, M. J., 63, 64, *74*

K

Kagan, H. B., 72, *74*
Kahovec, J., 50, *56*
Kajimoto, T., 238, *258,* 266, *273*
Kakáč, B., 122, *141*
Kakisawa, H., 43, *56*
Kaliya, O. L., 191, *211*
Kalvoda, J., 133, *144*
Kametani, T., 152, 153, *172*
Kanai, H., 180, *213*
Kanakkanatt, A. T., 263, *273*
Kaneko, Y., 7, *59*
Kang, J. W., 61, *76,* 191, 192, *211*
Karakhanov, R. A., 15, *58*
Karkowski, F. M., 160, *171*
Karol, F. J., 39, *56*
Karpenko, I., 18, *58*
Kasahara, A., 92, *117*
Kathawala, F., 85, *118*
Kato, T., 111, *120*
Katsui, N., 107, *119*
Katsuno, R., 176, *212*
Katsuyama, Y., 78, *118*
Katz, T. J., 169, *172,* 187, 188, *211, 212, 213*
Kaufmann, H. P., 124, *142*
Kawamoto, K., 91, *117,* 176, 178, *212*
Kawashima, Y., 183, *213*
Kaye, B., 51, *59*
Kazanskii, B. A., 151, 153, *171, 172*
Kealy, T. J., 245, 246, *257*
Keblys, K. A. 140, *142*
Kehoe, L. J., 215, 239, *257*
Keim, W., 187, 197, 198, 203, *212, 213, 214*
Keith, C. D., 163, *172*
Kemp, A. L. W., 61, *74*
Kende, F., 41, *54*
Kenyon, W. G., 47, *56*
Kepler, J. A., 7, *59*

Kern, R. J., 193, *212*
Ketley, A. D., 83, 87, *117,* 148, *172,* 176, 177, *212*
Ketter, D. C., 271, *273*
Khan, M. M. T., 61, *75*
Kierstead, R. C., 130, *142*
Kiff, B. W., 23, *56*
Kiji, J., 236, 239, 248, 249, 250, *258*
Kikuchi, S., 78, *119*
Kilroy, M., 4, 26, *58,* 100, 109, *119*
Kindler, K., 162, *172*
Kirk, D. N., 42, *54*
Kiss, J., 105, *117*
Kitahara, Y., 111, *120*
Kitamura, T., 112, *118*
Kitching, W., 82, 83, *117*
Klein, H. S., 176, 184, *212*
Klimova, N. V., 109, *117*
Kline, G. B., 29, *56*
Kloetzel, M. C., 44, *58*
Knight, J. C., 133, *142*
Knowles, W. S., 71, 73, *75*
Kobayashi, Y., 179, *212*
Kober, E. H., 253, *257*
Koch, W., 104, *117*
Kochetkov, N. K., 130, *142*
Koenig, K. E., 291, *293*
Kohll, C. F., 81, 82, *118, 120,* 233, 249, 250, *257*
Kohnle, J. F., 186, 201, *212,* 223, *256*
Kojer, H., 77, 78, *119*
Kolarikal, A., 133, *141*
Komarewsky, V. I., 145, *172*
Konrad, F. M., 32, *55*
Kornfeld, E. C., 29, *56*
Korshak, V. V., 283, *292*
Korte, F., 247, *256*
Koster, R. A., 276, *292*
Kottke, R. H., 113, *117*
Kraihanzel, C. S., 283, 290, *292*
Kratzer, J., 78, *117*
Kraus, S., 266, *272*
Kraus, T. C., 253, *257*
Krause, J. H., 23, 24, *59*
Krauth, C. A., 127, *142*
Kreidl, J., 68, *76*
Kreienbuhl, P., 267, *273*
Krekeler, H.,
Kressley, L. J., 89, *116*
Krewer, W. A., 209, *212*
Kripalani, K. J., 69, *74*
Kröner, M., 187, *214*
Kroll, L. C., 61, *74*
Kruse, W., 125, *142*
Kubota, M., 82, 84, 85, *116,* 224, *255*
Kubota, T., 36, *56*
Kuhn, R., 159, *172*
Kuivila, H. G., 275, *292*
Kumada, M., 278, 289, *293*
Kuo, C. H., 11, *59*
Kupchan, S. M., 107, *118, 119*
Kuper, D. G., 232, *257*
Kurematsu, S., 176, *212*
Kurkov, V. P., 85, 98, *118*
Kuroda, Y., 7, *59*
Kusumi, T., 43, *56*
Kusunoki, Y., 176, *212*
Kyrides, L. P., 23, *57*

L

Lafont, P., 149, *172*
Lago, R. M., 2, *57*
Lagowski, J. M., 37, *55*
Laing, S. B., 65, *75*
Lake, R. D., 8, *54*
Lambert, B. F., 53, *58*
Lamoreaux, H. F., 276, *292*
Langer, E., 260, *273*
Lapan, E. I., 7, *59*
Larkin, D. R., 23, *57*
Larson, J. K., 47, *56*
Lasky, J. S., 149, *173*
Lassau, C., 222, *255*
Laundon, R. D., 16, *54*
Lavigne, J. B., 85, 98, *118*
Lawrenson, M. J., 215, 227, *257*
Lawton, B. T., 137, *142*
Lazutkin, A. M., 187, *212*
Lazutkina, A. I., 187, *212*
Leder-Packendorff, L., 2, *57*
Lednicer, D., 2, 38, 47, *57*
Lee, H. B., 61, *76*
Lefort, M. J. C., 282, *292*
Legzdins, P., 73, *75*
Lehman, D. D., 61, *75*
Leites, L. A., 277, *292*
LeMaistre, J. W., 47, *57*
Lemberg, S., 150, *172*

Lemieux, R. U., 130, *143*
Leonard, N. J., 12, *57,* 167, 168, *172*
Leonova, A. I., 153, *172*
L'eplattenier, F., 254, *257*
Lesbre, M., 289, *292*
Levanda, O. G., 84, *118*
Levin, C., 79, *118*
Levina, R. Y., 151, *172*
Levine, S. G., 7, *59*
Lewy, G. A., 105, *118*
Li, G. S., 274, 275, *291*
Liao, C.-W., 5, *54*
Liberman, A. L., 160, *171*
Liesenfelt, H., 70, *74*
Linder, D. E., 50, *55, 58*
Lindsey, R. V., Jr., 60, *74,* 146, 148, *171,* 176, 177, 179, 186, 189, 191, *211, 212,* 239, 245, *254, 255, 257*
Lines, C. B., 218, 219, *257*
Ling, N. C., 24, *57*
Lini, D. C., 133, *143,* 166, *172*
Linn, W. J., 181, *212*
Linsen, B. G., 66, *76*
Linsk, J., 29, *57*
Linstead, R. P., 2, 3, 29, 30, 35, 38, 39, 45, 50, *54, 57*
Lipman, C., 51, *55*
Lipshes, Z., 269, *272*
Little, J. C., 167, *172*
Little, W. F., 191, *211*
Lloyd, W. G., 87, 88, 101, *118*
Lochte, H. L., 9, *57*
Locke, D. M., 9, 48, *57,* 168, *172*
Loew, O., 30, *57*
Lohse, F., 104, *117*
Long, R., 218, 219, 225, 248, 249, *255, 257*
Lorenc, W. F., 163, *172*
Losee, M. L., 283, *292*
Louvar, J. J., 24, *56*
Luberoff, B. J., 87, 88, *118*
Lucas, R. A., 53, *58*
Lukes, R. M., 49, *58*
Lumb, J. T., 184, *212*
Lunk, H. E., 69, *76*
Lun-Syan, T., 160, *171*
Luttinger, L. B., 192, *212*
Lutz, E. F., 100, 109, *118, 120*
Lyons, J. E., 60, 71, *75,* 84, *118,* 145, 146, 149, *173,* 274, 285, 287, *291, 292*

M

Mabry, T. J., 63, 64, *76,* 152, 159, *172, 173*
McCall, E. B., 265, 271, *272, 273*
McCasland, G. E., 128, *142*
McClain, D. M., 113, *118*
McClure, J. D., 34, *59,* 180, *213*
McCollum, J. D., 61, *74*
McCrindle, R., 156, *171*
McCurdy, O. L., 37, *55*
McDevitt, J. P., 28, *54*
McDonagh, P. M., 278, *292*
McDonald, W. S., 61, *75*
McEuen, J. M., 151, 160, *173*
McGarvey, B. R., 62, *74*
Mackay, F. P., 279, *292*
McKenzie, S., *254*
McKeon, J. E., 78, 88, 89, 90, 96, *116, 118,* 202, *211*
McKervey, M. A., 160, 164, *172, 173*
MacLean, A. F., 79, *118,* 129, 134, *142, 143*
MacMillan, J., 157, *171*
McQuillin, F. J., 3, *56,* 61, 62, 64, 71, 72, 73, *74, 75,* 178, *212*
McQuillin, J., 157, *173*
Macrae, T. F., 99, *118*
MacSweeney, D. F., 130, *142,* 151, *173*
Maddock, S. J., 154, *171*
Madison, N. L., 152, *172*
Madison, R. K., 253, *254*
Mador, I. L., 113, *118,* 225, *237,* 248, 249, *257, 258*
Maggiolo, A., 37, *55*
Mague, J. T., 61, *75,* 147, *171,* 267, *272*
Maier, C. A., 188, *211*
Maitlis, P. M., 61, 71, *75, 76,* 78, 82, *118,* 193, 194, *211, 212,* 215, *257*
Majima, R., 36, *57*
Makeev, A. G., 109, *117*
Malone, J. F., 150, *171*
Maloney, L. S., 126, *143*
Manassen, J., 169, *173*
Manchard, B., 122, *142*
Mangham, J. R., 46, *57,* 269, *273*
Mango, F. D., 168, *173*
Manly, D. G., 265, *273*
Mann, G., 160, *173*
Mann, M. J., 29, *56*
Manoharan, P. T., 61, *75*
Manyik, R. M., 185, 201, 202, 207, *211, 212, 214*

Marburg, S., 44, *56*
Marco, M., 151, *173*
Marechal, J., 30, *58*
Marggraff, I., 4, *55*
Marko, L., 73, *75*, 227, *256*
Marples, B. A., 225, *257*
Marschik, J. F., 42, *57*
Marsh, C. A., 101, 105, *118*
Marsico, W. E., 243, *255*
Martin, J. C., 133, *143*
Martin, M. M., 276, *292*
Masada, H., 230, *258*
Masamune, S., 169, *173*
Mason, T. J., 86, *115*
Masters, C., 61, *75*
Mastikhin, V. M., 187, *212*
Mathauser, G., 210, *211*
Mathew, C. T., 162, *173*
Matsuda, A., 238, *257*
Matsuda, M., 91, *116*
Matsuda, S., 112, *119*
Matsui, M., 39, *57*
Matsumura, H., 24, *57*
Matsuo, A., 51, *56*
Matsuura, S., 24, *57*
Matsuura, T., 51, *56*
Matthews, J. S., 271, *273*
Matthys, P., 254, *257*
Mattox, W. J., 216, *256*
Maurel, R., 151, *173*
Maxted, E. B., 60, *75*
Mayell, J. S., 133, *142*
Mayer, J. R., 3, *55*
Mayo, F. R., 210, *212*
Mays, M. J., 215, *255*
Mazerolles, P., 289, *292*
Mears, D. E., 216, *257*
Mechoulam, R., 139, *143*
Medema, D., 82, *120*, 187, *213*, 233, 249, 250, *257*
Mehltretter, C. L., 102, 104, *118*
Meinerts, U., 127, *141*
Meinwald, Y. C., 48, *54*
Meisels, A., 139, *144*
Meisinger, E. E., 29, *56*
Mellies, R. L., 102, *118*
Menzie, G. K., 280, *292*
Merker, R. L., 286, *292*
Merzoni, S., 218, *255*
Meschino, J. A., 107, *119*
Meuche, D., 19, *54*
Meystre, C., 133, *144*
Michael, K. W., 274, *292*
Michaelis, K. O. A., 3, 30, 39, 45, 50, *57*
Miescher, K., 127, *142*
Milas, N. A., 123, 124, 125, 126, 128, *142, 143*
Miles, D. H., 290, *292*
Miles, G. B., 127, *142*
Miller, L. A., 168, *172*
Millidge, A. F., 2, 30, *57*, 216, 217, *257*
Milstein, D., 269, *272*
Minamida, H., 186, *213*
Minato, J., 22, *59*
Minn, J., 49, *57*
Mirrington, R. N., 130, *143*
Misono, A., 180, *213*
Misono, M., 148, *173*
Mitchell, P. W. D., 38, *54, 57*
Mitchell, R. W., 73, *75*
Mitchell, T. R. B., 42, *56*, 61, 64, 65, *74, 75*
Mitra, R. B., 133, *142*
Mitsuhashi, H., 260, *273*
Mitsui, S., 64, *76*
Mitsuyasu, T., 204, 206, 207, *213*
Miyake, A., 199, 204, 205, 206, *212, 214*
Mizoroki, T., 211, *213*
Mleczak, W., 42, *57*
Modic, F. J., 276, 285, *292*
Moffat, J., 194, *212*
Mohan, A. G., 254, *257*
Moiseev, I. I., 78, 82, 83, 84, 86, 87, 101, *115, 118, 120*
Mokotoff, M., 107, *118*
Montelatici, S., 61, *75*
Montgomery, P. D., 225, *258*
Mooberry, J. B., 85, *118*
Moody, G. J., 127, *143*
Moore, D. W., 154, *174*
Morandi, J. R., 69, *75*
Moreau, J. J. E., 289, *291*
Morgan, C. R., 176, 177, *212*
Mori, K., 39, *57*, 211, *213*
Mori, T., 7, *57*
Mori, Y., 244, *258*
Moriarty, R. M., 133, 136, *142, 143*
Morikawa, M., 218, 226, 230, 236, 239, 248, 249, 250, *257, 258*
Moritani, I., 88, 90, 91, 92, *115, 116, 117, 118*, 154, *171*

Moro-oka, Y., 112, *118*
Morris, D. E., 70, *74*, 227, *258*
Morris, G. H., 99, *117*
Morrison, D. E., 29, *56*
Morrison, J. D., 73, *75*
Morrow, C. J., 73, *75*
Morton, M., 185, *213*
Moseley, K., 71, *75*
Mosettig, E., 2, 26, *57*
Moss, G. P., 105, *118*
Moss, J. B., 51, *56*, 268, *273*
Mozingo, R., 9, *57*
Mrowca, J. J., 187, *213*
Müller, E., 158, *173*
Mugdan, M., 125, 126, *143*
Muller, E., 99, *118*, 139, *143*, 226, *256*, 260, *273*
Muller, H. C., 47, *56*
Mullineaux, R. D., 226, *258*
Murahashi, S., 36, *57*
Murphy, F. X., 7, 51, *58*
Murphy, R. A., 276, *292*
Musolf, M. C., 277, *292*
Muxfeldt, H., 85, *118*

N

Nagao, Y., 176, *212*
Nagasaki, T., 22, *59*
Naito, T., 7, *57*
Nakagawa, T., 130, *141*
Nakai, T., 170, *172*
Nakajima, M., 210, *212*
Nakamaye, K. L., 186, 201, *212*
Nakamura, A., 186, *213*
Nakamura, N., 260, *273*
Nakamura, S., 81, *118*
Nakata, H., 136, *143*
Nakazawa, T., 112, *119*
Nanaumi, K., 130, *141*
Nebzydoski, J. W., 87, *115*
Nelson, J. A., 26, 44, *54*
Nelson, L. E., 274, 275, 289, *291*, *293*
Nerlekar, P. G., 274, 275, 283, *291*
Nettleton, D. E., Jr., 26, *59*
Neubauer, D., 215, 216, 218, 236, 238, 239, 240, 241, 242, 244, 247, *255*, *258*, *259*
Newman, M. S., 2, 3, 26, 31, 33, 38, 43, 46, 47, 50, *57*, 154, *173*, 261, 269, *273*
Ng, F. T. T., 68, *75*
Nicholson, J. K., 150, *171*
Nicolini, M., 61, *74*
Nielsen, A. T., 6, 49, *57*
Nielsen, J. M., 282, *292*
Nikiforova, A. V., 101, *118*
Ninomiya, I., 7, *57*,
Nishimura, S., 68, *75*, 97, *119*
Nitzsche, S., 280, *292*
Nixon, A. C., 4, *58*
Niznik, G., 262, *273*
Nogi, T., 219, 220, 221, 241, *257*, *258*
Nohe, H., 82, *118*
Nolan, J. T., Jr., 125, *143*
Noma, T., 78, *118*
Nomine, G., 127, *141*
Nord, F. F., 2, *56*
Norell, J. R., 166, *173*
Norman, R. O. C., 92, *116*
Norton, C. J., 125, 126, 127, *143*
Nutt, R. F., 133, *143*
Nyilasi, J., 129, *143*
Nyman, C. J., 262, *272*

O

Oberender, F. G., 128, 134, *143*
Oberkirch, W., 187, *214*
Obeschchalova, N. V., 151, *173*
Oblerg, R. C., 31, *58*
O'Brien, D. H., 274, *293*
Ochiai, E., 82, *117*
O'Connor, C., 61, *75*
Oda, O., 260, *273*
Odaira, Y., 97, *119*, 182, 194, *213*
Oehlschlager, A. C., 260, *273*
Oesterlin, R., 10, *54*
Oftedahl, M. L., 111, *119*
Oga, T., 78, *119*
Ogata, I., 61, *75*
Ogawa, S., 105, 107, *119*
O'Halloran, J. P., 265, *273*
Ohnishi, S., 92, *117*
Ohno, K., 207, 208, *213*, 222, 238, *258*, 262, 263, 264, 266, 267, *273*, 276, *292*
Ohno, M., 133, *142*
Oh-uchi, K., 112, *119*

Oka, S., 23, *57*
Okada, H., 78, *118*
Okinoshima, H., 278, 289, *293*
Okita, H., 278, *293*
Olah, G. A., 267, *273*
Oldroyd, D. M., 47, *57*
Olechowski, J. R., 149, 154, *174*
O'Leary, T. J., 46, 50, *57*
Olivé, S., 216, *256*
Olivier, K. L., 216, 228, 234, 239, 240, *256, 257*
Oppenheimer, E., 267, *272*
Orchin, M., 145, *173,* 215, 216, *257, 259*
Ordzhonikidze, S., 7, *59*
Osborn, J. A., 60, 61, 62, 73, *74, 75, 76,* 145, 147, *171, 172,* 215, 219, 227, *255, 256, 257,* 267, *272,* 276, *291*
Oshima, A., 251, *257*
Otsuka, S., 186, *213*
Ottman, G. F., 253, *257*
Ouellette, R. J., 79, *118*
Overend, W. G., 133, *141, 142*
Overton, K. H., 156, *171*
Ovsyannikova, I. A., 187, *212*
Owen, L. N., 126, *143*
Owyang, R., 180, *213*
Ozaki, A., 112, *118,* 211, *213*

P

Packendorff, K., 2, *57*
Pal, B. C., 9, *57*
Palkin, S., 39, *55*
Palladino, N., 192, *211*
Pallaud, R., 36, *57*
Palous, S., 128, *143*
Panchenkov, G. M., 163, *173*
Pannell, K. H., 283, *291*
Pappo, R., 130, 138, 140, *143*
Paquette, L. A., 168, 169, *173*
Parikh, I., 159, *173*
Parikh, V. M., 137, *143*
Parish, R. V., 276, 277
Parry, D. J., 276, 277, *292*
Parshall, G. W., 247, *257*
Parsons, J. L., 62, *75*
Pascal, Y. L., 159, *173*
Pasedach, H., 232, *256*
Pasky, J. Z., 85, 98, *118*
Pasternak, I. S., 112, *118*
Paterson, W., 113, *116*
Patrick, J. B., 99, *120*
Patterson, J. M., 13, *57*
Patton, D. S., 170, *172*
Paul, I. C., 188, *211, 212*
Paulik, F. E., 216, 217, 224, 228, *255, 257, 258*
Paulsen, H., 101, 105, 107, *117*
Pavlath, A. E., 272, *272*
Pearson, K. W., 253, *255*
Pek, Yu. G., 83, *115*
Pelletier, S. W., 9, 48, *57,* 140, *143,* 161, *173*
Pendleton, L., 12, *54*
Perichon, J., 128, *143*
Perras, P., 111, *115*
Peschke, W., 162, *172*
Peterson, R. C., 44, *58*
Petrov, A. D., 277, *292*
Petrov, D. A., 151, *172*
Pettit, R., 149, *171*
Pfefferle, W. C., 163, *174*
Phillips, C., 73, *75*
Phillips, D. D., 34, *58*
Phillips, F. C., 77, *118*
Piatak, D. M., 135, 139, 140, *141, 143*
Pickholtz, Y., 266, *272*
Pickles, V. A., 149, *172*
Pidcock, A., 288, *292*
Pierdet, A., 127, *141*
Piers, E., 61, 63, *75*
Pierson, W. G., 53, *58*
Pietra, S., 154, *173*
Pike, R. A., 278, *292*
Pines, H., 29, 30, 31, *56, 57,* 262, *273*
Pino, P., 217, 226, *255, 258*
Piszkiewicz, L. W., 62, 67, *74*
Pitkethly, R. C., *254*
Pittman, A. G., 9, *57*
Platt, E. J., 2, 8, 53, *56*
Platz, R., 82, *118*
Plonsker, L., 151, 160, *173*
Poist, J. E., 290, *292*
Pollitzer, E. L., 163, *173*
Polyakova, A. M., 283, *292*
Poos, G. I., 49, *58*
Pospisil, J., 50, *56*
Possanza, G., 69, *74*
Post, G. G., 106, *115, 118*

Potts, G. O., 10, *54*
Poulter, S. R., 65, *75*
Powell, J., 150, *171*, 206, *212*
Prater, C. D., 2, *57*
Pregaglia, G. F., 148, *171*, 208, 209, *213*
Prelog, V., 9, 21, *58*
Prince, R. H., 262, *273*
Pritchard, W. W., 253, *258*
Proctor, G. R., 113, *116*
Pruett, R. L., 180, *211*, 226, 227, 231, *258*
Pummer, W. J., 51, *55*, 268, *273*
Puterbaugh, W. H., 154, *173*
Puthenpurackel, T., 263, *273*

Q

Quigley, S. T., 269, *272*
Quinn, H. A., 160, *173*

R

Rabjohn, N., 2, *58*
Rae, D. S., 38, *54*
Ragaini, V., 38, *55*
Rahman, A., 6, *58*
Ramage, R., 130, *142*, 151, *173*
Ramey, K. C., 133, *143*, 166, *172*
Rank, J. S., 151, *171*
Raper, G., 61, *75*
Rapoport, H., 11, 14, 17, *58*
Rappoport, Z., 82, 83, *117*
Rasberger, M., 123, *144*
Raspin, K. A., 262, *273*
Rauch, F. C., 95, *115*
Ream, B. C., 89, 90, 96, *116*
Ree, B. R., 133, *143*
Reese, C. B., 105, *118*
Reggel, L., 43, *58*
Regnault, A., 20, *54*
Reimlinger, H., 145, *172*
Reiners, R. A., 102, *118*
Reinheimer, H., 194, *212*
Reinhold, D. F., 43, 51, *54*
Reis, H., 215, 216, 236, 238, 240, 241, 242, *255*
Rempel, G. L., 60, 68, 73, *75*, 223, *255*
Renner, W., 210, *211*
Rennick, L. E., 60, *75*
Reppe, W., 158, *173*
Rhodes, R. E., 60, *76*
Richards, L. M., 37, *54*
Richter, H. J., 155, *173*
Rick, E. A., 182, *211*
Riehl, J. J., 47, *56*
Rietz, E., 123, *142*
Rinehart, R. E., 149, 150, 154, *173*, 185, *213*
Rist, C. E., 102, 104, *118*
Ritchie, A. W., 4, *58*
Ritter, J. J., 7, 51, *58*
Roberts, H. L., 129, *141*
Robinson, D. A., 19, *53*
Robinson, J. M., 49, *58*
Robinson, K. K., 216, 217, *257*, *258*
Robison, B. L., 47, *58*
Robison, M. M., 47, *53*, *58*
Rodd, E. H., 157, *173*
Roderick, H. R., 108, *116*
Rodriguez, H. R., 131, *143*
Rodriguez, N. M., 6, *58*
Rogalski, W., 85, *118*
Rogers, J. L., 47, *56*
Rohwedder, W. K., 153, *172*
Rolfe, J. R. K., 163, *171*
Romeyn, H., Jr., 154, *173*, 185, *213*
Rona, P., 13, *55*
Rony, P. R., 61, *76*, 81, *119*, 216, *258*
Rooney, J. J., 160, 164, *172*, *173*
Rosenblatt, D. H., 114, 115, *116*
Rosenblum, M., 99, *120*
Rosenman, H., 98, *116*, 267, *272*
Rosenstock, P. D., 9, *58*
Rosenthal, A., 138, *143*
Roth, J. F., 216, 217, 224, 228, *255*, *257*, *258*
Roth, R. J., 188, *211*
Rottenberg, M., 99, *119*
Roychaudhuri, D. K., 36, *59*
Rucker, J. T., 210, *213*
Ruddick, J. D., 73, *75*
Rudzik, A. D., 8, *56*
Ruesch, H., 63, 64, *76*
Rupilius, W., 216, 229, 247, *256*, *257*
Ruttinger, R., 77, 78, *119*
Ryan, J. W., 274, 280, 281, 282, *291*, *292*
Rydjeski, D. R., 131, *142*

Rylander, P. N., 1, 4, 18, 26, 38, 42, *57, 58,* 60, 62, 63, 65, 68, *74, 76,* 100, 109, *119,* 124, 133, *141, 143,* 151, 152, 156, *173,* 260, *273*

S

Saam, J. C., 274, 277, *292*
Sabacky, M. J., 71, 73, *75*
Saegusa, T., 183, *213,* 226, *258*
Sagar, W. C., 31, 33, *57*
Saito, G., 92, *117*
Saito, R., 92, *117*
Saito, Y., 148, *173*
Sajus, L., 61, *74*
Sakai, K., 260, *273*
Sakai, M., 169, *173*
Sakai, S., 183, *213*
Sakakibara, T., 97, *119*
Sakuraba, M., 237, 240, *256*
Sam, T. W., 31, *54*
Samuelson, G. E., 124, *144*
Sander, W. J., 209, *212*
Sandermann, W., 130, *143*
Sanderson, T. F., 49, *57*
Sandhu, R. S., 107, *118*
Santambrogio, A., 192, *211*
Sarel, S., 139, *143*
Sarett, L. H., 49, *58*
Sargent, L. H., 8, *58*
Sargeson, A. M., 18, *58*
Sasse, W. H. F., 17, 18, *56, 58*
Sasson, Y., 269, *272*
Sato, J., 217, *259*
Sato, T., 103, *120*
Sauer, J. C., 218, *258*
Sauer, R. J., 33, *55*
Sauter, F. J., 47, *56*
Sauvage, G. L., 156, *171*
Saxena, B. L., 132, *143*
Sbrana, G., 217, *255, 258*
Scaros, M. G., 3, *58*
Schachtschneider, J. H., 168, *173*
Schaeffer, W. D., 176, *213*
Schaffner, K., 127, *142*
Schane, P., 10, *54*
Scharf, G., 271, *272*
Scheben, J. A., 225, 237, 248, 249, *257, 258*
Schell, R. A., 215, 239, *257*
Schenach, T. A., 39, *56*
Schenker, K., 21, *58*
Schiller, G., 217, *258*
Schleppnik, A. A., 111, *119*
Schlesinger, G., 188, *213*
Schlesinger, S. I., 152, *172*
Schleyer, P. von R., 163, *173*
Schmalzl, K. J., 130, *143*
Schmidlin, J., 127, 132, *142, 144*
Schmidt, W., 210, *211*
Schnabel, W. J., 253, *257, 258*
Schneider, O., 123, *142*
Schneider, R. F., 263, *273*
Schneider, W., 190, *213*
Schniepp, L. E., 23, *58*
Schnitzer, A. N., 22, *58*
Schoental, R., 128, 129, *141*
Schofield, K., 105, *118*
Scholfield, C. R., 153, *172*
Schoolenberg, J., 62, *75*
Schrauzer, G. N., 188, *213*
Schreyer, R. C., 252, *258*
Schrock, R. R., 61, 73, *76*
Schuller, W. H., 92, *119*
Schultz, R. G., 81, 83, 84, *119,* 225, *258*
Schwabe, K., 99, *118,* 139, *143*
Schweitzer, G. W., 286, *291*
Schwerdtel, W., 81, *119*
Scott, M. J., 286, *292*
Scotti, F., 43, *59*
Sedlak, M., 38, *58*
Sedlmeier, J., 77, 78, *117, 119*
Seelye, R. N., 130, *143*
Segnitz, A., 158, *173,* 260, *273*
Seibert, R. P., 165, *174*
Seligman, A. M., 101, *119*
Selin, T. G., 274, 286, *292*
Selke, E., 153, *172*
Selman, L. H., 70, *76*
Selwitz, C. M., 79, 80, 81, 95, *116, 119*
Sen, K., 15, *55*
Senda, Y., 64, *76*
Sen Gupta, G., 162, *173*
Sengupta, S. C., 32, *58*
Settimo, S. F. a., 217, *258*
Shamasundar, K. T., 44, *56*
Shamma, M., 9, *58,* 131, *143*
Shapiro, R., 105, *118*
Shapley, J. R., 61, 73, *76,* 178, *214*

Shaw, B. L., 61, *75*, 150, *171*
Sherman, A. H., 47, *57*
Sherstyuk, V. P., 17, *55*
Shibano, T., 186, 196, 200, 201, 203, *214*, 276, 279, *293*
Shields, T. C., 244, *255*
Shier, G. D., 186, 195, *213*
Shiihara, I., 278, *293*
Shima, K., 153, *172*
Shimizu, Y., 260, *273*
Shingu, T., 155, *174*
Shinohara, H., 180, *213*
Shriner, R. L., 100, *115*
Shriver, D. F., 61, *75*
Shryne, T. M., 185, 197, 198, 200, 202, 203, *211*, *213*
Shue, R. S., 90, 91, *119*
Shuikin, N. I., 15, 21, *58*, 165, *171*, 270, *273*
Sibert, J. W., 215, *258*
Sidgwick, N. V., 122, *143*
Sieber, R., 77, 78, 80, *119*
Siegel, H., 71, *75*, 147, *172*
Siegel, S., 64, *76*
Sih, C. J., 107, *119*
Silber, A. D., 4, 26, *55*
Silverstein, R. M., 279, *292*
Simpson, B. D., 22, *58*
Sims, J. J., 70, *76*
Singer, H., 193, *213*
Singh, H. S., 132, *143*
Singh, M. P., 132, *143*
Singh, N. P., 132, *143*
Singh, V. N., 132, *143*
Sisti, A. J., 44, *58*
Skapski, A. C., 62, *76*
Slates, H. L., 27, *59*
Slaugh, L. H., 180, 186, 201, *212*, *213*, 223, 226, *256*, *258*
Sletzinger, M., 11, *55*
Sloan, A. D. B., 49, *56*
Small, T., 18, *55*
Smidt, J., 77, 78, 80, *117*, *119*
Smith, A. G., 280, *292*
Smith, E., 253, *258*
Smith, H., 36, *54*
Smith, H. P., 154, *173*, 185, *213*
Smith, J. A., 226, 227, 231, *258*
Smith, P. N., 126, *143*
Smith, R. C., 36, *54*
Smith, R. D., 81, 83, *116*
Šmolik, S., 122, *141*
Smutny, E. J., 166, *173*, 186, 197, 198, 199, 200, 201, 203, *213*
Snatzke, G., 133, *143*
Sneeden, R. P. A., 101, 107, 108, 109, *119*
Snyder, L. R., 216, *257*
Sobti, R. R., 138, *143*
Sommer, L. H., 274, 279, 285, 287, *291*, *292*
Somogyi, P., 129, *143*
Sondheimer, F., 139, *143*
Sonoda, A., 92, *115*
Spaethe, H., 218, *256*
Sparke, M. B., 148, *173*
Spector, M. L., 80, 87, *119*, 201, *213*, 222, *258*
Speier, J. L., 274, 277, 280, 282, 284, *291*, *292*, *293*
Spialter, L., 274, *293*
Sprecher, M., 139, *143*
Springer, J. M., 43, 50, *55*, *58*
Sprinzl, M., 138, *143*
Stacey, M., 104, *115*
Stangl, H., 81, *119*
Stautzenberger, A. L., 79, *118*, 129, *143*, 176, *211*
Steadman, T. R., 176, 177, *212*
Steele, D. R., 1, *58*, 68, *76*, 124, *143*
Steinrücke, E., 187, *214*
Steinwand, P. J., 235, *256*
Stern, E. W., 78, 80, 83, 87, 98, *119*, 201, *213*, 222, *258*
Stern, R., 61, *74*
Sternbach, L. H., 133, *142*
Sternberg, H. W., 215, *259*
Stevens, C. L., 127, *144*
Steward, H. F., 165, *174*
Steward, O. W., 279, *292*
Stiles, A. B., 181, *212*
Stiles, M., 44, *58*
Stolberg, U. G., 60, *74*, 239, *255*
Stone, F. G. A., 73, *74*
Stone, H., 115, *119*
Stork, G., 139, *144*
Strätz, F., 210, *211*
Strojny, E. J., 49, *59*, 89, *116*
Strow, C. B., 180, *211*
Stuart, E. R., 51, *55*
Su, A. C. L., 189, *213*

Suami, T., 105, 107, *119*
Subba Rao, G. S. R., 98, *116,* 151, *171*
Subluskey, L. A., 49, *57*
Subramaniam, P. S., 157, *172*
Suchkova, M. D., 283, *292*
Suga, K., 86, *119*
Suginome, H., 36, *59*
Suld, G., 21, *59*
Sullivan, M. F., 181, 191, *211, 214*
Supanekar, V. P., 20, *55*
Sussman, S., 125, 126, 128, *143*
Susuki, T., 246, *258*
Sutherland, J. K., 31, *54*
Suzuki, T., 180, *213*
Svobada, P., 216, *255,* 276, *291*
Swallow, J. C., 84, *115*
Swisher, J. V., 274, 275, *291*
Sykes, P. J., 65, *75*
Syrkin, Ya. K., 86, 87, 101, *118, 120*
Szarek, W. A., 137, *142*
Szkrybalo, W., 160, *171*
Szmuszkovicz, J., 5, *54*
Szonyi, G., 80, *119*
Szpilfogel, S., 9, *58*

T

Taira, S., 179, *212*
Takahashi, K., 199, 204, 205, 206, *212, 214*
Takahashi, S., 186, 196, 199, 200, 201, 203, *214,* 276, 279, *293*
Takahashi, Y., 183, *213*
Takahasi, H., 278, *293*
Takano, I., 24, *57*
Takao, K., 82, 84, *119*
Takase, K., 130, *141*
Takebe, N., 223, *255*
Takedo, K., 22, *59*
Takegami, Y., 230, *258*
Takesada, M., 217, 233, *258, 259*
Takeuchi, Y., 69, *75*
Takiwa, K., 8, *56*
Tamura, F., 193, *212*
Tamura, M., 79, 81, 82, 83, *119*
Tanaka, H., 91, *117*
Tanaka, K., 11, *55,* 187, *214*
Tanaka, R., 88, *117*
Tani, K., 186, *213*
Tateishi, M., 43, *56*
Taub, D., 11, *59*
Tayim, H. A., 60, 66, *76,* 149, *174*
Taylor, E. C., 49, *59*
Taylor, H. A., 122, 123, *144*
Taylor, K. G., 127, *144*
Temkin, O. N., 191, *211*
Tener, G. M., 105, *120*
Teramoto, K., 78, *119*
Teranishi, S., 82, 84, 88, 90, 91, 92, *115, 116, 117, 118, 119,* 176, 178, *212*
Terashima, T., 7, *59*
Terry, E. M., 123, *143*
Tessie, P., 185, *212*
Tessier, J., 127, *141*
Tetenbaum, M. T., 115, *119*
Theissen, R. J., 112, *119*
Thiers, M., 180, *211*
Thomas, C. B., 92, *116*
Thomas, J. C., 238, *254*
Thomas, S. L. S., 2, 29, 30, 39, *57*
Thompson, J. B., 7, *59*
Thompson, W. H., 51, *55*
Thomson, R. H., 124, *141*
Thurkauf, M., 99, *119*
Thyret, H. E., 197, 198, 203, *212, 213*
Tinker, H. B., 227, *258*
Tipping, A. E., 283, *291*
Tishler, M., 27, *59*
Toda, F., 286, *292*
Todd, A. R., 105, *118*
Tomascewski, A. J., 51, *55,* 268, *273*
Tomimatsu, T., 155, *174*
Tori, K., 155, *174*
Trebellas, J. C., 149, 154, *174*
Trenner, N. R., 100, 104, *119*
Trenta, G. M., 177, *212*
Trepagnier, J. H., 125, *143*
Tretter, J. R., 17, *58*
Triggs, C., 80, 88, 89, 90, 101, *106, 116*
Trocha-Grimshaw, J., 73, *76*
Tronchet, J., 133, *144*
Tronchet, J. M. J., 133, *144*
Troughton, P. G. H., 70, *76*
Tsou, K.-C., 101, *119*
Tsuchiya, T., 210, *212*
Tsuda, T., 183, *213,* 226, *258*
Tsuji, J., 204, 206, 207, 208, *213,* 215, 218, 219, 220, 221, 222, 226, 236, 238, 239, 241, 243, 244, 245, 246, 248, 249, 250, 251, *256, 257, 258,* 262, 263, 264, 266, 267, *273,* 276, *292*

Tsuneda, K., 68, *75*
Tsutsumi, S., 176, 182, 194, *212, 213*
Tuck, B., 18, *55*
Tucker, L. C. N., 101, 106, *116*
Turley, R. J., 47, *56*
Turner, J. O., 84, *118*
Turner, L., 148, *173*
Turner, R. B., 26, *59,* 101, 107, 108, 109, *119*
Tute, M. S., 51, *59*
Tyman, J. H. P., 65, *76*

U

Uchida, A., 112, *119*
Uchida, Y., 180, *213*
Uchimaru, F., 12, *57*
Ueberwasser, H., 132, *144*
Ueda, S., 186, *213*
Uguagliati, P., 225, *256*
Uhlig, H. F., 163, *174*
Unger, M. O., 89, *120*
Uno, K., 286, *292*
Unrau, A. M., 260, *273*
Urry, W. H., 181, *214*
Usov, A. I., 130, *142*

V

Vadekar, M., 112, *118*
Van Allan, J. A., 154, *174*
van Bekkum, H., 66, *76*
Van Catledge, F. A., 160, *171*
Vandegrift, J. M., 49, *58*
van der Ent, A., 61, *75, 76*
Vanderwerf, C. A., 13, *55*
Vanderwerff, W. D., 90, *120*
van Gaal, H., 61, *76*
van Gogh, J., 66, *76*
van Helden, R., 81, 82, 88, 89, 90, *118, 120,* 187, *213,* 233, 249, 250, *257*
van Minnen-Pathuis, G., 66, *76*
Van Peppen, J. F., 146, *171*
van't Hof, L. P., 66, *76*
Van Winkle, J. L., 34, *59*
Vargaftik, M. N., 82, 86, 87, *118, 120*
Vaska, L., 60, 61, *74, 76,* 101, *120*
Vassilian, A., 149, *174*
Vatakencherry, P. A., 133, *141, 142*
Vdovin, V. M., 283, *292*
Vedejs, E., 85, *118,* 170, *174*
Veldkamp, W., 8, *56*
Venturello, C., 218, *255*
Venuto, P. B., 209, *212*
Verberg, G., 82, 88, 90, *120*
Vernier, F., 159, *173*
Vineyard, B. D., 71, 73, *75*
Viola, A., 6, *59*
Vivant, G., 149, *172*
Vizsolyi, J. P., 105, *120*
Voelter, W., 69, *76*
Volger, H. C., 168, 169, *172, 174*
Volokhova, G. S., 163, *173*
Voltman, A., 51, *55,* 268, *273*
Voltz, S. E., 23, 24, *59*
von Dresler, D., 127, *141*
von Kutepow, N., 215, 216, 218, 224, 236, 238, 239, 240, 241, 242, 244, 247, *255, 256, 258, 259*
von Schuching, S., 103, 104, *120*
Vuano, B. M., 6, *58*

W

Wade, R. H., 5, *54*
Wagner, G. H., 274, 276, *293*
Wakamatsu, H., 217, 227, *255*
Walborsky, H. M., 262, *273*
Walker, F. E., 44, *56*
Walker, G., 216, *256*
Walker, G. N., 27, 47, 51, *56, 59*
Walker, K. A. M., 62, 67, 68, 69, 70, 71, *74*
Walker, W. E., 185, 201, 202, 207, *211, 212, 214,* 244, *255*
Wall, M. E., 7, *59*
Walpole, A. L., 2, 30, *57*
Walsh, W. D., 86, *116*
Walter, D., 187, *214*
Walton, E., 133, *143*
Wani, M. C., 7, *59*
Wannowius, H., 122, *142*
Warner, C. R., 275, *292*
Warnhoff, E. W., 26, *55*
Watamatsu, H., 233, *258*
Watanabe, J., 180, *213*
Watanabe, S., 86, *119*
Watanabe, Y., 180, *213,* 230, *258*
Waters, W. A., 122, *144*
Watkins, W. B., 130, *143*
Wayaku, M., 84, *119*

Weber, J., 216, *256*
Weber, W., 210, *211*
Weber, W. P., 291, *293*
Weberg, B. C., 155, *173*
Webster, G. R. B., 130, *142*
Webster, J. A., 274, 277, *293*
Webster, S. K., 154, *174*
Wehrli, U., 127, *142*
Weichet, J., 122, *141*
Weigert, F. J., 178, *214*
Weigert, W., 233, *255*
Weiner, N., 158, *174*
Weinreb, S. M., 130, *141*
Weinstock, J., 133, *144*
Weiss, U., 158, *174*
Weisz, P. B., 2, *57*
Weitkamp, H., 247, *256*
Weller, S. W., 23, *59*
Wells, P. B., 151, *174*
Welsh, H. G., 133, *143*
Wender, I., 215, *257, 259*
Wendler, N. L., 11, 27, *59*
Wenham, A. J. M., 148, *173*
Wenkert, E., 36, *59*
Werner, A., 232, *256*
West, R., 274, 286, *292*
Westaway, M., 216, *256*
Wettstein, A., 132, 133, *144*
Weyenberg, D. R., 274, 289, *293*
Whalley, W. B., 52, *54*
Wharf, I., 61, *75*
White, C., 61, *76*
White, E. H., 9, *59*
White, R. E., 125, 126, *143*
Whitehead, W. E., 276, *293*
Whitehurst, D. D., 216, *256*
Whitesides, G. M., 191, *214*
Whitfield, G. H., 248, 249, *255*
Whittle, C. P., 17, 18, *56*
Whittle, C. W., 12, *54*
Whyte, D. R. A., 51, *55*
Wibberley, D. G., 40, *56*
Wicha, J., 135, 140, *143*
Wieland, H., 99, *120*
Wieland, P., 132, 133, *144*
Wieland, von P., 68, *76*
Wildman, W. C., 26, *55*
Wilds, A. L., 43, *54*
Wilke, G., 187, *214*
Wilkinson, G., 60, 61, 62, 73, *74, 75, 76,* 145, 147, *171, 172,* 185, 193, *213,* 215, 216, 219, 226, 227, *255, 256, 257, 259,* 262, 267, *272,* 276, *291*
Wilkinson, P. J., 148, *171*
Willard, A. K., 291, *293*
Williams, J. L. R., 154, *174*
Williams, N. R., 133, *141, 142*
Williams, P. H., 34, *59,* 100, 109, *118, 120*
Williams, R. O., 225, *255*
Williamson, J. B., 7, *59*
Willing, D. N., 276, *293*
Willis, B. J., 65, *76*
Willson, C. D., 11, *58*
Wilson, G. R., 151, *174*
Wilt, J. W., 262, *273,* 280, *293*
Wilt, M. H., 261, 264, *273*
Windgassen, R. J., Jr., 22, *54*
Winstein, S., 82, 83, *117*
Winter, R., 133, *141*
Wise, W. B., 166, *172*
Witkop, B., 99, *120*
Witt, H. S., 154, *173,* 185, *213*
Wolfe, S., 140, *144*
Wolff, M. E., 133, 135, *144*
Wolz, H., 127, *141*
Woodruff, S., 122, *142*
Woods, G. F., 6, 43, *59*
Woodward, R. B., 29, *56,* 169, *174*
Wooldridge, K. R. H., 35, 38, *54, 57*
Woolley, J. M., 253, *255*
Work, S. D., 274, 275, *291*
Wright, D., 83, *120,* 187, 203, *212*
Wyler, H., 159, *173*
Wyss, P. C., 105, *117*
Wythe, S. L., 37, *55*

Y

Yagupsky, G., 216, *259*
Yakhontov, L. N., 7, *59*
Yakshin, V. V., 86, *120*
Yamaguchi, M., 61, *76,* 278, *293*
Yamaguuchi, H., 169, *173*
Yamamoto, K., 278, 289, *293*
Yamamoto, M., 84, *119*
Yamazaki, H., 199, *214*
Yanagawa, H., 111, *120*
Yang, A. H., 21, *58*
Yanuka, Y., 139, *143*
Yasui, T., 79, 81, 82, 83, *118, 119*
Yates, K., 184, *214*
Yates, P., 15, *59*

Yodona, M., 92, *117*
Yokogawa, H., 210, *212*
Yoneda, Y., 148, *173*
Yoshimoto, H., 89, *117*
Yoshimura, J., 103, *120*
Young, D. P., 125, 126, *143*
Young, H. S., 78, *117*
Young, J. F., 60, 61, 62, *74, 75, 76,* 215, 219, *257*
Young, W. G., 82, 83, *117*
Youngman, E. A., 69, *76,* 185, *211*
Yuguchi, S., 204, *212*
Yurtchenko, E. N., 187, *212*

Z

Zachry, J. B., 243, 252, *259*
Zahm, H. V., 3, 47, *57,* 269, *273*
Zajcew, M., 152, 153, *174*
Zalay, A. W., 10, *54*
Zalkow, V. B., 160, *171*
Zanati, G., 135, *144*
Zaweski, E. F., 225, *256*
Zbiral, E., 123, *144*
Zelikoff, M., 122, 123, *144*
Zellner, R. J., 13, *59*
Zhorov, Yu. M., 163, *173*
Zienty, F. B., 23, *57*
Zilkha, A., 10, *54*
Zimmerman, H. E., 124, *144*
Zimmermann, H., 187, *214*
Zitzmann, K., 210, *211*
Zuech, E. A., 149, *174*
zu Reckendorf, W. M., 138, *144*
Žvaček, J., 122, *141*

Subject Index

A

Acenaphthene, 21, 97
4,5-Acenaphthenequinonedibenzenesulfonimide, 155
Acenaphthenone, 97, 270
Acenaphthylene, 21
Acetaldehyde, 81
α-Acetamidocinnamic acid, 72
Acetic acid, 81–83, 195, 201, 202, 224–225, 285, 286
Acetic anhydride, 81
Acetone, 83, 132, 209
4,5-Acetone-2,5-furanose-*d*-gluconic acid, 104
2α-Acetonyl-2β,6β-dimethyl-6α-phenylcyclohexanone, 131
Acetophenone, 79, 84, 97, 271
17β-Acetoxy-1,4-androstadiene-3-one, 68
3α-Acetoxy-24,24-diphenylchol-23-ene, 139
2-Acetoxyethanol, 82
1-(2-Acetoxyethyl)-2-formyl-5-nitroimidazole, 132
Acetoxylation, 81, 82, 89, 95–97
p-Acetoxymethylbenzylidene diacetate, 96
2-Acetoxy-3-methylene-7-methyl-oct-6-ene, 86
3-Acetoxymethyl-7-methyl-octa-2,6-diene, 86
2-Acetoxymethylpent-1-ene-3-yne, 195
3α-Acetoxynorcholanic acid, 139
2-Acetoxy-1-octene, 83
3β-Acetoxy-5β-pregnan-16-en-20-one, 140
1-Acetoxypropene, 83
2-Acetoxypropene, 83
β-Acetoxypropionic acid, 234
α-Acetoxy-*p*-tolualdehyde, 96
Acetylacetone, 80, 184, 195, 206
17β-Acetyl-1,4-androstadiene-3-one, 68
Acetylene, 217–219, 221, 249, 282
Acetylphenylalanine, 72
Acrolein, 17, 181
Acrolein dimethylacetal, 281
Acrylaldehyde, 92
Acrylamide, 179
Acrylic acid, 70, 234
Acrylonitrile, 88, 92, 179–180, 232, 279, 280
Acrylyl chloride, 217, 237
α-Acylaminoacrylic acid, 73
Adamantane, 163, 164
Adipaldehyde, 126, 130
Adipic acid, 126, 141, 244
Adiponitrile, 179–180
Adipoyl chloride, 266

Aflatoxin, M_1, 130
Agitation, effect of, 102
3-Alkoxypyridines, 16
Alkylation, 20
13-Alkylgona-1,3,5(10), 8-tetraene, 36
3-Alkylidenegrisens, 133
Allene, 185–186, 195–196, 245–246
Allo-inositol, 106
Allo-inosose-1, 106
Allomucic acid, 128
Alloxypyridine, 165
Allyl acetate, 83, 195
Allyl acetic acid, 244
3-Allylacetylacetone, 184
Allyl alcohol, 128, 132, 181, 184, 247, 280
Allylamine, 184, 247
Allyl chloride, 181, 248–250, 280
Allyl crotonate, 247
Allyl cyclohexane, 275
Allyl ethyl ether, 250
Allylic rearrangement, 251
o-Allylphenol, 24
Allyl phenyl ether, 150
Allylpyrrolidone, 223
Allyl vinylacetate, 247
Amination, 16, 17
Aminobiphenyls, 23, 53
4-Amino-3-cyanopyridine, 41
Aminocyclitols, 105, 107
4-Amino-2-ethylpyrimidine, 34
2-Aminomethyltetrahydrofuran, 16
Ammonia, 16, 204, 240
1,4-Androstadiene-3,17-dione, 67, 68
4,6-Androstadiene-3,17-dione, 68
Androstane-1,17-dione, 67, 69, 136
Androstane-3,17-dione, 67
5α-Androstane-3β-ol-17-one, 136
Androsta-4,6,8(14)-triene-3,17-dione, 42
5α-Androst-1-ene-3,17-dione, 68
Androst-1-ene-3,7-dione-1α,2α-d_2, 69
Androst-4-ene-3,17-dione, 67, 69
Angelica lactone, 245
Anilines, 47, 53, 286
Anilinodimethylhexylsilane, 284
Anilinodimethylsilane, 284
Anisole, 89
Anthracene, 3, 21, 49, 128
Antifoaming agents, 103
Apopinene, 262–263
Aromatization, 4–17, 26–47, 155–156, 166–167
Arylmercuric salts, 93
Asymmetric hydrogenation, 71
Asymmetric hydrosilylation, 278
Atropic acid, 72
Azides, 224
Azines, 53
Azobenzene, 73, 253
Azophenetole, 73
Azoxyanisole, 73
Azoxybenzene, 252
Azulenes, 21, 36, 170

B

Benzaldehyde, 84, 109, 126, 130, 206, 265
1,2-Benzanthracene, 128
Benzene, 4, 37, 71, 88, 91, 95, 112, 128
Benzenesulfonyl chloride, 271
Benzenesulfonyl fluoride, 271
Benzil, 207
Benzocyclobutenes, 6
Benzofuran, 24
Benzoic acid, 39, 89, 285
Benzoic anhydrides, 269
Benzonorbornadiene, 188, 276
1,2-Benzopyrrocoline, 23
2,3-Benzopyrrocoline, 23
Benzoquinone, 67, 79, 82, 111
Benzoylacetamide, 271
Benzoyl chloride, 225, 260, 265, 267
6-*O*-Benzoyl-1,2:4,5-di-*O*-isopropylidene dulcitol, 137
6-*O*-Benzoyl-1,2:4,5-di-*O*-isopropylidene-*threo*-glycero-3-hexulose, 137
Benzoyl fluoride, 267
Benzthiophene, 14
Benzyl 2-acetamino-2-deoxy-α-D-glucopyranoside, 103
Benzyl 2-acetamino-2-deoxy-α-D-glucopyranosiduronic acid, 103
Benzyl acetate, 89, 96
Benzyl alcohol, 109, 260, 286
1-Benzyl-5-carbomethoxy-1,2,3,4-tetrahydroisoquinoline, 11
Benzyl formate, 271
Benzyl methyl ketone, 79
1-Benzyl-11-phenylbicyclo[6.2.1]hendecane-9,10-dione, 167
1-Benzyl-2-phenylborazarene, 13
1-Benzyl-2-phenyltetrahydroborazarene, 13

N-Benzyl-3-piperidone, 10
2-Benzylpyridine, 23
Biacetyl, 207
Bibenzyl, 166
Bicyclo[4.2.2]deca-2,4,7,9-tetraene, 170
Bicyclofarnesol, 111
Bicycloheptadiene, 230, 275
Bicycloheptencarboxaldehyde, 232
Bicyclo[2.2.1]-heptene, 182
Bicyclo[5.2.0]nonane, 160
Bicyclopentadiene, 230
Bicyclopentane, 184
Biferrocenyls, 92
1,1′-Biisoquinoline, 6
Binor-S, 188
Biphenyl, 29, 45, 88, 210
3,3′-Bipyridazines, 210
Bipyridyl, 17
2,2′-Bipyrrole, 14
2,2′Biquinolyl, 17
2,3′-Biquinolyl, 17
4,4′-Biquinolyl, 17
3,3-Bis(allyl)acetylacetone, 184
Bis(4-aminocyclohexyl)methane, 162
1,2-Bis(3-cyclohexen-1-yl) ethylene, 166
2,6-Bis(dimethylaminomethyl)cyclohexanone, 28
Bis(hydroxymethyl)bicycloheptane, 230
Bis(hydroxymethyl)cyclooctane, 230
1,1-Bis(*p*-hydroxyphenyl)cyclohexane, 33
Bis(pentachlorocyclopentadienyl) 210
α,α-Bis(trifluoromethyl)benzyl alcohol, 200
Bis(trimethylsiloxy)bis(ethylhydrogensiloxane), 280
1,2-Bis(trimethoxysilyl)ethane, 282
Bitolyls, 89
Blocking groups, 290
p-Bromostyrene, 91
Bullvalene, 170
Butadiene, 7, 93, 186–187, 189, 191, 196–208, 229, 242–245, 250, 276, 279
Butadienyl acetate, 201
Butanal, 261
Butanol, 209, 285, 286
1-Butene, 91, 146, 241
2-Butene, 91
2-Butene-1-carboxylic acid, 244
2-Butene-1,4-diol, 247
But-3-enoyl chloride, 248
2-Butenyl phenyl sulfide, 199
3-Butenyl phenyl sulfide, 199
Butyl acrylate, 235
Butyl amine, 222, 284, 286
N-Butylaminodimethylsilane, 284
t-Butylbenzene, 262
Butyl β-butopropionate, 235
3-*t*-Butylcyclohexanone, 42
4-*n*-Butyl-1,2-dihydrophenanthrene, 32
Butylformamide, 223
2-*n*-Butyl-1-hexyl-1,1,3,3-tetramethyldisilazane, 284
cis-4-*t*-Butylmethylcyclohexane, 64
2-*n*-Butyl-2-methyl-1,3-dioxolane-4-methanol, 88
4-*t*-Butylmethylenecyclohexane, 64
2-Butyl-4-methylenepyran, 65
β-*n*-Butylnaphthalene, 21
2-Butyloctanol, 209
2-Butyl-1-octene, 140
p-*t*-Butylphenol, 134
2-Butyne, 282
But-1-yn-3-ol, 193
Butynone, 193
Butyric acid, 238
Butyrolactone, 183

C

2-Carbamoyl-4,5-dihydro-1-acenaphthenone, 270
Carbanilides, 253
Carbazoles, 8, 23
Carbethoxypropionaldehyde, 230
Carbohydrates, *see also* individual compounds, 101
5-Carbomethoxyquinoline, 11
Carbon dioxide, 186, 223
Carbon monoxide, 92, 201, 215–254
Carbon tetrachloride, 225
Carbonylation, 215–254
7-Carboxy-*p*-cymene, 47
Carboxylation, 97
7-Carboxy-1,8(9)-*p*-methadiene, 47
5-Carboxyuracil, 110
Carvomenthene, 84
Carvotanacetone, 84
Catalysts, 1–3, 60–61, 100–101, 146–148, 151, 169, 215–216, 260, 274–276,
  supported complex, 61, 148, 216–217, 276
Catalyst agglomeration, 108

Catalyst deactivation, 1, 7, 14, 67, 68, 100, 104, 105, 112, 124, 148, 176, 189, 198, 267, 277
Catalyst recovery, 216
Catalyst regeneration, 13, 112, 189, 216
Cetyl alcohol, 108
exo-2-Chloro-*syn*-7-acetoxynorborane, 86
p-Chlorobenzaldehyde, 264
Chlorobenzene, 91, 264–265, 271
4-Chlorobenzenesulfonyl chloride, 271
4-Chlorobenzonitrile, 268
4-Chlorobenzoyl cyanide, 268
3-Chloro-1-butene, 94
4-Chloro-*n*-butyl-3-butenoate, 249
4-Chlorobutyric acid, 183
Chloroethanol, 81
2-Chloroethyl-4-chlorobutyrate, 183
1-Chloronaphthalene, 267
4-Chlorophenol, 51, 198
1-Chloro-4-phenyl-3-butene, 291
Chloroprene, 245
1-Chloropropan-2-ol, 125
β-Chloropropionyl chloride, 217
Cholestane, 65
Cholestane-3,6-dione-4,5-diol, 127
Cholestanol, 107
3-Cholestanone, 107
Cholestan-3-one-1,2-diol, 127
Cholestan-3-one-4,5-diol, 127
4-Cholestene-3,6-dione, 127
1-Cholesten-3-one, 69, 127
4-Cholesten-3-one, 127
Cinnamaldehyde, 260, 263
Cinnamic acid, 8, 36, 37, 154
Cinnamolide, 111
Citral, 111, 263
Citronellal, 263
Citronellene, 263
Clindamycin, 114
Confertiflorin, 63
Configurational changes, 152
Coreximine, 152
Coronopilin, 63
Coumarins, 15
Crotonaldehyde, 80
Crotonic acid, 70, 123, 249
2-Crotonylthiophene, 68
2-Crotoxypyridine, 165
Crotyl chloride, 94
Crotyl methyl ether, 250
Crowded Compounds, 33
Cumene, 89
2-Cyanobicyclo[2.2.1]heptane, 182
2-(2-Cyanoethyl)cyclohexanone, 22
Cyano-2-ethylidenebicyclo[2.2.1]heptane, 182
2-Cyanotricyclo[4.2.1.0(3,7)]nonane, 182
Cyano-2-vinylbicyclo[2.2.1]heptane, 182
Cycl(3,2,2)azine, 18
Cyclitols, 105, 106
1,2-Cyclobutanedicarboxylic acid, 134
1α-5-Cyclo-5α-cholestane, 65
3α,5-Cyclo-5α-cholestane, 65
1α-5-Cyclo-5α-cholest-2-ene, 65
3α,5-Cyclo-5α-cholest-6-ene, 65
Cyclodecadienes, 154
Cyclodecane, 21
4,8-Cyclododecadiene-1-carboxylic ester, 241
Cyclododeca-1,5,9-triene, 66, 154, 241, 278
Cyclododecene, 66
Cyclohepta-1,3-diene, 169
Cycloheptane, 4
2,3-Cycloheptenoindole, 98
Cyclohexa-1,3-diene, 71
Cyclohexa-1,4-diene, 71
Cyclohexane, 4, 112
Cyclohexanecarbonamide, 241
Cyclohexanecarboxyaldehyde, 229
Cyclohexanecarboxylic acid, 134, 238
Cyclohexanedicarboxylic acid, 39
Cyclohexanediol, 126
Cyclohexanol, 3, 285
Cyclohexanone, 3, 26, 87
Cyclohexene, 70, 71, 85, 87, 126, 128, 129, 130, 238, 240, 274
Δ^4-Cyclohexene-1,2-dicarboxylic acid, 39
Cyclohexene oxide, 85
2-Cyclohexenone, 85
p-Cyclohexenylphenol, 32
Cyclohexenylpropionate, 242
Cyclohexylamine, 246
N-Cyclohexylformamide, 246
Cyclohexylidene succinic anhydride, 220
N-Cyclohexylmethacrylamide, 246
Cyclohexylmethylamine toluene-*p*-sulfonate, 71
Cyclohexylphenol, 24, 45
1,3-Cyclooctadiene, 149, 278
1,5-Cyclooctadiene, 66, 149, 230, 241, 278
Cyclooctanemethanol, 230
Cyclooctene, 66

2,3-Cyclooctenoindole, 98
Cyclopentadiene, 126
Cyclopentanone, 266
Cyclopentene, 85, 130, 184
Cyclopentene-2-diol-1,4, 126
2,3-Cyclopentenoindole, 98
Cyclopropylamine, 223
Cyclopropyl methyl ketone, 79
Cyclopropylpyrrolidone, 223
p-Cymene, 30, 36, 38, 47

D

Damsin, 63
Dealkylation, 29–35, 39, 113, 114
Deamination, 34
Debenzylation, 10
Decahydroquinoxalines, 12
Decalin, 2, 112
Decanoyl chloride, 265
Decarbonylation, 46, 47, 48, 51, 62, 239, 260–271
Decarboxylation, 49
1-Decene, 266
n-Decylamine, 222
Dehalogenation, 8, 210
Dehydroabietic acid, 92
Dehydrogenation, 1–53, 99, 261, 268, 269, 284–288
2,3-Dehydrohomopterocarpin, 52
Dehydrolinalool, 68
Dehydropodophyllotoxin, 49
1′-Demethylclindamycin, 114
Deoxyribopolynucleotide, 105
Desmethoxy-β-erythroidine, 156
O-Desmethylthebainone, 157
Desulfonylation, 260, 272
Deuteration, 69, 70, 82, 87, 91, 146, 147, 184, 191, 196, 262, 289
Dextromethorphan, 162
1,4-Diacetoxy-1,3-butadiene, 81
1,2-Diacetoxyethane, 82
1,2-Diacetoxy-2-methyl-3-butene, 93
1,3-Diacetoxypropane, 83
Di-*N*-acetyl-*myo*-inosodiamine-4,6, 107
Di-*N*-acetyl-2-oxo-*myo*-inosodiamine-4,6, 107
2,3-Dialkoxy-3,4-dihydro-1,2-pyrans, 16
Diallyl ether, 150, 250, 280
1,2-5,6-Dibenzanthracene, 128
4,5-Dibenzenesulfonamidoacenaphthene, 155
Dibenzofuran, 24
Dibenzyl, 18
o-Dibenzylbenzene, 48
3,10-Dibenzyl-1,2-cyclodecanedione, 167
3,10-Dibenzylidene-1,2-cyclodecanedione, 167
3,7-Dibenzylidene-1,2-cycloheptanedione, 167
2,6-Dibenzylidenecyclohexanone, 167
2,7-Dibenzylidenecycloheptanone, 168
3,5-Dibenzylidenetetrahydro-4*H*-pyran-4-one, 168
2,6-Dibenzylphenol, 168
3,5-Dibenzyl-4*H*-pyran-4-one, 168
2,7-Dibenzyltropolone, 168
Dibutylamine, 223
2,6-Di-*t*-butyl-1,4-benzoquinone, 110
Di-*tert*-butylethylene, 154
3,5-Di-*t*-butyl-4-hydroxybenzaldehyde, 110
3,5-Di-*t*-butyl-4-hydroxybenzyl alcohol, 110
Dibutylurea, 223
m-Dichlorobenzene, 265
p-Dichlorobenzene, 271
1,2-Dichloro-3-butene, 251
1,4-Dichloro-2-butene, 251
3,4-Dichlorobutroyl chloride, 248
Dichotine, 24
Dicyanobutene, 179–180
2,6-Di(1-cyclohexenyl)cyclohexanone, 50
N,N-Di(cyclohexylmethyl)formamide, 241
N,N-Di(cyclohexylmethyl)methylamine, 241
Didecyloxamide, 222
1,3-Didecylurea, 222
2,3-Dideuterosuccinic acid, 70
Di-*endo*-2,3-dimethylbicyclo[2.2.1]heptane, 160
1,4-Diethoxy-2-butene, 251
2,3-Diethoxy-3,4-dihydro-1,2-pyran, 17
1,2-Diethoxyethylene, 17
2,3-Diethoxy-4-methyl-3,4-dihydro-1,2-pyran, 17
3,3-Diethoxytetrahydrofuran, 88
Diethyl acetylenedicarboxylate, 18
Diethyl amine, 286
Diethyl ketone, 237–238
Diethyl malonate, 195
2,6-Diethyltetrahydropyran, 165
Diformylcyclohexane, 232
2,6-Diformyltetrahydropyran, 233

9,10-Dihydroanthracene, 3
2,5-Dihydrobenzylamine, toluene-*p*-sulfonate, 70
Dihydrocoronopilin, 63
Dihydrocoumarins, 15
Dihydrodamsin, 64
Dihydrodes-methoxy-β-erythroidine, 156
9,10-Dihydro-9,10-dihydroxyphenanthrene, 128
3,4-Dihydro-1,5-dimethylnaphthalene, 6
2,3-Dihydro-*p*-dioxin, 23
13,14-Dihydroeremophilone, 67
22-Dihydroergosteryl acetate, 69
2,3-Dihydro-2-furaldehyde, 150
Dihydrofurans, 15, 88
Dihydroisoquinolines, 9
Dihydromorphinone, 157
Dihydromyrcene, 150
1,2-Dihydronaphthalene, 94
Dihydroouabagenin, 107
3,4-Dihydropapeverine, 9
6,7-Dihydro-1,5-pyrindine, 9
1,4-Dihydrotetralin, 69
Dihydrothiazoles, 14
2,3-Dihydrothiophene-1,1-dioxide, 150
2,5-Dihydrothiophene-1,1-dioxide, 150
Dihydroxanthyletin, 16
3α,17β-Dihydroxy-5β-androst-9(11)-en-12-one, 108
1,2-Dihydroxybutyric acid, 123
1,2-Dihydroxycyclohexadiene-3,5, 128
1,4-Di(1-hydroxycyclohex-1-yl)but-1-en-3-yne, 193
1,17β-Dihydroxyestra-1,3,5(10)-triene, 135
2,3-Dihydroxymethylbuta-1,3-diene diacetate, 195
1,17β-Dihydroxy-4-methylestra-1,3,5(10)-triene, 135
3,6-Diisopropenyl-1-phenyl-2-piperidone, 208
1,6-Dimesitoyl-1-cyclohexene, 112
2,4-Dimesitoyltoluene, 33
2,3-Dimethoxybenzoquinone, 67
7,8-Dimethoxyisoquinoline, 44
(3,3-Dimethoxypropyl)pentamethyldisiloxane, 281
Dimethylacetal, 87
Dimethylacetylenedicarboxylate, 18, 191
1,3-Dimethyladamantane, 164
Dimethylaminodimethylsilane, 284
2,4-Dimethylbicyclobutanes, 169
2,3-Dimethylbuta-1,3-diene, 195, 201
Dimethylchlorosilane, 280
5,8-Dimethylcoumarin, 27
1,3-Dimethylcyclohexanes, 160
1,4-Dimethylcyclohexenes, 69
2,6-Dimethylfluorenone, 269
Dimethylformamide, 115
Dimethyl fumarate, 218, 252
2,6-Dimethyl-4-heptanone, 209
2,6-Dimethyl-1-hepten-6-ol, 232
Dimethyl 2-hexenedioate, 178
Dimethyl maleate, 218
Dimethyl 3-methoxy-1-buten-1,2-dicarboxylate, 220
2β,6β-Dimethyl-2α-(2′-methylallyl)-6α-phenyl cyclohexanone, 131
α,α-Dimethyl-α′-methylene glutarate, 245
Dimethyl muconate, 218
2,7-Dimethyloct-3-en-5-yne-2,7-diol, 192
1,5-Dimethylnaphthalene, 6
2,2-Dimethylnorpinane, 31
2,3-Dimethyl-1,4-pentadiene, 190
2,6-Dimethylphenol, 27
2,5-Dimethylphenyl acetate, 97
3,3-Dimethyl-4-phenylbutanal, 261
Dimethylphenylsilane, 279
Dimethyl pyrotartarate, 234
1,1-Dimethyl-1-silacyclobutane, 289
1,1-Dimethyl-1-silacyclopentane, 289
Dimethyl sulfoxide, 133
Dimethyl 2,2,4-trimethylglutarate, 252
Dimethylvinylsilane, 282
Diocta-2,7-dienylamine, 204
1,3-Dioxanes, 183
1,3-Dioxolane-2-acetonitrile, 88
1,4-Dioxospiro[4,5]decane, 88
Diphenic acid, 128
Diphenoquinone, 67
Diphenylacetaldehyde, 266
Diphenylacetyl chloride, 266
Diphenylacetylene, 194, 219, 282
Diphenylamine, 23
1,2-Diphenylbenzocyclobutene, 48
1,4-Diphenyl-1,3-butadiene, 290
1,3-Diphenyl-1-butene, 91, 178
1,4-Diphenylbutenyne, 193
Diphenylchloromethane, 267
Diphenyl ether, 24, 89
3,6-Diphenylhexahydrophthalic acid, 40
Diphenyl maleate, 219
Diphenylmethylsilane, 282

1,8-Diphenylnaphthalene, 34
Diphenylsilane, 283, 289
3,6-Diphenyltetrahydrophthalic acid, 39
Diphenylurea, 252, 254
1,2-Di(2-quinolyl)ethane, 18
Disproportionation, 15, 30, 38, 39, 40, 70, 71, 133, 146, 159, 289, 290
Di-*endo*-trimethylenenorbornane, 160
1,2-Divinylcyclohexane, 149
Divinylpiperidones, 208
2,5-Divinyltetrahydropyran, 207
2-Dodecanone, 79
Dodecatetraene, 187, 199
1-Dodecene, 79, 130, 140, 160
2-Dodecene, 160

E

3-Epiisotelekin, 157
Epoxidation, 82, 84
Equilenin, 28
Eremophilone, 66
Ergost-7-en-3β-ol-5α,6-d_2-3β-acetate, 69
Ergosterol, 68, 69
Estradiol diacetate, 135
Estrone, 135, 161
Ethanediol, *see* Ethylene glycol
Ethanol, 200, 221
1-Ethnylcyclohexan-1-ol, 193
4-Ethoxyaniline, 73
2-(Ethoxycarbonyl)cyclopentanone, 205
3-Ethoxy-4-methylpyridine, 17
1-Ethoxy-octa-2,7-diene, 244
3-Ethoxypyridine, 17
Ethyl acetoacetate, 195, 205
Ethyl acetylenecarboxylate, 221
Ethyl acetylenedicarboxylate, 221
Ethyl acrylate, 230, 231
1-Ethyladamantane, 164
o-Ethylaniline, 23
Ethylbenzene, 7, 97, 148
2-Ethylbutandial-1,5
Ethyl 2-butenoate, 249
Ethyl 3-butenoate, 247, 249, 250
Ethyl *trans*-but-2-enyl sulfone, 184
Ethyl carbamate, 254
Ethyl 4-chloro-3-pentenoate, 245
α-Ethylcinnamaldehyde, 263
Ethyl cinnamate, 21
Ethyl crotonate, 247
Ethyl cyanoacetate, 195
Ethylcyclohexane, 148
1-Ethylcyclohexene, 275
Ethyldichlorosilane, 277
2-Ethyl-3,5-dimethylpyridine, 247
4-Ethyl-3,5-dimethylpyridine, 247, 254
Ethylene, 80, 81, 82, 88, 89, 91, 128, 129, 175–176, 178, 183, 184, 189, 190, 225, 234, 235, 237, 238, 240, 291
Ethylenediamine, 22
Ethylene glycol, 23, 88, 109, 128, 245
Ethylene glycol diacetate, 82
Ethylene glycol monoacetate, 82
Ethyl 4-ethoxy-4-methylvalerate, 243
Ethyl β-ethoxypropionate, 238
Ethyl fumarate, 221
2-Ethylhexanol, 209
Ethylidene diacetate, 81
3-Ethylidene-1-phenyl-6-vinyl-2-piperidones, 208
Ethyl isobutyrate, 249
Ethyl itaconate, 246
Ethyl maleate, 221
Ethyl methyl ketone, 132
Ethyl 2-methyl-3-pentenoate, 243
Ethyl 4-methyl-3-pentenoate, 243
9-Ethyl-9-nitro-1,6,11,16-heptadecatetraene, 206
Ethyl nona-3,8-dienoate, 244
o-Ethylphenol, 24
Ethyl propiolate, 192
Ethyl propionate, 238
2-Ethyl-5-propionylfuran, 165
1-Ethylpyrene, 32
Ethyl succinate, 221
α-Ethylstyrene, 71
Ethyl vinyl sulfone, 184
1-Ethynylcyclohexanol, 220
Eudalene, 39
Exchange reactions, *see also* Hydrogen exchange, 88, 155, 176, 199, 206, 267, 276

F

Ferrocene, 92
Ferrocenylation, 92
Fluorene, 97
Fluorenones, 97
Fluorobenzene, 271, 272
6-Fluoroequilenin, 28
Formaldehyde, 183, 245

Formation of new bonds, 17–29
Formic acid, 202
N-Formylallylamine, 247
2-Formyl-2′-biphenylcarboxylic acid, 264
2-Formylcyclohexanone, 205
Formylcyclohexene, 232
2-Formyl-3,4-dihydro-2*H*-pyran, 233
N-Formylpiperidine, 115
Formylstearate, 229
5-Formyluracil, 110
Fumaric acid, 36, 70, 122, 123
Furfural, 124, 260, 264–265
2-Furoic acid, 124

G

Geraniol, 111
Geraniolene, 263
Geranyl acetate, 86
Germacyclobutane, 289
D-Glucosaccharic acid, 104
Glucosamine, 104
Glucosaminic acid, 104
Glucose, 128
α-D-Glucosylamine, 104
D-Glucuronic acid, 102
Glutaraldehyde, 130
Glycerol, 128, 132
Glycolaldehyde, 109
Glycolic acid, 109
Glyoxal, 81, 82, 128
Grisen-3-ones, 133
Guerbert reaction, 208

H

3-Halopyridazines, 210
2,5-Heptadiene, 190
Heptaldehyde, 219, 227, 261
Heptamethyltrisiloxane, 286
Heptanoic acid, 239
Heptanone-2, 164
3-Heptene, 147, 277
Hexachlorocyclopentadiene, 210
1,3,7,11,15-Hexadecapentaene, 199
1,4-Hexadiene, 189, 190
1,5-Hexadiene, 181, 241, 245
2,4-Hexadiene, 205
9,10,11,12,13,14-Hexahydroxystearic acid, 124
Hexamethylenediamine, 179
Hexamethylene glycol, 245
Hexamethylenimine, 4
Hexamethylmellitate, 191
Hexandial-1,6, 229
Hexanol, 209, 286
Hexanone-2, 79, 88
1-Hexene, 79, 83, 88, 146, 147, 227, 277, 284
2-Hexene, 147, 177
3-Hexene, 177
n-Hexyltriphenylsilane, 277
Hex-l-yne, 193, 219, 283
Homoangelica lactone, 237
Homogeneous hydrogenation, 60–76
Homopterocarpin, 52
Hydratropic acid, 71, 72
Hydrazoanisole, 73
Hydrazobenzene, 73
Hydroformylation, 215–254
Hydrogen chloride, 287
Hydrogen cyanide, 182
Hydrogen fluoride, 287
Hydrogenolysis, 1, 15, 27, 31, 52, 91, 210, 251
Hydrogen sulfide, 287
Hydrogen exchange, 2, 8, 21, 32, 35–43
Hydrogermylation, 289
α-Hydromuconamides, 179
Hydroquinone, 111, 124, 217
Hydrosilylation of olefins, 274–285
17β-Hydroxy-1,4-androstadiene-3-one, 68
4-Hydroxyazulene, 26
o-Hydroxybenzyl alcohol, 109
9-Hydroxy-7,10-dibenzylbicyclo[5.3.0]10-decen-9-one, 167
3-Hydroxydiphenide, 264
3-Hydroxy-5,7,9(10)-estratrien-17-one, 28
N-(β-Hydroxyethyl)allylamine, 248
N-(β-Hydroxyethyl)pyrrolidone, 248
2-Hydroxymethyl-6-chlorophenol, 109
2-Hydroxymethyl-3,4-dihydropyran, 233
2-Hydroxymethyl-6-ethoxyphenol, 109
2-Hydroxymethyl-4-methylphenol, 109
2-Hydroxymethyl-6-methylphenol, 109
2-Hydroxymethyltetrahydrofuran, 16
3-Hydroxymethyl-4,5,6,7-tetrahydroindazole, 47
5-Hydroxymethyluracil, 110
1-(*p*-Hydroxyphenyl)-5-methoxy-2-phenylindole, 10
3-Hydroxyphthalide, 264
Hydroxyquinolines, 37

I

Imidazole, 22, 240
Imidodiacetic acid, 115
4-Imino-3-cyanopiperidine, 41
1,11 α-Iminoestrone-3-methyl ether, 8
Indazole, 7, 47
3(1*H*)-Indazolone, 7
Indenes, 92, 94
Indole, 7, 9, 23
Isocoronopilin, 63
Isobutyl decanoate, 270
Isocyanates, 208
Isodamsin, 64
Isoequilen, 161
Isomerization, 6, 7, 14, 15, 24, 43, 63, 64, 69, 82, 93, 111, 145–173, 184, 190, 202, 218, 226, 228, 229, 236, 247, 248, 261, 263, 266, 270, 277, 278
Isomexicanol, 156
Isophthaloyl chloride, 265
Isoprene, 93, 189, 190, 200, 201, 205, 208, 243, 244
Isopropanol, 200, 209
α-Isopropylcinnamaldehyde, 263
1,2-Isopropylidene-3-benzyl-6-trityl-α-D-glucofuranose, 137
1,2-Isopropylidene-3-benzyl-6-trityl-5-keto-α-D-glucofuranose, 137
1,2-Isopropylidene-D-glucose, 102
1,2-Isopropylidene-D-glucuronic acid, 102
9-(3′,5′-*O*-Isopropylidene-2′-keto-D-xylofuranosyl) adenine, 138
9-(3′,5′-*O*-Isopropylidene-β-D-xylofuranosyl)-adenine, 138
Isoquinolines, 9
Isotetralin, 69
Itaconic acid, 219
Itaconic anhydrides, 220

J

Juglone, 67

K

Ketene, 95
3-Ketobenzo(*d*,*e*)steroids, 21
Ketones as hydrogen acceptors, 42
1-Keto-1,2,3,4-tetrahydrophenanthrene, 26
4-Keto-1,2,3,4-tetrahydrophenanthrene, 26
10-Ketoundecanoic acid, 80

L

d-Limonene, 36, 38
Linalyl acetate, 86
Linderane, 22

M

Maleic acid, 36, 37, 70, 122
Maleic anhydride, 184
Malonitrile, 195
Mandelic acid, 132
Menthane, 31
Mesitylene, 209
Mesityl oxide, 209
Methallyl chloride, 280
Methanethiol, 287
Methanol, 199, 200, 224–225, 244, 245, 285
1-(*p*-Methoxybenzyl)-2-methyl-1,2,3,4,5,6,7,8-octahydroisoquinoline, 162
1-Methoxycyclohexadienes, 151
1-Methoxy-6-deuterio-2,7-octadiene, 199
3-Methoxyestra,1,3,5(10)-trien-17-one, 3
2-Methoxy-4-hydroxymethyl phenol, 109
5-Methoxy-1-naphthol, 134
1-Methoxy-2,7-octadiene, 199
Methoxypentamethyldisiloxane, 281
p-Methoxyphenylacetone, 80
1-Methoxy-1-phenylethane, 71
3-Methoxyphthalic acid, 134
4-Methoxy-1-propenylbenzene, 80
α-Methoxystyrene, 71
Methylacetylene, 195
Methyl *N*-acetyl-α-D-glucosaminide, 104
Methyl aconitate, 219
Methyl acrylate, 92, 178, 211, 252, 279, 291
2-Methyladamantane, 164
Methyl atropate, 72
2-Methylbenzofuran, 24
m-Methylbenzoic acid, 39
p-Methylbenzyl acetate, 97
Methyl *N*-benzyloxycarbonyl-α-D-glucosaminide, 104
2-Methylbiphenyl, 29
2-Methyl-2-buten-l-ol, 111
2-Methyl-3-butyn-2-ol, 220
3-Methyl-1-but-l-yn-3-ol acetate, 192
Methyl butyrate, 234
α-Methylbutyrolactone, 231
2-Methyl-5-carbomethoxy-1,2,3,4-tetrahydroisoquinoline, 10
Methyl carbonate, 234

α-Methylcinnamaldehyde, 263
Methyl cinnamate, 211, 243
β-Methylcinnamic acid, 72
Methyl crotonate, 234
1-Methylcyclobutene, 78
Methylcyclohexene, 277
Methyl cyclopent-2-ene-l-carboxylate, 242
3-Methyl-2-cyclopentenone, 246
Methylcyclopropene, 177
Methyldichlorosilane, 274, 275, 276, 278, 279, 280, 282
2-Methyldichlorosilyl-2-butene, 282
Methyl 2,4-dihydro-2,2-dimethyl-5-oxo-3-furoate, 220
Methyl 3α,6α-dihydroxycholanate, 108
Methyl 3α,12α-dihydroxycholanate, 108
2-Methyl-4,5-dihydrofuran, 15
Methyl α,γ-dimethylenevalerate, 246
Methyl 3,4-dimethylpent-3-enoate, 243
2-Methyl-1,3-dioxolane, 88
1-Methyl-2,2-diphenylbicyclo[1.1.0]-butane, 170
3-Methylenecyclohexene, 169
1-Methylene-3,4-diphenyl-2-cyclohexene, 123
Methyleneglutaronitrile, 180
Methylestradioldiacetate, 135
1-Methyl-1-ethylcyclohexane, 30
α-Methyl-7-ethyl-2-naphthenacetic acid, 48
2-Methyl-6-ethyltetrahydropyran, 165
Methyl formate, 245
α-Methylfuran, 15
Methyl-α-D-glucopyranoside, 104
Methyl-α-D-glucosamineuronide, 105
α-Methyl-α-D-glucosaminide, 105
Methyl-α-D-glucoside, 102
α-Methylheptanoic acid, 239
2-Methyl-1,4-hexadiene, 190
3-Methyl-1,4-hexadiene, 190
4-Methyl-1,4-hexadiene, 190
2-Methylhexaldehyde, 219
α-Methylhexanoic acid, 239
Methyl homolevulinate, 237
Methyl hydratropate, 72
Methyl hydrocinnamate, 243
3-Methyl-2-hydroxymethylbuta-1,3-diene acetate, 195
2-Methyl-3-hydroxypropanenitrile, 232
2-Methyl-3-hydroxy-4-pyrone, 111
1-Methyl-1-isopropylcyclohexane, 30
1-Methyl-4-isopropylcyclohexenes, 69
1-Methylisoquinoline, 51
Methyl itaconate, 219
Methyl linolenate, 66, 69
Methyl methacrylate, 231, 234, 245, 252
Methyl 1-methoxycarbonylmethyl-6,8-dimethylindolizine-2-carboxylate, 19
Methyl 2-methoxymethylacrylate, 220
Methyl-4-methoxy-2-pentenoate, 220
Methyl α-(methyldichlorosilyl)propionate, 279
3-Methyl-2-methylene-3-butenylamine, 196
α-Methylnaphthalene, 30
Methyl α-naphthyl ether, 134
2-Methyl-3-nitropropionaldehyde, 233
9-Methyloctalin, 30
Methyl oleate, 69, 228
2-Methylpentandial-1,5, 229
2-Methylpenta-1,3-diene, 169
4-Methyl-2-pentanone, 209
3-Methyl-2-pentene, 153
2-Methylpent-1-ene-3-yne, 195
Methyl pent-3-enoate, 179, 242, 250
Methylphenanthrenes, 32
2-Methylphenol, 27
4-Methylphenol, 198
Methylphenylacetylene, 194
Methyl-3-phenylbutanoate, 71
Methyl 3-phenylbut-2-en-oate, 71
1-Methyl-1-phenylcyclohexane, 29
Methyl-*N*-phenylurethane, 253
N-Methylpiperidine, 115
Methyl 5α,10α-podocarpa-8(9)-en-15-oate, 161
Methyl propiolate, 19, 192
Methyl propionate, 237, 245
2-Methyl-5-propionylfuran, 165
1-Methyl-1-*n*-propylcyclohexane, 30
Methyl propyl ketone, 15
1-Methylpyrene, 32
3-(6-Methyl-2-pyridyl)-2-phenyl-1-propanol, 22
N-Methylpyrrole, 13
N-Methylpyrrolidine, 13
2-Methylquinoline, 18
2-Methylquinoxaline, 18
α-Methylstyrene, 32, 278
Methyl teraconate. 220
Methyl 1,4,5,8-tetrahydro-1-naphthoate, 70
Methyl 3α,7α,12α-trihydroxycholanate, 108
Methyl vinyl ether, 87
Mexicanol, 156

1,2-Mono-*O*-cyclohexylidene-L-xylofuranose, 103
1,2-Mono-*O*-cyclohexylidene-L-xyluronic acid, 104
Morphine, 157
Morpholine, 204
Muco-inositol, 106
Muco-inosose-1, 106
Muconyl chloride, 218
Myo-inositol, 105
Myo-inosose-2, 105
Myo-inosose-5, 106
Myrcene, 85, 150
Myrcenol, 150
Myristyl alcohol, 108
Myristaldehyde, 108
Myrtenal, 263

N

1-Naphthaldehyde, 264
Naphthalene, 21, 30, 128, 264
Naphthols, 44, 45, 198
1,4-Naphthoquinone, 67
2,6-Naphthoquinone, 67
1-Naphthoyl chloride, 267
1-Naphthoyl cyanide, 268
Naphthylacetates, 95
Naphthylamines, 53
α-Naphthylphenylmethylsilane, 285, 287
α-Naphthyltrifluoroacetate, 95
Neo-inositol, 106
Neopentylbenzene, 261
Neryl acetate, 86
Nitrilotriacetate acid, 115
5-Nitro-7-azaindoline, 47
Nitrobenzene, 32, 40–42, 252–254
1-Nitrobutane, 254
Nitrocyclohexane, 47, 206
Nitroethane, 206
Nitromethane, 206
4-Nitrophenylpropionaldehyde, 233
1-Nitropropane, 206, 254
3-Nitropropene, 233
Nitrosobenzene, 252
4-Nitrostyrene, 233
9-Nitroundeca-1,6-diene, 206
Nonanone-3, 165
n-Nonanal, 227
Nonene, 266, 270
Norbornadiene, 66, 168, 187, 188, 243
Norbornanols, 111
Norbornene, 86
syn-7-Norbornenol, 86
Norlaudanosine, 153
19-Nor-9β-10α-$\Delta^{4,17(20)}$-pregnadiene-3-one, 127
19-Nor-9β-10α-Δ^{4}-pregnene-17α-ol-3,20-dione, 127
Nucleosides, 105
Nucleotides, 105

O

Ocimene, 150
Ocimenol, 150
Octa-1,6-diene, 187, 202, 203
Octa-1,7-diene, 187, 203
Octa-2,7-dien-l-ol, 185, 201
3,7-Octadienoyl chloride, 250
2,7-Octadien-1-yl acetate, 201
o-[1-(2,7)Octadienyl)]phenol, 198
N-(2,7-Octadienyl)piperidine, 203
Octa-3,6-dione, 238
1,9-Octalin, 69
9,10-Octalin, 69
Octahydro-1,5-pyrindine, 9
Octamethylene glycol, 245
3,6-Octanedione, 237
Octanoic acid, 239
Octanone-2, 165
Octanone-3, 165
Octa-1,3,7-triene, 186, 198, 199, 207
1-Octene, 68, 83, 129, 227
Octoyl bromide, 266
Octyl alcohol, 271
Octyl formate, 271
Oligomerizations, 175–208
Osmium tetroxide, 121–133
Oxalic acid, 128
Oxazolidines, 24
Oxidation, 77–119
Oxidative addition of alcohols, 87–88
Oxidative carbonylation, 233–235
Oxidative coupling, 81, 88–94, 113
Oxidative dehydrogenation, 112–113
3-Oxo-11α-acetoxy-20-ethylenedioxy-Δ^{4}-pregen, 68
3-Oxo-4β,5β-dihydroxy-17β-acetoxyandrostane, 127
5-Oxo-1,5-diphenylpentadiyne, 158
3-Oxo-1,5-diphenylpentadiyne, 158
3,3′-Oxphthalide, 264

P

Palmitaldehyde, 108
Papeverine, 9
1,3-Pentadiene, 189, 190, 205, 230, 242
n-Pentanol, 245
1-Pentene, 147, 274
2-Pentene, 147, 274, 277
β-Perchloromethylpropionyl chloride, 225
Perfluoroacetone, 207
Perhydroacenaphthene, 164
Peroxides as hydrogen acceptors, 42
d-*α*-Phellandrene, 36
Phenanthrene, 6, 21, 128, 269
Phenanthrene quinone, 128
Phenanthrol, 2, 26
Phenol, 4, 26, 95, 197, 198, 204, 264, 285, 286
1-Phenoxy-2,7,11-dodecatriene, 198
1-Phenoxy-2,7-octadiene, 166, 197
3-Phenoxy-1,7-octadiene, 197
Phenylacetaldehyde, 78
α-Phenylacetamidoacrylic acid, 72
Phenyl acetate, 95
Phenylacetone, 185
Phenylacetonitrile, 185
Phenylacetylalanine, 72
Phenylacetylene, 193
α-Phenylacrylic acid, 71
Phenylacylaniline, 113
9-Phenylanthracene, 48
m-Phenylbenzoic acid, 39
2-Phenylborazarene, 13
2-Phenylbutane, 71
3-Phenylbutanoic acid, 73
1-Phenyl-1-butene, 91, 94, 178
1-Phenyl-2-butene, 91, 94, 149, 178
2-Phenyl-2-butene, 94, 149, 154
4-Phenyl-1-butene, 149
1-Phenyl-3-buten-2-yl acetate, 93
1-Phenyl-4-carbomethoxy-1,3-butadiene, 291
α-Phenylcinnamaldehyde, 263
2-Phenylcyclobutanecarboxylic acid, 134
Phenylcyclohexane, 134
1-Phenyl-2-cyclohexylethane, 166
Phenylcyclopropane, 79
2-Phenyl-4,9-decadienenitrile, 185
Phenyldichlorosilane, 277, 280
Phenyldimethylallyloxysilane, 280
Phenyldimethylbutoxysilane, 280
Phenyldimethylpropoxysilane, 280
Phenyldimethylsilane, 280, 286
Phenyldimethylsilane-*d*, 289
2-Phenyl-3,6-divinyltetrahydropyran, 207
Phenylene oxide, 89
Phenylethyl alcohol, 269
2-Phenylethyl chloride, 93
N-(*β*-Phenylethyl)indole, 159
Phenylfluoroformate, 272
Phenylfluorothioformate, 272
Phenyl *β*-D-glucopyranoside, 101
Phenyl *β*-D-glucopyruronside, 101
Phenylglyoxal anil, 113
Phenyl isocyanate, 208, 252
β-Phenylisovaleraldehyde, 262
Phenylmercuric acetate, 93
Phenylmercuric chloride, 93, 94
Phenylmethyldiethynylsilane, 283
2-Phenyl-5-methylpyrrocoline, 22
1-Phenylnaphthalene, 193
Phenylphenol, 24, 32
Phenyl propenyl ether, 150
β-Phenylpropionic acid, 141
2-Phenylpropylmethyldichlorosilane, 278
3-Phenyl-5,10-undecadiene-2-one, 185
1-Phenyl-2-vinyl-4,6-heptadiene-1-ol, 207
Phosgene, 225
Phthalaldehydric acid, 264
Phthalic acid, 128, 134
Pimelaldehyde, 230
Pinane, 30, 38
Pinene, 39, 262
Pinitol, 106
Pinonenone, 261
Pinonic aldehyde, 261
Pinonone, 261
Piperidine, 4, 8, 203
Piperitone, 84
Pivalic acid, 134
Poisoning, *see* Catalyst deactivation
Polydimethylsilethylene, 282
Polymerization, 185
Polysilethylenes, 282
Polysulfones, 69
Poncitrin, 155
Pressure, effect of, 67, 109, 129, 180, 217, 231
Propanol, 200
Propargyl alcohol, 219–220
Propargyl chloride, 219–220
N-Propen-1-ylpyrrolidone, 223
Propionaldehyde, 238
Propionic acid, 238

Propionitrile, 113, 232
Propiophenone, 79, 97
3-Propoxypyridine, 17
Propylamine, 113
n-Propylbenzene, 97, 265
α-*n*-Propylcinnamaldehyde, 263
2-Propyl-3,5-diethylpyridine, 254
4-*n*-Propyl-1,2-dihydrophenanthrene, 32
Propylene, 83, 124, 125, 126, 176–177, 191, 223, 234, 238, 239, 240, 249, 261
Propylene oxide, 124, 125
Propylidene diacetate, 83
2-Propylidenethiophene, 68
1-*n*-Propylphenanthrene, 43
2-Propylpyridine, 23
N-Propylpyrrolidone, 223
Psilostachyine, 63
Pyracene, 5
Pyrene, 128, 134
Pyrene-4,5-quinone, 128, 134
Pyridines, 4, 8, 16, 18
Pyridones, 9
3-(2′-Pyridyl)-1-propanol, 22
1,5-Pyrindine, 9
Pyrrocoline, 22
Pyrroles, 8, 13, 18
Pyrrolidines, 8, 13, 287
2-Pyrrolidinone, 14
Pyrrolidone, 223, 247
1-Pyrroline, 13
Pyrrolinylbipyrrole, 14
2,2′-(1′-Pyrrolinyl)-pyrrole, 14

Q

Quadricyclane, 168
Quaternary carbons, 29, 49, 51
Quebrachitol, 106
Quinhydrone, 67, 124
Quinol, 67
Quinolines, 10, 11, 22
Quinones, 67

R

Reductive alkylation, 223
Resin acids, 140
Ribopolynucleotides, 105
Rosenmund reduction, 266, 288
Ruthenium tetroxide, 121, 133–144

S

Safflower oil, 229
Salicylaldehyde, 109
Scyllitol, 105
β-Selinene, 39
Sesquiterpenes, 36
Seychellene, 63
Silacyclobutane, 289
Skeletal isomerization, 163–166
Solvent, effect of, 62, 69, 79, 80, 83, 84, 93, 98, 108, 110, 114, 125, 126, 176, 177, 180, 186, 192, 197, 207, 210, 234, 237, 264
Spiranes, 32
Steroids, *see also* individual compounds, 67 107, 135, 161, 168, 268
Stilbenes, 91, 126, 130, 153, 211
Strain relief, 30
Strophanthidin, 107
Styrene, 78, 90, 91, 92, 178, 211, 278
Styrene oxide, 84
α-Styrylferrocene, 92
N-Styrylindolin, 159
2-Styrylpyridine, 154
Succinic acid, 141
Succinoyl chloride, 217
Sucrose, 129
Sulfinic acids, 271
Sulfur dioxide, 184

T

Tartaric acid, 122, 123, 124, 128
Telomerization of olefins, 175–208
Temperature, effect of, 17, 20, 21, 22, 31, 34, 38, 47, 50, 51, 52, 84, 87, 100, 147, 180, 190, 226, 227, 228, 231, 242, 244, 247, 248, 260, 261, 267, 268
Terpenes, 178
Terphenyl, 40, 43, 50
Terpinolene, 36
Terpyrroles, 14
Testosterone acetate, 127
3,3′,5,5′-Tetra-t-butyldiphenoquinone, 110
Tetracycline, 85
Tetradeca-3,6,9-12-tetraone, 237
Tetradehydroyohimbic acid, 36
1,2,3,4-Tetrahydroanthracene, 3
Tetrahydrocarbazole, 8, 98

Tetrahydrodicyclopentadiene, 163, 164
1,2,3,4-Tetrahydro-6,7-dihydroxy-1-(3,4,5-trimethoxybenzyl)isoquinoline, 153
Tetrahydrofuran, 270
4,5,6,7-Tetrahydroindazole, 7
4,5,6,7-Tetrahydro-3(1H)-indazolone, 7
4,5,6,7-Tetrahydroindoles, 9
Tetrahydroisoquinolines, 9, 22
Tetrahydromethylfuran, 15
Tetrahydrophenanthridone, 98
Tetrahydroponcitrin, 155
Tetrahydropyran, 164
Tetrahydropyridine, 16
Tetrahydroquinoxaline, 12
Tetrahydroselinene, 39
Tetrahydroxanthyletin, 16
9,10,11,12-Tetrahydroxystearic acid 124
Tetralin, 2, 89, 97, 112
Tetramethyl-1,3-butanediamine, 202
1,1,2,3-Tetramethylcyclohexane, 31
1,1,2,5-Tetramethylcyclohexane, 31
1,1,3,3-Tetramethyl-1,3-disilacyclobutane, 289
1,1,4,4-Tetramethyl-1,4-disilacyclohexane, 282
Tetramethyldisilazane, 284
Tetramethyldisiloxane, 281
2,3-Tetramethylenenorborane, 164
1,3,5,7-Tetramethyl-1,3,5,7-tetrahydrocyclotetrasiloxane, 285
Tetraphenylethane, 267
Tetraphenylethylene, 267
1,4,5,8-Tetraphenyl-2-naphthaldehyde, 264
Tetraphenylnaphthalene, 264
Tetraphenylsilane, 289
Tetrasilylmethanes, 286
4-(2-Thenoyl)-1-butene, 68
Thiazoles, 14
2-(2′-Thiazolin-2′-yl)benzimidazole, 14
2,(2′-Thiazolyl)benzimidazole, 14
5-(-2-Thienyl)-1-pentene, 68
Thiophenol, 199
Thymidylyl-(3′,5′)-thymidine, 105
Toluene, 18, 89, 96
p-Toluic anhydride, 269
p-Toluidine, 52
2-(*o*-Tolyl)pyridine, 23
o-Tolysilane, 286
Triacetoxytolylsilane, 286
1,3,5-Triacetylbenzene, 80
Triallylisocyanurate, 281
Tribenzylsilane, 277
14,16,18-Tribora-13,15,17-triazatriphenylene, 12
2,4,5-Tributylimidazole, 240
Tri(carbomethoxy)benzenes, 192
Trichlorosilane, 275, 277, 279, 280, 282, 283
Tricyclo[4.1.0^{2,7}]heptane, 169
Tri(cyclohexylmethyl)amine, 241
exo-Tricyclo[4.2.1.0^{2,5}]non-3-ene, 87
13-Tridecanolactam, 241
Tri(dimethylsilyl)amine, 284
Triethoxysilane, 277
2,4,5-Triethylimidazole, 240
Triethylsilane, 277, 278, 279, 287, 288
5-(Triethylsilyl)cyclooctene,
Trifluoroacetic acid, 95
1-(*p*-Trifluoromethylphenyl)-4-methoxycyclohexene, 41
N-Trifluoromethyl-*O*-trimethylsilylhydroxylamine, 283
Trifluoronitrosomethane, 283
1,2,4-Tri(1-hydroxy-1-methylethyl)benzene, 192
Trimethoxysilane, 277, 281, 282
1-Trimethoxysilylpropyl-3,5-diallylisocyanurate, 281
Trimethylamine, 115
1,2,3-Trimethylbenzene, 30
1,1,3-Trimethylcyclohexane, 29
3,3,5-Trimethylcyclohexanone, 42
1,1,3-Trimethylcyclohexene, 30
4,5-Trimethylene pyrazole, 7
1,4,7-Trimethylenespiro[4.4]nonane, 185
2,2,4-Trimethylglutarate, 252
Trimethyl phosphite, 42
Trimethylsilane, 276, 278, 279, 283
3-(Trimethylsilyl)cyclooctene, 278
5-Trimethylsilylindole, 10
5-Trimethylsilylindoline, 10
β-Trimethylsilylstyrene, 290, 291
1,2,3-Trimethyl-4,5,6-triphenylbenzene, 194
Triocta-2,7-dienylamine, 204
1,2,4-Triphenylbenzene, 193
1,3,5-Triphenylbenzene, 193
1,3,5-Triphenylbiuret, 252
1,3,5-Triphenyl-1-hexyne-2,5-diene, 194
1,1,3-Triphenylindane, 29
1,2,3-Triphenylindene, 29
Triphenylsilane, 277, 282, 289
2,4,5-Tripropylimidazole, 240

U

Ujacazulene, 22
Undecanal, 130
Undecanoic acid, 140
Undecanone-5, 140
Undecanoyl chloride, 266
Undeca-3,6,9-trione, 237
10-Undecenoic acid, 80

V

Valence isomerization, 168–170
Valerolactams, 9
4-Valerolactone, 244
Vinylacetate, 80, 81, 82, 83
Vinylacetic acid, 249
o-Vinylaniline, 23
2-Vinylbicyclo[2.2.1]hept-5-ene, 182
Vinyl chloride, 181
Vinylcyclohexene, 7, 148, 166, 186, 242, 243
Vinylcyclopropanes, 65
4-Vinyl-*meta*-dioxane, 207
Vinyl fluoride, 181
Vinylsilanes, 282
Vinyltrichlorosilane, 282
Vinyltrimethoxysilane, 282

W

Wacker process, 77
Water, 201, 217, 285

X

Xanthotoxin, 15
Xanthotoxol, 15
Xanthyletin, 16
Xylenes, 30, 31, 96
Xylyl acetates, 96
Xylylene diacetates, 96

Y

Yohimbic acid, 36

ORGANIC CHEMISTRY

A SERIES OF MONOGRAPHS

1. Wolfgang Kirmse. CARBENE CHEMISTRY, 1964; 2nd Edition, 1971

2. Brandes H. Smith. BRIDGED AROMATIC COMPOUNDS, 1964

3. Michael Hanack. CONFORMATION THEORY, 1965

4. Donald J. Cram. FUNDAMENTALS OF CARBANION CHEMISTRY, 1965

5. Kenneth B. Wiberg (Editor). OXIDATION IN ORGANIC CHEMISTRY, PART A, 1965; Walter S. Trahanovsky (Editor). OXIDATION IN ORGANIC CHEMISTRY, PART B, 1973

6. R. F. Hudson. STRUCTURE AND MECHANISM IN ORGANO-PHOSPHORUS CHEMISTRY, 1965

7. A. William Johnson. YLID CHEMISTRY, 1966

8. Jan Hamer (Editor). 1,4-CYCLOADDITION REACTIONS, 1967

9. Henri Ulrich. CYCLOADDITION REACTIONS OF HETEROCUMULENES, 1967

10. M. P. Cava and M. J. Mitchell. CYCLOBUTADIENE AND RELATED COMPOUNDS, 1967

11. Reinhard W. Hoffman. DEHYDROBENZENE AND CYCLOALKYNES, 1967

12. Stanley R. Sandler and Wolf Karo. ORGANIC FUNCTIONAL GROUP PREPARATIONS, VOLUME I, 1968; VOLUME II, 1971; VOLUME III, 1972

13. Robert J. Cotter and Markus Matzner. RING-FORMING POLYMERIZATIONS, PART A, 1969; PART B, 1; B, 2, 1972

14. R. H. DeWolfe. CARBOXYLIC ORTHO ACID DERIVATIVES, 1970

15. R. Foster. ORGANIC CHARGE-TRANSFER COMPLEXES, 1969

16. James P. Snyder (Editor). NONBENZENOID AROMATICS, VOLUME I, 1969; VOLUME II, 1971

17. C. H. Rochester. ACIDITY FUNCTIONS, 1970

18. Richard J. Sundberg. THE CHEMISTRY OF INDOLES, 1970

19. A. R. Katritzky and J. M. Lagowski. CHEMISTRY OF THE HETEROCYCLIC *N*-OXIDES, 1970

20. Ivar Ugi (Editor). ISONITRILE CHEMISTRY, 1971

21. G. Chiurdoglu (Editor). CONFORMATIONAL ANALYSIS, 1971

22. Gottfried Schill. CATENANES, ROTAXANES, AND KNOTS, 1971

23. M. Liler. REACTION MECHANISMS IN SULPHURIC ACID AND OTHER STRONG ACID SOLUTIONS, 1971

24. J. B. Stothers. CARBON-13 NMR SPECTROSCOPY, 1972

25. Maurice Shamma. THE ISOQUINOLINE ALKALOIDS: CHEMISTRY AND PHARMACOLOGY, 1972

26. Samuel P. McManus (Editor). ORGANIC REACTIVE INTERMEDIATES, 1973

27. H.C. Van der Plas. RING TRANSFORMATIONS OF HETEROCYCLES, Volumes 1 and 2, 1973

28. Paul N. Rylander. ORGANIC SYNTHESES WITH NOBLE METAL CATALYSTS, 1973

In preparation

Stanley R. Sandler and Wolf Karo. POLYMER SYNTHESIS